T0174545

Leveraging Artificial Intelligence in Engineering, Management, and Safety of Infrastructure

Editor

M.Z. Naser
Assistant Professor
Clemson University, Clemson, South Carolina, USA

CRC Press
Taylor & Francis Group
Boca Raton London New York

CRC Press is an imprint of the
Taylor & Francis Group, an **informa** business
A SCIENCE PUBLISHERS BOOK

First edition published 2023
by CRC Press
6000 Broken Sound Parkway NW, Suite 300, Boca Raton, FL 33487-2742

and by CRC Press
4 Park Square, Milton Park, Abingdon, Oxon, OX14 4RN

© 2023 Taylor & Francis Group, LLC

CRC Press is an imprint of Taylor & Francis Group, LLC

Reasonable efforts have been made to publish reliable data and information, but the author and publisher cannot assume responsibility for the validity of all materials or the consequences of their use. The authors and publishers have attempted to trace the copyright holders of all material reproduced in this publication and apologize to copyright holders if permission to publish in this form has not been obtained. If any copyright material has not been acknowledged please write and let us know so we may rectify in any future reprint.

Except as permitted under U.S. Copyright Law, no part of this book may be reprinted, reproduced, transmitted, or utilized in any form by any electronic, mechanical, or other means, now known or hereafter invented, including photocopying, microfilming, and recording, or in any information storage or retrieval system, without written permission from the publishers.

For permission to photocopy or use material electronically from this work, access www.copyright.com or contact the Copyright Clearance Center, Inc. (CCC), 222 Rosewood Drive, Danvers, MA 01923, 978-750-8400. For works that are not available on CCC please contact mpkbookspermissions@tandf.co.uk

Trademark notice: Product or corporate names may be trademarks or registered trademarks and are used only for identification and explanation without intent to infringe.

Library of Congress Cataloging-in-Publication Data (applied for)

ISBN: 978-0-367-42210-3 (hbk)
ISBN: 978-1-032-30575-2 (pbk)
ISBN: 978-0-367-82346-7 (ebk)

DOI: 10.1201/9780367823467

Typeset in Times New Roman
by Radiant Productions

Dedication

To the hard-working engineers around the world,
To the future of civil and environmental infrastructure,
&
To the convergence of domains.

#TheFutureisAI

Preface

Artificial Intelligence (AI) continues to transform our lives on a daily basis. Advancements on the AI front can only be described as lightning-fast, especially when compared to what we are accustomed to in our civil and environmental engineering (CEE) discipline, one of the most classical and perhaps oldest engineering disciplines. Such rapid advancements, while admirable, their applications often struggle to find a home in CEE. This is possibly due to the fact that AI represents facets that can be foreign to us. For example, AI relies on coding—an exercise that is rarely covered in our curriculum. AI, and for the most part, remains a blackbox that can be hard to visualize—unlike many of the methods favored and practiced by our engineers. Most importantly, our discipline has been thriving and possibly, with presumably little credit to AI.[1] Thus, it is not surprising that we are a bit careful about embracing AI.

Thus, the prime motivation behind this edited book is to present our community with a series of successful stories to counter the above and set the stage for a foundation for AI. This book not only hopes to narrow the knowledge gap between AI and CEE but also to ease the transition into a future CEE; one where AI works hand-in-hand with engineers. It is our hope that this book charts a path to how AI can help re-shape our domain; to one that is modern, resilient, and intelligent. We anticipate the showcased chapters and case studies to be seen as evidence of the potential of AI. We hope that these studies become benchmarks that our readers can refer to and, most importantly, extend to more complex problems.

In parallel, this edited book also aims to address the following pivotal question; given that infrastructure is a massive investment that is expected to last for decades, then *how can we leverage AI in engineering, management, and safety of our infrastructure?*[2] Throughout a journey spanning 18 chapters and over 400 pages, we hope that we were able to show a glimpse of the potential of AI to our discipline. Each chapter is tailored to address a particular dimension to the aforenoted question. Each chapter is also crafted by experts in their niche area to highlight how various AI methods can be created, tweaked, and applied to problems relating to infrastructure. Most chapters are designed to be a standalone document that is comprehensive and

---

[1] For a more in-depth discussion on engineering, philosophical and educational aspects pertaining to these items, please refer to the following works as well as (Lin and Huang, 2015; Martínez-Barrera et al., 2015; Sacks et al., 2020; Thai, 2022) those by yours truly and co-authors (Naser and Ross, 2022; Naser, 2021; Naser, 2021).

[2] And hence, the title of this edited book!

encompass of a detailed cases study to the most extent, and some were dedicated as key references/reviews to give our readers the best of two worlds.

A primary objective of designing the internal structure of this book was to be as inclusive and diverse as possible in terms of the invited contributors, as well as topics covered—without any substantial compromises. We are proud to announce that we have contributions from Africa, Asia, Australia, Europe, and North America. A good portion of our chapters includes or is led by minorities, under representative groups, and early careers/Post-docs. While our contributors from South America could not make it to this edition, we are confident that future editions will be able to highlight a more inclusive and diverse body of works.

We start this edited book with a review on convolutional neural networks and their applications in civil infrastructure led by Onur Avci, Osama Abdeljaber, Serkan Kiranyaz, Turker Ince and Daniel J. Inman. This review outlines the latest advancements in AI in infrastructure. Then, Yifan Gao, Vicente A. González, Tak Wing Yiu and Guillermo Cabrera-Guerrero present their work on identifying construction workers' personalities. This chapter explores prominent AI models for the safety of construction workers in big projects.

Chapters 3 and 4 (led by Ahmad N. Tarawneh and Eman F. Saleh and Iman Mansouri, Jale Tezcan and Paul O. Awoyera, respectively) describe procedures to create novel AI models for designing structural elements and predicting properties of construction materials. Vafa Soltangharaei, Li Ai and Paul Ziehl in Chapter 5 further our discussion onto how AI can be adopted as a tool for condition assessment of structures and in Chapter 6, Diana Andrushia, Anand, N., Richard Walls, Daniel Paul T. and Prince Arulraj compliment this track with a case study on crack detection via AI.

We take a deep breath to continue our reviews on AI in Chapter 7 (where Islam H. El-adaway and Rayan H. Assaad lead the discussion on recent advancements with regard to construction engineering and management and safety). Their review is augmented with an equally impressive analysis with a scientometric spin by Zhanzhao Li and Aleksandra Radlińska on the use of AI as a design tool for concrete construction materials (Chapter 8).

Koosha Khorramian and Fadi Oudah provide us with an in-depth introduction and discussion on AI for attaining reliable structures via active learning kriging in Chapter 9. Then, William Locke, Stefani Mokalled, Omar Abuodeh, Laura Redmond and Christopher McMahan (the authors of Chapter 10) present a thorough Bayesian analysis for multilevel damage classification and identification of bridges.

Chapters 11–15 are dedicated to successful works that scaled AI to system-levels and infrastructure. For example, in Chapter 11, Andrew Fahim, Tahmid Mehdi, Ali Taheri, Pouria Ghods, Aali Alizadeh and Sarah De Carufel paint a vivid picture of how AI and Internet-Of-Things (IoT) can be valuable for international and large-scale monitoring of concrete across various site jobs. Haifeng Wang and Teng Wu (Chapter 12) present a detailed analysis of how deep learning can be enhanced through domain knowledge to predict the nonlinear response of structures. In Chapter 13, Hayder A. Rasheed, Ahmed Al-Rahmani and AlaaEldin Abouelleil present a novel synergistic approach that combines numerical simulations with AI to create a new tool for bridge girder damage detection. Zadid Khan, Sakib Mahmud Khan, Mizanur Rahman,

Mhafuzul Islam and Mashrur Chowdhury (Chapter 14) elevates the discussion by exploring the role of AI in transportation cyber-physical systems. Chapter 15 (co-authored by Amir H. Behzadan, Nipun D. Nath and Reza Akhavian) discusses the added value of AI to the construction area from the lens of the Future of Work.

The last three chapters are dedicated to the merit of adopting AI to overcome the unique hazard of fire. Unlike other hazards, fire can break out anywhere and anytime. Yet, building codes have not matured enough in this often forgotten area. Thus, these three chapters showcase the potential of AI in overcoming such a bottleneck from a practical, research, and futuristic look into the problem of fire in infrastructure. For example, in Chapter 16, Yavor Panev, Tom Parker and Panagiotis Kotsovinos articulate the use of AI in structural fire engineering design applications. Chapter 17, as led by Srishti Banerji, presents a case of using AI as a tool to establish the response of construction materials under elevated temperatures. Finally, this book ends with Chapter 18, where Xinyan Huang, Xiqiang Wu, Xiaoning Zhang and Asif Usmani highlight the application of AI in smart tunnels from a fire perspective.

Note to Readers and Instructors

We believe that this edited book will be valuable on a number of fronts. For example, the dedicated review chapters (Chapters 1, 7, 8, and 14[3]) can deliver the latest information on the frontier of AI and CEE. These chapters are a good starting point for readers of varying backgrounds on AI (e.g., little or advanced knowledge). Personally, I like to assign these chapters to senior undergraduate students interested in AI, as well as to graduate students who are planning/carrying out research on AI. Students who are looking for some exciting projects might find these chapters worthy of their time. Equally, I also like to refer my industry colleagues and practicing engineers to such chapters to introduce them to AI-based solutions and proofs of concepts that can be extended to other problems.

All other chapters are designed to present the reader with a collective background on their particular problems and adopted AI methods in sufficient detail. Admittedly, these chapters also home references to more complete works that can be visited for additional information. In some instances, some chapters might include links to codes and/or datasets that readers/students might like to view and apply. In all cases, our contributors have displayed interest in sharing their codes upon reasonable request. Please feel free to reach out to us shall you have any questions.[4]

This edited book emphasizes six areas within CEE. Namely, structural engineering, structural health monitoring, construction management and safety, construction materials, transportation engineering, and fire engineering. Other areas such as geotechnical engineering, environmental engineering, pavement engineering, and others were not left out of negligence but rather due to limitations and constraints imposed by the COVID-19 pandemic. We hope to showcase works

---

[3] Please note that Chapters 9 and 15 are two special hybrids in which they contain a good amount of introduction/review and a detailed discussion as well.

[4] In addition, please feel free to visit my website for additional codes, datasets, and AI-based applications [https://www.mznaser.com/fireassessmenttoolsanddatabases].

from the aforenoted areas, as well as others, in future editions of this book. On a more positive note, many, if not all, the explored AI methods in this edited book are equally applicable to other areas.

A Series of Thoughts and Advice from our Contributors

This is a collection of personal thoughts and advice collected from our contributors, as well as yours truly. For brevity, I will convey these thoughts in short sentences.

- Think of AI as a tool that can supplement and augment our existing knowledge and methods.
- While AI can help us on a variety of fronts, it, and just other methods, has its limitations.
- Keep an eye on details.
- Do not be afraid to try and use AI through coding or coding-less approaches.
- Get the fundamentals down. Teach yourself to see where AI can help you.
- Ask, how & why did AI arrive at such predictions?
- Ask, what does AI "see" that traditional methods do not?
- With technological advancements in sensors, cloud storage, and computing power, AI and data-driven approaches have become essential to determine the current structural condition and estimate the remaining service life of structures.
- Verify that the inputs to the ML model are representative of the dataset used for its training.
- Obtaining data under consistent environmental and operational conditions is probably the most important task when performing model updating with real system response data, as unknown sources of noise cannot easily be captured in a model.
- Sharing data and tools and standardizing AI application procedures will be a big step towards the development of a robust, data-informed construction intelligence ecosystem.
- Developing AI-based methods to form surrogate models in a reliability-based framework of analysis has opened the door for assessing the reliability of complex structural systems that otherwise was difficult or inefficient for us to quantify.
- AI offers unprecedented opportunities to retrieve and reveal remarkable patterns, trends, relationships, and knowledge from big data that can better help in managing civil infrastructure systems, construction engineering operations, and associated safety practices.
- The culture of openness, explainability/interpretability, and uncertainty quantification are important research gaps that need to be addressed in the AI-based civil engineering community.
- One of the challenges facing machine learning is over fitting. Make sure your model is generalized by selecting the appropriate training algorithm, dividing the database into training and validation datasets, and visualizing the effect of each of the variables.

- The confined tunnel is fatal in case of any fire event, so it needs an intelligent fire safety management system that combines AI, IoT, and digital twin to enable smart firefighting.
- Artificial Intelligence (AI) techniques allow the investigation of natural patterns between variables without assuming any preconception in terms of the mathematical structure of the data, which can have a considerable impact on the way that workers' safety behaviour can be predicted.
- Efforts toward sharing data and tools and standardizing AI application procedures are essential to foster the development of a robust, data-informed construction intelligence ecosystem.
- Finite element analysis is one of the most reliable tools to predict behavior of structural engineering systems. On the other hand, AI framework is one of the most promising tools to generalize efficient pattern recognition of complex phenomena. Therefore, the synergy of the two approaches holds great potential to accomplish effective solutions to a variety of open-ended problems in civil engineering.
- AI and big data should be viewed as accelerators of (rather than replacements to) human ingenuity and creativity.

A Look into How this Edited Book came to be

Finally, I would like to take the following few lines to reminisce and acknowledge the kind support of our contributors and CRC staff – for which, without them, this edited book would not have been possible. A special thanks go to my family, friends, students, colleagues, and my home school at Clemson University.

Let me start with a brief history of this project. The idea for this edited book came to light after a brief discussion with Vijay Primlani, an Acquisitions Editor at CRC. We set sail on this journey at the end of 2019 with a goal of completion within 18–24 months. Little did we know that a pandemic was on the horizon. I must admit that in the first few months of the pandemic, this project felt far away from seeing the light. Nevertheless, CRC's support and the trust of our contributors were tremendous. Despite this pandemic,[5] which unfortunately continues to exist as of the writing of this preface, our contributors and CRC staff managed to complete this book on time. Their hard work and persistence are acknowledged within the pages of this book. This is, and always will be, a true community effort. I am humbled to be part of this team and indebted to all of you.

M.Z. Naser

April 7th, 2022   Clemson, SC

---

[5]   If you happen to read this edited in a few, or many, many years from now, you may/may not have witnessed the COVID-19 pandemic. A quick Google search can result in a brief look at our lives during the 2019–2022.

References

Lin, S. and Huang, Z. 2015. Comparative design of structures: Concepts and methodologies. *Comp. Des. Struct. Concepts Methodol.*, 1–404. https://doi.org/10.1007/978-3-662-48044-1.

Martínez-Barrera, G., Beycioğlu, A., Gencel, O., Subaşi, S. and González-Rivas, N. 2015. Artificial intelligence methods and their applications in civil engineering. *In: Handb. Res. Recent Dev. Mater. Sci. Corros. Eng. Educ.*, 2015. https://doi.org/10.4018/978-1-4666-8183-5.ch009.

Naser, M.Z. 2021. An engineer's guide to eXplainable Artificial Intelligence and Interpretable Machine Learning: Navigating causality, forced goodness, and the false perception of inference. *Autom. Constr.*, 129: 103821. https://doi.org/10.1016/J.AUTCON.2021.103821.

Naser, M.Z. 2021. Demystifying ten big ideas and rules every fire scientist & engineer should know about blackbox, whitebox & causal artificial intelligence. https://arxiv.org/abs/2111.13756v1 (accessed January 26, 2022).

Naser, M.Z. and Ross, B. 2022. An opinion piece on the dos and don'ts of artificial intelligence in civil engineering and charting a path from data-driven analysis to causal knowledge discovery. https://Doi.Org/10.1080/10286608.2022.2049257. https://doi.org/10.1080/10286608.2022.2049257.

Sacks, R., Girolami, M. and Brilakis, I. 2020. Building information modelling. *Artificial Intelligence and Construction Tech, Dev. Built Environ.*, 4: 100011. https://doi.org/10.1016/J.DIBE.2020.100011.

Thai, H.T. 2022. Machine learning for structural engineering: A state-of-the-art review. *Structures*, 38: 448–491. https://doi.org/10.1016/J.ISTRUC.2022.02.003.

Contents

Chapter 1

Convolutional Neural Networks and Applications on Civil Infrastructure

Onur Avci,[1,]* Osama Abdeljaber,[2] Serkan Kiranyaz,[3] Turker Ince[4] and Daniel J. Inman[5]

1. Background on Artificial Intelligence, Machine Learning, and Deep Learning

Artificial Intelligence (AI) is the intelligence ability showcased by computers/ machines to conduct tasks and solve problems that might be laborious for humans but simple for computer algorithms (Sajja, 2021; Goel and Davies, 2019). AI has evolved as an interdisciplinary field where its subset topics of Machine Learning (ML) and Deep Learning (DL) are being implemented in areas of multiple sciences and engineering at an increasing pace (Simeone, 2018; Hutter, 2019). The AI algorithm for running tasks and solving problems could be coded by a programmer (Goodfellow et al., 2017) so that this knowledge-based approach could be utilized in various applications (Hsu, 2002). Even though the knowledge-based AI systems sometimes fail at extremely simple human tasks (object recognition, face recognition, speech recognition), researchers have been working on developing different ways of teaching common-sense and intuitive knowledge/information to computers. ML was introduced as an application in AI to overcome the drawbacks of knowledge-based approaches (Guo et al., 2016; Tiwari, 2017). It is predominantly defined as the capability to automatically learn from different size groups of data, enhance from

[1] Assistant Professor, Department of Civil and Environmental Engineering, West Virginia University, Morgantown, WV, 26506.
[2] Department of Building Technology, Linnaeus University, Sweden-35195.
  Email: osama.abdeljaber@lnu.se
[3] Professor, Department of Electrical Engineering, Qatar University, Qatar. Email: mkiranyaz@qu.edu.qa
[4] Professor, Electrical & Electronics Engineering Department, Izmir University of Economics, Izmir, Turkey-35330. Email: turker.ince@ieu.edu.tr
[5] Professor, Department of Aerospace Engineering, University of Michigan, Ann Arbor, MI, USA-48109.
  Email: daninman@umich.edu
* Corresponding author: onur.avci@mail.wvu.edu

experience, and adapt without any explicit programming or instructions. In ML, algorithms and statistical models are the major components to draw a conclusion from data patterns (Jordan and Mitchell, 2015). In simple ML algorithms, the data needs to be presented with regard to a certain number of features and this requires preprocessing to extract distinctive pieces, also known as feature extraction (Guyon and Elisseeff (n.d.)). The following stage is the system training where the processed data is taught by an ML algorithm to learn how the features correlate with each other and/or with other data. ML algorithms can be mainly classified as unsupervised and supervised learning algorithms (Avci et al., 2021). Supervised algorithms operate with an input-output kind of format where the objective of supervised learning is to determine an optimized mapping based on the inputs to targeted outputs which means that the supervised algorithms demand a 'supervisor' individual to assign every data point with an accurate label before training (Ghahramani, 2015; Shalev-Shwartz and Ben-David, 2013). However, unsupervised algorithms operate with input-only data without any supervision on assigning targets. As such, the purpose of unsupervised learning is to determine the data distribution to gain useful information (Yegnanarayana, 2006).

ML applications are extensively used in almost every field of engineering. The ML processes can be categorized into classification, regression, prediction, and clustering tasks. The objective of 'classification' is to find out the category the input belongs to whereas the goal of the 'regression' is to determine the relationship between inputs and outputs. Regression is similar to classification, yet the difference is the format of the output. On the other hand, 'prediction' is a particular kind of regression where the goal is to foretell the upcoming values in a time series. In 'clustering' the input dataset is divided into groups of similar instances. It is performed with unsupervised ML techniques (e.g., Self-Organizing Maps (Abdeljaber and Avci, 2016; Abdeljaber et al., 2016; Avci and Abdeljaber, 2016; Avci et al., 2019)); meanwhile supervised methods are utilized for classification, regression, and prediction operations.

The standard unsupervised and supervised ML algorithms need manual feature extraction procedures performed in advance so that the input data can be represented with a certain number of hand-crafted features. This is why the success of these algorithms depends on the extracted feature type. As such, it is pertinent to use the appropriate feature set that can accurately represent the characteristics of the input data. Following the feature extraction stage, it is simpler for the ML algorithm to form a relationship between the output and the characteristics extracted. For many tasks though, it is not very easy to determine an appropriate set of features "good enough" to be utilized for training. To eliminate the use of hand-crafted features in complicated AI applications, DL methods have been introduced. DL algorithms are a particular type of ML tools that can extract the optimized input representation without any user intervention, directly from the raw data. From another perspective, DL methods can learn to correlate attributes to the targeted output in addition to extracting features. As a result, an AI system trained by a DL procedure is capable of direct mapping without feature extraction. As such, DL methods are capable of simplifying relatively complicated features into relatively simpler *learned features* (Goodfellow et al., 2017) since the DL algorithms simply break few large tasks into multiple relatively easier/smaller functions. This has been validated by the latest research which resulted in enhanced performance for difficult problems like heart-beat classification (Kiranyaz et al., 2016), image classification (Kiranyaz et al., 2016), and object detection (Scherer et al., 2010). Within the AI context, the relationship between the DL, ML, and knowledge-based AI paradigms are shown in Fig. 1.1, while a comparison between the three approaches is presented in Fig. 1.2.

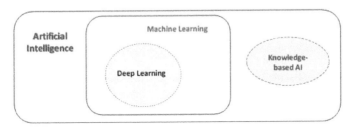

Fig. 1.1: Relationship between various AI systems.

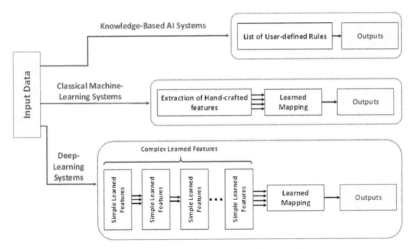

Fig. 1.2: Knowledge-based, ML-based, and DL-based systems.

2. Artificial Neural Networks (ANNs)

The mammalian brain consists of billions of *neurons* interconnected by trillions of synapses or connections (Yegnanarayana, 2006). This massive network of processing units allows the brain to carry out complex and nonlinear processes in a seamless and highly parallel manner. Thus, the brain is known for its remarkable efficiency in perception and pattern recognition, unmatched even by the most powerful digital computers. For instance, it takes an average human about 100 ms to recognize a familiar person in an unfamiliar environment, while a fast computer requires much more computational time to carry out a similar simple task (Goodfellow, 2017). This ability has motivated researchers to develop artificial computational models inspired by the interactions, communications, and interconnections between neurons in a human brain.

Artificial Neural Networks (ANNs), or predominantly used as Neural Networks (NNs), are ML models that are loosely based on the framework of neurons in the human central nervous system. A typical ANN consists of nonlinear processing units (i.e., Artificial Neurons), usually arranged in layers, interconnected by numerous pathways. As with any other ML or DL method, ANNs learn the required knowledge from a training dataset. The learned experience is stored in the interconnections between the ANN neurons.

ANNs have been widely utilized in numerous disciplines of science and engineering such as speech pronunciation, pattern recognition, computer vision, and medical diagnoses. In civil engineering, ANNs have been used in several applications such as structural control (Abdeljaber et al., 2016; Avci et al., 2020), finite element mesh generation (Bohn, 2003), structural health monitoring (Arangio and Bontempi, 2014), structural damage detection (Meruane and Heylen, 2012; Radzieński et al., 2011; Bayissa et al., 2008; Cha and Buyukozturk, 2015), and design optimization (Abdeljaber et al., 2015; Abdeljaber et al., 2016; Avci et al., 2017; Abdeljaber et al., 2017). The widespread use of ANNs is due to their superior capabilities which can be summarized as the following (Yegnanarayana, 2006; Bani-Hani et al., 1999):

1. ANNs are highly nonlinear, and therefore efficient in predicting the behavior of complex nonlinear systems.
2. ANNs are adaptive models with the ability to learn difficult tasks especially when the available data is inconsistent or fuzzy.
3. ANNs are capable of 'generalization', which means they can assign accurate outputs to inputs that have not been used in the training process.
4. Constructing a model for a system with a large number of inputs and outputs using an ANN is a relatively easy task compared to the other methods.
5. Due to their unique parallel architecture, ANNs, implemented in hardware form, are robust even if some of the processing units fail.

2.1 Biological Analogy

The human nervous system consists of three main components: receptors, effectors, and the central nervous system. The receptors translate stimuli affecting the body into electrical impulses which are sent to the brain in the central nervous system. The "biological neural network" of the brain continually processes the input impulses and sends the resulting output impulses to the effectors to produce the physical response (Haykin, 2008).

As shown in Fig. 1.3, neurons, which are the building blocks of this biological neural network, consist mainly of the "cell body" which houses the nucleus. The cell body receives the signals from the other cell through several nerve fibers referred to as the dendrites. The output of the neuron is conveyed through a relatively long nerve fiber called the 'axon' which branches into a large number of fibers leading to thousands of other cells. The branches of the axon are connected to the dendrites and cell bodies of other neurons at junctions called synapses. The function of these synapses is to regulate the interactions between the neurons.

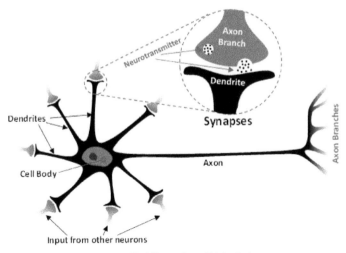

Fig. 1.3: A simplified illustration of biological neurons.

Neurons typically produce their output as a series of voltage impulses or "action potentials". The transmission of the output signal to another neuron at a synapse is a highly sophisticated chemical process carried out by special substances called neurotransmitters. Initially, the protoplasm inside the cell body is negatively charged at a potential of –70 mv against the surrounding neural liquid, which contains positive sodium ions. In the resting state, the membrane around the cell body prevents the ions from penetrating its protoplasm, which maintains the potential level. This level gradually increases as the cell body receives electrical signals from other neurons (Ming and Song, 2011; Allen and Eroglu, 2017). When the potential level increases beyond –60 mv, the membrane suddenly loses its ability to hold the positive ions, resulting in an abrupt change in the potential that causes the cell body to discharge or 'fire'.

The cell's discharge is transferred through the axon of the neuron as a sequence of pulses having a frequency of 1–100 pulses (Yegnanarayana, 2006). The signal propagates across the axon and its branches until it stops at the synaptic junction. The synapse then releases a small amount of its neurotransmitter substance and sends it along the neuron fibers leading to the neuron at the other end of the junction. The neurotransmitter alters the conductance of the synapse at its post-synaptic end which affects the way the signals are transmitted to the post-synaptic neuron. This process results in an 'excitatory' or 'inhibitory' synaptic end. Excitatory synaptic ends cause the passing impulse to activate the receiving neuron (i.e., bringing the potential of its cell body closer to the discharge voltage), while the inhibitory ends inhibit the receiving neuron by bringing it closer to the resting state.

As will be detailed in the next section, ANNs are analogous to biological neural networks in the sense that both of them are made up of interconnected processing units (i.e., neurons), and both of them use synaptic connections to manage the interactions between neurons and store the learned knowledge. However, there are notable differences between the biological neural networks and their artificial counterparts (Goodfellow et al., 2017; Yegnanarayana, 2006; Haykin, 2008):

- Biological neurons are significantly slower when compared to silicon chips. Neural events in biological neurons take place in the range of microseconds, while it takes silicon chips only a few nanoseconds to carry out a single step of a program.
- However, the human brain compensates for the slow speed of its components by having a massive network composed of approximately 10^{11} neurons interconnected by nearly 10^{15} synapses. On the other hand, the most complicated ANNs ever designed have only several millions of interconnections.
- The enormous parallel structure of the brain enables it to routinely carry out complex classification and pattern recognition tasks that are challenging for computers.
- As a result of its ability to execute a massive number of operations in parallel, the brain is highly efficient in terms of energy consumption compared to ANNs.
- Usually, ANNs are designed to carry out a single specialized task, while a biological neural network is capable of learning numerous different tasks.

2.2 *Artificial Neurons*

Artificial neurons are crude mathematical models of biological neurons that constitute the building units of ANNs. There are several models for artificial neurons available in the literature. Perceptrons are among the most used neuron models in the feed-forward, fully-connected ANNs, otherwise known as multilayer perceptrons (MLPs). As displayed in Fig. 1.4, simple perceptron consists of the following components:

1. *Connection links* that connect the neuron to other neurons in the network. Each link is assigned a number called the *connection weight* w_{ik}, where the subscript i denotes the neuron at the input end of the link, while the subscript k represents the neuron at the receiving end (i.e., the current neuron). The role of the connection weight is to adjust the input y_i, coming from the ith neuron, before being processed by the current neuron. As will be detailed in the next section, ANNs store the knowledge acquired by training in their connection weights.

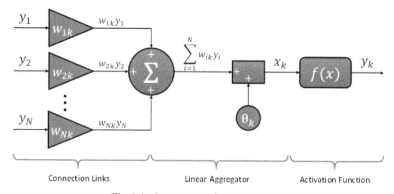

Fig. 1.4: Components of an MLP neuron.

2. A linear aggregator is that which sums the weighted inputs from the other neurons together with a *threshold* θ_k associated with the current neuron. This can be described by Eq. (1.1):

$$x_k = \theta_k + \sum_{i=1}^{N} w_{ik} y_i \qquad (1.1)$$

where N is the number of neurons connected to the current neuron. The threshold θ_k applies an affine transformation on the weighted inputs to the neuron. Similar to the connection weights, θ_k is also a parameter that can be adjusted through ANN training.

3. An *activation function* $f(\cdot)$ is that which processes the output of the aforementioned linear combination process x_k to produce the final output of the neuron y_k. The activation process can be expressed as follows Eq. (1.2):

$$y_k = f(x_k) \qquad (1.2)$$

The task of the activation function is to constrain the output of the neuron within a certain finite range. Generally, activation functions can be categorized into unipolar and bipolar functions. Unipolar functions map the output within the interval [0,1], while their bipolar counterparts produce an output within the range of [−1, 1]. Sigmoid functions are among the most widely used activation functions in ANNs. These functions are smooth, differentiable, and strictly increasing (Bani-Hani, 1999). They exhibit a smooth transition between linear and nonlinear behavior. Logistic function as well as hyperbolic tangent function, $y = \tanh(x)$ are common examples of sigmoid activation functions.

3. Multi-Layer Perceptrons

MLPs are a widely used class of ANNs. As illustrated in Fig. 1.5, MLPs are made up of interconnected MLP neurons structured within the layers. Within the MLP structure, the first layer is called the "input layer", while the last layer is called the "output layer". The "hidden layers" are lying in between the input and the output layers. The input signals are received by the input neurons and then processed by the hidden neurons and then sent to the output neurons which broadcast the final

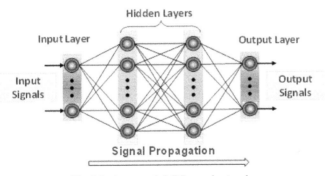

Fig. 1.5: An example MLP neural network.

output of the network. Note that MLP networks can have one or more hidden layers. The network shown in Fig. 1.5 has two hidden layers only for illustration purposes. The number of neurons in input and output layers is determined by the number of the inputs and outputs of the modeled system, respectively. On the other hand, the number of hidden layers and the number of neurons in the hidden layers are the 'hyperparameters' that should be defined before running the training process. They are predominantly chosen by trial-and-error.

It is worth noting that MLPs fall in the category of multilayer, fully-connected, and feedforward neural networks since the input signal is only allowed to propagate in the forward direction. Also, every neuron in any MLP layer is connected to the neurons in the preceding layer, which is why MLPs are viewed as fully-connected networks.

3.1 Training of MLPs

MLPs belong to the class of supervised neural networks, and therefore, they are trained using a dataset containing several input samples with the corresponding target outputs. As stated earlier, the connection links between the neurons are assigned by scalar values called the weights, which are responsible for determining how signals propagate throughout the network. Initially, the weights are random but are then tuned through the training process so that the error is minimized between the actual output of the neural network and the desired target. Training of MLPs is a systematic and iterative process that involves two operations: forward-propagation, and back-propagation.

As the name implies, the input sample is propagated in the forward direction (forward-propagation) until the output emerges from the output layer. This output can be viewed as a nonlinear function of the input in terms of the current weights and thresholds. A certain cost function is then used to calculate the error between the actual output of the neural network and the targeted output corresponding to that input sample. Next, the error is backpropagated from the output layer through the hidden layers and finally to the input layer. During the backpropagation process, the 'sensitivity' of each weight and threshold in the network to the error is calculated. These sensitivities are utilized to update the weights and thresholds of the network according to the "gradient descent method". This process of forward-propagation (FP), backpropagation, and weights updating is iterated until the neural parameters (i.e., weights and thresholds) converge to optimal values that minimize the error between the actual output and the target values. The entire procedure explained in this section is an optimization technique called the "backpropagation algorithm", which is the predominant algorithm used in neural network training.

3.1.1 Backpropagation Algorithm in MLPs

The MLP network depicted in Fig. 1.6 consists of L layers (an input layer, an output layer, and $L - 2$ hidden layers). Each layer consists of N_v neurons, where $v = 1,2, ..., L$ is the index of the layer. Figure 1.6 displays a closer look at the kth neuron of layer l which lies between layers $l - 1$ and $l + 1$.

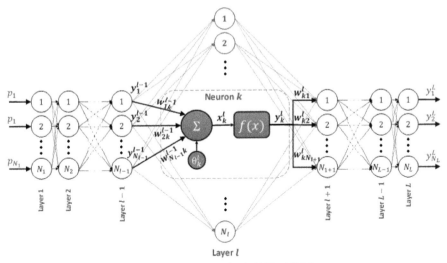

Fig. 1.6: Forward-propagation in hidden MLP layers.

In FP, the neuron collects the weighted outputs of layer $l - 1$, sums them to a threshold θ_k^l, and then applies an activation function $f(\cdot)$ on the result of the summation to calculate the output of the neuron y_k^l. This FP operation can be expressed by the following Eqs. (1.3, 1.4):

$$x_k^l = \theta_k^l + \sum_{i=1}^{N_{l-1}} w_{ik}^{l-1} y_i^{l-1} \tag{1.3}$$

$$y_k^l = f(x_k^l) \tag{1.4}$$

Note that the superscripts in the above equations denote the layer to which the quantities belong. Accordingly, the input propagates throughout the entire network (neuron by neuron) until an output vector $Y = y_1^L, y_2^L, ..., y_{N_L}^L$ emerges from the last layer L. Now suppose that the input-output example used in this training iteration consists of an input $P = p_1^1, p_2^1, ..., p_{N_1}^1$ associated with the desired target vector $T = t_1, t_2, ..., t_{N_L}$. The error E between the actual and desired outputs can be then computed according to the following cost function as given in Eq. (1.5):

$$E = \frac{1}{2}\sum_{i=1}^{N_L}(y_i^L - t_i)^2 \tag{1.5}$$

The factor $1/2$ is introduced here for convenience, just to cancel out the exponent when the cost function is differentiated with respect to y.

The objective of the backpropagation algorithm is to be optimizing the network's weights and thresholds so that the cost function is minimized. For this, the first step is to obtain the sensitivity factors. For the kth neuron of layer l, the sensitivity factors to be calculated are the partial derivatives of the error E with respect to the weights

$w_{1k}, w_{2k}, ..., w_{N_{l-1},k}$ along with the threshold θ_k^l, which can be defined based on the chain rule as follows Eqs. (1.6, 1.7):

$$\frac{\partial E}{\partial w_{ik}^{l-1}} = \frac{\partial E}{\partial x_k^l} \frac{\partial x_k^l}{\partial w_{ik}^{l-1}} = \frac{\partial E}{\partial x_k^l} y_i^{l-1} \tag{1.6}$$

$$\frac{\partial E}{\partial \theta_k^l} = \frac{\partial E}{\partial x_k^l} \frac{\partial x_k^l}{\partial \theta_k^l} = \frac{\partial E}{\partial x_k^l} \tag{1.7}$$

It is clear from the formulas above that the sensitivity factors are dependent on the partial derivative of the error as for x_k^l. The partial derivative is often referred to as the "delta error" of the neuron Δ_k^l Eq. (1.8):

$$\Delta_k^l = \frac{\partial E}{\partial x_k^l} \tag{1.8}$$

According to the chain rule, the delta error of any neuron can be expressed as Eq. (1.9):

$$\Delta_k^l = \frac{\partial E}{\partial y_k^l} \frac{\partial y_k^l}{\partial x_k^l} = \frac{\partial E}{\partial y_k^l} f'(x_k^l) \tag{1.9}$$

For a neuron k in the output layer L, Eq. (1.9) can easily be simplified as Eq. (1.10):

$$\Delta_k^L = (y_k^l - t_k) \cdot f'(x_k^l) \tag{1.10}$$

However, for the neurons in layers 1 to $L - 1$, we need to derive a formula for computing the derivative $\frac{\partial E}{\partial y_k^l}$. As illustrated in Fig. 1.6, the output of the kth neuron of layer l (i.e., y_k^l) contributes to the input of neurons of the next layer $l + 1$. Accordingly, it can be proven that the derivative of the error with respect to y_k^l is Eq. (1.11):

$$\frac{\partial E}{\partial y_k^l} = \sum_{i=1}^{N_{l+1}} \frac{\partial E}{\partial x_i^{l+1}} \frac{\partial x_i^{l+1}}{\partial y_k^l} = \sum_{i=1}^{N_{l+1}} \Delta_i^{l+1} w_{ki}^l \tag{1.11}$$

Substituting Eq. (1.11) in Eq. (1.9), the delta error corresponding to any neuron k in layers 1 to $L - 1$ can be expressed as Eq. (1.12):

$$\Delta_k^l = f'(x_k^l) \sum_{i=1}^{N_{l+1}} \Delta_i^{l+1} w_{ki}^l \tag{1.12}$$

Equation (1.12) suggests that the delta errors, and hence the sensitivity factors, of a certain layer l depend on the delta errors of the next layer $l + 1$. Therefore, in the backpropagation algorithm, we start by calculating the delta errors of the output layer L, then we move in the backward direction and calculate the delta errors starting from the last hidden layer $L - 1$ until arriving at the input layer. Once the delta errors and sensitivity factors are computed for all neurons, the weights and thresholds can be updated according to the gradient descent method as expressed by the following Eqs. (1.13, 1.14):

$$w_{ik}^l(t+1) = w_{ik}^l(t) - \varepsilon \frac{\partial E}{\partial w_{ik}^{l-1}} \qquad (1.13)$$

$$\theta_k^l(t+1) = \theta_k^l(t) - \varepsilon \frac{\partial E}{\partial \theta_k^l} \qquad (1.14)$$

where ε is called the learning factor. The standard backpropagation algorithm for training of MLP networks are summarized below:

1. Select the hyperparameters that determine the architecture of the ANN (number of hidden layers and number of neurons in each hidden layer).
2. Initialize the weights and thresholds by uniformly distributed random numbers $U(-a, a)$.
3. Pick an input-output sample (or a batch of input-output samples) and forward-propagate its input through the network according to Eqs. (1.3) and (1.4).
4. Compute the error between the actual output and the target value associated with the forward-propagated input-output sample(s) as expressed by Eq. (1.5).
5. Calculate the delta errors of the output layer according to Eq. (1.10).
6. Calculate the delta errors of the other layers starting from the hidden layer $L-1$ until the input layer according to Eq. (1.12).
7. Obtain the sensitivity factors as explained by Eqs. (1.6), (1.7), and (1.8).
8. Apply the gradient descent method to update the weights and thresholds according to Eqs. (1.13) and (1.14).
9. Repeat steps 3 to 8 for each sample (or batch of samples) in the training dataset.
10. Repeat steps 3 to 9 for a sufficient number of *epochs* until a certain stopping criterion is satisfied. An epoch is defined as one complete cycle of the described training dataset.

3.1.2 Batch Training vs. Online Training

Two general backpropagation approaches can be used for training ANNs. The difference between these two approaches lies only in the way the weights and thresholds of the ANN are updated. In the first approach, referred to as online training, the ANN parameters are updated for every single input-output sample. Therefore, for a dataset composed of n training samples, the network is updated n times for each training epoch. This can be represented by the following piece of pseudocode:

> *for each training epoch, DO:*
>   *for each input-output sample, DO:*
>   *compute the sensitivity factors for the current sample*
>   *update the weights and thresholds using the sensitivity factors*
>   *end for*
> *end for*

In the second approach, usually called "batch training", the entire dataset is forward and backpropagated through the network, and all sensitivity factors are accumulated and then used to update the network. Accordingly, the network's parameters are updated only once per epoch as explained by the following pseudocode:

for each training epoch, DO:
  for each input-output sample, DO:
      compute the sensitivity factors for the current sample
      accumulate the sensitivity factors
    end for
    update the weights and thresholds using the accumulated sensitivity factors
  end for

Online training is generally easier to implement and more efficient than batch training in dealing with challenging classification problems (Haykin, 2008). On the other hand, batch training parallelizes the ANN training process since the sensitivity factors for each sample can be computed in parallel (i.e., independent from the sensitivities of the other samples). This can increase the training speed dramatically when using multi-core CPUs.

3.1.3 Random Initialization

As mentioned earlier, the first step toward training an MLP network is to randomly assign its connection weights. The weights are then adjusted by applying the backpropagation algorithm on a training dataset. Investigations on deep MLP networks (i.e., MLPs with many hidden layers) showed that the training process can be negatively affected by the phenomenon of "vanishing gradients" (He et al., 2015). This phenomenon greatly depends on the initial random weights and the type of activation function used in the MLP neurons. To understand this phenomenon, consider the sigmoid activation function shown in Fig. 1.7. A neuron is called 'saturated' when the input to its activation function approaches the extreme ends. As the neuron gets saturated, it is clear from the plot that the derivative of the activation function approaches zero. When training an improperly initialized deep network, it is very likely that the neurons at the last hidden layer get saturated quickly, leading

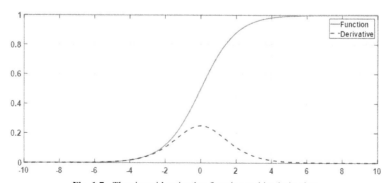

Fig. 1.7: The sigmoid activation function and its derivative.

to small gradients. When backpropagating these small gradients through the layers of the network, the gradients become increasingly smaller until they completely disappear at the first few layers. This can lead to poor training performance since the first hidden layers of deep ANNs are usually responsible for the feature extraction process.

Glorot and Bengio (Glorot and Bengio, 2010) investigated the relationship between weight initialization and the phenomenon of vanishing gradients for several types of activation functions. They showed statistically that such a phenomenon can be avoided by initializing deep networks according to the formulas given in Table 1.1 (Glorot and Bengio, 2010).

Table 1.1: Recommended formulas for weight initialization.

Activation Function	Uniform Distribution $U(-a, a)$
tanh	$a = 4\sqrt{\dfrac{6}{n_i + n_o}}$
Logistic function	$a = \sqrt{\dfrac{6}{n_i + n_o}}$

Note: The notation n_i denotes the number of neurons in the preceding layer, while n_o is the number of neurons in the next layer (Glorot and Bengio, 2010).

3.1.4 Overfitting vs. Generalization

Training of neural networks by backpropagation can be viewed as a nonlinear curve-fitting problem in which the objective is to obtain the correct mapping from the inputs of the training dataset to their associated targets. In some cases, when the selected architecture of an ANN is overly complex or when few training samples are used, the training process might lead to an overfitted neural network. Overfitting, which is the opposite of generalization, implies that the ANN can only deal with the exact input samples used during the training process. Using an overfitted ANN to predict the output of an input sample that is slightly different from the training samples might result in erroneous results. Overtraining causes ANNs to curve fit the noise in the training data rather than learning the underlying process itself.

Therefore, it is critical to check the ability of trained ANNs to generalize. To do so, the data is usually divided into training and validation datasets. Among these datasets, the training dataset is utilized to train the network, while the validation set is utilized afterward to test the network against overfitting.

4. Convolutional Neural Networks (CNNs)

CNNs are another type of supervised multilayer feed-forward ANNs. As mentioned earlier, the structural body and organization of CNNs are predominantly inspired by the simple and complex cells in the primary cortex of the mammalian brain (Kiranyaz et al., 2017). CNNs have recently become the standard for DL procedures such as the classification of large image databases and face recognition (Kiranyaz et al., 2016; KrizhKrizhevsky et al., 2012). This can be attributed to their inherent ability to merge the feature extraction and feature classification operations into a single learning body (Kiranyaz et al., 2016). An important characteristic of CNNs is their capability to classify images independent of their scale and orientation.

4.1 The Convolution Operation

As the name suggests, convolutional neural networks rely mainly on a mathematical operation called convolution. In CNNs terminology, an input *I* is convolved with a 2D *filter* or *kernel w* to produce a *feature map S*. For a 2D input (e.g., image), a standard convolution operation in CNNs, denoted by the operator (∗), can be written as Eq. (1.15):

$$I * w = S(i, j) = \sum_m \sum_n I(i+m, j+n)w(m, n) \tag{1.15}$$

The equation above is illustrated graphically in Fig. 1.8. As shown in the figure, standard convolution is carried out by sliding the filter within the boundaries of the input image in a pixel-by-pixel manner. Therefore, each pixel in the output feature map corresponds to a specific location of the kernel over the input image. Since the kernel is not allowed to cross the edges of the image, standard convolution reduces the image dimensions by $Kx -1$, $Ky -1$ where Kx and Ky are the dimensions of the applied kernel. In Fig. 1.8 for example, the 4×4 input image was reduced to a 3×3 feature map.

In some cases, the size of the image must remain unaltered after convolution. This is why zero-padding is sometimes applied by adding $(K_x -1, K_y -1)$ zeros to the boundaries of the image before convolution. This operation, depicted in Fig. 1.9, is usually called "full convolution".

4.2 Structure of CNNs

To explain the structure of CNNs, Fig. 1.10 shows an example CNN used to classify a 24×24-pixel grayscale image into two categories. The first layer (i.e., the input layer)

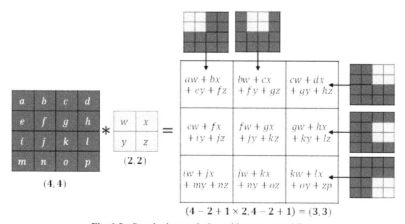

Fig. 1.8: Standard convolution without zero-padding.

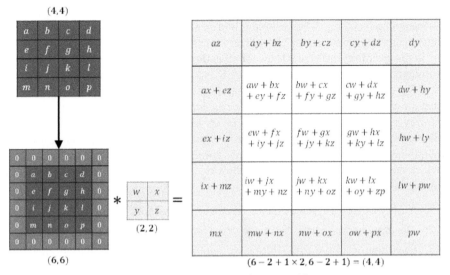

Fig. 1.9: Full convolution with zero-padding.

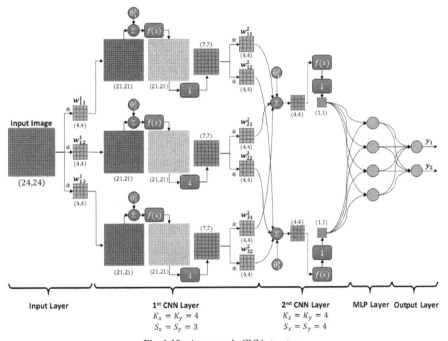

Fig. 1.10: An example CNN structure.

is made of a single neuron accepting 24×24 images. The input layer is followed by a number of interconnected convolutional neurons arranged in *two hidden CNN layers*. The output of the CNN layers is processed by a single MLP hidden layer and finally an output layer.

Each convolutional neuron is assigned by a certain kernel size (K_x, K_y), subsampling factor (S_x, S_y), threshold θ, and an activation function f. Also, each of the interconnections between the CNN layers is assigned a $K_x \times K_y$ kernel (i.e., weight), w. In this example, the kernel sizes for all neurons were set to $K_x = K_y = 4$, while the subsampling factors were chosen as $S_x = S_y = 3$ for the first convolutional layer, and $S_x = S_y = 4$ for the second one. Note that these values were deliberately selected so that the output of the last CNN layer (i.e., the input to the first MLP layer) is a scalar. However, the number of MLP hidden layers along with the number of neurons in the CNN and MLP layers were arbitrarily chosen just for illustration purposes. The output layer consists of two MLP neurons corresponding to the number of classes to which the image is categorized. The following steps describe a complete FP process in this example CNN:

1. A 24×24-pixel grayscale image is received by the input neuron.
2. The input neuron sends the image to the three neurons in the first CNN layer.
3. Each neuron in the first CNN layer applies a standard convolution operation (i.e., without zero-padding) between the image and the associated weighting kernel, resulting in a 21×21 feature map.
4. The CNN neuron then sums the feature map to a threshold and then applies the activation function.
5. The output of the activation function (21×21) is decimated by a subsampling factor of 3, resulting in a 7×7 map.
6. The output of the three neurons in the first hidden CNN layer is sent to the second one. Therefore, each neuron in the second CNN layer receives three input maps.
7. Similarly, each neuron applies standard convolution between each input map and its corresponding kernel, resulting in three 4×4 feature maps.
8. The three feature maps are summed together with a threshold. The output of the summation is activated by the activation function.
9. The activated output (4×4) is decimated by a subsampling factor of 4, resulting in a 1×1 (i.e., scalar) output.
10. The scalar outputs of the last CNN layer are forward propagated through the following MLP and output layers as explained in Section 2.3 to produce the final output that represents the classification of the input image.

From this example, the following conclusions can be drawn regarding the structure and components of CNNs:

- CNNs are composed of a number of hidden CNN layers usually, but not necessarily, followed by some MLP layers.
- The convolutional neurons in CNN layers are capable of performing standard convolution, aggregation, activation, and subsampling operations.
- The structure of the CNN is determined by the following hyperparameters:

○ The number of CNN and MLP hidden layers.

○ The number of neurons in each hidden layer.

○ The kernel sizes (K_x, K_y) and subsampling factors (S_x, S_y) for the convolutional neurons.

• In FP, the size of the input image gets gradually reduced by successive convolution and subsampling operations. The number of hidden CNN layers along with the kernel size and subsampling factor of each layer should be selected in a way that forces scalar outputs at the last CNN layer.

4.3 Advantages of CNNs

In Fig. 1.10, each convolutional neuron in the first hidden CNN layer was connected to the input image by a single kernel composed of only 3×3=9 weights. Now suppose that a regular MLP network was used to classify the same image instead of the CNN. In this case, the input layer would have 24×24 neurons for receiving the pixels of the image. As explained in Section 2.3, since MLP networks are fully connected, each neuron in the first hidden layer would have 24×24=576 connections to the input layer assigned by 576 scalar weights.

This example demonstrates the significant difference between CNN and fully connected MLP networks in terms of the number of parameters that need to be tuned during the training process. This can be attributed to the fact that CNN layers are sparsely connected, which means that the outputs of a CNN neuron are only correlated to small portions of the input map. Furthermore, it is obvious that in a standard convolution operation inside a CNN neuron, the same kernel is used over the entire image. Hence, rather than learning an individual weight for each pixel of the image (as in MLPs), a CNN neuron only learns a small set of parameters and applies it over the whole input map. Having shared parameters, or in other words, "tied weights" significantly reduce the memory and computational time required for carrying out the training process and allows the parallel implementation of CNNs (Haykin, 2008).

Moreover, the subsampling process allows CNN neurons to extract features that are invariant to small translations and distortions. This is particularly important when the objective is to detect certain features rather than finding their exact locations (Goodfellow et al., 2017). The subsampling process also allows CNNs to process images of different sizes just by changing the subsampling factors accordingly.

To summarize, the main advantages of using CNNs in DL applications are:

1. CNNs have the capability of fusing the feature extraction and feature classification stages into a single model. They can learn what features to extract and how to classify these features directly from the raw input.

2. Because the CNN neurons are sparsely connected with tied weights, they can operate with large inputs with superior computational efficiency.

3. CNNs are immune to small transformations in the input data including distortion, skewing, scaling, and translation.

4. CNNs can adapt to different input sizes.

4.4 Backpropagation Algorithm in CNNs

Similar to MLPs, CNNs are trained in a supervised manner using a dataset of input-output samples. The input is an image (or color channels of the image), while the output is the corresponding class vector of the input image. In the example shown in Fig. 1.11, the output vector can be either [1 0] indicating that the image belongs to class 1, or [0 1] if the image belongs to class 2. This section describes the backpropagation algorithm by which the training dataset is used to train CNNs (Kiranyaz et al., 2016).

Consider a CNN consisting of L layers (input layer + output layer + L_a hidden CNN layers + L_b hidden MLP layers). Each layer consists of N_v neurons, where $v = 1,2, ..., L$ is the index of the layer. Figure 1.12 shows a closer look at the hidden CNN layer l which lies between two CNN layers, $l - 1$ and $l + 1$. In FP, the input image is firstly propagated through the hidden CNN layers. An FP operation from CNN layer $l - 1$ to the kth neuron of layer l can be written as Eqs. (1.16, 1.17, 1.18):

$$x_k^l = b_k^l + \sum_{i=1}^{N_{l-1}} \text{Conv}(w_{ik}^{l-1}, s_i^{l-1}) \tag{1.16}$$

$$y_k^l = f(x_k^l) \tag{1.17}$$

$$s_k^l = \text{down}_{(s_x, s_y)}(y_k^l) \tag{1.18}$$

where the operator $\text{Conv}(\cdot)$ Indicates a standard convolution operation (i.e., without zero-padding), w_{ik}^{l-1} denotes the kernel weight connecting the kth neuron of layer l to the ith neuron of layer $l - 1$, s_i^{l-1} is the output of the ith neuron of layer $l - 1$, s_k^l is the final output of the kth neuron of layer l, and b_k^l is the threshold associated with the current neuron. The operator $\text{down}_{(s_x, s_y)}$ represents a down-sampling operation

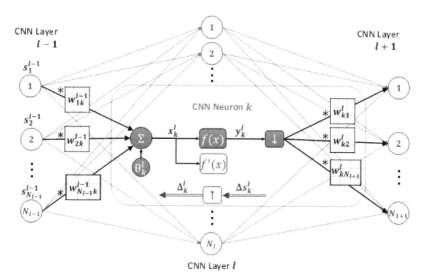

Fig. 1.11: Forward-propagation and back-propagation in hidden CNN layers.

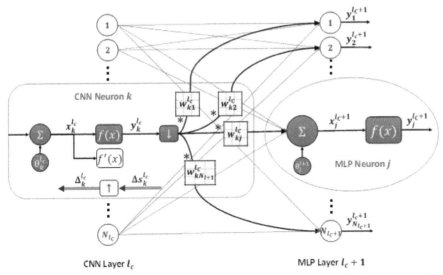

Fig. 1.12: Forward-propagation and back-propagation between the last hidden CNN layer and the first hidden MLP layer.

by average pooling according to the subsampling factors S_x and S_y. Note that the superscripts in the above equation denote the layer to which the quantities belong.

As mentioned earlier, the kernel sizes (K_x, K_y) and subsampling factors (S_x, S_y) of the hidden CNN layers should be carefully chosen such that the outputs of the last CNN layer are scalars. These scaler outputs are then propagated through the MLP according to Eqs. (1.3) and (1.4) until an output vector $Y = y_1^L, y_2^L, ..., y_{N_L}^L$ emerges from the output layer L. Supposing that the input image used in this FP operation is associated with the desired target vector $T = t_1, t_2, ..., t_{N_L}$, the error between the actual and desired outputs E can be computed according to the cost function defined in Eq. (1.5).

The next step is to backpropagate (BP) this error from the output layer through the hidden MLP layers then the CNN layers and finally to the input layer. During this BP process, the sensitivity factor of each weight and threshold in the network to the error is computed. The delta errors and sensitivity factors of the output and the hidden MLP layers can be easily calculated as explained previously in Eqs. (1.6), (1.7), (1.8), (1.10), (1.12).

After that, the error is backpropagated from the first MLP layer $l_c + 1$ to the last CNN layer l_c. Figure 1.12 shows the connection between the jth neuron of the layer $l_c + 1$ and the kth neuron of the layer l_c. The *down-sampled delta error* $\Delta s_k^{l_c}$ of the CNN neurons of the last hidden CNN layer l_c can be computed in terms of the associated kernels $w_{ki}^{l_c}$ along with the delta errors of the first MLP layer $l_c + 1$ as follows Eq. (1.19) (Kiranyaz et al., 2016):

$$\Delta s_k^{l_c} = \frac{\partial E}{\partial s_k^{l_c}} = \sum_{i=1}^{N_{l_c+1}} \frac{\partial E}{\partial x_i^{l_c+1}} \frac{\partial x_i^{l_c+1}}{\partial s_k^{l_c}} = \sum_{i=1}^{N_{l_c+1}} \Delta_i^{l_c+1} w_{ki}^{l_c} \qquad (1.19)$$

Then, as given in Eq. (1.20) this down-sampled delta error can be used to compute the delta error $\Delta_k^{l_c}$:

$$\Delta_k^{l_c} = \frac{1}{S_x S_y} \mathrm{up}_{(S_x, S_y)}(\Delta s_k^{l_c}) f'(x_k^{l_c}) \tag{1.20}$$

where the operator $\mathrm{up}_{(S_x, S_y)}$ represents an up-sampling operation by the factors S_x and S_y. Note that the exponent $1/S_x S_y$ is introduced here since, according to Eq. (1.17), each pixel in the output $s_k^{l_c}$ was computed by averaging $S_x S_y$ pixels of the output of the activation function $y_k^{l_c}$.

For the CNN neurons in the other CNN layers (layers 1 to $l_c - 1$), it can be shown that the down-sampled delta error Δs_k^l and the delta error Δ_k^l of a neuron in the CNN hidden layer l are given in Eqs. (1.21, 1.22) (Kiranyaz et al., 2016):

$$\Delta s_k^l = \sum_{i=1}^{N_{l+1}} \mathrm{ConvZ}\left(\Delta_i^{l+1}, \mathrm{rot}180(w_{ki}^l)\right) \tag{1.21}$$

$$\Delta_k^l = \frac{1}{S_x S_y} \mathrm{up}_{(S_x, S_y)}(\Delta s_k^l) f'(x_k^l) \tag{1.22}$$

where $\mathrm{ConvZ}(\cdot)$ represents a full convolution operation (i.e., with $K_x - 1$, $K_y - 1$ zero-padding), Δ_i^{l+1} is the delta error of the next CNN layer, w_{ki}^l are the kernels associated with the current neuron, f' is the derivative of the activation function, and the operator $\mathrm{rot}180(\cdot)$ denotes a $180°$ matrix rotation operation.

Using the delta errors computed by Eqs. (1.20) and (1.22), the sensitivity factors of all neurons in the network (including the neurons of the last CNN layer l_c) can be obtained as Eqs. (1.23, 1.24):

$$\frac{\partial E}{\partial w_{ki}^l} = \mathrm{conv}(s_k^l, \Delta_i^{l+1}) \tag{1.23}$$

$$\frac{\partial E}{\partial \theta_{ki}^l} = \sum_m \sum_n \Delta_k^l(m, n) \tag{1.24}$$

where m and n are the dimensions of the matrix Δ_k^l. Finally, once the delta errors and sensitivity factors are computed for all neurons, the weights and thresholds can be updated according to the gradient descent method as expressed in Eqs. (1.13) and (1.14).

The backpropagation algorithm for training of CNNs can be summarized in the following steps:

1. Choose the hyperparameters that determine the architecture of the CNN:
    a. The number of hidden CNN layers.
    b. The number of hidden MLP layers.
    c. The number of neurons in each hidden layer.
    d. Kernel sizes of each layer (K_x, K_y).
    e. Subsampling factors of each layer (S_x, S_y).

2. Initialize the kernels, weights, and thresholds by uniformly distributed random numbers $U(-a, a)$.

3. Pick an input-output sample (or a batch of input-output samples for the case of batch training).

4. Forward-propagate the input of the selected sample through the CNN layers according to Eqs. (1.16–1.18).

5. Forward-propagate the output of the last CNN layer through the MLP layers according to Eqs. (1.3) and (1.4).

6. Compute the error between the actual output and the desired target associated with the forward-propagated input-output sample(s) as expressed by Eq. (1.5).

7. Calculate the delta errors of the output layer according to Eq. (1.10).

8. Calculate the delta errors of the MLP layers starting from layer $L-1$ until the first MLP layer $l_c + 1$ according to Eq. (1.12).

9. Calculate the delta errors of the last CNN layer l_c according to Eqs. (1.19) and (1.20).

10. Calculate the delta errors of the remaining CNN layers starting from layer $l_c - 1$ until the input layer according to Eqs. (1.21) and (1.22).

11. Obtain the sensitivity factors of the MLP layers as explained by Eqs. (1.6), (1.7), and (1.8).

12. Obtain the sensitivity factors of the CNN layers as explained by Eqs. (1.23) and (1.24).

13. Apply the gradient descent method to update the weights and thresholds according to Eqs. (1.13) and (1.14).

14. Repeat steps 3 to 13 for each sample (or batch of samples for the case of batch training) in the training dataset.

15. Repeat steps 3 to 14 for a sufficient number of *epochs* until a certain stopping criterion is satisfied. An epoch is defined as one complete pass of the whole training dataset.

4.5 One-Dimensional Convolutional Neural Networks

The CNNs explained so far are designed to run solely on 2D data like videos and images. Therefore, they are known as 2D CNNs. Nevertheless, there have been several attempts to apply 2D CNNs to deal with naturally 1D signals. Different techniques were proposed to represent 1D data in a 2D manner including simply reshaping the input array into a matrix (Wei et al., 2017) and concatenation of multiple 1D signals into a 2D 'image' (Janssens et al., 2016).

Alternatively, a modified version of 2D CNNs has been developed to run for one-dimensional applications. The modified version is called One-Dimensional Convolutional Neural Networks (1D CNNs) (Kiranyaz et al., 2016). 1D CNNs are generally similar to their 2D version except for some structural differences. The weighting kernels of 1D CNNs are 1D arrays (instead of the 2D kernels in 2D CNNs). Therefore, all the 2D FP and BP operations explained in Section 4.3 should be replaced by their 1D counterparts. Also, the 2D hyperparameters (K_x, K_y) and (S_x, S_y) are substituted by single scalers K and S. As a result, the Eqs. (1.25–1.27) for FP operation in 1D CNNs can be expressed as follows:

$$x_k^l = b_k^l + \sum_{i=1}^{N_{l-1}} \text{conv1D}(w_{ik}^{l-1}, s_i^{l-1}) \tag{1.25}$$

$$y_k^l = f(x_k^l) \tag{1.26}$$

$$s_k^l = \text{down}_s(y_k^l) \tag{1.27}$$

where $\text{conv1D}(\cdot)$ represents a standard 1D convolution operation conducted by sliding the kernel array w_{ik}^{l-1} over the array s_i^{l-1} without zero-padding, and $\text{down}_s(\cdot)$ denotes a down-sampling operation by average pooling according to a subsampling factor of S.

Also, the BP Eqs.(1.28–1.31) for computing the deltas and sensitivity factors of the CNN layers in 1D CNNs are as follows:

$$\Delta s_k^l = \sum_{i=1}^{N_{l+1}} \text{conv1Dz}[\Delta_i^{l+1}, \text{rev}(w_{ki}^l)] \tag{1.28}$$

$$\Delta_k^l = \frac{1}{S} \text{up}_s(\Delta s_k^l) f'(x_k^l) \tag{1.29}$$

$$\frac{\partial E}{\partial w_{ki}^l} = \text{conv1D}(s_k^l, \Delta_i^{l+1}) \tag{1.30}$$

$$\frac{\partial E}{\partial \theta_{ki}^l} = \sum_n \Delta_k^l(n) \tag{1.31}$$

where $\text{conv1Dz}(\cdot)$ represents a full 1D convolution operation with $K-1$ zero-padding, $\text{rev}(\cdot)$ denotes an array reversing operation, $\text{up}_s(\cdot)$ is an up-sampling operation by to a factor of S, and n is the number of elements in the array Δ_k^l.

It can be noticed from Eqs. (1.25–1.31) that 1D CNNs are computationally far more efficient than their 2D counterparts in terms of operating on 1D data due to various factors:

- Current research revealed that 1D CNNs with relatively shallow architectures (i.e., a small number of hidden layers and neurons) are capable of learning complex tasks via 1D signals such as the classification of electrocardiogram (ECG) signals (Kiranyaz et al., 2017). While the networks with shallow architectures are relatively easier to implement and train, for the tasks about the same difficulty, 2D CNNs usually require deeper architectures.
- The special hardware setup is required for the training of deep 2D CNNs whereas even low-power computers are sufficient for 1D CNN training.
- 1D CNNs are well-suited for real-time applications thanks to their low computational requirements. The use of 1D CNNs in real-time structural damage detection has been shown to result in successful performance.

5. 1D CNN Use in Civil Infrastructure Applications

As mentioned earlier, 1D CNNs are more efficient than their 2D CNNs in terms of operating on 1D data, and this is the reason why compact 1D CNNs have been preferred over 2D CNNs in various science and engineering applications. 1D CNNs have been extensively used in various fields of engineering (Kiranyaz et al., 2019; Kiranyaz et al., 2021). There has been successful use of 1D CNNs in Real-time electrocardiogram (ECG) monitoring (Kiranyaz et al., 2017); automatic speech recognition, condition monitoring of rotating machinery (Abdeljaber et al., 2019; Avci et al., 2021), and on high-power multilevel converters (Kiranyaz et al., 2018). For civil infrastructure, CNNs have been used predominantly for damage detection in structural health monitoring applications (Avci et al., 2021; Avci et al., 2021). Damage is an important indicator for the integrity and safety of civil engineering structures, and it has always drawn attention in structural health monitoring research (Chaabane et al., 2017; Mansouri et al., 2015; Mansouri et al., 2015). Vibration response of structures has always been a useful tool utilized for modal testing (Mosallam et al., 2018; Chinka et al., 2021; Avci and Murray, 2012; Han et al., 2020; Avci, 2012; Avci, 2017; Ewins and Saunders, 1986; Avci et al., 2010; Avci, 2015; Avci, 2005), vibrations serviceability (Barrett et al., 2006; Younis et al., 2017; Muhammad et al., 2018), condition assessment (Haji Agha Mohammad Zarbaf et al., 2018; Khodabandehlou et al., 2019; Sony et al., 2019; Delgadillo and Casas 2020; Jablonski, 2021), and monitoring (Celik et al., 2016; Catbas et al., 2017) purposes for civil infrastructure. Vibration response is also used in structural damage detection with 1D CNNs resulting in successful performance. The complete vibration-based structural damage detection process includes detecting the damage as a first step, localizing the damage as a second step, and quantifying the damage as the third and final step. In the literature, the studies usually focus on only one or more of these steps.

The first use of 1D CNNs in vibration-based damage detection was by Abdeljaber et al. (Abdeljaber et al., 2017). They used a large-scale steel laboratory structure for experimentation using one accelerometer per node, for 30 nodes (Abdeljaber et al., 2016). It was reported that the steel frame structure was arguably the largest instrumented mock-up stadium structure built in a laboratory (Fig. 1.13). The

Fig. 1.13: Mock-up steel grid of laboratory stadium structure with 4.2 m×4.2 m footprint dimensions.

Fig. 1.14: Wireless sensors used on the laboratory grid structure.

artificial 'damage' introduced to the structure was purely loosening the steel bolts at a filler beam connection (Avci et al., 2021). While this an insignificant change on the rotational stiffness of the connection, all steps of the structural damage detection process were accomplished with the use of 1D CNNs (Avci et al., 2017). For the damage detection process, individual 1D CNNs were trained for 31 damage conditions (one damage condition per node, and one undamaged condition where none of the bolts were loosened) by recording the vibration response via shaker excitation for each condition. Each 1D CNN was assigned to a node for processing the local data via the accelerometer placed at the node. The system was tested for multiple single and double damage conditions and succeeded in all of them. Based on the accompanying complexity analysis, it was demonstrated that using a standard PC, the damage detection speed was 45 times faster than the real-time requirement, which was published as a patent (Kiranyaz et al., 2019). After the wired accelerometer tests with shaker excitations, the same group of authors carried on with additional tests based on a modified 1D CNN methodology with wireless sensors on the same laboratory structure (Fig. 1.14) and achieved successful damage detection results, again (Avci et al., 2018; Avci et al., 2020). According to computational complexity analysis, with the wireless sensors, the damage detection processing speed was 37 times faster than the real-time requirement. With two CNN layers (each layer with four neurons), the tests with wireless sensors resulted in superior performance even though the training was conducted in ambient vibration conditions without any shaker excitations. The same group of authors continued with the use of 1D CNNs on an existing Benchmark data (Dyke et al., 2003) created by others. The authors developed an updated damage detection system that requires less effort for training the CNN classifiers. It was demonstrated in (Abdeljaber et al., 2017; Avci et al., 2018) that the updated procedure was again successful for damage detection and quantification using the benchmark data.

In a recent article, 1D CNNs were used to evaluate damage severity with a simple Finite Element (FE) cantilever model with non-rigid rotational spring support (Almutairi et al., 2020; Almutairi et al., 2021). Damage was introduced at the support location by reducing rotational spring stiffnesses and 1D CNNs were reported to be working efficiently in the assessment of damage severity. In another recent study, 1D-CNNs were utilized for autonomous defect detection on bridge decks via ground penetrating radar (Ahmadvand et al., 2021).

6. Conclusions

Convolutional Neural Networks (CNNs) are a type of feed-forward ANNs used extensively in various ML applications that perform consecutive convolution and pooling operations in a hierarchical architecture. The 1D CNNs have recently achieved superior performance in various engineering fields with an elegant computational efficiency. In this chapter, the authors presented the evolution of ANNs into CNNs demonstrating the predominant concepts and step-by-step procedures to display the way 1D CNNs operate. The chapter started with the motivation behind the AI systems and continued with an overview of ANNs. Then, MLPs were explained; and training of 2D and 1D CNNs was presented. Finally, 1D CNN applications on civil infrastructure were briefly discussed.

Based on the structural damage detection case study results utilizing compact 1D CNNs, it was reported that 1D CNNs can precisely distinguish uncorrelated and complex acceleration time history data for detecting, locating, and quantifying structural damage on a large-scale laboratory structure, and on an existing benchmark dataset. While shaker excitations were used for wired sensor tests and ambient vibration conditions were used for wireless sensor tests, the computational complexity analysis results showed that the total time required for the classification of a 1-sec signal was 45 times and 37 times faster than the real-time requirements, respectively for the wired and wireless sensor tests.

References

Abdeljaber, O. and Avci, O. 2016. Nonparametric structural damage detection algorithm for ambient vibration response: Utilizing artificial neural networks and self-organizing maps. *J. Archit. Eng.*, 22. https://doi.org/10.1061/(ASCE)AE.1943-5568.0000205.

Abdeljaber, O., Avci, O. and Inman, D.J. 2015. Optimization of chiral lattice based metastructures for broadband vibration suppression using genetic algorithms. *J. Sound Vib.* https://doi.org/10.1016/j.jsv.2015.11.048.

Abdeljaber, O., Avci, O. and Inman, D.J. 2016. Active vibration control of flexible cantilever plates using piezoelectric materials and artificial neural networks. *J. Sound Vib.* p. 363. https://doi.org/10.1016/j.jsv.2015.10.029.

Abdeljaber, O., Avci, O. and Inman, D.J. 2016. Genetic algorithm use for internally resonating lattice optimization: Case of a beam-like metastructure. *In: Conf. Proc. Soc. Exp. Mech. Ser.* https://doi.org/10.1007/978-3-319-29751-4_29.

Abdeljaber, O., Avci, O., Do, N.T., Gul, M., Celik, O. and Necati Catbas, F. 2016. Quantification of structural damage with self-organizing maps. *Conf. Proc. Soc. Exp. Mech. Ser.* https://doi.org/10.1007/978-3-319-29956-3_5.

Abdeljaber, O., Avci, O., Kiranyaz, M.S., Boashash, B., Sodano, H. and Inman, D.J. 2017. 1-D CNNs for structural damage detection: Verification on a structural health monitoring benchmark data. *Neurocomputing.* https://doi.org/10.1016/j.neucom.2017.09.069.

Abdeljaber, O., Avci, O., Kiranyaz, S. and Inman, D.J. 2017. Optimization of linear zigzag insert metastructures for low-frequency vibration attenuation using genetic algorithms. *Mech. Syst. Signal Process.* 84. https://doi.org/10.1016/j.ymssp.2016.07.011.

Abdeljaber, O., Avci, O., Kiranyaz, S., Gabbouj, M. and Inman, D.J. 2017. Real-time vibration-based structural damage detection using one-dimensional convolutional neural networks. *J. Sound Vib.*, 388. https://doi.org/10.1016/j.jsv.2016.10.043.

Abdeljaber, O., Sassi, S., Avci, O., Kiranyaz, S., Ibrahim, A.A. and Gabbouj, M. 2019. Fault detection and severity identification of ball bearings by online condition monitoring. *IEEE Trans. Ind. Electron.* https://doi.org/10.1109/TIE.2018.2886789.

Abdeljaber, O., Younis, A., Avci, O., Catbas, N., Gul, M., Celik, O. and Zhang, H. 2016. Dynamic testing of a laboratory stadium structure. *In: Geotech. Struct. Eng. Congr.* 2016: pp. 1719–1728. https://doi.org/10.1061/9780784479742.147.

Ahmadvand, M., Dorafshan, S., Azari, H. and Shams, S. 2021. 1D-CNNs for autonomous defect detection in bridge decks using ground penetrating radar. https://doi.org/10.1117/12.2580575.

Allen, N.J. and Eroglu, C. 2017. Cell biology of astrocyte-synapse interactions. *Neuron.* https://doi.org/10.1016/j.neuron.2017.09.056.

Almutairi, M., Avci, O. and Nikitas, N. 2020. Efficiency of 1d cnns in finite element model parameter estimation using synthetic dynamic responses. *In: Proc. Int. Conf. Struct. Dyn., EURODYN.* https://doi.org/10.47964/1120.9009.19640.

Almutairi, M., Nikitas, N., Abdeljaber, O., Avci, O. and Bocian, M. 2021. A methodological approach towards evaluating structural damage severity using 1D CNNs. *Structures*, 34: 4435–4446. https://doi.org/10.1016/j.istruc.2021.10.029.

Arangio, S. and Bontempi, F. 2014. Structural health monitoring of a cable-stayed bridge with Bayesian neural networks. *Struct. Infrastruct. Eng.*, 11: 575–587. https://doi.org/10.1080/15732479.2014.951867.

Avci, O. 2005. *Effects of Bottom Chord Extensions on the Static and Dynamic Performance of Steel Joist Supported Floors.* Virginia Polytechnic Institute and State University.

Avci, O. 2012. Retrofitting steel joist supported footbridges for improved vibration response. *In: Struct. Congr. 2012—Proc. 2012 Struct. Congr.* https://doi.org/10.1061/9780784412367.041.

Avci, O. 2015. Modal parameter variations due to joist bottom chord extension installations on laboratory footbridges. *J. Perform. Constr. Facil.* https://doi.org/10.1061/(asce)cf.1943-5509.0000635.

Avci, O. 2017. Nonlinear damping in floor vibrations serviceability: Verification on a laboratory structure. *In: Conf. Proc. Soc. Exp. Mech. Ser.* https://doi.org/10.1007/978-3-319-54777-0_18.

Avci, O. and Abdeljaber, O. 2016. Self-organizing maps for structural damage detection: A novel unsupervised vibration-based algorithm. *J. Perform. Constr. Facil.* 30. https://doi.org/10.1061/(ASCE)CF.1943-5509.0000801.

Avci, O. and Murray, T.M. 2012. Effect of bottom chord extensions on the static flexural stiffness of open-web steel joists. *J. Perform. Constr. Facil.* https://doi.org/10.1061/(asce)cf.1943-5509.0000262.

Avci, O., Abdeljaber, O. and Kiranyaz, S. 2022. An overview of deep learning methods used in vibration-based damage detection in civil engineering. *In:* Grimmelsman, K. (ed.). *Dynamics of Civil Structures*, Volume 2. Conference Proceedings of the Society for Experimental Mechanics Series. Springer, Cham. https://doi.org/10.1007/978-3-030-77143-0_10.

Avci, O., Abdeljaber, O. and Kiranyaz, S. 2022. Structural damage detection in civil engineering with machine learning: Current state of the art. *In:* Walber, C., Stefanski, M. and Seidlitz, S. (eds.). *Sensors and Instrumentation, Aircraft/Aerospace, Energy Harvesting & Dynamic Environments Testing*, Volume 7. Conference Proceedings of the Society for Experimental Mechanics Series. Springer, Cham. https://doi.org/10.1007/978-3-030-75988-9_17.

Avci, O., Abdeljaber, O., Kiranyaz, S. and Inman, D. 2017. Structural damage detection in real time: implementation of 1D convolutional neural networks for shm applications. pp. 49–54. *In:* Niezrecki, C. (ed.). *Struct. Heal. Monit. Damage Detect.* Vol. 7. *Proc. 35th IMAC, A Conf. Expo. Struct. Dyn.* 2017. Cham: Springer International Publishing. https://doi.org/10.1007/978-3-319-54109-9_6.

Avci, O., Abdeljaber, O., Kiranyaz, S. and Inman, D. 2020. Control of plate vibrations with artificial neural networks and piezoelectricity. *Conf. Proc. Soc. Exp. Mech. Ser.* https://doi.org/10.1007/978-3-030-12676-6_26.

Avci, O., Abdeljaber, O., Kiranyaz, S. and Inman, D. 2020. Convolutional neural networks for real-time and wireless damage detection. *In: Conf. Proc. Soc. Exp. Mech. Ser.* https://doi.org/10.1007/978-3-030-12115-0_17.

Avci, O., Abdeljaber, O., Kiranyaz, S. and Inman, D.J. 2017. Vibration suppression in metastructures using zigzag inserts optimized by genetic algorithms. *In: Conf. Proc. Soc. Exp. Mech. Ser.* https://doi.org/10.1007/978-3-319-54735-0_29.

Avci, O., Abdeljaber, O., Kiranyaz, S. and Inman, D. 2020. Structural health monitoring with self-organizing maps and artificial neural networks. *In*: Mains, M.L. and Dilworth, B.J. (eds.). *Topics in Modal Analysis & Testing*, Volume 8. Conference Proceedings of the Society for Experimental Mechanics Series. Springer, Cham. https://doi.org/10.1007/978-3-030-12684-1_24.

Avci, O., Abdeljaber, O., Kiranyaz, S., Boashash, B., Sodano, H. and Inman, D.J. 2018. Efficiency validation of one-dimensional convolutional neural networks for structural damage detection using an SHM benchmark data. *International Institute of Acoustics and Vibration (IIAV)*, 2018: 4600–4607.

Avci, O., Abdeljaber, O., Kiranyaz, S., Hussein, M. and Inman, D.J. 2018. Wireless and real-time structural damage detection: A novel decentralized method for wireless sensor networks. *Journal of Sound and Vibration*, 424: 158–172.

Avci, O., Abdeljaber, O., Kiranyaz, S., Hussein, M., Gabbouj, M. and Inman, D.J. 2021. A review of vibration-based damage detection in civil structures: From traditional methods to machine learning and deep learning applications. *Mech. Syst. Signal Process.* https://doi.org/10.1016/j.ymssp.2020.107077.

Avci, O., Abdeljaber, O., Kiranyaz, S., Hussein, M., Gabbouj, M. and Inman, D. 2022. A new benchmark problem for structural damage detection: Bolt loosening tests on a large-scale laboratory structure. *In Dynamics of Civil Structures*, Volume 2, pp. 15–22. Springer, Cham.

Avci, O., Abdeljaber, O., Kiranyaz, S., Sassi, S., Ibrahim, A. and Gabbouj, M. 2022. One-dimensional convolutional neural networks for real-time damage detection of rotating machinery. *In Rotating Machinery, Optical Methods & Scanning LDV Methods*, Volume 6, pp. 73–83. Springer, Cham.

Avci, O., Setareh, M. and Murray, T.M. 2010. Vibration testing of joist supported footbridges. *In*: *Struct. Congr.* 2010. https://doi.org/10.1061/41130(369)80.

Bani-Hani, K. 1999. *Analytical and Experimental Study of Nonlinear Structural Control Using Neural Networks*. University of Illinois at Urbana-champaign.

Bani-Hani, K., Ghaboussi, J. and Schneider, S.P. 1999. Experimental study of identification and control of structures using neural network. *Part 1: Identification, Earthq. Eng. Struct. Dyn.*, 28. https://doi.org/10.1002/(SICI)1096-9845(199909)28:9<995::AID-EQE851>3.0.CO;2-8.

Barrett, A.R., Avci, O., Setareh, M. and Murray, T.M. 2006. Observations from vibration testing of *in situ* structures. *In*: *Struct. Congr.* 2006. https://doi.org/10.1061/40889(201)65.

Bayissa, W.L., Haritos, N. and Thelandersson, S. 2008. Vibration-based structural damage identification using wavelet transform. *Mech. Syst. Signal Process.* https://doi.org/10.1016/j.ymssp.2007.11.001.

Bohn, C.-A. 2003. A neural network based finite element grid generation algorithm for the simulation of light propagation. *Appl. Numer. Math.*, 46: 263–277.

Catbas, F.N., Celik, O., Avci, O., Abdeljaber, O., Gul, M. and Do, N.T. 2017. Sensing and monitoring forstadium structures: A review of recent advances and a forward look. *Front. Built Environ.*, 3: 38. https://doi.org/10.3389/fbuil.2017.00038.

Celik, O., Do, N.T., Abdeljaber, O., Gul, M., Avci, O. and Catbas, F.N. 2016. Recent issues on stadium monitoring and serviceability: A review. *In*: *Conf. Proc. Soc. Exp. Mech. Ser.* https://doi.org/10.1007/978-3-319-29763-7_41.

Cha, Y.J. and Buyukozturk, O. 2015. Structural damage detection using modal strain energy and hybrid multiobjective optimization. *Comput. Civ. Infrastruct. Eng.* https://doi.org/10.1111/mice.12122.

Chaabane, M., Ben Hamida, A., Mansouri, M., Nounou, H.N. and Avci, O. 2017. Damage detection using enhanced multivariate statistical process control technique. *In*: *17th Int. Conf. Sci. Tech. Autom. Control Comput. Eng. STA 2016—Proc.* https://doi.org/10.1109/STA.2016.7952052.

Chinka, S.S.B., Putti, S.R. and Adavi, B.K. 2021. Modal testing and evaluation of cracks on cantilever beam using mode shape curvatures and natural frequencies. *Structures*. https://doi.org/10.1016/j.istruc.2021.03.049.

Delgadillo, R.M. and Casas, J.R. 2020. Non-modal vibration-based methods for bridge damage identification. *Struct. Infrastruct. Eng.* https://doi.org/10.1080/15732479.2019.1650080.

Dyke, S., Bernal, D., Beck, J. and Ventura, C. 2003. Experimental phase II of the structural health monitoring benchmark problem. *In*: *Proc. 16th ASCE Eng. Mech. Conf.*, 2003: 1–7.

Ewins, D.J. and Saunders, H. 1986. Modal testing: Theory and practice. *J. Vib. Acoust.* https://doi.org/10.1115/1.3269294.

Ghahramani, Z. 2015. Probabilistic machine learning and artificial intelligence. *Nature*. https://doi. org/10.1038/nature14541.

Glorot, X. and Bengio, Y. 2010. Understanding the difficulty of training deep feedforward neural networks. *Pmlr.*, 9: 249–256. https://doi.org/10.1.1.207.2059.

Goel, A.K. and Davies, J. 2019. Artificial intelligence. *Cambridge Handb. Intell.* https://doi. org/10.1017/9781108770422.026.

Goodfellow, I., Yoshua Bengio and Aaron Courville. 2017. *Deep Learning.* Cambridge, USA: MIT Press. https://doi.org/10.1561/2000000039.

Guo, Y., Liu, Y., Oerlemans, A., Lao, S., Wu, S. and Lew, M.S. 2016. Deep learning for visual understanding: A review. *Neurocomputing*. https://doi.org/10.1016/j.neucom.2015.09.116.

Guyon, I. and Elisseeff, A. (n.d.). An introduction to feature extraction. *Featur. Extr.*, pp. 1–25. https://doi. org/10.1007/978-3-540-35488-8_1.

Haji Agha Mohammad Zarbaf, S.E., Norouzi, M., Allemang, R., Hunt, V., Helmicki, A. and Venkatesh, C. 2018. Vibration-based cable condition assessment: A novel application of neural networks. *Eng. Struct.* https://doi.org/10.1016/j.engstruct.2018.09.060.

Han, Z., Brownjohn, J.M.W. and Chen, J. 2020. Structural modal testing using a human actuator. *Eng. Struct.* https://doi.org/10.1016/j.engstruct.2020.111113.

Haykin, S.O. 2008. *Neural Networks and Learning Machines*. https://doi.org/978-0131471399.

He, K., Zhang, X., Ren, S. and Sun, J. 2015. Delving deep into rectifiers: Surpassing human-level performance on imagenet classification. *Proc. IEEE Int. Conf. Comput. Vis. 2015 Inter.*, pp. 1026–1034. https://doi.org/10.1109/ICCV.2015.123.

Hsu, F.-H. 2002. *Behind Deep Blue*. Princeton, NJ, USA: Princeton University Press.

Hutter, F., Kotthoff, L. and Vanschoren, J. 2019. *Automated Machine Learning: Methods, Systems, Challenges* (p. 219). Springer Nature.

Jablonski, A. 2021. Vibration-based condition assessment methods. *In: Springer Tracts Mech. Eng.* https://doi.org/10.1007/978-3-030-62749-2_5.

Janssens, O., Slavkovikj, V., Vervisch, B., Stockman, K., Loccufier, M., Verstockt, S., Van de Walle, R. and Van Hoecke, S. 2016. Convolutional neural network based fault detection for rotating machinery. *J. Sound Vib.*, 377: 331–345. https://doi.org/10.1016/j.jsv.2016.05.027.

Jordan, M.I. and Mitchell, T.M. 2015. Machine learning: Trends, perspectives, and prospects. *Science* (80-). https://doi.org/10.1126/science.aaa8415.

Khodabandehlou, H., Pekcan, G. and Fadali, M.S. 2019. Vibration-based structural condition assessment using convolution neural networks. *Struct. Control Heal. Monit.* https://doi.org/10.1002/stc.2308.

Kiranyaz, S., Avci, O. and Jaber, Q.A. 2019. Real-time structural damage detection by convolutional neural networks, US16031519. https://patents.google.com/patent/US20190017911A1/en.

Kiranyaz, S., Avci, O., Abdeljaber, O., Ince, T., Gabbouj, M. and Inman, D.J. 2021. 1D convolutional neural networks and applications: A survey. *Mech. Syst. Signal Process.* 151. https://doi.org/10.1016/j. ymssp.2020.107398.

Kiranyaz, S., Gastli, A., Ben-Brahim, L., Alemadi, N. and Gabbouj, M. 2018. Real-time fault detection and identification for MMC using 1D convolutional neural networks. *IEEE Trans. Ind. Electron.* https://doi.org/10.1109/TIE.2018.2833045.

Kiranyaz, S., Ince, T. and Gabbouj, M. 2016. Real-time patient-specific ECG classification by 1-D Convolutional Neural Networks. *IEEE Trans. Biomed. Eng.* https://doi.org/10.1109/TBME.2015.2468589.

Kiranyaz, S., Ince, T. and Gabbouj, M. 2017. Personalized monitoring and advance warning system for cardiac arrhythmias. *Sci. Rep.*, 7. https://doi.org/10.1038/s41598-017-09544-z.

Kiranyaz, S., Ince, T. and Gabbouj, M. 2017. Personalized monitoring and advance warning system for cardiac arrhythmias. *Sci. Rep.*, 7. https://doi.org/10.1038/s41598-017-09544-z.

Kiranyaz, S., Ince, T., Abdeljaber, O., Avci, O. and Gabbouj, M. 2019. 1-D convolutional neural networks for signal processing applications. *In: ICASSP, IEEE Int. Conf. Acoust. Speech Signal Process. Proc.*, 2019. https://doi.org/10.1109/ICASSP.2019.8682194.

Kiranyaz, S., Waris, M.A., Ahmad, I., Hamila, R. and Gabbouj, M. 2016, September. Face segmentation in thumbnail images by data-adaptive convolutional segmentation networks. In *2016 IEEE International Conference on Image Processing (ICIP)*, pp. 2306–2310. IEEE.

KrizhKrizhevsky, A., Sutskever, I. and Hinton, G.E. 2012. ImageNet classification with deep convolutional neural networks: Advances in neural information processing systems. *Adv. Neural Inf. Process. Syst.*, pp. 1–9.

Mansouri, M., Avci, O., Nounou, H. and Nounou, M. 2015. A comparative assessment of nonlinear state estimation methods for structural health monitoring. *In: Conf. Proc. Soc. Exp. Mech. Ser.* https://doi.org/10.1007/978-3-319-15224-0_5.

Mansouri, M., Avci, O., Nounou, H. and Nounou, M. 2015. Iterated square root unscented Kalman filter for nonlinear states and parameters estimation: Three DOF damped system. *J. Civ. Struct. Heal. Monit.*, 5. https://doi.org/10.1007/s13349-015-0134-7.

Meruane, V. and Heylen, W. 2012. Structural damage assessment under varying temperature conditions. *Struct. Heal. Monit.* https://doi.org/10.1177/1475921711419995.

Ming, G.L. and Song, H. 2011. Adult neurogenesis in the mammalian brain: Significant answers and significant questions. *Neuron*, 70(4): 687–702. https://doi.org/10.1016/j.neuron.2011.05.001.

Mosallam, A., Zirakian, T., Abdelaal, A. and Bayraktar, A. 2018. Health monitoring of a steel moment-resisting frame subjected to seismic loads. *J. Constr. Steel Res.* https://doi.org/10.1016/j.jcsr.2017.10.023.

Muhammad, Z., Reynolds, P., Avci, O. and Hussein, M. 2018. Review of pedestrian load models for vibration serviceability assessment of floor structures. *Vibration.* https://doi.org/10.3390/vibration2010001.

Radzieński, M., Krawczuk, M. and Palacz, M. 2011. Improvement of damage detection methods based on experimental modal parameters. *Mech. Syst. Signal Process.* https://doi.org/10.1016/j.ymssp.2011.01.007.

Sajja, P.S. 2021. Introduction to artificial intelligence. *Stud. Comput. Intell.* https://doi.org/10.1007/978-981-15-9589-9_1.

Scherer, D., Müller, A. and Behnke, S. 2010. Evaluation of pooling operations in convolutional architectures for object recognition. *In: Proc. 20th Int. Conf. Artif. Neural Networks Part III.* Berlin, Heidelberg: Springer-Verlag, pp. 92–101. https://doi.org/https://doi.org/10.1007/978-3-642-15825-4_10.

Shalev-Shwartz, S. and Ben-David, S. 2013. *Understanding Machine Learning: From Theory to Algorithms.* https://doi.org/10.1017/CBO9781107298019.

Simeone, O. 2018. A brief introduction to machine learning for engineers. *Found. Trends Signal Process.* https://doi.org/10.1561/2000000102.

Sony, S., Laventure, S. and Sadhu, A. 2019. A literature review of next-generation smart sensing technology in structural health monitoring. *Struct. Control Heal. Monit.* https://doi.org/10.1002/stc.2321.

Tiwari, A.K. 2017. Introduction to machine learning. *Ubiquitous Machine Learning and its Applications.* https://doi.org/10.4018/978-1-5225-2545-5.ch001.

Wei, Z., Gaoliang, P. and Chuanhao, L. 2017. Bearings fault diagnosis based on convolutional neural networks with 2-D representation of vibration signals as input. *MATEC Web of Conferences*, 13001: 1–5.

Yegnanarayana, B. 2006. *Artificial Neural Networks.* New Delhi.

Younis, A., Avci, O., Hussein, M., Davis, B. and Reynolds, P. 2017. Dynamic forces induced by a single pedestrian: A literature review. *Appl. Mech. Rev.*, 69. https://doi.org/10.1115/1.4036327.

Chapter 2

Identifying Non-linearity in Construction Workers' Personality
Safety Behaviour Predictive Relationship Using Neural Network and Linear Regression Modelling

Yifan Gao,[1,*] *Vicente A. González,*[2] *Tak Wing Yiu*[3]
and *Guillermo Cabrera-Guerrero*[4]

1. Introduction

The construction sector, in recent years, has recorded a high accident rate (Baradan et al., 2019; Feng et al., 2018). In New Zealand, for instance, this sector has been responsible for the largest portion of all occupational injuries in 2019, which caused a substantial economic loss of more than 100 million USD (ACC, 2018). Investigating the root causes of occupational accidents has been a pivotal focus of contemporary safety research (Zhang et al., 2020). Research has found that many occupational accidents are foreseeable, being the result of people's unsafe behaviour from a retrospective point of view (Abdelhamid and Everett, 2000; Feng et al., 2018; Gangwar and Goodrum, 2005; Gao et al., 2019a; Zohar and Luria 2004). Accordingly, it has been suggested that gaining priori insights into workers' intended safety behaviours will be useful in the context of safety management (Gao et al., 2019b; Patel and Jha, 2015; Zhang et al., 2020). Safety management practices are proactive interventions and incentives established to prevent the occurrence of accidents

---

[1] College of Civil Engineering and Architecture, Zhejiang University, Hangzhou, China-310058.
[2] Department of Civil and Environmental Engineering, University of Auckland, Auckland, New Zealand-1010. Email: v.gonzalez@auckland.ac.nz
[3] School of Built Environment, Massey University, Auckland, New Zealand-0745.
  Email: tyiu@massey.ac.nz
[4] Escuela de Ingeniería Informática, Pontificia Universidad Católica de Valparaíso, Valparaíso, Chile-2950.
  Email: guillermo.cabrera@pucv.cl
* Corresponding author: yfgao91@zju.edu.cn

(Patel and Jha, 2015). The prediction of workers' intended behaviours in advance can help identify at-risk workers who intend to undertake unsafe behaviours in the future (Merigó et al., 2019). Such insights will be useful in the design of effective management practices to minimise the occurrence of accidents and contribute to the reduction of injury rates (Low et al., 2019). In addition, the workers' behaviour prediction can improve their self-awareness and knowledge on their own intended behavioural responses and associated risks, which will be helpful to motivate workers to work safely and take their own safety more seriously (Patel and Jha, 2015).

The latest literature reveals that there is within-population diversity that leads individuals' intended safety behaviours in the workplace, which are found to vary among individuals as a function of their inner drives—personality traits (Chu et al., 2019; Hasanzadeh et al., 2019). The predictor role of personality traits in workers' safety behaviours has been evidenced by the prevailing behavioural theories such as self-determination theory (Ryan and Deci, 2000) and theory of purposeful work behaviour (Barrick et al., 2013). According to these theories, an intended safety behaviour is the function of individuals to attain certain goals at work (e.g., communion, status, self-control, and achievement). For example, in the dimensions that constitutes an individual's personality structure, the trait agreeableness is associated with the goal of communion, and individuals with a high level of agreeableness were found to perform safely in the workplace as unsafe behaviours could place other colleagues' health and safety at risk and then result in damaged interpersonal relationships (Jirjahn and Mohrenweiser, 2019). The trait extraversion is associated with the goal of status, and individuals with a high level of extraversion were found to work productively but sometimes in an unsafe manner (e.g., ignoring safety rules) in order to sustain competitive advantages over their teammates (Gao et al., 2019b). Neuroticism is associated with the goal of self-control, and individuals with a high level of neuroticism were found to have poor self-control and more negative emotions, which usually result in distracted thinking and then affect safety behaviour (Hidayat et al., 2015). Conscientiousness is associated with the goal of achievement, and individuals with a high level of conscientiousness were found to perform safely during work as unsafe behaviours could incur hazards to people or equipment in the work environment and delay the completion of work as a result (Beus et al., 2015).

Previous studies on the ascertainment of construction workers' personality-safety behaviour predictive relationship are found to rely on the linear regression (LR) method (Al-Shehri, 2015; Gao et al., 2019b; Hasanzadeh et al., 2019; Jackson, 2009; Manjula, 2017; Rau et al., 2018; Sing et al., 2014; Zhang et al., 2020). However, it has been found that nonlinear effect exists in the relationship between individuals' personality and intended safety behaviour according to meta-analyses published in recent years (Beus et al., 2015; Clarke and Robertson, 2008; Frazier et al., 2017; Yuan et al., 2018). LR is a linear rule-based method which maps the relationship among variables by imposing a linearity assumption on the data (Jorgensen, 2019). The linearity assumption does not hold good for nonlinear settings, which transforms the nonlinear nature into a linear character and can lead to inaccurate prediction mappings (Schoukens et al., 2016). This raises doubts about the validity of the LR models developed in previous studies.

On the other side, neural network (NN) has gained significant popularity in the field of statistical prediction (Goetz et al., 2015). In contrast to the rule-driven nature of LR, NN is a data-driven machine learning approach that allows identifying natural patterns between variables without assuming any preconception in terms of the mathematical structure of the data and uses algorithms to build models based on data in order to make inferences about likely future outcomes (Velasco et al., 2020). NN has been proven to achieve high accuracy levels in the determination of complex relationships in various fields such as medical diagnosis (Liu et al., 2019), financial forecasting (Yan and Zhao, 2019), speech recognition (Alloghani et al., 2020), and behaviour prediction (MolaAbasi et al., 2019). Given that NN allows the ascertainment of nature patterns between variables (which can be either nonlinear or linear), the use of NN can overcome the linearity limitation of the LR models developed in previous studies as mentioned above. To contribute to the growing research literature in this field, this chapter aims to apply NN to develop a more reliable ascertainment of the predictive relationship between construction workers' personality traits and intended safety behaviour. To ascertain whether the relationship can be described best with linear or nonlinear models, the authors develop an LR model and compare the prediction accuracy against the NN model.

This chapter is organised as follows. *The Dataset* introduces the data collected for carrying out the numerical simulations of NN and LR. The development of the NN and LR models is presented in *The Neural Network and Linear Regression Models*. Next, the results, which involve the evaluation and comparison of the prediction performance of the NN and LR models, are discussed in *Results and Discussion*. Finally, *Conclusions* summarises the results, interprets the practical implications, lists the limitations, and recommends future research directions.

2. The Dataset

Collecting relevant data for model development is a critical first step in this research. As mentioned in Section *Introduction*, this chapter aims to ascertain the predictive relationship between construction workers' personality traits and intended safety behaviour. Using the personality and behaviour measures as determined below, data were collected from 268 construction workers which constitutes the dataset for this research.

2.1 Personality Measurement

As previous studies have pointed out (Costa et al., 2019; Kunnel John et al., 2019), the Big Five Inventory (BFI) by John et al. (1991), the NEO Five-Factor Inventory (NEO-FFI) by Costa and McCrae (1989), and the Revised NEO Personality Inventory (NEO-PI-R) by Costa and McCrae (1992) are the three well accepted and widely used instruments for personality measurement in the research domain of psychology. The psychometric properties (i.e., reliability and validity) of these instruments have been assessed extensively by numerous researchers and have been proven to be highly reliable and valid (e.g., Hamby et al., 2016; Kerry and Murray, 2018; Kunnel John et al., 2019). From a practical point of view, the BFI has been recommended to be more useful than NEO-FFI and NEO-PI-R for being brief (Salgado and Táuriz, 2014). Huang (2019) has estimated that the BFI, NEO-FFI, and NEO-PI-R are composed of 44, 60, and 240 items, respectively, and take approximately 5, 10, and 45 minutes, respectively, to complete. Lengthy instruments can cause respondents to generate negative attitudes towards the survey (e.g., becoming fatigued, refusing to participate, and responding in a careless manner), which may influence the validity of the data gathered (Anglim and O'Connor, 2018). Considering the above discussion, the BFI (John et al., 1991) was selected as the instrument for personality measurement (Appendix A2.1).

In addition, it has been pointed out that the BFI as well as NEO-FFI and NEO-PI-R was developed purposely for the general measure of personality (Anglim and O'Connor, 2018), and general personality items can cause individuals to perceive the context of the items differently (Meryem, 2018). For example, in response to the item 'Q18: I see myself as someone who tends to be disorganised' in the BFI (Appendix A2.1), people's perceptions of the context can differ from one another (e.g., workplace or non-workplace) (Anglim and O'Connor, 2018). This may influence their responses to the item as individuals can act differently in different contexts, e.g., highly organised in the workplace (e.g., maintaining a tidy workspace) but disorganised at home (e.g., accumulating dirty laundry) (Wagner et al., 2019). It has been pointed out that the BFI items can usefully be adapted to derive contextualised measures (Desson et al., 2014). To enhance the specificity of the BFI for personality measurement in the work context, Schmit et al. (1995) have developed the work-specific BFI by appending a reference to work (i.e., 'at work') to each item in the BFI. As also pointed out by Schmit et al. (1995), the openness sub-scale (see Appendix A2.1) was not included in the work-specific BFI because they found that: 1. The appendage of 'at work' could not meaningfully fit into many items in the openness sub-scale (as suggested by the safety experts participated in their study to evaluate the content validity of the work-specific BFI); and 2. No correlation could be identified between openness and people's intended safety behaviour at work, which has also been confirmed by meta-analyses conducted in recent years (Beus et al., 2015; Frazier et al., 2017).

The psychometric properties (i.e., reliability and validity) of the work-specific BFI have been examined extensively by numerous researchers and have been proven to be highly reliable and valid (Desson et al., 2014; Gao et al., 2019b; Smith and DeNunzio, 2020; Swift and Peterson, 2019; Wang 2011). Taking the above considerations into account, the work-specific BFI was used as the instrument for personality measurement, which is a 34-item psychological measure of four personality traits: extraversion, agreeableness, conscientiousness, and neuroticism (Appendix A2.2).

2.2 Safety Behaviour Measurement

In the literature, it has been suggested that behaviour observation would expose researchers to workers' instant work environment, which is risky for the researchers given the hazardous nature of construction sites (Gao et al., 2019b; Hasanzadeh et al., 2019; Rau et al., 2018; Zhang et al., 2020). Moreover, behaviour observation could interfere with the construction progress, which is found to be a method that is unwelcome in most workplaces (Liu et al., 2020). A review of recent publications reveals that self-reporting is a favourable technique being widely utilised by today's behavioural research in the construction sector (e.g., Gao et al., 2019b; Guo et al., 2016; Hasanzadeh et al., 2019; Liu et al., 2020; Rau et al., 2018; Sing et al., 2014; Zhang et al., 2020). Taking the above into consideration, participants in this study were requested to self-report their intended safety behaviour at work. The Safety Behaviour Scale (SBS) by Hayes et al. (1998) (Appendix A2.3) was utilised due to the following advantages.

First, the SBS is built on relevant behaviour theories that were ascertained through decades of research with hundreds of thousands of individuals involved— theory of purposeful work behaviour (Barrick et al., 2013) and theory of planned behaviour (Armitage and Conner, 2001). According to these theories, a thorough understanding of workers' intended safety behaviour consists of two dimensions, namely, task performance and contextual performance. Task performance refers to compliance-related behaviour that individuals carry out to keep themselves safe such as not taking shortcuts, using safety equipment, and following safety procedures and rules. Contextual performance refers to voluntary safety activities such as reporting safety problems, keeping workplace clean, and caring for colleagues' safety, which may not directly contribute to one's own safety but help to develop an environment that supports safety. The SBS was favourably developed by its authors (Hayes et al., 1998) to contain a wide variety of behavioural topics to measure task performance and contextual performance such as taking shortcuts, using safety equipment, following safety procedures and rules, reporting safety problems, keeping workplace clean, and caring for colleagues' safety (see Appendix A2.3). Thus, to the best of the authors' knowledge, the use of SBS can provide a more adequate assessment of workers' intended safety behaviour than some of the instruments used in previous studies in the field of construction management. For example, Patel and Jha (2015) assessed construction workers' safety behaviour by implementing a single item 'I (self) follow all of the safety procedures for the jobs that I perform'. Guo et al. (2016) used four items to assess construction workers' safety behaviour in terms of using safety equipment and other behavioural topics such as following safety procedures were not included in their survey instruments.

Second, the SBS has additional advantages of being brief (11 items) and possessing very good reliability (0.85) given the threshold (above 0.70) (Wagner III, 2016), as reported in the original paper (Hayes et al., 1998). The quality of being brief makes SBS an easy and cost-saving instrument to administer. The SBS has been widely utilised in occupational safety research over the years, where its reliability has been well-scrutinised by many researchers afterwards and proven to be highly reliable (e.g., Gao et al., 2019b; Singh and Misra, 2020).

3. The Neural Network and Linear Regression Models

In this section, the NN and LR models are developed to numerically simulate the predictive relationship between construction workers' personality traits and intended safety behaviour.

3.1 The Neural Network Model

An NN is a mathematical model consisting of artificial "biological neurons" in multiple conductive layers that imitates the way the human brain functions such as the process of information processing and adaptation (Tahavvor and Nazari, 2019) (Fig. 2.1). The neurons in the input, the hidden and output layers are responsible, respectively, for receiving input signals, processing the received signals, and decoding the processed signals into the outputs (Azizi et al., 2019). The output of an NN depends on the weights between neurons in different layers and the biases of neurons in the hidden and output layers (Rafiei and Adeli, 2018) (Fig. 2.1). Weight indicates the strength of a particular connection between two neurons (Tahavvor and Nazari, 2019). For example, the weight index $w_{j,k}^{IH}$ in Fig. 2.1 refers to the strength of the connection between the jth input neuron and the kth hidden neuron. Bias can be considered as an additional input to each neuron, which is a constant and is used to adjust the input sum to each neuron to increase the computational capability of the NN (Ghritlahre and Prasad, 2018). For example, the bias index $bias_k^H$ in Fig. 2.1 refers to the bias of the kth hidden neuron. During a learning process applying the LM-BP algorithm, the NN iteratively adjusts the weight and bias values to minimise the error between the predicted and actual outputs (Azizi et al., 2019).

The development of a NN involves specifying the number of hidden layers, selecting the combination of activation functions, determining the number of neurons in the input, hidden, and output layers, determining the data split ratio, and training the NN (Azizi et al., 2019). Each step is described in greater detail below.

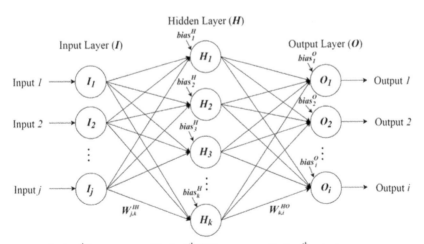

I_j: the jth input neuron H_k: the kth hidden neuron O_i: the ith output neuron

$W_{j,k}^{IH}$: weight between the jth input neuron and k^{th} hidden neuron

$W_{k,i}^{HO}$: weight between the kth hidden neuron and i^{th} output neuron

$bias_k^H$: bias of the kth hidden neuron $bias_i^O$: bias of the ith output neuron

Fig. 2.1: The architecture of an example neural network.

3.1.1 The Number of Hidden Layers

In general, an NN has only one input layer and one output layer, and there can be one or more hidden layers in between (Ghritlahre and Prasad, 2018). It has been pointed out by many researchers that a single hidden layer is sufficient for an NN to approximate any complex nonlinear mappings with desired accuracy (e.g., Dong et al., 2018; Liu et al., 2019; Peters et al., 2019). According to above considerations, the three-layer NN (i.e., one input layer, one hidden layer, and one output layer) was used in this study.

3.1.2 The Combination of Activation Functions

The next step is to select the combination of activation functions (Scardapane et al., 2019). In an NN, activation functions run on neurons in the hidden and output layers, transforming a neuron's input signal into an output signal (Mason et al., 2018). The commonly used activation functions are *tansig*, *logsig*, and *purelin* (Scardapane et al., 2019). The *tansig* and *logsig* are nonlinear functions and the *purelin* is a linear function (Pham and Hadi, 2014). Shen et al. (2018) pointed out that nonlinear functions (*tansig* and *logsig*) are capable of more complex computational rules than the linear function (*purelin*). The hidden layer of the NN performs computations on the network inputs and transfers the computed results to the output layer, and nonlinear functions (*tansig* and *logsig*) are commonly used in the hidden layer to increase the computational capability of the NN (Hajian and Styles, 2018). Particularly, among the nonlinear functions, *tansig* has previously proven to have performed better than *logsig* for achieving higher prediction accuracy with less working memory and training time required (e.g., Abbas et al., 2018; Jongprasithporn et al., 2018). Considering the above discussion, the nonlinear function *tansig* was used in the hidden layer.

In addition, Zarei and Behyad (2019) pointed out that the output layer of the NN decodes the computed results obtained in the hidden layer to provide the final output, and the activation functions used in the output layer can determine the output range of the NN. The output ranges are different among the activation functions, where the output of the linear function (*purelin*) can take on any value along a continuum from negative infinity to positive infinity and the output of nonlinear functions (*tansig* and *logsig*) is limited to the ranges of –1 to 1 and 0 to 1, respectively (Deng et al., 2019). In this study, participants were asked to self-report their safety behaviour using a 5-point Likert scale (1 = strongly disagree, 2 = disagree, 3 = neither agree nor disagree, 4 = agree, 5 = strongly agree) (Appendix A2.3). Their overall safety behaviour score as the output of the NN is calculated by summing the scores on all items (Appendix A2.3) and then averaging it and can thus take on any value in the range from 1 to 5. Considering the above discussion, the linear function *purelin* was used in the output layer to widen the output range of the NN.

3.1.3 The Number of Neurons in the Input, Hidden, and Output Layers

In addition to specifying the number of hidden layers and selecting the combination of activation functions, another important task is to determine the number of neurons in the input, hidden, and output layers (Azizi et al., 2019). In this paper, the network has four input neurons (one for each personality trait) and two output neurons (one for each safety behaviour indicator). In order to determine the number of hidden neurons, a method widely recommended and used in the literature was followed: $m \leq n \leq 2m$ where n is the number of hidden neurons and m is the number of input neurons (e.g., Desai et al., 2019; Lafif Tej and Holban, 2019; Velasco et al., 2020). As noted in Patel and Jha (2015), the process of determining the number of hidden neurons is also the process of identifying the network with the best performance. Therefore, several training trials were conducted in MATLAB®, varying the number of hidden neurons from four to eight ($m \leq n \leq 2m$), to identify the best performing network by examining two statistical parameters (i.e., mean squared error (MSE) and coefficient of determination (R^2)), as suggested by many researchers (e.g., Ramkumar et al., 2019; Tümer and Edebali, 2019).

Mean Squared Error (MSE) and Coefficient of Determination (R^2) are the most common measures of network performance as they indicate the global goodness-of-fit (Rafiei and Adeli, 2018). For MSE, a lower value indicates a better goodness-of-fit; for R^2, a higher value indicates a better goodness-of-fit (Rafiei and Adeli, 2018). The datasets used in the training and testing procedures are described in the following Section, *The Dataset Split Ratio*.

3.1.4 The Dataset Split Ratio

In order to enable the evaluation of network performance, the whole dataset is usually divided into three groups: the training dataset, the validating dataset, and the testing dataset (Ramkumar et al., 2019). The training dataset is used to adjust the network parameters (i.e., weights and biases) to best map the input-output relationships (Klaassen et al., 2016). The validating dataset is used to guarantee that the network is not overfitted while training (Manikandan and Subha, 2016). Overfitting refers to the phenomenon that the learning system overly adapts to the training dataset and even the noise in data, which affects the prediction accuracy of the trained model for new input (Ramkumar et al., 2019). As noted in Ramkumar et al. (2019), it is possible for a network to overfit the training dataset when the MSE value for the validating dataset starts to increase, and the training will be stopped at this point to prevent the network from overfitting. Finally, the testing dataset is established to evaluate the performance of the network after its development (Tümer and Edebali, 2019).

To determine the dataset split ratio, it has been recommended in the literature (Manikandan and Subha, 2016; Reddy and Juliet, 2019) that: 1. The training set should not be less than two-thirds of the whole dataset; and 2. The validation and testing sets should be one-fourth to one-eighth of the training set. Given this, the widely utilised data split ratios for training, validation, and testing sets are found to be 70:15:15 and 80:10:10 (Krzykowska and Krzykowski, 2019). In addition, 70:15:15 is recommended as a more balanced ratio than 80:10:10 for preserving as many data portions for training as possible and also including sufficient data points for validation and testing (Reddy and Juliet, 2019). In this study, the 70:15:15 ratio was therefore selected, where 188 (70 percentage), 40 (15 percentage), and

40 (15 percentage) samples were randomly assigned for training, validating, and testing of the proposed network.

3.1.5 The Training of the Neural Network

As mentioned in Section *The Number of Neurons in the Input, Hidden, and Output Layers*, the training of the proposed network was carried out repeatedly with different numbers of neurons in the hidden layer (4 —). To determine the best performing network, two statistical parameters, MSE and R^2, were estimated and compared for the training and validating datasets. The results are presented in Table 2.1. The authors found that the hidden layer with eight neurons provided the best performance among all the alternatives, showing the least MSE and highest R^2 values. As a result, the 4-8-2 (4 input neurons-8 hidden neurons-2 output neurons) was considered the best performing configuration and determined for the proposed network.

While training the network, its weights between neurons in different layers and biases of hidden as well as output neurons were adjusted to best map the input-output relationships (Klaassen et al., 2016). The weights and biases of the best performing configuration (4-8-2) are provided in Table 2.2.

Table 2.1: Statistical parameters of neural network models.

Number of Hidden Neurons	Data Sets	Statistical Parameters	
		MSE	R²
4	Training	0.203	0.771
	Validating	0.257	0.651
5	Training	0.234	0.724
	Validating	0.192	0.753
6	Training	0.176	0.811
	Validating	0.163	0.823
7	Training	0.099	0.882
	Validating	0.103	0.873
8	Training	0.041	0.942
	Validating	0.058	0.931

3.1.6 The Architecture of the Developed Neural Network

An overview of the architecture of the developed network is presented in Fig. 2.2. The input *I* is a 4x1 matrix, which represents a unit of four personality traits, namely, extraversion, agreeableness, conscientiousness, and neuroticism. The size of the first weight matrix W_1 is 8x4, which connects 4 input neurons to 8 hidden neurons. The second weight matrix W_2 is a 2x8 matrix, which indicates the connections between 8 hidden neurons and 2 output neurons. The bias matrices b_1 (8x1) and b_2 (2x1) in Fig. 2.2 indicates the biases of 8 hidden neurons and 2 output neurons. As shown in Fig. 2.2, the developed network has its hidden and output layers built on the nonlinear function *tansig* and linear function *purelin*, respectively. The matrix a_1 is the output

Table 2.2: Weights and biases.

Input /Output Neurons (I_j/O_i)	Weight/ Bias	Hidden Neurons (H_k)							
		H_1	H_2	H_3	H_4	H_5	H_6	H_7	H_8
I_1	$w_{1,k}^{IH}$	0.128	0.219	−0.452	0.325	2.325	1.831	1.017	−4.231
I_2	$w_{2,k}^{IH}$	−1.232	3.588	−1.433	5.222	2.176	−2.362	3.255	2.465
I_3	$w_{3,k}^{IH}$	3.247	−1.357	3.682	−2.425	−3.683	−6.433	−2.675	2.791
I_4	$w_{4,k}^{IH}$	1.342	−1.243	3.369	−1.435	−2.355	−2.131	2.323	1.266
	$bias_k^H$	−2.612	−0.322	1.991	−2.345	1.378	2.879	4.356	−2.338
O_1	$w_{k,1}^{HO}$	−2.364	3.256	0.111	−3.254	1.467	2.645	2.122	3.211
O_2	$w_{k,2}^{HO}$	−2.213	−1.121	−3.122	−3.122	3.322	2.322	−1.113	5.382
	$bias_i^O$	0.234 (for O_1)				0.162 (for O_2)			

Note: I_j = the jth input neuron; H_k = the kth hidden neuron; O_i = the ith output neuron; $w_{j,k}^{IH}$ (i.e., $w_{1,k}^{IH}$, $w_{2,k}^{IH}$, $w_{3,k}^{IH}$, $w_{4,k}^{IH}$) = the weight between the jth input neuron and the kth hidden neuron; $w_{k,i}^{HO}$ (i.e., $w_{k,1}^{HO}$, $w_{k,2}^{HO}$) = the weight between the kth hidden neuron and the ith output neuron; $bias_k^H$ = the bias of the kth hidden neuron; $bias_i^O$ = the bias of the ith output neuron.

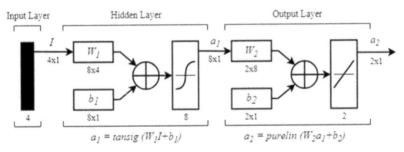

Fig. 2.2: Architecture of the developed network.

of the hidden layer, where the input matrix I is first multiplied by the weights W_1 and then the multiplication outcome is added with the biases b_1 and forwarded to the *tansig* function. The matrix a_2 (2x1) is the output of the network, which represents a unit of two safety behaviour dimensions, namely, task performance, and contextual performance. To obtain the matrix a_2, the matrix a_1 is multiplied by the weights W_2 and then the multiplication outcome is added with the biases b_2 and forwarded to the *purelin* function.

As a result, the output matrix a_2 can be calculated using the following formulas:

$$W_1 I + b_1 = \begin{pmatrix} \sum_{j=1}^{N_j} w_{j,1}^{IH} I_j + bias_1^H \\ \vdots \\ \sum_{j=1}^{N_j} w_{j,k}^{IH} I_j + bias_k^H \\ \vdots \\ \sum_{j=1}^{N_j} w_{j,8}^{IH} I_j + bias_8^H \end{pmatrix} \qquad (2.1)$$

$$a_1 = \tan sig(W_1I + b_1) = \frac{2}{1 + e^{-2(W_1I + b_1)}} - 1 \tag{2.2}$$

$$W_2a_1 + b_2 = \begin{pmatrix} w_{1,1}^{HO} \cdots w_{k,1}^{HO} \cdots w_{8,1}^{HO} \\ w_{1,2}^{HO} \cdots w_{k,2}^{HO} \cdots w_{8,2}^{HO} \end{pmatrix} \times a_1 + \begin{pmatrix} bias_1^{O} \\ bias_2^{O} \end{pmatrix} \tag{2.3}$$

$$a_2 = purelin(W_2a_1 + b_2) = W_2a_1 + b_2 \tag{2.4}$$

where I = input matrix; W_1 = weight matrix (including weights between input and hidden neurons); W_2 = weight matrix (including weights between hidden and output neurons); b_1 = bias matrix (including biases of hidden neurons); b_2 = bias matrix (including biases of output neurons); a_1 = output of the hidden layer; a_2 = output of the network; N_j = the total number of input neurons (N_j = 4); the definitions and values of $w_{j,k}^{IH}$ (e.g., $w_{j,1}^{IH}$, $w_{j,8}^{IH}$), $w_{k,i}^{HO}$ (e.g., $w_{k,1}^{HO}$, $w_{k,2}^{HO}$), $bias_k^{H}$ (e.g., $bias_1^{H}$, $bias_8^{H}$), and $bias_i^{O}$ (e.g., $bias_1^{O}$, $bias_2^{O}$) can be found in Table 2.2.

3.2 The Linear Regression Model

LR is a statistical method used to model the relationship between an output variable and one or more input variables by fitting all data points to the linear form (Jorgensen, 2019): $y = b_1x_1 + b_2x_2 + ... + b_ix_i + c$, where y is the output variable, x_i is the ith input variable, b_i is the weight of the ith input variable, and c is a constant which is used to adjust the sum of weighted input variables to best approximate the output variable. Using the dataset which includes the 188 training samples and 40 validating samples as assigned in the Section *The Data Split Ratio*, the LR calculation was performed using the computer programme SPSS (Schoukens et al., 2016), and the following LR formulas were obtained for the prediction of construction workers' intended safety behaviour:

$$y_1 = -0.086x_1 + 0.554x_2 + 0.543x_3 - 0.166x_4 + 0.658 \tag{2.5}$$

$$y_2 = -0.132x_1 + 0.755x_2 + 0.343x_3 - 0.170x_4 + 1.011 \tag{2.6}$$

where y_1 = task performance; y_2 = contextual performance; x_1 = Extraversion; x_2 = Agreeableness; x_3 = Conscientiousness; and x_4 = Neuroticism.

In addition, it has been pointed out that an LR generally has a relatively simple mathematical structure compared with nonlinear techniques such as NN (Velasco et al., 2020). Such a simple structure may lead to instability of the regression equations, being sensitive to potential structural breaks in the input signals (Peters et al., 2019). A structural break refers to an abrupt shift in the slope of a trend line over a series of data points. To ascertain the structural stability of the LR formulas (2.5) and (2.6), a widely used approach—Chow test—was adopted. The Chow test can estimate whether the parameters of an LR model are structurally stable, and runs as follows (Song et al., 2019): 1. Identifying the structural breaks in the dataset used to develop the model and splitting the dataset into subsets at the breakpoints; 2. Performing separate regressions on the entire dataset and each subset of the data;

3. Retrieving the residual sum of squares for each regression; and 4. Computing the Chow statistic using the formula:

$$F = \frac{(N-2k)[R_w - (\Sigma_{j=1}^n R_j)]}{k(\Sigma_{j=1}^n R_j)} \tag{2.7}$$

where F = the Chow statistic; R_w = residual sum of squares of the regression for the whole dataset; n = the number of sub-datasets split according to structural breaks; j = the jth sub-dataset; R_j = residual sum of squares of the regression for the jth sub-dataset; N = the number of samples in the whole dataset (N = 228, including 188 training samples and 40 validating samples as assigned in Section *The Data Split Ratio*); and k = the number of input variables (k = 4, including four personality traits).

The Chow test was conducted using the computer programme SAS (Song et al., 2019). The results showed that there was one structural break in the whole dataset, resulting in two split sub-datasets and the following R_w, R_1 and R_2 for the LR formulas (2.5) and (2.6) respectively: 33.497, 18.143, and 15.136 ($p < 0.001$); 82.325, 39.693, and 41.667 ($p < 0.001$). Applying the formula (2.7) above, the F values were computed respectively for the LR formulas (5) and (6) as follows: 0.37 ($p < 0.001$) and 0.65 ($p < 0.001$). According to the criteria of Chow test (Song et al., 2019), the parameters of an LR model are considered structurally stable when the calculated F value is less than the F-critical value with (k, N-2k) degrees of freedom which can be retrieved from the F-distribution table (available online at: www.stat.purdue.edu/~jtroisi/STAT350Spring2015/tables/FTable.pdf). As predefined, the values of k and N equal to 4 and 228, respectively. The degrees of freedom for the F-critical value are thereby determined as (4, 220). According to the F-distribution table, the F-critical value with degrees of freedom (4, 220) and p-value less than 0.001 is 4.81. The calculated F values, 0.37 ($p < 0.001$) and 0.65 ($p < 0.001$) are therefore well below the F-critical value 4.81 ($p < 0.001$), which indicates that the parameters of the LR formulas (2.5) and (2.6) are structurally stable.

4. Results and Discussion

After determining the architecture of the NN and LR models, the prediction accuracy was evaluated on the testing dataset. The results are presented in Fig. 2.3, showing a comparison between the predicted results (NN and LR) and actual results on both task performance and contextual performance for all the testing samples. The horizontal axis of the plot outlines 40 samples in the testing dataset, and the vertical axis represents one's score on task performance and contextual performance, where a higher score indicates a greater performance of an individual in terms of intending to behave safely at work.

Based on the predicted-actual plot, the prediction accuracy was evaluated using the statistical parameter generalisation error (E_{gen}), which is a measure of how accurately a mathematical model is able to make predictions for previously unseen data (De Jong et al., 2019). The E_{gen} can be calculated using the formula below, where

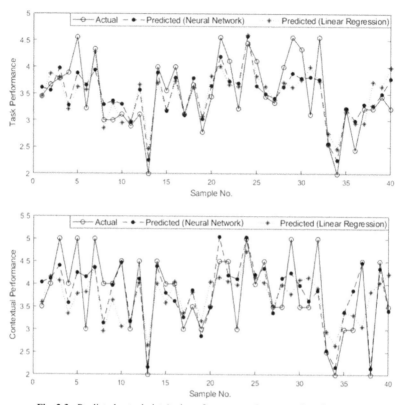

Fig. 2.3: Predicted-actual plot (task performance and contextual performance).

a value less than 0.25 is considered a low level of error as well as a satisfactory level of prediction accuracy (Singaravel et al., 2018):

$$E_{gen} = \frac{1}{n} \Sigma_{m=1}^{n} (Y_m - \hat{Y}_m)^2 \tag{2.8}$$

where E_{gen} = generalisation error; n = the number of samples in the testing dataset; Y_m = the actual result for the mth sample; $\hat{Y}_m$ = the predicted result for the mth sample.

In this study, the E_{gen} rates were 0.05 and 0.09 for NN predictions on task performance and contextual performance, respectively. According to the threshold (below 0.25), the prediction accuracy of the developed NN model is highly satisfactory. In addition, the E_{gen} rates were 0.32 and 0.41 for LR predictions on task performance and contextual performance, respectively. Given the threshold (below 0.25), the prediction accuracy of the developed LR formulas is unsatisfactory for task and contextual performance. In a comparison of the NN and LR performance as presented above, it is found that the predictive relationship between construction workers' personality and intended safety behaviour can be described better with NN. This implies that LR was unable to successfully capture the personality-safety behaviour relationship and a nonlinear effect thereby exists.

5. Conclusions

This chapter presents the utilisation of two mathematical models—neural network (NN) and linear regression (LR)—to ascertain whether the predictive relationship between construction workers' personality traits and intended safety behaviour can be described best with linear or nonlinear models. The developed NN and LR models can be represented by its mathematical expressions (1) (2) (3) (4) and (5) (6) as respectively shown in *The Architecture of the Developed Neural Network* and *The Linear Regression Model*. According to the results obtained on the testing dataset (see Section *Results and Discussion*), the NN model is shown to predict highly satisfactorily and yielded minimal error rates on the predictions. The results also showed that the LR method demonstrated greater prediction errors than NN, which resulted in unsatisfactory prediction accuracy (see *Results and Discussion*). This implies that LR was unable to successfully capture the predictive relationship between construction workers' personality traits and intended safety behaviour and a nonlinear effect thereby exists. As previous studies in this field all utilised the LR method to examine the personality-safety behaviour relationship (see *Introduction*), the empirical evidence (i.e., nonlinear effect) as identified in this chapter allows the authors to cast doubt on the validity of the LR models developed in previous studies and underpin the usefulness of the NN model developed.

The findings of this chapter imply practical implications for on-site safety management. A prior prediction on employees' intended safety behaviour can help identify at-risk workers who are prone to undertake unsafe behaviours. The NN model developed is proven to have a highly satisfactory prediction accuracy and is therefore recommended as a potential projection tool to assist project decision-makers to assess how prone workers are to carry out unsafe behaviours in the workplace. Site managers can input data on workers' personality traits and the NN model will estimate and output the results on safety behaviour for consideration. The NN model is developed to make predictions on the dimensions that constitute individuals' intended safety behaviour, which include task performance and contextual performance. This can be useful not only in identifying undesirable aspects in different individuals but also in the design of management practices prior to the occurrence of accidents. For example, individuals predicted low on task performance are more likely to conduct non-compliance work behaviours in the future such as taking shortcuts and violating safety rules. With these, site managers may adopt interventions to enhance their sense of discipline and compliance. Those who predicted low on contextual performance tend to be reluctant to conduct voluntary safety activities in the future such as keeping workplace clean and caring for co-workers' safety. Site managers may therefore employ incentive methods to promote their desire to make contributions to organisational safety.

Due to financial and resource constraints, the data collection was mainly focused on New Zealand construction projects with several samples obtained from overseas. This might leave out sections of the population (i.e., construction workforce) that could be meaningful to this study. For example, region-based cultural differences might influence the form of people's personality traits across countries and result in region-based diversity in people's intended safety behaviours in the workplace.

It would thus be interesting for future research to investigate regional differences in construction workers' personality traits and intended safety behaviour.

Ethics Approval

The ethics approval for this research was granted by University of Auckland Human Participants Ethics Committee. Reference number: 019515.

Data Availability Statement

The data that support the findings of this study are available from the corresponding author, YG, upon reasonable request.

References

Abbas, M.H., Norman, R. and Charles, A. 2018. Neural network modelling of high pressure CO2 corrosion in pipeline steels. *Process Safety and Environmental Protection*, 119: 36–45 doi: 10.1016/j.psep.2018.07.006.

Abdelhamid, T.S. and Everett, J.G. 2000. Identifying root causes of construction accidents. *Journal of Construction Engineering and Management*, 126: 52–60. doi: 10.1061/(ASCE)0733-9364(2000)126:1(52.)

ACC. 2018. The number of new accepted work related claims by the industry and injury cause. *New Zealand*. https://www.stats.govt.nz/information-releases/injury-statistics-work-related-claims-2017.

Al-Shehri, Y. 2015. Relationship between personality trait and multi-national construction workers safety performance in Saudi Arabia. Loughborough University. Thesis. https://hdl.handle.net/2134/18041.

Alloghani, M., Al-Jumeily, D., Mustafina, J., Hussain, A. and Aljaaf, A.J. 2020. A systematic review on supervised and unsupervised machine learning algorithms for data science. pp. 3–21. *In*: Berry, M.W., Mohamed, A. and Yap, B.W. (eds.). *Supervised and Unsupervised Learning for Data Science*. Cham: Springer International Publishing. doi: 10.1007/978-3-030-22475-2_1.

Anglim, J. and O'Connor, P. 2018. Measurement and research using the Big Five, HEXACO, and narrow traits: A primer for researchers and practitioners. *Australian Journal of Psychology*, 71: 16–25. doi: https://doi.org/10.1111/ajpy.12202.

Armitage, C.J. and Conner, M. 2001. Efficacy of the theory of planned behaviour: A meta-analytic review. *British Journal of Social Psychology*, 40: 471–499. doi: 10.1348/014466601164939.

Azizi, N., Rezakazemi, M. and Zarei, M.M. 2019. An intelligent approach to predict gas compressibility factor using neural network model. *Neural Computing and Applications*, 31: 55–64. doi: 10.1007/s00521-017-2979-7.

Baradan, S., Dikmen, S.U. and Akboga Kale, O. 2019. Impact of human development on safety consciousness in construction. *International Journal of Occupational Safety and Ergonomics*, 25: 40–50. doi: 10.1080/10803548.2018.1445069.

Barrick, M.R., Mount, M.K. and Li, N. 2013. The theory of purposeful work behavior: The role of mpersonality, higher-order goals, and job characteristics. *Academy of Management Review*, 38: 132–153. doi: 10.5465/amr.2010.0479.

Beus, J.M., Dhanani, L.Y. and McCord, M.A. 2015. A meta-analysis of personality and workplace safety: Addressing unanswered questions. *Journal of Applied Psychology*, 100: 481–498. doi: 10.1037%2Fa0037916.

Chu, F., Fu, Y. and Liu, S. 2019. Organization is also a "life form": Organizational-level personality, job satisfaction, and safety performance of high-speed rail operators. *Accident Analysis & Prevention*, 125: 217–223. doi: 10.1016/j.aap.2019.01.027.

Clarke, S. and Robertson, I. 2008. An examination of the role of personality in work accidents using meta-analysis. *Applied Psychology*, 57: 94–108. doi: 10.1111/j.1464-0597.2007.00267.x.

Costa, P.T. and McCrae, R.R. 1989. NEO Five-Factor Inventory (NEO-FFI) Odessa. *FL: Psychological Assessment Resources*, 3: 1–10. https://www.parinc.com/Products/Pkey/274.

Costa, P.T. and McCrae, R.R. 1992. Revised NEO Personality Inventory (NEO PI-R) and NEO five-factor inventory (NEO-FFI): Professional Manual. *Psychological Assessment Resources, Incorporated.* https://www.scienceopen.com/document?vid=2163aff2-9fd9-470e-8ad8-8bfa40ef035f.

Costa, P.T., McCrae, R.R. and Löckenhoff, C.E. 2019. Personality across the life span. *Annual Review of Psychology*, 70: 423–448. doi: 10.1146/annurev-psych-010418-103244.

De Jong, A.W., Rubrico, J.I.U., Adachi, M., Nakamura, T. and Ota, J. 2019. A generalised makespan estimation for shop scheduling problems, using visual data and a convolutional neural network. *International Journal of Computer Integrated Manufacturing*, 32: 559–568. doi: 10.1080/0951192X.2019.1599430.

Deng, J., Sun, J., Peng, W., Hu, Y. and Zhang, D. 2019. Application of neural networks for predicting hot-rolled strip crown. *Applied Soft Computing*, 78: 119–131. doi: 10.1016/j.asoc.2019.02.030.

Desai, S.D., Giraddi, S., Narayankar, P., Pudakalakatti, N.R. and Sulegaon, S. 2019. Back-propagation neural network versus logistic regression in heart disease classification. pp. 133–144. *In*: Mandal, J., Bhattacharyya, D. and Auluck, N. (eds.). *Advanced Computing and Communication Technologies. Advances in Intelligent Systems and Computing*, 702. Springer, Singapore. doi: 10.1007/978-981-13-0680-8_13.

Desson, S., Stephen, B. and John, G. 2014. Lumina spark—development of an integrated assessment of big 5 personality factors, type theory and overextension. *In*: *International Conference on Psychotechnology Abstract Book.* URL: https://corryrobertson.com/lumina-spark-development-of-an-integrated-assessment-of-big-5-personality-factors-type-theory-overextension/.

Dong, C., Dong, X., Jiang, Q., Dong, K. and Liu, G. 2018. What is the probability of achieving the carbon dioxide emission targets of the Paris Agreement? Evidence from the top ten emitters. *Science of the Total Environment*, 622-623: 1294–1303. doi: 10.1016/j.scitotenv.2017.12.093.

Feng, Z., González, V.A., Amor, R., Lovreglio, R. and Cabrera-Guerrero, G. 2018. Immersive virtual reality serious games for evacuation training and research: A systematic literature review. *Computers & Education*, 127: 252–266. doi: 10.1016/j.compedu.2018.09.002.

Frazier, M.L., Fainshmidt, S., Klinger, R.L., Pezeshkan, A. and Vracheva, V. 2017. Psychological safety: A meta-analytic review and extension. *Personnel Psychology*, 70: 113–165. doi: 10.1111/peps.12183.

Gangwar, M. and Goodrum, P.M. 2005. The effect of time on safety incentive programs in the US construction industry. *Construction Management and Economics*, 23: 851–859. doi: 10.1080/01446190500184527.

Gao, Y., Gonzalez, V.A. and Yiu, T.W. 2019a. The effectiveness of traditional tools and computer-aided technologies for health and safety training in the construction sector: A systematic review. *Computers & Education*, 138: 101–115. doi: 10.1016/j.compedu.2019.05.003.

Gao, Y., Gonzalez, V.A. and Yiu, T.W. 2019b. Exploring the relationship between construction workers' personality traits and safety behaviour. *Journal of Construction Engineering and Management*, 146: 04019111. doi: 10.1061/(ASCE)CO.1943-7862.0001763.

Ghritlahre, H.K. and Prasad, R.K. 2018. Application of ANN technique to predict the performance of solar collector systems: A review. *Renewable and Sustainable Energy Reviews*, 84: 75–88. doi: 10.1016/j.rser.2018.01.001.

Goetz, J.N., Brenning, A., Petschko, H. and Leopold, P. 2015. Evaluating machine learning and statistical prediction techniques for landslide susceptibility modeling. *Computers & Geosciences*, 81: 1–11. doi: 10.1016/j.cageo.2015.04.007.

Guo, B.H.W., Yiu, T.W. and González, V.A. 2016. Predicting safety behavior in the construction industry: Development and test of an integrative model. *Safety Science*, 84: 1–11. doi: 10.1016/j.ssci.2015.11.020.

Hajian, A. and Styles, P. 2018. Artificial neural networks. pp. 3–69. *In*: Hajian, A. and Styles, P. (eds.). *Application of Soft Computing and Intelligent Methods in Geophysics.* Cham: Springer International Publishing. doi: 10.1007/978-3-319-66532-0_1.

Hamby, T., Taylor, W., Snowden, A.K. and Peterson, R.A. 2016. A meta-analysis of the reliability of free and for-pay big five scales. *The Journal of Psychology*, 150: 422–430. doi: 10.1080/00223980.2015.1060186.

Hasanzadeh, S., Dao, B., Esmaeili, B. and Dodd Michael, D. 2019. Role of personality in construction safety: Investigating the relationships between personality, attentional failure, and hazard identification under fall-hazard conditions. *Journal of Construction Engineering and Management*, 145: 04019052. doi: 10.1061/(ASCE)CO.1943-7862.0001673.

Hayes, B.E., Perander, J., Smecko, T. and Trask, J. 1998. Measuring perceptions of workplace safety: Development and validation of the work safety scale. *Journal of Safety Research*, 29: 145–161. doi:10.1016/S0022-4375(98)00011-5.

Hidayat, S., Suwandi, T. and Qomarudin, M.B. 2015. The analysis of factors related to unsafe acts on welders in XYZ Ltd. *International Journal of Advanced Engineering, Management and Science*, 2: 760–766. https://www.academia.edu/download/46821890/36_The_Analysis_of_Factors_Related_to_Unsafe_Acts_on_Welders_in_XYZ_Ltd.pdf.

Huang, C. 2019. Social network site use and Big Five personality traits: A meta-analysis. *Computers in Human Behavior*, 97: 280–290. doi: https://doi.org/10.1016/j.chb.2019.03.009.

Jackson, C.J. 2009. Using the hybrid model of learning in personality to predict performance in the workplace. In: *8th Industrial & Organisational Psychology Conference, Conference Proceedings, 2009. Sydney, Australia: Manly*, pp. 75–79. https://www.researchgate.net/profile/Peter-Hayes-13/publication/230633967_Anticipation_in_context_Factors_that_support_or_hinder_anticipation_in_Incident_Management_Teams/links/54fd7bc80cf2c3f52424e6a2/Anticipation-in-context-Factors-that-support-or-hinder-anticipation-in-Incident-Management-Teams.pdf#page=76.

Jirjahn, U. and Mohrenweiser, J. 2019. Performance pay and applicant screening. *British Journal of Industrial Relations*, 57: 540–575. doi: 10.1111/bjir.12443.

John, O.P., Donahue, E.M. and Kentle, R.L. 1991. *The Big Five Inventory: Versions 4a and 54 Berkeley, CA: University of California, Berkeley, Institute of Personality and Social Research*. URL: https://www.ocf.berkeley.edu/~johnlab/bfi.htm.

Jongprasithporn, M., Yodpijit, N., Guerra, G. and Khawnuan, U. 2018. Evaluation of activation function capability for intent recognition and development of a computerized prosthetic knee. *In: 2018 IEEE International Conference on Industrial Engineering and Engineering Management (IEEM)*, 16–19 Dec. 2018, pp. 178–182. doi: 10.1109/IEEM.2018.8607594.

Jorgensen, B. 2019. *Theory of Linear Models*. New York: Routledge. doi: 10.1201/9780203718971.

Kerry, N. and Murray, D.R. 2018. Strong personalities: Investigating the relationships between grip strength, self-perceived formidability, and Big Five personality traits. *Personality and Individual Differences*, 131: 216–221. doi: 10.1016/j.paid.2018.05.003.

Klaassen, E.A.M., Frunt, J. and Slootweg, J.G. 2016. Experimental validation of the demand response potential of residential heating systems. *In: 2016 Power Systems Computation Conference (PSCC)*, 20–24 June 2016, pp. 1–7. doi: 10.1109/PSCC.2016.7540825.

Krzykowska, K. and Krzykowski, M. 2019. Forecasting parameters of satellite navigation signal through Artificial Neural Networks for the purpose of civil aviation. *International Journal of Aerospace Engineering*, 2019: 7632958. doi: 10.1155/2019/7632958.

Kunnel John, R., Xavier, B., Waldmeier, A., Meyer, A. and Gaab, J. 2019. Psychometric evaluation of the BFI-10 and the NEO-FFI-3 in Indian adolescents. *Frontiers in Psychology*, 10: 1057. doi: 10.3389/fpsyg.2019.01057.

Lafif Tej, M. and Holban, S. 2019. Determining optimal multi-layer perceptron structure using linear regression. pp. 232–246. *In*: Abramowicz, W. and Corchuelo, R. (eds.). *Business Information Systems*, Cham: 2019. Springer International Publishing. doi: 10.1007/978-3-030-20485-3_18.

Liu, Q., Xu, N., Jiang, H., Wang, S., Wang, W. and Wang, J. 2020. Psychological driving mechanism of safety citizenship behaviors of construction workers: Application of the theory of planned behavior and norm activation model. *Journal of Construction Engineering and Management*, 146: 04020027. doi: 10.1061/(ASCE)CO.1943-7862.0001793.

Liu, X., Zhou, Y. and Wang. Z. 2019. Recognition and extraction of named entities in online medical diagnosis data based on a deep neural network. *Journal of Visual Communication and Image Representation*, 60: 1–15. doi: 10.1016/j.jvcir.2019.02.001.

Low, B.K.L., Man, S.S., Chan, A.H.S. and Alabdulkarim, S. 2019. Construction worker risk-taking behavior model with individual and organizational factors. *International Journal of Environmental Research and Public Health*, 16: 1335, doi: 10.3390/ijerph16081335.

Manikandan, N. and Subha, S. 2016. Software design challenges in time series prediction systems using parallel implementation of artificial neural networks. *The Scientific World Journal*, 2016: 10. doi: 10.1155/2016/6709352.

Manjula, N.H.C. 2017. *Developing a Model to Predict Unsafe Behaviour of Construction Workers in Sri Lanka*. University of Moratuwa Thesis. http://dl.lib.mrt.ac.lk/handle/123/12862.

Mason, K., Duggan, J. and Howley, E. 2018 Forecasting energy demand, wind generation, and carbon dioxide emissions in Ireland using evolutionary neural networks. *Energy*, 155: 705–720. doi: 10.1016/j.energy.2018.04.192.

Merigó, J.M., Miranda, J., Modak, N.M., Boustras, G. and de la Sotta, C. 2019. Forty years of safety science: A bibliometric overview. *Safety Science*, 115: 66–88. doi: 10.1016/j.ssci.2019.01.029.

Meryem, R. 2018. Gender domesticity reinterpreted: Housework division in dual-work Moroccan families. *European Scientific Journal*, 14: 131. doi: 10.19044/esj.2018.v14n17p131.

MolaAbasi, H., Saberian, M., Kordnaeij, A., Omer, J., Li, J. and Kharazmi, P. 2019. Predicting the stress-strain behaviour of zeolite-cemented sand based on the unconfined compression test using GMDH type neural network. *Journal of Adhesion Science and Technology*, 33: 945–962. doi: 10.1080/01694243.2019.1571659.

Patel, D.A. and Jha, K.N. 2015. Neural network model for the prediction of safe work behavior in construction projects. *Journal of Construction Engineering and Management*, 141: 04014066. doi: 10.1061/(ASCE)CO.1943-7862.0000922.

Peter, S.C., Dhanjal, J.K., Malik, V., Radhakrishnan, N., Jayakanthan, M. and Sundar, D. 2019. Quantitative structure-activity relationship (QSAR): Modeling approaches to biological applications. pp. 661–676. *In*: Ranganathan, S., Gribskov, M., Nakai, K. and Schönbach, C. (eds.). *Encyclopedia of Bioinformatics and Computational Biology*. Oxford: Academic Press. doi: 10.1016/B978-0-12-809633-8.20197-0.

Peters, S.O., Sinecen, M., Gallagher, G.R., Pebworth, L.A., Jacob, S., Hatfield, J.S. and Kizilkaya, K. 2019. Comparison of linear model and artificial neural network using antler beam diameter and length of white-tailed deer (*Odocoileus virginianus*) dataset. *PLOS ONE*, 14: e0212545. doi: 10.1371/journal.pone.0212545.

Pham, T.M. and Hadi, M.N.S. 2014. Predicting stress and strain of FRP-confined square/rectangular columns using artificial neural networks. *Journal of Composites for Construction*, 18: 04014019. doi: 10.1061/(ASCE)CC.1943-5614.0000477.

Rafiei, M.H. and Adeli, H. 2018. Novel machine-learning model for estimating construction costs considering economic variables and indexes. *Journal of Construction Engineering and Management*, 144: 04018106. doi: 10.1061/(ASCE)CO.1943-7862.0001570.

Ramkumar, P.N. et al. 2019. Deep learning preoperatively predicts value metrics for primary total knee arthroplasty: Development and validation of an artificial neural network model. *The Journal of Arthroplasty*, 34: 2220–2227. doi: https://doi.org/10.1016/j.arth.2019.05.034.

Rau, P.-L.P., Liao, P.-C., Guo, Z., Zheng, J. and Jing, B. 2018. Personality factors and safety attitudes predict safety behaviour and accidents in elevator workers. *International Journal of Occupational Safety and Ergonomics*, 26: 719–727. doi: 10.1080/10803548.2018.1493259.

Reddy, A.S.B. and Juliet, D.S. 2019. Transfer learning with ResNet-50 for malaria cell-image classification. *In*: 2019 International Conference on Communication and Signal Processing (ICCSP), IEEE, pp. 0945–0949. doi: 10.1109/ICCSP.2019.8697909.

Ryan, R.M. and Deci, E.L. 2000. Self-determination theory and the facilitation of intrinsic motivation, social development, and well-being. *American Psychologist*, 55: 68–78. doi: 10.1037/0003-066X.55.1.68.

Salgado, J.F. and Táuriz, G. 2014. The five-factor model, forced-choice personality inventories and performance: A comprehensive meta-analysis of academic and occupational validity studies. *European Journal of Work and Organizational Psychology*, 23: 3–30. doi: 10.1080/1359432X.2012.716198.

Scardapane, S., Van Vaerenbergh, S., Totaro, S. and Uncini, A. 2019. Kafnets: Kernel-based non-parametric activation functions for neural networks. *Neural Networks*, 110: 19–32. doi: 10.1016/j.neunet.2018.11.002.

Schmit, M.J., Ryan, A.M., Stierwalt, S.L. and Powell, A.B. 1995. Frame-of-reference effects on personality scale scores and criterion-related validity. *Journal of Applied Psychology*, 80: 607–620. doi: 10.1037/0021-9010.80.5.607.

Schoukens, J., Vaes, M. and Pintelon, R. 2016. Linear system identification in a nonlinear setting: Nonparametric analysis of the nonlinear distortions and their impact on the best linear approximation. *IEEE Control Systems Magazine*, 36: 38–69. doi: 10.1109/MCS.2016.2535918.

Shen, Y., He, C., Cao, J. and Zhang, B.A. 2018. Relationship construction method between lifecycle cost and indexes of RMSST based on BP neural network. pp. 227–233. *In*: Zuo, M.J., Ma, L., Mathew, J. and Huang, H.-Z. (eds). *Engineering Asset Management* 2016. Cham, 2018. Springer International Publishing. doi: 10.1007/978-3-319-62274-3_20.

Sing, C., Love, P., Fung, I. and Edwards, D. 2014. Personality and occupational accidents: Bar benders in Guangdong Province, Shenzhen, China. *Journal of Construction Engineering and Management*, 140: 05014005. doi: 10.1061/(ASCE)CO.1943-7862.0000858.

Singaravel, S., Suykens, J. and Geyer, P. 2018. Deep-learning neural-network architectures and methods: Using component-based models in building-design energy prediction. *Advanced Engineering Informatics*, 38: 81–90. doi: 10.1016/j.aei.2018.06.004.

Singh, A. and Misra, S.C. 2020. A Dominance based rough set analysis for investigating employee perception of safety at workplace and safety compliance. *Safety Science*, 127: 104702. doi: 10.1016/j.ssci.2020.104702.

Smith, R.W. and DeNunzio, M.M. 2020. Examining personality: Job characteristic interactions in explaining work outcomes. *Journal of Research in Personality*, 84: 103884. doi: 10.1016/j.jrp.2019.103884.

Song, H.Y., Moon, T.W. and Choi. S.J. 2019. Impact of antioxidant on the stability of β-carotene in model beverage emulsions: Role of emulsion interfacial membrane. *Food Chemistry*, 279: 194–201. doi: 10.1016/j.foodchem.2018.11.126.

Swift, V. and Peterson, J.B. 2019. Contextualization as a means to improve the predictive validity of personality models. *Personality and Individual Differences*, 144: 153–163. doi: 10.1016/j.paid.2019.03.007.

Tahavvor, A.R. and Nazari, M. 2019. Numerical and neural network analysis of natural convection from a cold horizontal cylinder above an Adiabatic Wall. *Journal of Applied Fluid Mechanics*, 12: 369–377. doi: 10.29252/jafm.12.02.28779.

Tümer, A.E. and Edebali, S. 2019. Modeling of trivalent chromium sorption onto commercial resins by Artificial Neural Network. *Applied Artificial Intelligence*, 33: 349–360. doi: 10.1080/08839514.2019.1577015.

Velasco, L.C.P., Estoperez, N.R., Jayson, R.J.R., Sabijon, C.J.T. and Sayles, V.C. 2020. Performance analysis of artificial neural networks training algorithms and activation functions in day-ahead base, intermediate, and peak load forecasting. pp. 284–298. *In*: Arai, K. and Bhatia, R. (eds.). *Advances in Information and Communication*. Cham: 2020. Springer International Publishing. doi: 10.1007/978-3-030-12385-7_23.

Wagner III, W.E. 2016. *Using IBM® SPSS® Statistics for Research Methods and Social Science Statistics Vol.* California State University, Channel Islands, USA: Sage Publications. URL: https://us.sagepub.com/en-us/nam/using-ibm%C2%AE-spss%C2%AE-statistics-for-research-methods-and-social-science-statistics/book258010.

Wagner, J., Lüdtke, O. and Robitzsch, A. 2019. Does personality become more stable with age? Disentangling state and trait effects for the big five across the life span using local structural equation modeling. *Journal of Personality and Social Psychology*, 116: 666–680. doi: 10.1037/pspp0000203.

Wang, Q. 2011. *A Comparison of Work-Specific and General Personality Measures as Predictors of OCBs and CWBs in China and the United States*. Wright State University Thesis. https://corescholar.libraries.wright.edu/etd_all/504/.

www.stat.purdue.edu/~jtroisi/STAT350Spring2015/tables/FTable.pdf.

Yan, X. and Zhao, J. 2019. Application of improved convolution neural network in financial forecasting. *In: 2019 IEEE 4th International Conference on Cloud Computing and Big Data Analysis (ICCCBDA)*, 12–15 April 2019. pp. 321–326. doi: 10.1109/ICCCBDA.2019.8725661.

Yuan, X., Li, Y., Xu, Y. and Huang, N. 2018. Curvilinear effects of personality on safety performance: The moderating role of supervisor support. *Personality and Individual Differences*, 122: 55–61. doi: 10.1016/j.paid.2017.10.005.

Zarei, T. and Behyad, R. 2019, Predicting the water production of a solar seawater greenhouse desalination unit using multi-layer perceptron model. *Solar Energy*, 177: 595–603. doi: 10.1016/j.solener.2018.11.059.

Zhang, J., Xiang, P., Zhang, R., Chen, D. and Ren, Y. 2020. Mediating effect of risk propensity between personality traits and unsafe behavioral intention of construction workers. *Journal of Construction Engineering and Management*, 146: 04020023. doi: 10.1061/(ASCE)CO.1943-7862.0001792.

Zohar, D. and Luria, G. 2004. Climate as a social-cognitive construction of supervisory safety practices: Scripts as proxy of behavior patterns. *Journal of Applied Psychology*, 89: 322–333. doi: 10.1037/0021-9010.89.2.322.

Appendix A2.1: The Big Five Inventory

Please indicate the extent to which you agree with the following statements (*disagree strongly, disagree a little, neither agree nor disagree, agree a little, agree strongly*).

I see myself as someone who...

Q1: is talkative. [a]

Q2: tends to find fault with others. [b]

Q3: does a thorough job. [c]

Q4: is depressed. [d]

Q5: is original, comes up with new ideas. [e]

Q6: is reserved. [a]

Q7: is helpful and unselfish with others. [b]

Q8: can be somewhat careless. [c]

Q9: handles stress well. [d]

Q10: is curious about many different things. [e]

Q11: is full of energy. [a]

Q12: starts quarrels with others. [b]

Q13: is a reliable worker. [c]

Q14: can be tense. [d]

Q15: is ingenious, a deep thinker. [e]

Q16: generates a lot of enthusiasm. [a]

Q17: has a forgiving nature. [b]

Q18: tends to be disorganised. [c]

Q19: worries a lot. [d]

Q20: has an active imagination. [e]

Q21: tends to be quiet. [a]

Q22: is generally trusting. [b]

Q23: tends to be lazy. [c]

Q24: is emotionally stable. [d]

Q25: is inventive. [e]

Q26: has an assertive personality. [a]

Q27: can be cold and aloof. [b]

Q28: perseveres until the task is finished. [c]

Q29: can be moody. [d]

Q30: values artistic, aesthetic experiences. [e]

Q31: is sometimes shy. [a]

Q32: is considerate and kind to almost everyone. [b]

Q33: does things efficiently. [c]

Q34: remains calm in tense situations. [d]

Q35: prefers work that is routine. [e]

Q36: is outgoing. [a]

Q37: is sometimes rude to others. [b]

Q38: makes plans and follows through with them. [c]

Q39: gets nervous easily. [d]

Q40: likes to reflect, play with ideas. [e]

Q41: has few artistic interests. [e]

Q42: likes to cooperate with others. [b]

Q43: is easily distracted. [c]

Q44: is sophisticated in art, music, or literature. [e]

[a] Extraversion; [b] Agreeableness; [c] Conscientiousness; [d] Neuroticism; [e] Openness.

Appendix A2.2: The Work-specific Big Five Inventory

Please indicate the extent to which you agree with the following statements (*disagree strongly*, *disagree a little*, *neither agree nor disagree*, *agree a little*, *agree strongly*).

I see myself as someone who…

Q1: is talkative at work. [a]

Q2: tends to find fault with others at work. [b]

Q3: does a thorough job at work. [c]

Q4: is depressed at work. [d]

Q5: is reserved at work. [a]

Q6: is helpful and unselfish with others at work. [b]

Q7: can be somewhat careless at work. [c]

Q8: handles stress well at work. [d]

Q9: is full of energy at work. [a]

Q10: starts quarrels with others at work. [b]

Q11: is a reliable worker at work. [c]

Q12: can be tense at work. [d]

Q13: generates a lot of enthusiasm at work. [a]

Q14: has a forgiving nature at work. [b]

Q15: tends to be disorganised at work. [c]

Q16: worries a lot at work. [d]

Q17: tends to be quiet at work. [a]

Q18: is generally trusting at work. [b]

Q19: tends to be lazy at work. [c]

Q20: is emotionally stable at work. [d]

Q21: has an assertive personality at work. [a]

Q22: can be cold and aloof at work. [b]

Q23: perseveres until the task is finished at work. [c]

Q24: can be moody at work. [d]

Q25: is sometimes shy at work. [a]

Q26: is considerate and kind to almost everyone at work. [b]

Q27: does things efficiently at work. [c]

Q28: remains calm in tense situations at work. [d]

Q29: is outgoing at work. [a]

Q30: is sometimes rude to others at work. [b]

Q31: makes plans and follows through with them at work. [c]

Q32: gets nervous easily at work. [d]

Q33: likes to cooperate with others at work. [b]

Q34: is easily distracted at work. [c]

[a] Extraversion; [b] Agreeableness; [c] Conscientiousness; [d] Neuroticism.

Appendix A2.3: The Safety Behaviour Scale

Please indicate the extent to which you agree with the following statements (*disagree strongly, disagree a little, neither agree nor disagree, agree a little, agree strongly*).

I see myself as someone who…

Q1: overlooks safety procedures in order to get my job done more quickly. [a]

Q2: follows all safety procedures regardless of the situation I am in. [a]

Q3: handles all situations as if there is a possibility of having an accident. [a]

Q4: wears safety equipment required by practice. [a]

Q5: keeps workplace clean. [b]

Q6: helps co-workers to be safe. [b]

Q7: keeps my work equipment in safe working condition. [a]

Q8: takes shortcuts to safe working behaviours in order to get the job done faster. [a]

Q9: does not follow safety practices that I think are unnecessary. [a]

Q10: reports safety problems to my supervisor when I see safety problems performed by co-workers. [b]

Q11: corrects safety problems to ensure accidents will not occur. [a]

[a] Task Performance; [b] Contextual Performance.

Chapter 3

Machine Learning Framework for Predicting Failure Mode and Flexural Capacity of FRP-Reinforced Beams

*Ahmad N. Tarawneh** and *Eman F. Saleh*

1. Introduction

Corrosion of steel bars is a major concern in reinforced concrete (RC) structures (Wight, 2016). Fiber-reinforced polymers (FRP) bars provide a promising substitution to steel bars in RC structures, particularly in harsh environments, due to their high resistance to corrosion, high strength-weight ratio, and excellent fatigue resistance. In practice, FRP can be made with different types of fibers including a glass-fiber-reinforced polymer (GFRP), carbon fiber-reinforced polymer (CFRP), and aramid fiber-reinforced polymer (AFRP) (Wight, 2016).

FRP bars cannot readily replace steel reinforcement bars due to their different mechanical characteristics. Most FRP bars provide a relatively low elastic modulus compared to steel, with a linear stress-strain relation until failure. Thus, flexural and shear behavior of reinforced concrete members internally reinforced with FRP (i.e., FRP-RC members) have been extensively investigated (Wight, 2016), and different guidelines (ACI Committee 440, 2015; Canadian Standards Association (CSA), 2012) were developed to assist engineers in designing FRP-RC members in flexure and shear.

Despite the number of investigations conducted to study the behavior of FRP-RC beams, developed design codes and guidelines of FRP-reinforced beams exhibit different inconsistencies. For example, ACI 440.1R design guideline allows the design of FRP-RC beams as tension-controlled or compression-controlled members, while CSA-S806 provisions only allow the FRP-RC beams to be designed as compression-controlled members. Moreover, the strength reduction factors adopted in different design guidelines differ significantly. This indicates the need to fully

Assistant Professor, Civil Engineering Department, Faculty of Engineering, The Hashemite University, P.O. box 330127, Zarqa 13133, Jordan. Email: emanf_yo@hu.edu.jo
* Corresponding author: ahmadn@hu.edu.jo

characterize the behavior of FRP beams so that an agreement on the design process can be reached.

Recently, a paradigm shift toward the utilization of machine learning-based techniques in different aspects of engineering has been witnessed. Specifically, machine learning (ML) has been applied in structural engineering applications including earthquake engineering (Reddy et al., 2011; González and Zapico, 2008), structural health monitoring (Salehi et al., 2019; Salehi et al., 2019), prediction of failure modes, capacity, and mechanical properties of the structural materials (Ni and Wang, 2000; Vakhshouri and Nejadi, 2018; Demir, 2008). ML can break down complex mathematical relations into simple operations and recognize patterns through learning from input (training) data. Particularly, artificial neural network (ANN), support vector machine (SVM), and gene expression programming (GEP) are among the most widely used ML algorithms used in structural engineering applications (Reddy et al., 2011).

The concept of ANN simulates the human central nervous system, where all input and output data are fully connected with neurons that transmit signals (information) to each other. Each neuron has a weight (factor) that is calibrated through algorithms to minimize the error between input and output data. While GEP mimics the Darwinian natural selection in the evolution of prediction expressions, where a mathematical operation that provides a higher correlation to the output data (less error between input and output data) will take place as compared to the ones with lower correlation. Similarly, the SVM algorithm is used in prediction and classification problems such as identifying failure mode. The algorithm develops a separating hyperplane among two classes as will be discussed in a later section.

Tarawneh et al. (2021) utilized gene expression programming (GEP) with a comprehensive experimental database of 241 tested steel-fiber reinforced concrete (SFRC) beams to develop a prediction model for the shear capacity of the beams. Independent variables utilized in the GEP model development were fiber factor, longitudinal reinforcement ratio, concrete cylinder compressive strength, and shear span-to-depth ratio. The proposed model provided superior accuracy compared to the existing models developed using linear and nonlinear regression, resulting in a higher reliability index (lower failure probability).

Kodur and Naser (2021) successfully utilized ML to provide a risk-based classification for the fire hazard of bridges. The study's importance arises from the lack of an approach to identify the vulnerability of bridges to fire hazards. Three ML classification algorithms, namely, Random Forest (RF), Support vector machine (SVM), and the generalized additive model (GAM), was trained with collected database which comprises 80 steel bridges and 38 concrete bridges to provide a risk-based classification for a fire hazard. The study also highlights and evaluates the significance of features that affect the fire risk on a bridge including fuel, span, age, geographical significance, material type, structural system, and fire position.

Moreover, Roya et al. (2020) developed an ML framework to develop a prediction model for failure mode and capacity of ultra-high performance concrete beams. The framework begins with developing a comprehensive database of 360 tested beams. Afterward, the affecting features were selected, and the data was analyzed using the SVM, ANN, and k-nearest neighbor algorithms to identify the failure model of the beams. Finally, GEP was used to develop a simplified prediction model for the capacity for each failure mode. The framework was tested against values obtained using design codes and other proposed equations and showed an accuracy of 82% in predicting the failure mode and a capacity prediction accuracy of $R^2 = 0.92$.

Vu and Hoang (2016) developed a model employing the SVM algorithm for regression to predict the capacity of FRP-reinforced concrete slabs. Yan and Shi (2010) have also employed the SVM algorithm to predict the elastic modulus of high and normal strength concrete.

A simplified and robust design methodology of FRP-reinforced concrete beams is essential to optimize the use of FRP bars in reinforced concrete flexural members. To develop such a methodology, an ML framework is employed here to identify the mode of failure and predict the flexural capacity of FRP-reinforced concrete beams. A wide range of experimental databases was surveyed and used as input to the ML framework to learn and be able to identify patterns that control failure mode and flexural capacity. The SVM, and k-nearest neighbor (K-NN) algorithms were used to classify failure modes observed in the experiments, i.e., concrete crushing and FRP rupture. Finally, GEP prediction models linking the identified failure mode were developed to predict the flexural capacity of the FRP-reinforced beam. The proposed expressions are a function of geometric properties of the beams and material properties of concrete and FRP bars.

2. Experimental Database

Generally, ML requires input or training data for learning and patterns recognition. To classify the failure mode and develop prediction models using GEP, existing literature was extensively surveyed and experimental data of a total of 164 tested FRP-reinforced beams in flexure were collected. Data collection is followed by identifying the critical parameters governing flexural behavior. The collected database comprises geometrical properties of the tested beam including beam width (b_w), depth (h), effective depth (d), and reinforcement ratio (ρ); in addition to material properties including concrete compressive strength ($f'c$), FRP elastic modulus (E_{FRP}), and FRP tensile strength ($f_{ult,FRP}$). In addition, the experimental database includes the identified failure mode in the experiment. A summary of the collected experimental data and a range of variables are provided in Table 3.1. The distribution of the variables and the failure mode are shown in Fig. 3.1.

3. Data-driven Machine Learning Framework for Response Prediction

In this study, the ML algorithm is applied to predict the failure mode and flexural capacity of FRP-reinforced concrete beams. To accomplish this, the ML framework is illustrated in Fig. 3.2 was employed. This framework comprises three stages that include data collection, failure mode classification, and capacity prediction (Roya et al., 2020). The data collection stage involves the collection of experimental data that cover a wide range of parameters (features) that govern the behavior as discussed previously (Table 3.1).

Table 3.1: Summary of surveyed experimental database.

Ref.	# of Tests	Width (mm)	Depth (mm)	f_c' (MPa)	Rnft. Ratio	E, GPa	f_{ult}, MPa	Failure Mode
Almusallam et al., 1997	4	200	240	3536	1.33	43	885	C
Theriault and Benmokrane, 1998	8	130	180	46–97	1.19–2.79	38	773	C
Toutanji and Saafi, 2000	6	180	300	35	0.53–1.1	40	695	C
Kassem et al., 2011	12	200	300	39–41	0.55–2.24	36–122	617–1988	C
Benmokrane et al., 1995	9	200	300–550	43–52	0.57–1.12	42–49	641–689	C, T
Ashour, 2006	6	150	200–300	28–50	0.14–0.28	38	650	T
Benmokrane and Masmoudi, 1996	3	200	300–550	43	0.57–1.13	45	600	C, T
Yost et al., 2001	12	191–381	152–216	28	0.12–1.35	41	830	C, T
Masmoudi et al., 1998	8	200	300	45–52	0.68–2.23	38	773	C
Brown and Bartholomew, 1993	5	152	152	36	0.37	45	760	T
Duranovic et al., 1997	3	150	250	25–32	1.34	45	1000	C
Alsayed et al., 2000	6	200	210–300	31–41	1.15–3.6	36	700	C
Bischoff, 2005	2	152	152	49–52	0.32–0.50	140	1900	C
El-Nemr et al., 2013	12	200	400	29–73	0.38–1.82	49–69	1639	C, T
Faza and Gangarao, 1993	6	152	305	29–45	0.87–2.49	46–51	552–896	C
Wang and Belarbi, 2005	3	178	229	48	0.67–3.39	41–124	552–2069	C
Lau and Pam, 2010	5	280	380	34–43	0.35–2.07	38–40	582–603	C, T
Gao and Benmokrane, 2001	13	200	270–294	39–54	0.59–2.95	38–49	513–773	C, T
El Refai et al., 2015	3	230	300	40	0.4–1.08	50	1000	C, T
Alkhraisha et al., 2020	6	180	230	30	0.34–1.29	43–47	1075–1121	C
Abed et al., 2021	10	180	230	47–70	0.51–1.94	43–130	1029–2068	C
Maranan et al., 2015	17	200	400	29–48	0.38–1.87	47–69	666–1639	C
Habeeb and Ashour, 2008	5	200	300	24–28	0.16–0.42	200	1061–2000	T

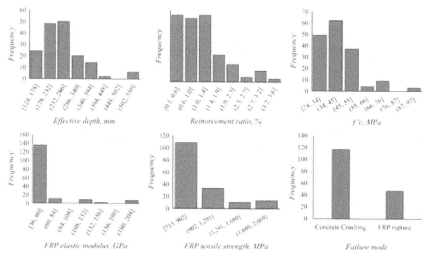

Fig. 3.1: Distribution of the variables in the collected experimental database.

After features selection, a classification algorithm, herein SVM and K-NN are utilized, is used to classify failure modes observed in the experiments. In the last step, a predictive equation for each failure mode classification is developed using the GEP algorithm. A detailed description of these steps is provided in the below subsections.

3.1 Failure Mode Classification

In this work, the SVM, and k-nearest neighbor (K-NN) algorithms were used to classify failure modes of FRP-reinforced beams observed in the experiments. A short description of these methods is presented herein, while a detailed description of these methods can be found in (Noble, 2006; Zhang and Zhou, 2007).

3.1.1 Support Vector Machine

The SVM is one of the most well-recognized ML algorithms that was originally proposed for binary classification in which the solution involves the developing of a separating hyperplane among two classes (Zhang and Zhou, 2007). This is done by simultaneously maximizing the width of the margin between the decision hyperplanes and the labeled data (i.e., with known failure mode) and minimizing the misclassification of the data. To illustrate this, consider the separating hyperplane is defined as (Zhang and Zhou, 2007)

$$w^T x + b = 0 \tag{3.1}$$

where w is a weighting vector that defines the direction of the margin boundaries, whereas b represents the bias. Following this, the decision function could be defined as follows:

$$f(x) = sgn(w^T x + b) = \begin{cases} 1 & w^T x + b \geq 0 \\ -1 & w^T x + b < 0 \end{cases} \tag{3.2}$$

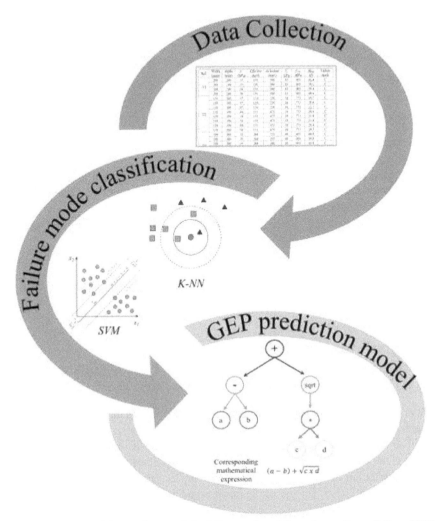

Fig. 3.2: Schematic of the ML framework for failure mode and flexural capacity prediction of FRP-reinforced beams.

The SVM algorithm tries to maximize the margin by minimizing the L_2-norm of w (i.e., $\|w\|$), this results in the following constrained optimization problem (Zhang and Zhou, 2007)

$$min_{w,\xi}\left[\frac{1}{2}\|w\|^2 + C\Sigma_{i=1}^{M}\xi_i\right]$$

$$s.t. \quad y_i(w^T x + b) \geq 1 - \xi_i \tag{3.3}$$

$$\xi_i > 0$$

$$C > 0$$

where ξ_i is the slack variable that is introduced to allow for some violation in the constraints (some data points inside the margin). If a linear separation of the data

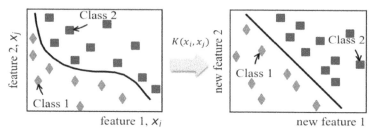

Fig. 3.3: An example of the application of the kernel trick to nonlinearly separable data.

is not possible, an alternative solution involves the employ of a kernel trick that projects the data points into a higher dimension where data can be made linearly separable (Hofmann, 2006) (see Fig. 3.3).

The kernel function $K(x_i, x_j)$ is defined as the dot product of functions as follows (Hofmann, 2006; Çevik et al., 2015),

$$K(x_i, x_j) = \varnothing(x_i)^T \varnothing(x_j) \tag{3.4}$$

where x_i and x_j are the features of the data. In SVM, the commonly used kernel functions are:

- Linear kernel, $K(x_i, x_j) = x_i^T x_j$
- Polynomial kernel, $K(x_i, x_j) = (\gamma x_i^T x_j + r)^d$
- Radial basis function (RBF) or Gaussian kernel, $K(x_i, x_j) = \exp(-||x_i - x_j||^2/2\sigma^2)$
- Sigmoid kernel, $K(x_i, x_j) = \tanh(\gamma x_i^T x_j + r)$

where γ, r, σ, and d are kernel parameters. The kernel parameter γ must be greater than zero.

The linear kernel is the simplest and has only one parameter C (see Eq. 3.3) which is the regularization or penalty parameter that controls the trade-off between the maximization of the width of the margin and the minimization of classification error (Tharwat, 2019). The use of the linear kernel is equivalent to the algorithm that does not use any kernel.

Radial basis function (RBF) or Gaussian kernel has only one parameter σ that controls the performance of SVM (Tharwat, 2019). RBF has many variants that have different sensitivity behaviors to the change of the parameter σ. For example, the Laplacian kernel which is defined as $K(x_i, x_j) = \exp(-||x_i - x_j||/2\sigma)$ is less affected by the change of the parameter σ. Another variant to the RBF kernel is the exponential kernel which is defined as $K(x_i, x_j) = \exp(-||x_i - x_j||/2\sigma^2)$, the only difference between the exponential kernel and the RBF kernel is that the RBF kernel includes the square of the norm.

Normally the SVM algorithm includes the following steps (Tharwat, 2019):

1. Data transformation and scaling as required by the SVM package.
2. Consider a certain kernel (e.g., RBF kernel).
3. Obtain the optimal kernel's parameters by cross-validation (e.g., for RBF the parameters are σ and C).

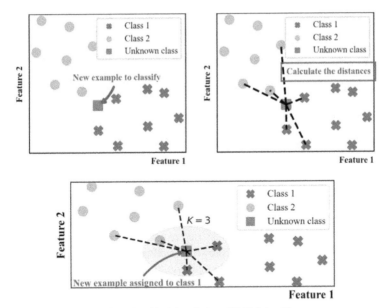

Fig. 3.4: Graphical description of K-NN algorithm.

4. Use the optimal parameters to train the entire training set.
5. Test the generated model using the test dataset.

3.1.2 K-nearest Neighbor

K-NN is a classification algorithm. The process of K-NN involves computing the distance between every two feature vectors (x_i, x_j) of the training dataset. Then this distance and the class that corresponds to x_i is added to an ordered collection. This ordered collection is arranged in an ascending order based on distances (Yu et al., 2001). After that, the first K entries of the sorted ordered collection are selected. Then, the class that corresponds to the feature vector x_i is determined as the mode of the K classes. The algorithm aims to evaluate the optimal K that reduces the error of misclassification by trying several K and choosing this optimal one. A graphical description of the K-NN concept is illustrated in Fig. 3.4. As shown in the figure, to identify the class of an unknown example, the distances between the unknown example and all other neighboring points are calculated first. The K-NN classification rule is to assign the unknown example to the class to which most of the K nearest training examples belong.

The Euclidean distance between a testing example and the specified training examples is the most used in the K-NN classification algorithm (Zhang and Zhou, 2007). Let x_i be a testing example with k features $(x_{i1}, x_{i2}, ..., x_{ik})$ and n is the total number of specified training examples, each has k features. The Euclidean distance between x_i and $x_j = (x_{j1}, x_{j2}, ..., x_{jk})$ where $j = 1, 2, ..., n$ is defined as (Zhang and Zhou, 2007; Yu et al., 2001):

$$D(x_i, x_j) = \sqrt{(x_{i1} - x_{j1})^2 + (x_{i2} - x_{j2})^2 + \cdots + (x_{ik} - x_{jk})^2} \qquad (3.5)$$

3.1.3 Classification Results

The collected dataset described in Table 3.1 was employed as input to the ML framework. This dataset was randomly divided into three sets: training set, testing set, and validation set. The training set was used to train the classification algorithm (to develop a predictive model), the validation set was used to calibrate the parameters of the model, and the test set was used to evaluate the performance of the developed model. To implement the SVM and K-NN algorithms, the dataset was classified into two classes: Tension-controlled class, and compression-controlled class whereas the input features $x_1, x_1, \dots x_6$ are the variables that control the behavior of FRP-reinforced concrete beams.

The MATLAB® software was utilized to implement the SVM and K-NN algorithms. The classification accuracy of each classifier is shown in Tables 3.2 and 3.3 for SVM and K-NN algorithms, respectively. The classification accuracy represents the number of correctly classified examples in the training set to the total number of examples in this set. As shown in Table 3.2, the quadratic kernel has the best performance; thus, it was used to evaluate the performance of the SVM model on the test set. Further, it is obvious from Table 3.3 that the weighted K-NN type led to the best classification performance; thus, it is used to assess the test set.

To assess the performance of each classifier on the test set, a confusion matrix that contains information on predicted and observed classifications was developed as shown in Fig. 3.5a and Fig. 3.5b for SVM with the quadratic kernel, and weighted K-NN, respectively. Herein, the observed classification represents the actual failure mode observed in the experiments on the FRP-reinforced concrete beam while the predicted classification represents the one predicted by the developed classifiers.

Table 3.2: SVM classification accuracy.

SVM Type	Classification Accuracy
Linear	92.7%
Quadratic	97.0%
Cubic	93.9%
Fine Gaussian	92.1%
Medium Gaussian	95.1%
Coarse Gaussian	78.7%

Table 3.3: K-NN classification accuracy.

K-NN Type	Classification Accuracy
Fine	97.0%
Medium	83.5%
Coarse	71.3%
Cosine	78.7%
Cubic	81.7%
Weighted	98.2%

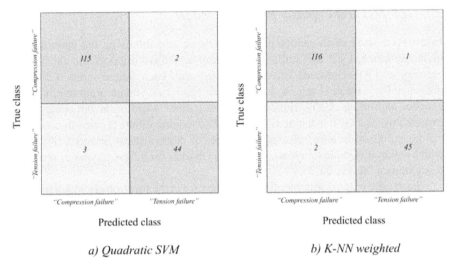

a) *Quadratic SVM*   b) *K-NN weighted*

Fig. 3.5: Classification confusion matrix (a) Quadratic SVM (b) K-NN weighted.

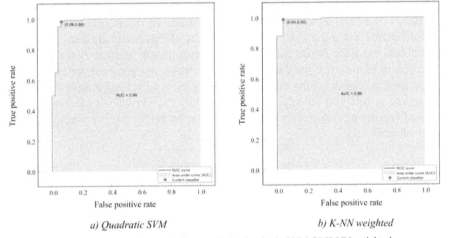

a) *Quadratic SVM*   b) *K-NN weighted*

Fig. 3.6: Classification confusion matrix (a) Quadratic SVM (b) K-NN weighted.

The diagonal terms (colored in green) of a confusion matrix represent the correctly classified test examples (observed failure mode is the same as the predicted one) while the off-diagonal terms (colored in red) represent the misclassified ones. As shown from Fig. 3.5b, the weighted K-NN performed slightly better than the quadratic SVM in predicting the failure modes in the test set. To further assess each algorithm, the receiver operating characteristics (ROC) curve was plotted and presented in Fig. 3.6a and Fig. 3.6b for the SVM and K-NN algorithm, respectively. The area under the ROC curve represents a metric that can be used to assess the performance of the classifier on the test. As shown in Fig. 3.6, the area under the ROC curve that corresponds to the weighted K-NN is higher compared to the area under the ROC curve that corresponds to the SVM algorithm; thus, the K-NN has a better performance.

3.2 GEP Flexural Capacity Prediction Model

In the third stage of the adopted ML framework and following the classification of failure mode, GEP was utilized to develop a prediction model for the flexural capacity of FRP-reinforced beams associated with each failure mode. GEP was presented by Ferreira (Ferreira, 2001), which combines genetic programming with genetics algorithms. GEP mimics Darwinian natural selection in the evolution of expressions. The GEP machine is capable of generating robust prediction codes that can be expressed in terms of expression trees or mathematical formulas (Fig. 3.7), which makes it suitable for creating prediction models for different types of problems (Gandomi and Roke, 2015).

Each GEP process involves five components: (1) the function set which includes the mathematical operation that will be used; (2) the terminal set which indicates the symbolic representation of the variables; (3) the fitness function which evaluates how fit is the chromosome with respect to the rest of the population (chromosomes) such as root mean square error (RMSE) Eq. 3.6, (4) control parameters; and (5) a termination condition. Figure 3.8 illustrates the GEP algorithm that evolves the expressions tree. The algorithm starts by selecting the five components mentioned earlier. The initial population of chromosomes is generated randomly using the predefined function set and the terminals. The produced chromosomes are executed and converted into a tree expression. The results of expression are compared to the actual data using the fitness function and if the results are satisfactory the process is terminated. However, if the desired level of fitness is not achieved, variation in the population is introduced by conducting one or more genetic operators (crossover, mutation, or rotation) on selected chromosomes. The individuals are selected and copied into the next generation according to the fitness by roulette wheel sampling with elitism.

$$RMSE = \sqrt{\frac{1}{N} \Sigma_{K=1}^{N} \left(v_{test} - v_{predicted} \right)^2} \qquad (3.6)$$

Features extracted from the database and included in the development of the prediction model are beam width b, beam effective depth d, concrete cylinder

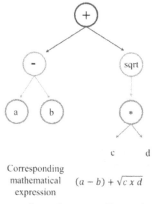

Corresponding mathematical expression $(a - b) + \sqrt{c \times d}$

Fig. 3.7: Tree expression and corresponding mathematical expression.

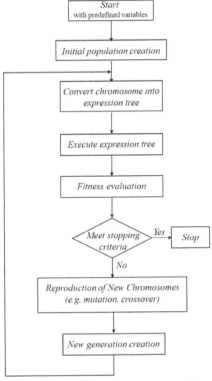

Fig. 3.8: Gene expression evolution algorithm.

Table 3.4: GEP selected parameters.

Parameter	Selected Value
Dependent variable (shear stress)	1
Independent variables	5
Genes	3, 4
Function set	$-, +, \times, \div, \sqrt{\square}, \sqrt[3]{\square}, \square^{\square}, \exp$
Head size	6, 7
The linking function between ETs	Multiplication

compressive strength f'_c, reinforcement ratio ρ, FRP elastic modulus E, and FRP tensile strength f_{ult}. In addition, various parameters were involved in GEP model development; these parameters affect the generalization capability of the developed model. Parameters selected for the GEP algorithm are shown in Table 3.4. The selection of the parameters was based on multiple trials. In each trial, one variable varies while the others remain constant. Trials were compared based on the coefficient of variation (COV) and the average of the tested-to-predicted ratio ($M_{u,test}/M_{u,predicted}$). In developing the prediction models, basic mathematical operations were selected in the function set to produce a simple model.

The surveyed database was devised into two sets: A concrete crushing failure set (117 tests) and an FRP-rupture failure set (47 tests). Each dataset was divided into a training set and validation set to ensure the generalization of the model. The developed GEP models for each failure mode and the associated coefficient of determination (R^2) are listed in Table 3.5 and shown in Fig. 3.9. The GEP model for compression-controlled failure resulted in RMSE of 9.45 *kN.m*, while the tension-controlled model resulted in 8.56 *kN.m*.

Table 3.5: Derived GEP model to predict the flexural capacity of FRP-reinforced beams.

Failure Mode	GEP Prediction Model	R^2
Concrete crushing	$$Y = \frac{1}{\big(((b+8.21)-(2\,f'c))+(b+5.05)\big)}$$ $$\times \left(\left(\left(\frac{d}{1.134}\right)+e^{6.27}\right)-\big((11.16+b)\times 5.89\big)\right)$$ $$\times \frac{d}{(5.98\times 5.98)^2 + \dfrac{E}{-0.22}}\times\big((\rho\times 8.08)+40.06-d\big)\times(\rho)^{0.5}$$	0.96
FRP-rupture	$$y = \left(\sqrt[3]{f'c}\times(3.6\times 2.684)-(d-f'c)\times 0.416\right)$$ $$\times \frac{\sqrt[6]{10^{-9.28}}}{(-1.614\times -5.225)+(-5.22^2)}\times(E$$ $$-\big(d+((2\rho)\times(b+f_{ult}))\big)\big)\big)$$	0.98

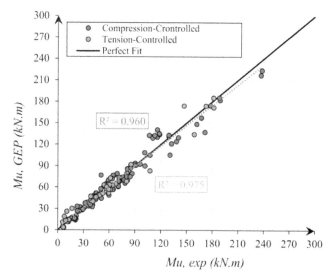

Fig. 3.9: Experimental vs. predicted ultimate flexural capacity for concrete crushing and FRP-rupture failure modes.

4. Comparison Study

To assess the prediction accuracy and validate the proposed framework, the flexural capacity obtained using the prediction models was compared to the design guidelines ACI 440-15 and CSA S806-12. Classifying the failure mode and determining the flexural capacity procedure for both design guidelines are summarized in Table 3.6. As mentioned earlier, CSA S806-12 allows only concrete crushing failure mode. ACI 440-15 and CSA S806-12 classification guidelines, which are based on the balanced reinforcement ratio, resulted in an accuracy of 94% and 95% in predicting the failure mode, compared to an accuracy of 99% when using the K-NN algorithm.

Figure 3.10 shows the experimental versus the predicted flexural capacity using CSA S806, ACI 440, and the proposed machine learning-based method. Although the results do not show a significant difference in accuracy, the ML-based model provides the highest accuracy with an overall RMSE of 9.2 kN.m compared to RMSE of 11.6 and 11.1 kN.m for CSA S806 and ACI 440, respectively. Also, in

Table 3.6: Classification and flexural capacity prediction procedure for ACI 440 and CSA S806 guidelines.

ACI 440-15	CSA S806-12
$\rho_f = \dfrac{A_f}{bd}$	$\rho_f = \dfrac{A_f}{bd}$
$\beta_1 = \begin{cases} 17 \le f_c' \le 28, & \beta_1 = 0.85 \\ 28 < f_c' < 55, & \beta_1 = 0.85 - \dfrac{0.05(f_c' - 28)}{7} \\ f_c' \ge 55, & \beta_1 = 0.65 \end{cases}$	$\alpha_1 = 0.85 - 0.0015 f_c'$
When $\rho_f > \rho_b$ (Concrete crushing)	$\beta_1 = 0.85 - 0.0025 f_c'$
$f_f = \sqrt{\dfrac{(E_f \varepsilon_{cu})^2}{4} + \dfrac{0.85 \beta_1 f_c'}{\rho_f} E_f \varepsilon_{cu}} - 0.5 E_f \varepsilon_{cu} \le f_{fu}$	*When $\rho_f > \rho_b$ (Concrete crushing)*
$M_u = \rho_f f_f bd^2 \left(1 - 0.59 \dfrac{\rho_f f_f}{f_c'}\right)$	$\alpha_1 \beta_1 f_c' bc - A_f E_f \varepsilon_{cu} \dfrac{d-c}{c} = 0$
When $\rho_f > \rho_b$ (FRP rupture)	$f_f = E_f \varepsilon_{cu} \dfrac{d-c}{c} < f_{fu}$
$c_b = \left(\dfrac{\varepsilon_{cu}}{\varepsilon_{cu} + \varepsilon_{fut}}\right) d$	$M_n = \rho_f f_f bd^2 \left(1 - \dfrac{\rho_f f_f}{2\alpha_1 f_c'}\right)$
$M_n = A_f f_{fw} \left(d - \dfrac{\beta_1 c_b}{2}\right)$	*When $\rho_f > \rho_b$ (FRP rupture)* The controlling limit state in this standard is a compressive failure and is only viable for concrete crushing failure mode beams
$\varepsilon_{cu} = 0.003$	$\varepsilon_{cu} = 0.0035$

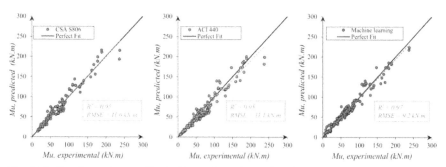

Fig. 3.10: Experimental vs. predicted ultimate flexural capacity using CSA S806, ACI 440, and the adopted ML framework.

terms of the Coefficient of determination (R^2), the proposed model provides an $R^2 = 0.97$ compared to $R^2 = 0.95$ for both CSA S806 and ACI 440.

5. Conclusions

This study is of a hybrid machine learning framework to predict the failure mode and flexural capacity of FRP-reinforced concrete beams. The framework was broken down into three stages: data collection, failure mode classification, and gene expression-prediction model development. Based on the study results, the following conclusions can be drawn:

1. Machine learning-based classification methods, SVM and K-NN, can be considered effective and reliable in predicting the failure mode of FRP-reinforced concrete beams with high accuracy. The K-NN method yielded a classification accuracy of 99% compared to 94% and 95% for the ACI 440 and CSA S806 procedures, which are based on the balanced reinforcement ratio.

2. Despite the high accuracy of ACI 440 and CSA S806 procedures, the proposed GEP flexural capacity prediction model provided higher accuracy with an overall R^2 of 0.97 and RMSE of 9.2 kN.m.

3. The ML framework provides a flexible, reliable, and fast approach to classify and predict the flexural capacity of FRP-RC beams. The adopted framework can be applied to a wide range of structural engineering applications.

4. Although the CSA S806 standard does not allow the designing of the FRP-reinforced beam with FRP rupture failure mode, the proposed framework has shown that FRP rupture failure mode and the corresponding moment capacity can be predicted with high accuracy.

References

Abed, F., Al-Mimar, M. and Ahmed, S. 2021. Performance of BFRP RC beams using high strength concrete. *Composites Part C: Open Access*, 4: 100–107.

ACI Committee 440. 2015. *Guide for the Design and Construction of Structural Concrete Reinforced with FRP Bars (ACI 440.1R-15)*, American Concrete Institute, Farmington Hills, MI, 83 pp.

Alkhraisha, H., Mhanna, H., Tello, N. and Abed, F. 2020. Serviceability and flexural behavior of concrete beams reinforced with basalt fiber-reinforced polymer (BFRP) bars exposed to harsh conditions. *Polymers*, 12(9): 21–10.

Almusallam, T.H., Al-Salloum, Y.A., Alsayed, S.H. and Amjad, M.A. 1997, October. Behavior of concrete beams doubly reinforced by FRP bars. *In: Proceedings of the Third International Symposium on Non-metallic (FRP) Reinforcement for Concrete Structures (FRPRCS-3), Japan* (Vol. 2), pp. 471–478.

Alsayed, S.H., Al-Salloum, Y.A. and Almusallam, T.H. 2000. Performance of glass fiber reinforced plastic bars as a reinforcing material for concrete structures. *Composites Part B: Engineering*, 31(67): 555–567.

Ashour, A.F. 2006. Flexural and shear capacities of concrete beams reinforced with GFRP bars. *Construction and Building Materials*, 20(10): 1005–1015.

Benmokrane, B. and Masmoudi, R. 1996. Flexural response of concrete beams reinforced with FRP reinforcing bars. *Structural Journal*, 93(1): 46–55.

Benmokrane, B., Chaallal, O. and Masmoudi, R. 1995. Glass fibre reinforced plastic (GFRP) rebars for concrete structures. *Construction and Building Materials*, 9(6): 353–364.

Bischoff, P.H. 2005. Reevaluation of deflection prediction for concrete beams reinforced with steel and fiber reinforced polymer bars. *Journal of Structural Engineering*, 131(5): 752–767.

Brown, V.L. and Bartholomew, C.L. 1993. FRP reinforcing bars in reinforced concrete members. *Materials Journal*, 90(1): 34–39.

Canadian Standards Association (CSA). 2012. *Design and Construction of Building Components with Fiber Reinforced Polymers (CAN/CSAS806-12)*, CSA Group, Rexdale, ON, Canada, 208 pp.

Çevik, A., Kurtoğlu, A.E., Bilgehan, M., Gülşan, M.E. and Albegmprli, H.M. 2015. Support vector machines in structural engineering: A review. *Journal of Civil Engineering and Management*, 21(3): 261–281.

Demir, F. 2008. Prediction of elastic modulus of normal and high strength concrete by artificial neural networks. *Construction and Building Materials*, 22(7): 1428–1435.

Duranovic, N., Pilakoutas, K. and Waldron, P. 1997. Tests on concrete beams reinforced with glass fibre reinforced plastic bars. *Non-metallic (FRP) Reinforcement for Concrete Structure*, 2: 479–486.

El Refai, A., Abed, F. and Al-Rahmani, A. 2015. Structural performance and serviceability of concrete beams reinforced with hybrid (GFRP and steel) bars. *Construction and Building Materials*, 96: 518–529.

El-Nemr, A., Ahmed, E.A. and Benmokrane, B. 2013. Flexural behavior and serviceability of normal-and high-strength concrete beams reinforced with glass fiber-reinforced polymer bars. *ACI Structural Journal*, 110(6).

Faza, S.S. and Gangarao, H.V. 1993. Theoretical and experimental correlation of behavior of concrete beams reinforced with fiber reinforced plastic rebars. *Special Publication*, 138: 599–614.

Ferreira, C. 2001. Gene expression programming: a new adaptive algorithm for solving problems. *arXiv preprint* cs/0102027.

Gandomi, A.H. and Roke, D.A. 2015. Assessment of artificial neural network and genetic programming as predictive tools. *Advances in Engineering Software*, 88: 63–72.

Gao, D. and Benmokrane, B. 2001. Calculation method of flexural capacity of GFRP-reinforced concrete beams. *J. Hydraul. Eng.*, 73–80.

González, M.P. and Zapico, J.L. 2008. Seismic damage identification in buildings using neural networks and modal data. *Computers & Structures*, 86 (3-5): 416–426.

Habeeb, M.N. and Ashour, A.F. 2008. Flexural behavior of continuous GFRP reinforced concrete beams. *Journal of Composites for Construction*, 12(2): 115–124.

Hofmann, M. 2006. Support vector machines-kernels and the kernel trick. *Notes*, 26(3): 1–16.

Kassem, C., Farghaly, A.S. and Benmokrane, B. 2011. Evaluation of flexural behavior and serviceability performance of concrete beams reinforced with FRP bars. *Journal of Composites for Construction*, 15(5): 682–695.

Kodur, V.K. and Naser, M.Z. 2021. Classifying bridges for the risk of fire hazard via competitive machine learning. *Advances in Bridge Engineering*, 2(1): 1–12.

Lau, D. and Pam, H.J. 2010. Experimental study of hybrid FRP reinforced concrete beams. *Engineering Structures*, 32(12): 3857–3865.

Maranan, G.B., Manalo, A.C., Benmokrane, B., Karunasena, W. and Mendis, P. 2015. Evaluation of the flexural strength and serviceability of geopolymer concrete beams reinforced with glass-fibre-reinforced polymer (GFRP) bars. *Engineering Structures*, 101: 529–541.

Masmoudi, R., Theriault, M. and Benmokrane, B. 1998. Flexural behavior of concrete beams reinforced with deformed fiber reinforced plastic reinforcing rods. *Structural Journal*, 95(6): 665–676.

Ni, H.G. and Wang, J.Z. 2000. Prediction of compressive strength of concrete by neural networks. *Cement and Concrete Research*, 30(8): 1245–1250.

Noble, William S. 2006. What is a support vector machine?. *Nature Biotechnology*, 24(12): 1565–1567.

Reddy, T.A., Devi, K.R. and Gangashetty, S.V. 2011, December. Multilayer feedforward neural network models for pattern recognition tasks in earthquake engineering. *In: International Conference on Advanced Computing, Networking and Security*. Berlin, Heidelberg: Springer, pp. 154–162.

Salehi, H., Biswas, S. and Burgueno, R. 2019. Data interpretation framework integrating machine learning and pattern recognition for self-powered data-driven damage identification with harvested energy variations. *Engineering Applications of Artificial Intelligence*, 86: 136–153.

Salehi, H., Das, S., Chakrabartty, S., Biswas, S. and Burgueño, R. 2019. An algorithmic framework for reconstruction of time-delayed and incomplete binary signals from an energy-lean structural health monitoring system. *Engineering Structures*, 180: 603–620.

Solhmirzaei, Roya, Salehi, H., Kodur, V. and Naser, M.Z. 2020. Machine learning framework for predicting failure mode and shear capacity of ultra-high-performance concrete beams. *Engineering Structures*, 224: 111–221.

Tarawneh, A., Almasabha, G., Alawadi, R. and Tarawneh, M. 2021. August. Innovative and reliable model for shear strength of steel fibers reinforced concrete beams. *In: Structures* (Vol. 32), Elsevier, pp. 154–162.

Tharwat, A. 2019. Parameter investigation of support vector machine classifier with kernel functions. *Knowledge and Information Systems*, 61(3): 1269–1302.

Theriault, M. and Benmokrane, B. 1998. Effects of FRP reinforcement ratio and concrete strength on flexural behavior of concrete beams. *Journal of Composites for Construction*, 2(1): 7–16.

Toutanji, H.A. and Saafi, M. 2000. Flexural behavior of concrete beams reinforced with glass fiber-reinforced polymer (GFRP) bars. *Structural Journal*, 97(5): 712–719.

Vakhshouri, B. and Nejadi, S. 2018. Prediction of compressive strength of self-compacting concrete by ANFIS models. *Neurocomputing*, 280: 13–22.

Vu, D.T. and Hoang, N.D. 2016. Punching shear capacity estimation of FRP-reinforced concrete slabs using a hybrid machine learning approach. *Structure and Infrastructure Engineering*, 12(9): 1153–1161.

Wang, H. and Belarbi, A. 2005. Flexural behavior of fiber-reinforced-concrete beams reinforced with FRP rebars. *ACI Structural Journal, SP230*, 51(230): 895–914.

Wight, J.K. 2016. *Reinforced Concrete Mechanics and Design*. (7th Edn.). Hoboken, New Jersey 07030: Pearson Education, Inc.

Yan, K. and Shi, C. 2010. Prediction of elastic modulus of normal and high strength concrete by support vector machine. *Construction and Building Materials*, 24(8): 1479–1485.

Yost, J.R., Goodspeed, C.H. and Schmeckpeper, E.R. 2001. Flexural performance of concrete beams reinforced with FRP grids. *Journal of Composites for Construction*, 5(1): 18–25.

Yu, C., Ooi, B.C., Tan, K.L. and Jagadish, H.V. 2001. September. Indexing the distance: An efficient method to K-NN processing. *In: Vldb* (Vol. 1), pp. 421–430.

Zhang, M.L. and Zhou, Z.H. 2007. ML-KNN: A lazy learning approach to multi-label learning. *Pattern Recognition*, 40(7): 2038–2048.

Appendix 3A.1

Ref.	Width (mm)	Depth (mm)	f_c' (MPa)	Effective Depth	As Bottom (mm²)	E, GPa	f_{ult}, MPa	M_{exp}, kN	Failure Mode
Salehi et al., 2019	200	240	35	191	508	43	885	41.4	C
	200	240	35	191	508	43	885	39.1	C
	200	240	36	191	508	43	885	39.4	C
	200	240	36	191	508	43	885	40.6	C
Ni and Wang, 2000	130	180	57	154	238	38	773	19.7	C
	130	180	57	154	238	38	773	20.6	C
	130	180	97	154	238	38	773	22.7	C
	130	180	46	131	475	38	773	20.6	C
	130	180	54	131	475	38	773	21.0	C
	130	180	54	131	475	38	773	21.4	C
	130	180	94	131	475	38	773	28.4	C
	130	180	94	131	475	38	773	29.5	C
Vakhshouri and Nejadi, 2018	180	300	35	268	253	40	695	60.0	C
	180	300	35	268	253	40	695	59.0	C
	180	300	35	268	380	40	695	65.0	C
	180	300	35	268	380	40	695	64.3	C
	180	300	35	255	507	40	695	71.0	C
	180	300	35	255	507	40	695	70.5	C
Demir, 2008	200	300	40	236	284	114	1506	71.2	C
	200	300	39	236	425	114	1506	83.1	C
	200	300	39	236	567	114	1506	90.4	C
	200	300	40	230	254	122	1988	78.8	C
	200	300	41	230	382	122	1988	80.9	C
	200	300	41	230	509	122	1988	89.4	C
	200	300	39	226	760	40	617	77.5	C
	200	300	39	226	1013	40	617	86.8	C
	200	300	39	227	679	36	747	71.0	C
	200	300	39	227	905	36	747	84.5	C
	200	300	39	236	425	52	1800	70.9	C
	200	300	39	236	567	52	1800	71.8	C

contd.... .

...Appendix 3A.1 contd.

Ref.	Width (mm)	Depth (mm)	f_c' (MPa)	Effective Depth	As Bottom (mm²)	E, GPa	f_{ult}, MPa	M_{exp}, kN	Failure Mode
Tarawneh et al., 2021	200	300	44	255	573	42	689	80.4	C
	200	300	44	255	573	49	641	50.6	C
	200	300	44	255	573	49	641	63.8	C
	200	450	52	405	573	49	641	106.6	C
	200	450	52	405	573	49	641	113.0	C
	200	550	43	505	573	42	689	181.5	T
	200	550	43	505	573	42	689	181.5	T
	200	550	43	505	573	49	641	146.9	T
	200	550	43	505	573	49	641	172.5	T
Kodur and Naser, 2021	150	200	28	164	57	38	650	5.9	T
	150	250	28	222	57	38	650	7.8	T
	150	300	28	269	57	38	650	10.8	T
	150	200	50	164	57	38	650	5.9	T
	150	250	50	222	57	38	650	9.5	T
	150	300	50	269	113	38	650	16.8	T
Solhmirzaei et al., 2020	200	300	43	254	573	45	600	80.4	C
	200	550	43	504	573	45	600	181.7	T
	200	550	43	504	573	45	600	181.7	T
Vu and Hoang, 2016	381	203	28	179	80	41	830	11.5	T
	381	203	28	179	80	41	830	12.7	T
	381	203	28	179	80	41	830	11.5	T
	318	216	28	192	80	41	830	13.6	T
	318	216	28	192	80	41	830	13.3	T
	318	216	28	192	80	41	830	13.1	T
	203	152	28	124	320	41	830	15.8	C
	203	152	28	124	320	41	830	15.6	C
	203	152	28	124	320	41	830	16.3	C
	191	152	28	124	320	41	830	16.4	C
	191	152	28	124	320	41	830	16.7	C
	191	152	28	124	320	41	830	15.8	C

contd... .

...Appendix 3A.1 contd.

Ref.	Width (mm)	Depth (mm)	f_c' (MPa)	Effective Depth	As Bottom (mm^2)	E, GPa	f_{ult}, MPa	M_{exp}, kN	Failure Mode
Yan and Shi, 2010	200	300	52	258	349	38	773	57.9	C
	200	300	52	258	349	38	773	59.8	C
	200	300	52	258	523	38	773	66.0	C
	200	300	52	258	523	38	773	64.8	C
	200	300	45	235	697	38	773	75.4	C
	200	300	45	235	697	38	773	71.7	C
	200	300	45	235	1046	38	773	84.8	C
	200	300	45	235	1046	38	773	85.4	C
Almusallam et al., 1997	152	152	36	127	71	45	760	7.0	T
	152	152	36	127	71	45	760	6.6	T
	152	152	36	127	71	45	760	7.2	T
	152	152	36	127	71	45	760	7.4	T
	152	152	36	127	71	45	760	6.8	T
Theriault and Benmokrane, 1998	150	250	25	213	429	45	1000	40.3	C
	150	250	32	213	429	45	1000	39.7	C
	150	250	32	213	429	45	1000	39.5	C
Toutanji and Saafi, 2000	200	210	31	158	1134	36	700	34.2	C
	200	260	31	211	507	43	886	45.1	C
	200	300	41	248	567	36	700	59.2	C
	200	250	41	198	1134	36	700	57.0	C
	200	300	24	265	88	200	2000	44.8	T
	200	300	27	265	226	200	1061	60.7	T
Kassem et al., 2011	152	152	52	130	63	140	1900	12.6	C
	152	152	49	130	99	140	1900	17.1	C
Benmokrane et al., 1995	200	400	34	344	261	67	1639	82.8	C
	200	400	34	344	385	49	817	81.3	C
	200	400	59	344	261	67	1639	101.6	C
	200	400	59	344	385	49	817	85.6	T
	200	400	29	319	970	69	1362	129.3	C
	200	400	34	319	1162	50	762	118.7	C
	200	400	73	319	970	69	1362	178.0	C
	200	400	73	319	1162	50	762	177.7	C
	200	400	34	319	970	60	1245	110.6	C
	200	400	34	338	1019	60	906	115.9	C
	200	400	73	342	1040	60	1245	188.4	C
	200	400	73	338	1019	60	906	189.1	C

contd... .

...Appendix 3A.1 contd.

Ref.	Width (mm)	Depth (mm)	f_c' (MPa)	Effective Depth	As Bottom (mm²)	E, GPa	f_{ult}, MPa	M_{exp}, kN	Failure Mode
Ashour, 2006	152	305	29	267	1013	46	552	54.2	C
	152	305	34	268	774	51	552	56.5	C
	152	305	45	267	1006	46	552	74.2	C
	152	305	45	267	1006	46	552	81.4	C
	152	305	45	273	381	48	738	42.2	C
	152	305	45	269	355	48	896	50.9	C
Benmokrane and Masmoudi, 1996	178	229	48	185	723	41	690	47.0	C
	178	229	48	179	1077	41	552	51.0	C
	178	229	48	185	219	124	2069	51.0	C
Yost et al., 2001	280	380	37	344	804	40	593	158.8	T
	280	380	41	340	1964	38	582	237.9	C
	280	380	42	346	339	40	603	80.4	T
	280	380	43	346	452	40	603	107.3	T
	280	380	34	340	1964	38	582	236.8	C
Masmoudi et al., 1998	200	294	43	254	299	45	552	38.5	T
	200	294	54	254	299	45	552	41.0	T
	200	294	43	254	299	49	641	52.8	T
	200	294	43	254	449	49	641	61.6	T
	200	270	42	230	1356	38	773	85.4	C
	200	294	39	254	381	45	552	57.2	T
	200	294	51	254	381	45	552	59.7	T
	200	294	40	254	381	49	513	59.7	T
	200	294	40	254	381	49	513	61.6	T
	200	294	45	231	508	45	552	61.0	T
	200	294	45	231	508	45	552	54.1	T
	200	294	40	231	508	49	513	59.7	T
	200	294	40	231	508	49	513	67.8	T
Brown and Bartholomew, 1993	230	300	40	244	226	50	1000	49.0	T
	230	300	40	244	339	50	1000	53.8	C
	230	300	40	242	603	50	1000	69.6	C
Duranovic et al., 1997	180	230	30	186	115	43	1121	21.3	C
	180	230	30	186	115	43	1121	21.7	C
	180	230	30	184	243	47	1119	28.7	C
	180	230	30	184	365	47	1119	36.1	C
	180	230	30	182	424	46	1075	37.9	C
	180	230	30	182	424	46	1075	40.7	C

contd... .

...Appendix 3A.1 contd.

Ref.	Width (mm)	Depth (mm)	f_c' (MPa)	Effective Depth	As Bottom (mm²)	E, GPa	f_{ult}, MPa	M_{exp}, kN	Failure Mode
Alsayed et al., 2000	180	230	47	186	172	43	1075	23.0	C
	180	230	47	185	171	43	1029	22.8	C
	180	230	47	184	243	47	1119	31.1	C
	180	230	47	182	636	46	1121	38.3	C
	180	230	47	184	243	130	2068	41.6	C
	180	230	70	186	172	43	1075	26.0	C
	180	230	70	185	171	43	1029	24.8	C
	180	230	70	184	243	47	1119	31.9	C
	180	230	70	182	636	46	1121	44.5	C
	180	230	70	184	243	130	2068	50.3	C
Bischoff, 2005	200	400	34	344	387	49	817	81.3	C
	200	400	39	320	645	49	817	130.6	C
	200	400	34	344	258	67	1639	82.8	C
	200	400	39	343	597	48	751	101.3	C
	200	400	39	343	796	48	751	138.2	C
	200	400	29	343	398	69	1362	95.9	C
	200	400	34	343	398	60	1245	91.3	C
	200	400	34	320	1194	48	751	118.3	C
	200	400	29	320	995	69	1362	129.3	C
	200	400	34	320	995	60	1245	110.6	C
	200	400	39	340	568	48	728	107.4	C
	200	400	42	340	852	48	728	140.4	C
	200	400	39	340	774	47	693	132.3	C
	200	400	48	340	852	53	1082	171.4	C
	200	400	48	338	1020	53	666	161.7	C
	200	400	48	338	1020	66	1132	167.2	C
	200	400	34	338	1020	60	906	115.9	C
El-Nemr et al., 2013	200	300	27	265	223	200	1061	64.1	T
	200	300	28	265	85	200	2000	44.3	T
	200	300	24	265	85	200	2000	44.8	T
	200	300	27	265	223	200	1061	60.7	T
	200	300	28	265	223	200	1061	56.0	T

Chapter 4

A Novel Formulation for Estimating Compressive Strength of High Performance Concrete Using Gene Expression Programming

Iman Mansouri,[1,*] *Jale Tezcan*[2] *and Paul O. Awoyera*[3]

1. Introduction

Many design problems in civil engineering can be described as finding the best option among a set of alternatives. One such problem is the concrete mix design, where the proportions of the ingredients are to be selected in such a way so as to optimize a certain mechanical property of the concrete. This process involves estimation of a regression function which models the relationship between the target variable and a set of predictors consisting of mix ingredients and other relevant variables such as temperature and concrete age. In general, the model that shows the closest agreement with the experimental observations is considered the optimal solution. Identifying the best model can be complicated in cases where the mathematical form of the regression function is unknown, or the objective function contains multiple local optima. In such cases, mathematically motivated algorithms such as gradient descent are either inapplicable or inefficient. Metaheuristic algorithms which seek to find a near-optimal solution in a reasonable amount of time have recently emerged as a viable approach to solving complicated optimization problems. Many metaheuristic algorithms are nature-inspired; they mimic animal behavior or natural processes such as evolution to solve optimization problems where the objective function is discontinuous, nonlinear, or non-differentiable. Metaheuristic algorithms can be classified in multiple ways. One approach is to characterize the algorithm as either population-based or trajectory-based (Gandomi et al., 2013; Yang, 2010) depending

[1] Dept. of Civil and Environmental Engineering, Princeton University, NJ 08544, United States.
[2] Professor, School of Civil, Environmental and Infrastructure Engineering, Southern Illinois University, Carbondale, IL 62901, United States.
[3] Department of Civil Engineering, Covenant University, 112233 Ota, Nigeria.
* Corresponding author: imansouri@princeton.edu

on the number of candidate solutions being considered at each iteration. Population-based algorithms use a set of strings or agents (Kennedy and Eberhart, 1995). Genetic algorithms and particle swarm optimization are considered population-based. On the other hand, trajectory-based algorithms use a single candidate solution that moves through the search space. Typical examples of trajectory-based algorithms are simulated annealing (Kirkpatrick et al., 1983), iterated local search, and guided local search. In addition to being used as the sole optimization strategy, metaheuristic algorithms have also been blended with other methods to combine the benefits of different strategies, resulting in hybrid metaheuristics. Over the last two decades, various metaheuristic algorithms have been used in the training of artificial neural networks (ANNs) (Haykin, 1999), especially those of feedforward type (Ojha et al., 2017). Different ANN components such as the number of layers, the number of nodes at the hidden layers, and the connection weights can be determined using metaheuristic optimization. This approach overcomes the limitations of the gradient-descent algorithms (e.g., backpropagation) that are commonly used in conventional ANNs. The use of evolutionary algorithms in the design of ANNs is a common form of hybrid metaheuristics where the local search ability of the backpropagation algorithm is combined with the global search ability of evolutionary algorithms. As ANNs have been successful in solving a wide range of civil/structural engineering problems (e.g., (Alavi and Gandomi, 2011; Ahmad et al., 2021; Asteris et al., 2021; Pizarro and Massone, 2021; Tarawneh et al., 2021; Abuodeh et al., 2020)), the use of GP and its extensions, both as an alternative to ANNs and as part of the ANN training procedure is being explored by researchers.

In this chapter, we explore the use of a linear variant of gene programming (GP) called gene expression programming (GEP) (Ferreira, 2001) in predicting the strength properties of high performance concrete (HPC). In particular, using a dataset (https://archive.ics.uci.edu/ml/datasets/concrete+compressive+strength) consisting of 1030 experimental test results, we derive a mathematical expression describing the relationship between the compressive strength of HPC and a set of eight predictors consisting of mix ingredients and the age at testing.

As the name implies, HPC refers to concrete designed to have better performance than conventional concrete. ACI Concrete Terminology (CT-18) defines HPC as "concrete meeting special combinations of performance and uniformity requirements that cannot always be achieved routinely using conventional constituents and normal mixing, placing, and curing practices". As the desired characteristics of concrete are problem-specific, proper interpretation of the meaning of HPC relies on the particular application and the environment. It is generally understood that HPC offers at least one of the following properties (Kosmatka et al., 2002): high strength, high early strength, high elastic modulus, low permeability, low shrinkage, high resistance to abrasion, high resistance to damaging chemicals, high durability in severe environments, high toughness and impact resistance, ease of placement and compaction without segregation.

Although HPC is often synonymous with high strength concrete, which is preferred in bridges and tall buildings, high strength is not necessary for HPC designation. High resistance to chloride penetration and an enhanced air-void system can be achieved in a normal strength concrete, reducing material costs in bridge design (Bickley and Fung, 2001).

To achieve the desired performance measures, in addition to the basic ingredients of traditional concrete (i.e., Portland cement, water, fine and coarse aggregates), HPC contains cementitious materials and chemical admixtures. Cementitious materials such as blended cement, fly ash, slag, and silica fume are added to the mix to increase the strength and durability of the concrete. Water reducers enable reducing the water content by up to 12% while maintaining a certain level of consistency. Up to 30% reduction in water content is possible using superplasticizers. Durability, corrosivity, and alkali-silica reactivity of concrete can also be controlled using specific admixtures.

These additional ingredients, however, often modify various properties of concrete in addition to the targeted property. For example, water reducers often affect the setting time. Some water reducers accelerate the initial setting time, and others do the opposite. The interactions between the ingredients and their effect on different properties are not fully understood. While laboratory testing can be used to investigate these properties, experiments are expensive and time-consuming. As HPC is increasingly being used in various civil engineering applications, understanding and modeling the behavior of HPC remains a challenging task.

The remainder of this chapter is organized as follows: The next section summarizes the state of knowledge on the use of metaheuristic optimization and machine-learning approaches for predicting concrete strength. Genetic algorithm and its extensions are briefly discussed in Section 3, and the main components of the GEP algorithm are described in Section 4. The GEP model for the compressive strength of HPC is developed in Section 5. Finally, Section 6 presents the conclusions of this study.

2. Artificial Intelligence-based Models of HPC Properties

Concrete is a highly complex material. Its strength and durability properties are determined by various factors including the mix ingredients and their proportions, preparation technique, curing conditions, and age. As high-performance mixes contain various admixtures and cementitious materials, predicting the behavior of HPC is more challenging than traditional concrete. This difficulty motivates the use of artificial intelligence techniques in modeling HPC behavior. As the functional relationships linking these factors to strength and durability properties of concrete are not fully understood, there is a need for empirically developed predictive models. Due to the persistent use of HPC, it has become necessary to develop an appropriate soft computing method for predicting HPC properties.

Compressive strength, being one of the most important properties of concrete, has been the subject of various studies since the early 1900s. One of the earliest and well-known models of compressive strength is Abrams law, which suggests that the strength of the concrete increases as the water-to-cement ratio decreases. Since the introduction of Abrams' law, additional variables have been found to influence the compressive strength. These include cement to sand ratio, amount of cementitious material, and type of cement. In addition, experimental studies contradicting Abrams' rule also exist. For example, Popovics (1990) observed that when two concrete mixtures with the same water-cement ratios were tested, the mixture having higher cement content yielded the higher strength. Despite the rich literature reporting results from thousands of experimental tests representing different mixes and curing conditions, it is difficult to generalize these results to other cases.

The quality of coarse aggregate significantly influences the performance of HPC, unlike conventional concrete. Silica fume is used in HPC, and aggregate types are carefully selected, as they must have a smaller maximum size (10–14 mm) than those used in conventional concrete. The smaller maximum size of aggregate in HPC aids reduction of the differential stresses that could cause micro-cracking at the aggregate-cement paste inter-face. Also, smaller aggregate particles are known to be stronger than larger ones.

Over the last few decades, various predictive models have been developed to model strength characteristics of both traditional (Golafshani and Behnood, 2018; Naderpour et al., 2018; Sobhani and Najimi, 2014) and high performance concrete (Bui et al., 2018; Tayfur et al., 2014; Yu et al., 2018). De Larrard (1999) developed one of the first models for concrete containing pozzolanic admixtures and lime, which required manual adjustment of parameters and weighting factors. The strength properties of HPC have also been estimated based on a Gaussian process-powered framework (Ke and Duan, 2021). In another study, Nguyen et al. (2021) used four different predictive algorithms, support vector regression, multilayer perceptron, gradient boosting regressor, and extreme gradient boosting to develop predictive models for compressive and tensile strengths of HPC. Mousavi et al., developed prediction models for the compressive strength of HPC mixes using gene expression programming (Mousavi et al., 2012), and a combination of gene expression programming and orthogonal least squares (Mousavi et al., 2010). ANN models have been used to predict the performance of composites made with steel, recycled aggregate, and other types of concrete (Abdalla and Hawileh 2011; Abdalla and Hawileh, 2013; Durodola et al., 2018). Choudhary et al. (2021) developed an ANN-based model to predict the compressive strength of ultra HPC. They observed an improvement in model accuracy when features are selected sequentially. Latif (2021), used deep learning method to estimate the compressive strength of HPC. Higher-order response surface methodology and multivariate adaptive regression splines approach have also been found used to model the performance of concrete mixture (Hameed et al., 2021; Kaloop et al., 2020). Solhmirzaei et al. (2020) proposed a machine learning framework to estimate the shear capacity and mode of failure of ultra HPC. The authors analyzed the database using various machine learning techniques to identify the major factors affecting the HPC shear capacity and failure mode. Genetic algorithm and its variants and extensions have also been used in modeling various properties of concrete (Al-Ghrery et al., 2021; Ketabdari et al., 2020; Mansouri et al., 2021; Mansouri et al., 2021; Murad et al., 2020; Naser et al., 2021; Sarir et al., 2021; Tarawneh et al., 2021; Saad and Malik, 2018).

Genetic algorithms and ANNs are Artificial Intelligence methods, belonging to evolutionary algorithms and machine learning methods, respectively. Compared to the traditional ANNs, which are essentially black boxes, and can only deal with a small class of optimization problems, genetic algorithms and its variants provide a powerful global optimization strategy and more transparency.

3. Overview of Genetic Algorithms and its Extensions

The Genetic Algorithm (GA) is an evolutionary optimization algorithm introduced by Holland (Holland, 1975). Inspired by Darwin's theory of evolution, GA uses operators mimicking biological processes such as natural selection, crossover, and mutation (Mitchell, 1996; Zhengjun et al., 1998). GA is a population-based algorithm which solves an optimization problem by manipulating populations of candidate solutions, where individuals are selected using a fitness function, and new generations are obtained using genetic operators.

Compared to traditional, gradient-based optimization algorithms, GAs offer two main advantages. The first is the ability to deal with complex problems where the objective function is discontinuous, nonlinear, noisy, or nonstationary. The second is parallelism, where the search space is explored by multiple agents simultaneously, allowing the manipulation of different parameters at the same time. GAs often use a combination of local and global search strategies which help ensure convergence to the global optimum. The term population refers to the set of candidate solutions currently being considered in the search algorithm. Each member of the population is called an individual and is represented by a chromosome, which refers to a string of genes. Given an optimization problem, the GA uses the following procedure to find a good solution: (1) encoding the objective function; (2) defining a selection criterion; (3) defining an initial population; (4) iteratively updating the population towards better solutions; (5) obtaining the solution to the problem by decoding the results. Each iteration of updating the population leads to a new generation.

The GA algorithm requires a set of user-defined parameters and functions, which must be carefully selected considering the specific problem and the goal of optimization. The optimal values for the parameters depend on the specifics of the optimization problem, e.g., mathematical properties of the objective function, in addition to the quality (representativeness) and the quantity of the available data. The selected set of parameters greatly impact the efficiency and performance of the algorithm.

In canonical GA, each member of the population is a character string of fixed length, called a chromosome, which encodes a candidate solution to the problem being considered. Genetic Programming (GP) is an extension of GA where the population consists of parse trees of varying shapes and sizes, where each parse tree represents a computer program or an equation which produces a candidate solution when executed (Gandomi and Alavi, 2011; Koza, 1992; Ferreira, 2006). GP can be seen as an evolutionary algorithm which generates computer programs by searching the program space and not the data space. The benefit of GP, compared to the canonical GA or other metaheuristic algorithms is that, being a symbolic optimization technique, it provides an analytical expression corresponding to the model, making the replica more transparent, and potentially easier to interpret.

Gene expression programming (GEP) is a more recent GA variant, where each member of the population is a chromosome encoded as a linear, symbolic string of fixed length. Despite their fixed length, GEP chromosomes can encode expression trees of different shapes and sizes. According to Ferreira (2006), GA and GP both suffer from the tradeoff between functional complexity and ease of genetic modification, while GEP does not, due to the genotype/phenotype separation, as described in the next section. Researchers from various disciplines have utilized the GEP techniques to solve civil engineering problems (Gandomi et al., 2021; Iqbal et al., 2021; Tadayon et al., 2021; Yao et al., 2021; Zou et al., 2021).

4. Gene Expression Programming (GEP) Algorithm

4.1 GEP Chromosome Structure

GEP is a genotype/phenotype-based genetic algorithm, where chromosomes and expression trees (ETs) function as the genotype and phenotype, respectively. The mapping from genotype to phenotype space is called decoding, and the reverse is called encoding.

In GEP, each chromosome is composed of one or more genes of equal length. The number of genes is selected for each problem. Each gene can be compactly represented as a K-expression, using the Karva language created by Ferreira specifically for reading and expressing the information contained in the GEP chromosomes. The K-expression is obtained by reading the expression tree from left to right and from top to bottom, as one would read a book. Given a K-expression, the corresponding ET representation is uniquely determined in a straightforward way.

A distinguishing feature of GEP is that each gene is composed of a head and a tail. The head encodes the functions and terminals, while the tail encodes the terminals. The length of the head is preselected for each problem, while the length of the tail (t) is determined using

$$t = h \times (n-1) + 1, \tag{4.1}$$

where h is the length of the head and n is the maximum arity, defined as the number of arguments of the function with the most arguments. The length of the gene is sum of the length of the head and that of the tail:

$$l = n \times h + 1 \tag{4.2}$$

The ET corresponding to each gene can be constructed considering the following rules: (1) The first line (topmost position) of the ET is occupied by the first position of the gene, which is called the root; (2) The number of nodes in each line is determined by the number of arguments of the function in the line above; (3) The nodes in each line are filled consecutively, reading the remaining elements of the K-expression; (4) The process is complete when a line which does not contain any function is obtained. As the termination point of the ET does not necessarily coincide with the last position of the gene, GEP genes often contain regions that do not appear in the ET. If location e of the gene marks the termination point of the ET, locations $e + 1$ through l forms the noncoding region of the gene. The presence of noncoding regions gives the GEP chromosomes the ability to code ETs of different shapes and sizes, despite having a fixed length. When the first location of the gene is occupied by a terminal, the ET consists of a single node. On the other extreme, when all elements of the head are functions with n arguments, the ET contains l nodes, achieving its maximum possible size.

As an example, Fig. 4.1 shows a chromosome composed of two genes with head length 5. Shaded cells represent the tail.

Figures 4.2a and 4.2b show the sub-ET encoded by Gene 1 and Gene 2, respectively. The sub-ET of Gene 1 is equivalent to the mathematical expression $a + (b/a - b)$, while the sub-ET of Gene 2 is equivalent to $(b - a)b$. Note that although both genes are 11 characters long, the sub-ETs they encode are of different sizes. The sub-ET of Gene 1 has 7 nodes, while sub-ET of Gene 2 has 5 nodes.

The sub-ETs can be combined into a single ET using a linking function which takes more than one argument. When the ETs of the genes represent algebraic equations, the linking function is a mathematical function such as addition or multiplication. On the other hand, ETs representing Boolean expressions can be linked using a Boolean function such as AND or OR. Figure 4.3 shows the ET obtained by linking the ETs shown in Fig. 4.2 through addition.

					Gene 1											Gene 2					
0	1	2	3	4	5	6	7	8	9	0	0	1	2	3	4	5	6	7	8	9	0
+	a	-	/	b	b	a	b	a	b	b	*	-	b	b	a	b	b	a	a	b	a

Fig. 4.1: K-expression of a two-genic chromosome.

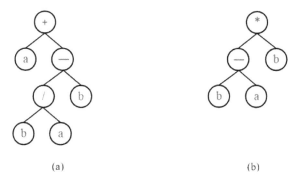

(a)   (b)

Fig. 4.2: Sub-ETs encoded by two genes (a) Gene 1, (b) Gene 2.

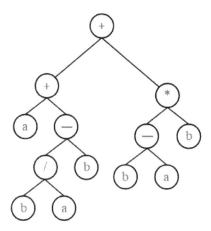

Fig. 4.3: The ET resulting from linking the two ETs by addition.

4.2 Fitness Function

A critical step in all conventional evolutionary algorithms is the evaluation of each candidate solution using a predefined fitness function to quantify the degree to which it meets the optimization goal. A fitness function can be defined as a mapping from a chromosome to a scalar value. It is closely related to the objective function used in mathematical optimization, and in the case of unconstrained optimization, the two terms can often be used interchangeably. As the optimization algorithm is guided by the fitness scores of the candidate solutions, proper selection of the fitness function is very important to the success of any genetic algorithm. A good fitness function should have a clear mathematical definition, and the values it generates should be easy to interpret in the context of the desired optimization goal. Traditionally, the fitness function is selected such that it produces a positive value, higher values corresponding to better solutions. Another important consideration in selecting a fitness function is its efficient implementation. As the fitness function is evaluated many times during the course of optimization, its computation should be sufficiently fast.

In regression problems, where the dependent variable is numeric, and often continuous, the fitness of a chromosome can be evaluated based on an error measure. Although the goal is to find a mathematical expression that produces a value close to the target value for each chromosome and each fitness case, using an excessively narrow selection range is problematic as it slows down the evolution of populations, which in turn may prevent the algorithm to find a good solution. Using a very broad selection is also problematic, as in this case, high fitness values do not necessarily correspond to good solutions. For symbolic regression problems, the fitness function takes the form of Eq. (4.3) and Eq. (4.4), respectively, using the absolute error and the relative error (Ferreira, 2001):

$$f_i = \Sigma_{j=1}^{C_t} (M - |C_{(i,j)} - T_j|) \qquad (4.3)$$

$$f_i = \Sigma_{j=1}^{C_t} \left(M - \left| 100 \times \frac{C_{(i,j)} - T_j}{T_j} \right| \right), \qquad (4.4)$$

where M is a constant representing the range of selection, $C_{(i,j)}$ is the value returned by the individual chromosome i for fitness case j, and T_j is the target value for fitness case j, and C_t is the number of fitness cases. In the case of a perfect fit, the $C_{(i,j)}$ equals the target value T_j for all fitness cases, leading to the fitness value $f_i = M \times C_t$ using either equation. The fitness function can be defined in many other ways, using a variety of similarity metrics, such as squared correlation coefficient or mean squared error. In classification problems, where a categorical or nominal target variable is being predicted, the fitness function is designed considering the rates of correct and incorrect classification. For simple classification problems with balanced datasets, the number of correctly classified examples can be used as the fitness function. This approach may not work well for complex classification problems, especially for those involving unbalanced datasets. In such cases, an efficient fitness function can be designed using the counts of true positives, true negatives, false positives, and false negatives, assigning different weights to these four counts. A similar approach can be used in problems where the target variable represents a binary or Boolean prediction.

4.3 Genetic Operators

Genetic operators are functions that operate on one or two chromosomes, potentially modifying their genetic composition. Genetic operators determine the genetic composition of the offspring and help efficient exploration of the search space. In its most basic form, GEP uses three main types of genetic operators: selection, crossover, and mutation. Selection refers to the identification of the fit individuals to create the next generation. Instead of using a greedy algorithm, where the selection is done based solely on the fitness scores, a probabilistic approach is preferred where a high fitness score increases the likelihood of being selected, but does not guarantee it. Recombination, also called crossover, refers to combining two parent chromosomes to produce two offsprings. In GEP algorithm, three types of recombination are defined: one-point recombination, two-point recombination, and gene recombination. In the one-point recombination, two-parent chromosomes are split up at the same point, and the segment downstream of this point is exchanged. The location of this point is randomly determined. In the two-point recombination, two points are randomly selected, and two-parent chromosomes exchange the material between these two points. In gene recombination, two-parent chromosomes exchange the entire genes. Mutation refers to a random change in the chromosome, to introduce and maintain diversity in the population. Mutation provides a mechanism by which the population can escape from a local optimum, helping efficient exploration of the search space. As opposed to the crossover, which is mainly a local action which helps exploitation of the current subspace, mutation may lead to a solution outside the current subspace. This property helps the algorithm prevent getting trapped in a local optimum.

In addition to these basic operators, various architecture-altering genetic operators have been developed. The inversion operator works by randomly selecting a segment of the gene head and reversing the order of the characters occupying this segment. Since the tail region is not altered during this operation, the tail of the new gene does not contain any function, and therefore there is no risk of creating syntactically incorrect genes. Another genetic operation is transposition. In GEP, transposition takes three different forms. The first is Insertion Sequence (IS) transposition, which refers to a short fragment that transposes to the head of gene, avoiding the first element of the head, which corresponds to the root of the ET encoded by the gene. The second is Root Insertion Sequence (RIS) transposition, which refers to a short fragment that starts with a function that transposes to the root of the gene. The third is gene transposition which refers to the total genes that transpose to the beginning of chromosomes.

5. GEP Model for HPC Compressive Strength

In this section, a relationship is derived for the compressive strength of HPC based on age and mix ingredients. GeneXproTools (GeneXproTools. Version 4.0) was used for the GEP model development.

5.1 Database

The GEP model was developed using the Concrete Compressive Strength database obtained from the UCI (UC Irvine) Machine Learning Repository (https://archive. ics.uci.edu/ml/datasets/concrete+compressive+strength). The database contains 1030 HPC compressive strength test results, representing different mix proportions and ages at testing. Each observation consists of nine variables: eight input variables and the compressive strength as the output variable. Table 4.1 shows the descriptive statistics of the input and output variables.

Table 4.1: Descriptive statistics of the variables.

	Cement (kg/m³)	Blast Furnace Slag (kg/m³)	Fly Ash (kg/m³)	Water (kg/m³)	Superplasticizer (kg/m³)	Coarse Aggregate (kg/m³)	Fine Aggregate (kg/m³)	Age (day)	Compressive strength(MPa)
Average	281.17	73.90	54.19	181.57	6.20	972.92	773.58	45.66	35.82
Standard deviation	104.51	86.28	64.00	21.36	5.97	77.75	80.18	63.17	16.71
Median	272.90	22.00	0.00	185.00	6.35	968.00	779.51	28.00	34.44
Mode	425.00	0.00	0.00	192.00	0.00	932.00	594.00	28.00	33.40
Min	102.00	0.00	0.00	121.75	0.00	801.00	594.00	1.00	2.33
Max	540.00	359.40	200.10	247.00	32.20	1145.00	992.60	365.00	82.60

The database used was randomly divided into training (80%) and test (20%) datasets. The training data was used in generating the GEP model. The test data was used to evaluate the performance of the model.

5.2 User-defined Parameters and Settings

A general expression for the compressive strength (s) is given as follows:

$$s = f(d_0, d_1, d_2, d_3, d_4, d_5, d_6, d_7) \qquad (4.5)$$

where d_0, d_1, d_2, d_3, d_4, d_5, d_6, and d_7 represent age at testing (day), blast furnace slag (kg/m³), cement (kg/m³), coarse aggregate (kg/m³), fly ash (kg/m³), fine aggregate (kg/m³), superplasticizer (kg/m³), and water (kg/m³), respectively.

It was assumed that a GEP chromosome consists of three genes with a head length of twelve. An addition operator was selected as the linking function.

The function set consists of the following eight operations: addition, subtraction, multiplication, division, power, square root, exponential, and natural logarithm.

The terminal set consists of the seven predictors (d_0 through d_7), in addition to two numerical constants per gene.

The fitness of individuals was evaluated based on the root-mean-squared-error (RMSE), using the following function:

$$f_i = \frac{1000}{1 + RMSE_i} \tag{4.6}$$

where

$$RMSE_i = \sqrt{\frac{1}{C_t} \sum_{j=1}^{C_t} (C_{(i,j)} - T_j)^2}, \tag{4.7}$$

C_t is the number of fitness cases, $C_{(i,j)}$ is the value returned by the individual chromosome i for fitness case j, and T_j is the target value for fitness case j. As the GEP aims to maximize the fitness, RMSE itself is not a valid fitness function. To obtain a valid fitness function, where lower RMSE corresponds to a higher fitness score, the RMSE was used in the denominator of the fitness function. The scalar 1 was added to the RMSE to prevent division by zero. Finally, 1000 used in the numerator indicates the maximum possible fitness value. Note that $f_i = 1000$ if chromosome i reproduces the target value perfectly for all fitness cases.

One of the most important parameters that must be selected by the user is the population size. When the population size is small, the algorithm may converge to a poor solution. On the other hand, a large population may require excessive search time. A population of 10–160 individuals is suggested for typical applications (Fogel, 2002). In this study, the population size was set to 100. Selection of the other control parameters was determined using trial-and-error, considering the findings reported in similar studies (e.g., (Baykasoğlu et al., 2008; Cevik and Cabalar, 2009)). Table 4.2 shows the control parameters used in the execution of the GEP algorithm. The process is repeated until the solution has reached a plateau, and no significant improvement is achieved in the average fitness scores of the chromosomes.

Table 4.2: Control parameters used in the GEP algorithm.

Parameter	Value
Population size	100
Number of genes	3
Head length	12
Linking function	Addition
Mutation rate	0.044
Inversion rate	0.1
IS transposition rate	0.1
RIS transposition rate	0.1
Gene transposition rate	0.1
One-point recombination rate	0.3
Two-point recombination rate	0.3
Gene recombination rate	0.1

5.3 Results

The equation for HPC compressive strength obtained using the GEP algorithm is:

$$s = s_1 + s_2 + s_3 \qquad (4.8)$$

where s is the compressive strength, and s_1, s_2, and s_3 are the expressions represented by the three genes of the fittest chromosome selected by the GEP algorithm, which are defined as follows:

$$s_1 = \sqrt{d_2 \times ln\left[\dfrac{d_0}{\left(\dfrac{-0.109d_3 - 2.77 + d_7}{-0.361d_7 + d_2}\right)}\right]}$$

$$s_2 = \left[4.58(d_0 + \sqrt{d_6} - 2.94)(\dfrac{d_1^2}{d_7})\right]^{0.25} + d_5^{0.1}$$

$$s_3 = -9.91 + \sqrt{(d_4 + d_6)^{0.25}\left(\dfrac{d_0(d_2 - d_6)}{d_0 + d_7}\right)}$$

Figure 4.4 compares of the compressive strength predictions of the GEP model to the experimentally measured values obtained using the training and test sets.

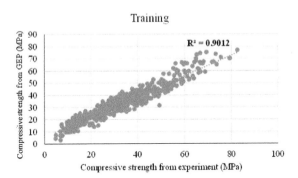

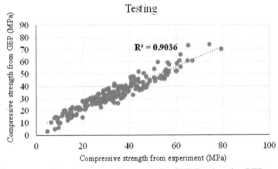

Fig. 4.4: Measured versus predicted compressive strength of HPC using the GEP model: (a) training set and (b) test set.

Table 4.3 shows the performance metrics of the resulting GEP-based model for the compressive strength, which is represented by Eq. (4.8). In this table, R denotes the correlation coefficient, RMSE is the root mean square error, and MAE stands for mean absolute error.

Table 4.3: Performance metrics of the GEP model.

	R	R^2	RMSE	MAE
Training	0.9493	0.9012	4.8182	3.9996
Test	0.9506	0.9036	4.7499	4.0844

According to Smith (1986), a regression model with correlation coefficient $|R| > 0.8$ implies good performance. While being a good indicator of the degree of linearity, correlation coefficient between the predicted and target values is not a reliable indicator of the model accuracy. Indeed, when one or both of the variables undergo a linear transformation, where all values of the variables are modified by adding a constant and/or multiplication by a constant, the correlation coefficient remains the same. Thus, additional metrics are needed to assess the performance of any regression model. The mean absolute error (MAE) and RMSE are useful metrics in this regard. The low values of MAE and RMSE, as shown in Table 4.3, indicates the ability of the model to produce predictions close to the measured values. Further, the similarity of the values obtained for these metrics for the training and test dataset indicates good generalization performance.

6. Conclusions

Regression techniques are widely used in many disciplines, to describe the relationship between a target variable and one or more predictor variables, based on a set of empirical observations. The resulting mathematical expression produces a prediction of the target variable for different values of input variables, reducing the need to collect real-world data for different settings of the predictor variables. The use of a regression model to predict various mechanical properties of concrete reduces the need for laborious, time-consuming, and expensive laboratory tests.

This study used GEP, a variant of genetic programming, to develop a new regression model for the compressive strength of HPC using a database of 1030 HPC compressive test results. The prediction performance of the model was evaluated using a test set using correlation and error measures and found to be satisfactory. The similarity of the performance values over the training and test data is a sign that the model was not overfit if the observations are assumed to be independent. As opposed to ANN, the GEP is a gradient-free method able to deal with complex, multi-objective optimization problems, without getting trapped in local minima.

There are two limitations of the proposed approach. The first is related to the database used in this study which has been used by many other researchers. Strictly speaking, the observations listed in the database are not independent. For example, the database contains multiple test results from the same mix, tested at different times (7 days, 28 days, etc.). The lack of independence of the observations means that the resulting model is not immune to overfitting, despite the similarity of the performance metrics computed for training and test sets. The second is in regards to the interpretability of the GEP models. One of the purported advantages GEP over the machine learning approach is the resulting model's interpretability, as opposed to the black-box approach of ANNs. While this may indeed be the case for simple models consisting of one or two predictors, the mathematical expression produced by GEP gets complicated as the number of predictors increase. Further, as the user is required to pre-define the function set, the terminal set, the number and length of genes, and the linking function, the mathematical form of the resulting expression is in part pre-determined by the user. Thus, while the GEP appears to produce accurate predictions, it may be difficult to gain insight into complex relationships.

References

Abdalla, J.A. and Hawileh, R. 2011. Modeling and simulation of low-cycle fatigue life of steel reinforcing bars using artificial neural network. *Journal of the Franklin Institute*, 348(7): 1393–1403. https://doi.org/10.1016/j.jfranklin.2010.04.005.

Abdalla, J.A. and Hawileh, R.A. 2013. Artificial neural network predictions of fatigue life of steel bars based on hysteretic energy. *Journal of Computing in Civil Engineering*, 27(5): 489–496. https://doi.org/10.1061/(ASCE)CP.1943-5487.0000185.

Abuodeh, O.R., Abdalla, J.A. and Hawileh, R.A. 2020. Assessment of compressive strength of ultra-high performance concrete using deep machine learning techniques. *Applied Soft Computing Journal*, 95. https://doi.org/10.1016/j.asoc.2020.106552.

Ahmad, A., Elchalakani, M., Elmesalami, N., El Refai, A. and Abed, F. 2021. Reliability analysis of strength models for short-concrete columns under concentric loading with FRP rebars through Artificial Neural Network. *Journal of Building Engineering*, 42: https://doi.org/10.1016/j.jobe.2021.102497.

Alavi, A.H. and Gandomi, A.H. 2011. Prediction of principal ground-motion parameters using a hybrid method coupling artificial neural networks and simulated annealing. *Computers and Structures*, 89(23-24): 2176–2194. https://doi.org/10.1016/j.compstruc.2011.08.019.

Al-Ghrery, K., Kalfat, R., Al-Mahaidi, R., Oukaili, N. and Al-Mosawe, A. 2021. Prediction of concrete cover separation in reinforced concrete beams strengthened with FRP. *Journal of Composites for Construction*, 25(4). https://doi.org/10.1061/(ASCE)CC.1943-5614.0001130.

Asteris, P.G., Skentou, A.D., Bardhan, A., Samui, P. and K. Pilakoutas. 2021. Predicting concrete compressive strength using hybrid ensembling of surrogate machine learning models. *Cement and Concrete Research*, 145. https://doi.org/10.1016/j.cemconres.2021.106449.

Baykasoğlu, A., Güllü, H., Çanakçi, H. and Özbakir, L. 2008. Prediction of compressive and tensile strength of limestone via genetic programming, *Expert Systems with Applications*, 35(12): 111–123. https://doi.org/10.1016/j.eswa.2007.06.006.

Bickley, J.A. and Fung, R. 2001. *Optimizing the Economics of High-performance Concrete*. A Concrete Canada and Canadian Cement Industry Joint Research Project. Ottawa, Ontario, Cement Association of Canada.

Bui, D.K., Nguyen, T., Chou, J.S., Nguyen-Xuan, H. and Ngo, T.D. 2018. A modified firefly algorithm-artificial neural network expert system for predicting compressive and tensile strength of high-performance concrete. *Construction and Building Materials*, 180: 320–333. https://doi.org/10.1016/j.conbuildmat.2018.05.201.

Cevik, A. and Cabalar, A.F. 2009. Modelling damping ratio and shear modulus of sand-mica mixtures using genetic programming. *Expert Systems with Applications*, 36(4): 7749–7757. https://doi.org/10.1016/j.eswa.2008.09.010.

Choudhary, D., Keshari, J. and Khan, I.A. 2021. Prediction of compressive strength of ultra-high-performance concrete using machine learning algorithms—SFS and ANN. *Advances in Intelligent Systems and Computing*, pp. 17–27.

De Larrard, F. 1999. *Concrete Mixture Proportioning.* CRC Press.

Durodola, J.F., Ramachandra, S., Gerguri, S. and Fellows, N.A. 2018. Artificial neural network for random fatigue loading analysis including the effect of mean stress. *International Journal of Fatigue*, 111: 321–332. https://doi.org/10.1016/j.ijfatigue.2018.02.007.

Ferreira, C. 2001. Gene expression programming: A new adaptive algorithm for solving problems. *Complex Systems*, 13(2): 87–129.

Ferreira, C. 2006. *Gene Expression Programming: Mathematical Modeling by an Artificial Intelligence (2nd Edn.).* Germany: Springer-Verlag.

Fogel, D.B. 2002. *Evolutionary Computing.* NY: IEEE Press.

Gandomi, A.H. and Alavi, A.H. 2011. Multi-stage genetic programming: A new strategy to nonlinear system modeling. *Information Sciences*, 181(23): 5227–5239. https://doi.org/10.1016/j.ins.2011.07.026.

Gandomi, A.H., Yang, X.S., Talatahari, S. and A.H. Alavi. 2013. *Metaheuristic Applications in Structures and Infrastructures.* Elsevier.

Gandomi, M., Kashani, A.R., Farhadi, A., Akhani, M. and Gandomi, A.H. 2021. Spectral acceleration prediction using genetic programming based approaches. *Applied Soft Computing*, 106. https://doi.org/10.1016/j.asoc.2021.107326.

GeneXproTools. Version 4.0.

Golafshani, E.M. and Behnood, A. 2018. Automatic regression methods for formulation of elastic modulus of recycled aggregate concrete. *Applied Soft Computing Journal*, 64: 377–400. https://doi.org/10.1016/j.asoc.2017.12.030.

Hameed, M.M., AlOmar, M.K., Baniya, W.J. and AlSaadi, M.A. 2021. Prediction of high-strength concrete: high-order response surface methodology modeling approach. *Engineering with Computers.* https://doi.org/10.1007/s00366-021-01284-z.

Haykin, S. 1999. *Neural Networks: A Comprehensive Foundation.* Englewood Cliffs, NJ: Prentice Hall.

Holland, J.H. 1975. *Adaptation in Natural and Artificial Systems: An Introductory Analysis with Applications to Biology, Control, and Artificial Intelligence.* University of Michigan Press.

https://archive.ics.uci.edu/ml/datasets/concrete+compressive+strength.

Iqbal, M., Zhang, D., Jalal, F.E. and Faisal Javed, M. 2021. Computational AI prediction models for residual tensile strength of GFRP bars aged in the alkaline concrete environment. *Ocean Engineering*, 232. https://doi.org/10.1016/j.oceaneng.2021.109134.

Kaloop, M.R., Kumar, D., Samui, P., Hu, J.W. and Kim, D. 2020. Compressive strength prediction of high-performance concrete using gradient tree boosting machine. *Construction and Building Materials*, 264. https://doi.org/10.1016/j.conbuildmat.2020.120198.

Ke, X. and Duan, Y. 2021. A Bayesian machine learning approach for inverse prediction of high-performance concrete ingredients with targeted performance. *Construction and Building Materials*, 270. https://doi.org/10.1016/j.conbuildmat.2020.121424.

Kennedy, J. and Eberhart, R. 1995. Particle swarm optimization. *In: Proceedings of the IEEE International Conference on Neural Networks*, Piscataway, NJ.

Ketabdari, H., Karimi, F. and Rasouli, M. 2020. Shear strength prediction of short circular reinforced-concrete columns using soft computing methods. *Advances in Structural Engineering*, 23(14): 3048–3061. https://doi.org/10.1177/1369433220927270.

Kirkpatrick, S., Gellat, C.D. and Vecchi, M.P. 1983. Optimization by simulated annealing. *Science*, 220: 671–680.

Kosmatka, S.H., Kerkhoff, B. and Panarese, W.C. 2002. *Design and Control of Concrete Mixtures.* (14th Edn.), Skokie, Illinois, USA: Portland Cement Association.

Koza, J.R. 1992. *Genetic Programming: on the Programming of Computers by means of Natural Selection.* Cambridge, MA: MIT Press.

Latif, S.D. 2021. Concrete compressive strength prediction modeling utilizing deep learning long short-term memory algorithm for a sustainable environment. *Environmental Science and Pollution Research.* https://doi.org/10.1007/s11356-021-12877-y.

Mansouri, I., Güneyisi, E.M. and Mosalam, K.M. 2021. Improved shear strength model for exterior reinforced concrete beam-column joints using gene expression programming, *Engineering Structures*, 228. https://doi.org/10.1016/j.engstruct.2020.111563.

Mansouri, I., Ostovari, M., Awoyera, P.O. and Hu, J.W. 2021. Predictive modeling of the compressive strength of bacteria-incorporated geopolymer concrete using a gene expression programming approach. *Computers and Concrete*, 24(4): 319–332. https://doi.org/10.12989/cac.2021.27.4.319.

Mitchell, M. 1996. *An Introduction to Genetic Algorithms*. Cambridge: MIT Press.

Mousavi, S.M., Aminian, P., Gandomi, A.H., Alavi, A.H. and Bolandi, H. 2012. A new predictive model for compressive strength of HPC using gene expression programming. *Advances in Engineering Software*, 45(1): 105–114. https://doi.org/10.1016/j.advengsoft.2011.09.014.

Mousavi, S.M., Gandomi, A.H., Alavi, A.H. and Vesalimahmood, M. 2010. Modeling of compressive strength of HPC mixes using a combined algorithm of genetic programming and orthogonal least squares. *Structural Engineering and Mechanics*, 36(2): 225–241. https://doi.org/10.12989/sem.2010.36.2.225.

Murad, Y.Z., Tarawneh, B.K. and Ashteyat, A.M. 2020. Prediction model for concrete carbonation depth using gene expression programming. *Computers and Concrete*, 26(6): 497–504. https://doi.org/10.12989/cac.2020.26.6.497.

Naderpour, H., Rafiean, A.H. and Fakharian, P. 2018. Compressive strength prediction of environmentally friendly concrete using artificial neural networks. *Journal of Building Engineering*, 16: 213–219. https://doi.org/10.1016/j.jobe.2018.01.007.

Naser, M.Z., Thai, S. and Thai, H.T. 2021. Evaluating structural response of concrete-filled steel tubular columns through machine learning. *Journal of Building Engineering*, 34. https://doi.org/10.1016/j.jobe.2020.101888.

Nguyen, H., Vu, T., Vo, T.P. and Thai, H.T. 2021. Efficient machine learning models for prediction of concrete strengths. *Construction and Building Materials*, 266. https://doi.org/10.1016/j.conbuildmat.2020.120950.

Ojha, V.K., Abraham, A. and Snášel, V. 2017. Metaheuristic design of feedforward neural networks: A review of two decades of research. *Engineering Applications of Artificial Intelligence*, 60: 97–116. https://doi.org/10.1016/j.engappai.2017.01.013.

Pizarro, P.N. and Massone, L.M. 2021. Structural design of reinforced concrete buildings based on deep neural networks. *Engineering Structures*, 241. https://doi.org/10.1016/j.engstruct.2021.112377.

Popovics, S. 1990. Analysis of the concrete strength versus water-cement ratio relationship. *ACI Materials Journal*, 87(5): 517–529.

Saad, S. and Malik, H. 2018. Gene expression programming (GEP) based intelligent model for high performance concrete comprehensive strength analysis. *Journal of Intelligent and Fuzzy Systems*, 35(5): 5403–5418. https://doi.org/10.3233/JIFS-169822.

Sarir, P., Chen, J., Asteris, P.G., Armaghani, D.J. and Tahir, M.M. 2021. Developing GEP tree-based, neuro-swarm, and whale optimization models for evaluation of bearing capacity of concrete-filled steel tube columns. *Engineering with Computers*, 37(1). https://doi.org/10.1007/s00366-019-00808-y.

Smith, G.N. 1986. *Probability and Statistics in Civil Engineering*. London: Collins.

Sobhani, J. and Najimi, M. 2014. Numerical study on the feasibility of dynamic evolving neural-fuzzy inference system for approximation of compressive strength of dry-cast concrete. *Applied Soft Computing Journal*, 24: 572–584. https://doi.org/10.1016/j.asoc.2014.08.010.

Solhmirzaei, R., Salehi, H., Kodur, V. and Naser, M.Z. 2020. Machine learning framework for predicting failure mode and shear capacity of ultra high performance concrete beams. *Engineering Structures*, 224. https://doi.org/10.1016/j.engstruct.2020.111221.

Tadayon, B., Dehghani, H. and Ershadi, C. 2021. Proposing new breaking wave height prediction formulae using gene expression programming. *Ocean Engineering*, 228. https://doi.org/10.1016/j.oceaneng.2021.108952.

Tarawneh, A., Almasabha, G., Alawadi, R. and Tarawneh, M. 2021. Innovative and reliable model for shear strength of steel fibers reinforced concrete beams. *Structures*, 32: 1015–1025. https://doi.org/10.1016/j.istruc.2021.03.081.

Tarawneh, A., Momani, Y. and Alawadi, R. 2021. Leveraging artificial intelligence for more accurate and reliable predictions of anchors shear breakout capacity in thin concrete members. *Structures*, 32: 1005–1014. https://doi.org/10.1016/j.istruc.2021.03.074.

Tayfur, G., Erdem, T.K. and Kirca, Ö. 2014. Strength prediction of high-strength concrete by fuzzy logic and artificial neural networks. *Journal of Materials in Civil Engineering*, 26(11). https://doi.org/10.1061/(ASCE)MT.1943-5533.0000985.

Yang, X.S. 2010. *Engineering Optimization: An Introduction with Metaheuristic Applications.* Hoboken, NJ: John Wiley and Sons.

Yao, L., Leng, Z., Jiang, J., Ni, F. and Zhao, Z. 2021. Nondestructive prediction of rutting resistance of in-service middle asphalt layer based on gene expression programing. *Construction and Building Materials*, 293. https://doi.org/10.1016/j.conbuildmat.2021.123481.

Yu, Y., Li, W., Li, J. and Nguyen, T.N. 2018. A novel optimised self-learning method for compressive strength prediction of high performance concrete. *Construction and Building Materials*, 184: 229–247. https://doi.org/10.1016/j.conbuildmat.2018.06.219.

Zhengjun, P., Lishan, K. and Yuping, C. 1998. *Evolutionary Computation.* Beijing: Tsinghua University Press.

Zou, W.L., Han, Z., Ding, L.Q. and Wang, X.Q. 2021. Predicting resilient modulus of compacted subgrade soils under influences of freeze–thaw cycles and moisture using gene expression programming and artificial neural network approaches. *Transportation Geotechnics*, 28. https://doi.org/10.1016/j.trgeo.2021.100520.

Chapter 5

Implementation of Data-Driven Approaches for Condition Assessment of Structures and Analyzing Complex Data

Vafa Soltangharaei,[1,*] *Li Ai*[1] and *Paul Ziehl*[2]

1. Introduction

Both concrete and steel structures are exposed to internal and external degradation and damaging agents during their service life. For example, concrete structures are susceptible to alkali-silica reaction, rebar corrosion, salt crystallization, sulfate attack, freeze and thaw cycles, delayed ettringite formation, etc. These types of degradation and damage endanger the durability and serviceability of structures. Cracks are formed due to the agents mentioned above, and extension and coalescence of existing cracks will inevitably impair the integrity of concrete structures. Steel structures are also exposed to corrosion and brittle cracking. Reconstructing damaged infrastructure is very expensive and requires a large amount of resources and energy. Therefore, structural inspection and monitoring have become more significant in recent years. Structural health monitoring of essential infrastructure can lead to a reasonable extension of structural service life.

Acoustic emission (AE) is one method for structural health monitoring. AE is a phenomenon wherein a material emits stress waves due to a sudden release of energy, such as cracking (ASTM, 2006). Piezoelectric sensors are used to acquire these mechanical waves and transfer them to digital signals. The signals have valuable information about the crack formation and the internal condition of materials in which cracks are formed. Many studies have focused on the use of AE to localize

[1] Department of Civil and Environmental Engineering, University of South Carolina, USA.
[2] Associate Dean for Research, College of Engineering and Computing, University of South Carolina, USA.
* Corresponding author: vafa@email.sc.edu

crack formation, assess damage, and monitor structural health (ElBatanouny et al., 2014; Lokajíček et al., 2017; Shiotani, 1994; Shiotani et al., 2017; Soltangharaei et al., 2021a; Soltangharaei et al., 2019b).

Acoustic emission data typically includes parametric features extracted from the signals. Managing, analyzing, understanding, and interpreting data may be challenging due to the complexity of the data attributed to structural detailing, the duration of monitoring, and the temporal evolution of the acoustic property of the material. Today, several data-driven methods are available to overcome challenges attributed to data and the prediction of data patterns, including methods based on statistical concepts and machine learning algorithms.

2. AE Wave Sources in Different Structures

AE waves are emitted due to different mechanisms in structures. In the following section, some degradation mechanisms leading to AE wave generation are discussed.

2.1 Stress Corrosion Cracking

Cracks may be initiated and propagated in steel structures under corrosive conditions. Cracking in a steel structure is referred to as stress corrosion cracking (SCC) when it is caused by applying stress under corrosive conditions. Steel alloys used in structures are commonly considered to be ductile or flexible in nature. When steel alloys are exposed to high temperatures, they can become sensitive to corrosion. The corrosive products formed in the steel are more brittle than the original ductile steel and vulnerable to cracking. Therefore, in the presence of stress, cracking may initiate. This degradation process can also be observed in austenitic stainless-steel alloys when the steel chemical species are altered due to exposure to high temperatures (e.g., welding). The carbon in the material bonds with the chromium and forms carbides in the grain boundaries, which reduces the concentration of chromium near the grain boundaries and makes the steel more susceptible to intergranular SCC (Soltangharaei et al., 2020c). Figure 5.1 shows how heat treatment affects the crystal structure of 304 L stainless steel.

One potential risk of SCC in the structure is related to dry cask storage systems (DCSS). DCSSs are temporary nuclear waste disposal systems which have been utilized since the 1970s. The structures are stainless-steel canisters (vertical cylinders) which encase used nuclear fuel and high-level waste. The DCSSs are covered by a steel lid, which is welded to the canister. The welding on the storage containers causes austenitic materials to become sensitized and more susceptible to SCC. Microcracks are expected to form at the welded region due to the exposure of DCSSs to the environment. Microcracks will widen, extend, and become macro-cracks. The cracking endangers the serviceability of DCSSs and may eventually lead to concerns with radiation leakage to the environment.

Visually inspecting the cracks in the steel canisters is challenging. Furthermore, these storage structures are usually under restricted access, which prohibits inspections on a regular-basis. AE technology can be employed in this case, taking

(a) Sensitization after 2 hours   (b) Sensitization after 14 hours

Fig. 5.1: Evolution of crystal structure of steel during heat treatment (Soltangharaei et al., 2020c).

advantage of the high sensitivity of the AE sensors as well as the non-destructive nature of the method. Microcracks are initiated at the welded region of the canisters. Then, stress waves are generated and travel through the material to reach an AE sensor. Minimal numbers of sensors may be attached at the bottom of the canister or on the base.

2.2 Cracking due to Alkali-silica Reaction

Many concrete structures such as dams, bridges, nuclear structures, and hydraulic structures are exposed to alkali-silica reaction (ASR). ASR is a chemical reaction which occurs between alkali hydroxides and siliceous minerals in some aggregates used in concrete structures (Rajabipour et al., 2015; Villeneuve et al., 2012). Reaction components are available in concrete structures. The reaction product is a hygroscopic gel, which is formed around and/or inside aggregates (Bažant and Steffens, 2000; Dron and Brivot, 1992; Hanson, 1944; Jun and Jin, 2010; Ponce and Batic, 2006). This gel tends to absorb moisture from concrete pores or premature cracks and then expands. The volume of the gel is larger than the original components of the reaction. Therefore, the gel exerts pressure on the surrounding concrete components, such as aggregates, cement paste, and interfacial transition zones. If the gel finds a way (cracks or void) to flow, its pressure will drop, otherwise, the pressure will accumulate and surpass the material fracture strength causing microcracks and crack extensions in the concrete. ASR in a concrete structure can be accompanied by other degradation agents such as freeze—thaw cycles, chloride penetration in concrete and steel corrosion, and delayed ettringite formation, which expedites degradation.

The reaction rate depends on several factors such as reactivity of aggregates, the sodium and potassium ion content in the cement, the availability of alkali-hydroxides in the pore solution, high humidity (more than 80%), high temperature (Saouma and Hariri-Ardebili, 2014), and confinement from different directions (Allard et al., 2018; Barbosa et al., 2018; Karthik et al., 2016).

The ASR microcracks inside a concrete structure merge and widen through the ASR process. The reaction starts with a lower rate, then the rate increases followed by a decrease (Saouma and Perotti, 2006). The ASR progress rate can be observed by measuring the expansion of concrete. The expansion strain has a trend similar to the reaction. Figure 5.2a shows the volumetric strain of concrete blocks exposed to high temperature and humidity for 560 days.

Another important factor in cracking due to ASR is internal or external confinement. Concrete structures can be internally confined against a specific direction due to existing reinforcements perpendicular to that direction. Furthermore, the confinement may be externally imposed on a structure. One example is a rigid structure, which may restrain an adjacent structure from ASR expansion in a specific direction. Confinement in one direction does not necessarily prevent the concrete from cracking. According to Fig. 5.2b, the confined concrete block was cracked mostly in one direction, while cracking in the unconfined block was more randomly distributed (Fig. 5.2c). The direction of the crack opening in the confined specimen is perpendicular to the plane with reinforcements as shown in Fig. 5.2b.

Stress waves are generated due to ASR crack formation in a concrete structure and are recorded as digital signals. Signal features may be affected by the fracture mechanism of active cracking in different ASR stages, the potential evolution of internal acoustic properties of concrete during ASR, sensor coupling, structure geometry, and concrete components (rebars, aggregate size and type, void ratio).

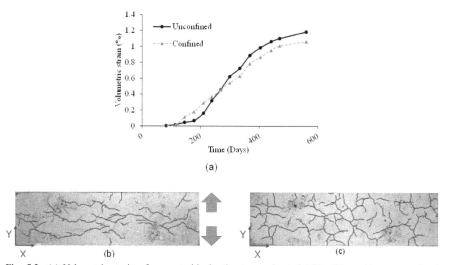

Fig. 5.2: (a) Volumetric strain of concrete blocks due to accelerated ASR; (b) cracking on confined specimen; (c) cracking on unconfined specimen.

2.3 Loading and Impact

AE signals can be generated by the crack formation induced by external loading or impacts. In terms of material science, AE can be used to investigate the different cracking mechanisms and fracture mechanics. The crack mechanisms can be identified by moment tensor analysis (Ohtsu, 1995) or using machine learning methods (Anay et al., 2018; Soltangharaei et al., 2021b). Further, more detailed information is obtained about the failure mechanism of structural components using AE during a full-scale test (Abdelrahman et al., 2014).

Load testing is also used as a procedure to evaluate the condition of structures under service loads. In the United States, load tests are conducted using standard truck loads to assess the condition of bridges and remove the load postings. One method to evaluate the superstructural components of a bridge is to calculate displacement distribution factors between the slabs using displacement and strain sensors during load tests. However, using conventional load testing is time-consuming and costly.

Concrete superstructural components of bridges (bridge decks or girders) have cracks and defects due to premature concrete conditions or degradation factors. Existing cracks in bridge decks or girders open or extend as a truck passes over the bridge, generating stress waves. These waves can be captured by AE sensors placed on bridge components. The advantage of AE over stain gauges or displacement sensors is its higher sensitivity, which can be helpful for bridge evaluation (Anay et al., 2020).

AE monitoring is not limited to civil engineering infrastructure, but can also be utilized in aerospace structures. AE waves are generated by the impact of debris or hail striking an airplane component, such as an elevator, during flight. The impacts might cause minor damage, referred to as barely visible impact damage (BVID), on a surface manufactured with fiber composite material (Soltangharaei et al., 2019b). The location and level of damage is correlated with the evolution of AE signal features. Figure 5.3 illustrates impacting on an airplane elevator used to collect AE data for a machine learning algorithm.

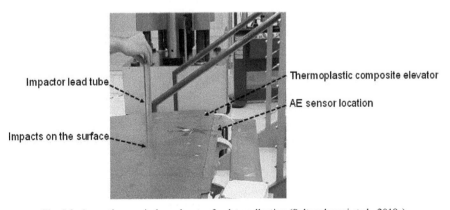

Fig. 5.3: Impacting an airplane elevator for data collection (Soltangharaei et al., 2019a).

3. Data Analysis Approaches

Data-driven techniques aim to find inherent features and structures of complex phenomena by evaluating data from different aspects (Hayashi, 1998). The techniques are based on mathematical relationships, which are derived from experimental data (Solomatine et al., 2009) and can be used for damage detection and prediction when a physical model is not available, but the historical damage behavior is accessible.

Although the data-driven methods are efficient for a noisy dataset and complex behavior, they may also be computationally expensive. However, the former drawback is diminishing due to progressive advances in computational hardware and cluster computers.

In physics-based methods, the availability of a physical model and loading conditions are essential. Each model should be validated and its parameters determined before application, which is not always feasible and straightforward (An et al., 2015). Computational intelligence, machine learning, and regression are three examples of data-driven methods (Solomatine et al., 2009). Pattern recognition can be categorized under the machine learning field, which focuses on discovering the regularities inside a data set by employing computer algorithms. There are two main categories associated with this method: unsupervised and supervised pattern recognition. Unsupervised pattern recognition is utilized to find and cluster unlabeled data when no predefined pattern is available. On the other hand, supervised pattern recognition is employed when a large labeled dataset is accessible for training. Subsequently, the trained algorithm can be engaged to classify new data (Bishop, 2006).

3.1 Data Preparation before Analysis

Although the high sensitivity of AE is employed for early detection of deterioration or damage, this advantage might lead to collecting a large amount of data during acquisition, which brings more challenges to data management and analysis. Therefore, different approaches should be taken to recognize more valuable data (essential data) within the raw data. Extraneous AE data may be collected for several reasons in a data acquisition system. Background noise, electrical noise, mechanical noise, boundary reflection data, and irrelevant environmental data are among the non-genuine data which may potentially be recorded during structural monitoring. Background noise is usually handled by setting an appropriate threshold in a system before running a test. The first step before using an AE system is to run a background noise test for a few hours to make sure the test environment is not acoustically contaminated. Using this information, an appropriate amplitude threshold for the system can be set. Furthermore, it is recommended that an AE sensor be attached to a component in a control condition during the test. Mechanical noise may be generated by friction between the structural components during loading of the structures or between the structure and the loading instruments. The solution for reducing the mechanical noise is isolating the loading instrumentation as much as possible from the structure by using materials with minimal acoustic impedance, such as neoprene pads, and/or a guard sensor in the vicinity of components with a friction potential.

Electrical noise usually appears in electrical components such as connections, cables, channels, or faulty sensors. Water penetration in connections is also one reason for electrical noise that may occur during long-term monitoring of structures exposed to environmental precipitation or high-humidity conditions. Fortunately, electrical noise can be clearly recognized from genuine data sets (Fig. 5.4). Electrical noise has a low average frequency and peak frequency and can be removed by applying frequency-based filters. The hit rates for electrical noise are high, which may cause

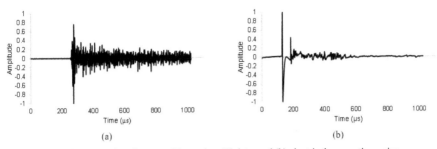

Fig. 5.4: Comparison between (a) genuine AE data; and (b) electrical connection noise.

acquisition and buffering issues. Therefore, it is recommended to check electrical components and protect them from water by utilizing shrinkage tubes.

Stress waves are reflected when they reach a media with different impedance or structure boundaries. In small-scale specimen tests, receiving a large amount of reflections can be confusing in terms of data processing. A simple method to filter the reflections is using front-end filters (Ziehl and ElBatanouny, 2015). This is usually set in a data acquisition system by defining timing setups such as hit definition time (HDT), peak definition time (PDT), and hit lockout time (HLT) (MISTRASGroup, 2011). HLT is a parameter which is directly associated with removing a signal tailing or reflections. It refers to a time at the end of a signal where any threshold crossing is neglected. The amount of tailing data after an event can be adjusted using HLT.

AE raw data should go through several filters to be useful for data analysis and damage assessment procedures. One of the most traditional filtering methods is the Swansong II filter (Fowler et al., 1989; Soltangharaei et al., 2018). In this method, distributions of log (duration) versus amplitude (in decibels (dbAE)) or log (rise time) versus amplitude (in dbAE) are used to determine non-genuine data. Genuine AE data with low-amplitude generally has low-duration, and data with high-amplitude generally has high-duration. The other data, which does not follow the stated rule, is referred to as non-genuine data and is deleted. In this method, filtering limits are determined by observing the AE waveforms one by one in different amplitude intervals. Therefore, this is time-consuming and inconsistent. The latter disadvantage means that the filtering method depends on experience and preferences, i.e., the data filtered by two people may be different. Thus, research has been conducted using statistical and signal processing methods for data filtering (Abdelrahman et al., 2019; Sagasta et al., 2014; Sagasta et al., 2013). Abdelrahman et al. (2019) employed the continuous wavelet transform to recognize genuine data from noisy data. Despite using different methods for filtering, accuracy, and performance in different applications is still under question and it appears that the Swansong II filter remains one of the most reliable methods for AE filtering. However, as mentioned, this method is time-consuming and challenging for long-term monitoring and low-level AE, where acquired data might be as large as a hundred gigabytes or a few terabytes. For this case, attaching several sensors on a structure in the sensitivity distance range of sensors can be beneficial for fast filtering. As an example, a script has been developed to filter AE data based on event definition (Soltangharaei et al., 2020a, Soltangharaei et al., 2020c). In the code, the hits registered by a specific number of sensors in specific time differences are recognized and retained as a genuine data set. The number of sensors depends on the number of locators used in the test and the source location scope of project. Two-dimensional planes and three-dimensional spaces require at least three and four sensors, respectively, for source localization using triangulation. This filtering method has been found to be faster and more consistent than traditional methods. The algorithm is applicable to both parametric data and waveforms.

In the end, the pre-mentioned filtering methods are efficient to some extent but none may filter all extraneous data; hence random visual inspection of filtered data is recommended to check the efficiency of the filtering method and remove remaining potential non-geniune data.

After filtering, AE data should be organized and prepared for analyzing purposes. For example, the time attributed to data is modified to represent real-time monitoring, and the data is usually organized by the time.

3.2 Statistical Analysis and Distributions

AE waveforms are non-stationary signals (Suzuki et al., 1996a). Therefore, different statistical and signal processing methods can be used to analyze data and derive essential information from the complex data. The outliers in signal features can be identified using statistical methods. In statistics, an outlier is defined as an observation with a large deviation from other observations (Hawkins, 1980). Outliers in a dataset may result in confusion and errors in some data-driven techniques. For instance, k-means methods are expected to misclassify a dataset with outliers (Tan et al., 2013). Therefore, in some data-driven techniques, identification and removing outliers from data before conducting any analyses is significant. One of the simplest statistical methods is end-trimming data. In this method, a percentage (i.e., 5% or 10%) of the highest and lowest data is considered as outlier data and removed from the data set (Ott and Longnecker, 2015). Another common statistical method is using box-and-whiskers plots. By calculating a median, lower quartile, and upper quartile, a boxplot can be drawn for a dataset. Then an inner fence and an outer fence are determined using quartiles and an interquartile range. Finally, mild and extreme outliers are identified by observing the data in the inner and outer fences (Ott and Longnecker, 2015).

Deriving a statistical distribution from AE data can be helpful for AE feature analyses. For example, distribution events along a specific dimension can be representative of cracking distribution (Soltangharaei et al., 2020a). In Fig. 5.5, event medians, first, and third quartiles are shown. The median line separates the lowest 50% and highest 50% of data, and the first and third quartiles split the lowest 25% of the data from the highest 75%, and the highest 25% of data from the lowest 75%, respectively. The vertical axis of the Fig. 5.5 is the relative frequency of events. For example, in Fig. 5.5a, the value for the first bin from the left is approximately equal to 0.04, which means only 4% of data is between 0 to 1.6 cm. In other words, the probability of event occurrence between 0 to 1.6 cm is 4%. The distribution for the confined specimen is completely different from the unconfined specimen.

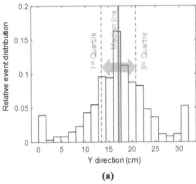 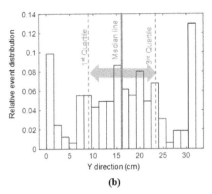

(a)   (b)

Fig. 5.5: Relative event distribution for concrete specimens exposed to ASR deterioration, (a) Confined specimen; (b) unconfined specimens (Soltangharaei et al., 2020a).

Another application of statistical methods is using a relative frequency histogram of signals to calculate Shannon entropy, which is explained in the next section.

The Gutenberg-Richter equation is the basis for b-value analysis (Rao and Lakshmi, 2005), which is used in seismology and risk analysis to determine the relationship between the frequency of an event occurrence and the event magnitude. The modified version of this method has been used for AE data as a damage index (Colombo et al., 2003; Jung et al., 2017; Rao and Lakshmi, 2005; Shiotani, 1994). The relationship between AE amplitude (in dBAE) and the number of AE signals with amplitudes larger than A (signal amplitude) is presented as follows:

$$LogN = a - b(\frac{A}{20}) \tag{5.1}$$

Linear regression can be used to estimate a and b because the relationship between $logN$ and A is close to linear. Two approaches are recommended to calculate b-values, incremental b-value and global b-value (Soltangharaei et al., 2020c). In the former method, the entire data is separated according to intervals and b-values are calculated for the data in each interval. In the global b-value, b-values are calculated for each time, by considering the AE data that occurred before that time.

In addition to b-values, the coefficient of determination (R^2-value) can be calculated to identify the deviation of data from the linear relationship (Soltangharaei et al., 2020c). The parameter can be used as a damage identification criterion. AE signals are emitted with a higher rate and larger energy or amplitude when severe damage or cracking occurs in a structure. AE signals with higher amplitude cause AE data to deviate from the b-value relationship decreasing the coefficient of determination for the system. R^2-value is calculated from the following equation:

$$R^2 = 1 - \frac{\sum_{i=1}^{n}(\hat{y}_i - y_i)^2}{\sum_{i=1}^{n}(y_i - \bar{y})^2} \tag{5.2}$$

where $\hat{y}_i$ is the estimated value of $logN$ for an ith point by using the fitted line and y_i is the real value of $logN$ for the ith point. $\bar{y}$ is the average value of $logN$. Index n denotes the last desired data to calculate the b-value.

3.3 Signal Processing Methods

Several features and important information are embedded in AE signals. Using signal processing methods, the inherent information can be extracted and used for the temporal assessment of features.

3.3.1 Fast Fourier Transform

The Fourier transform transfers a time-series signal to a frequency domain cue by convoluting the digitized message with sinusoidal signals ($e^{-\frac{j2\pi kn}{N}}$) with different frequencies. The following equation represents the formulation for a Fourier transform:

$$X_k = \sum_{n=0}^{N-1} x_n e^{-\frac{j2\pi kn}{N}}$$ (5.3)

where N is the number of samples, x_n is a signal in a time domain, and X_k is the Fourier transform coefficients for the kth frequency.

3.3.2 Continuous Wavelet Analysis

The wavelet transform (WT) is utilized to present the energy distributions of signals in a time-frequency domain. Contrary to the windowed Fourier transform (WFT), the WT presents data in a high-time resolution for the high-frequency components and a high-frequency resolution for the low-frequency components. The CWT is defined according to the following equation:

$$CWT(a,b) = \int S(t) * |a|^{-0.5_\psi} \left(\frac{t-b}{a} \right) dt$$ (5.4)

where a and b are scale and shift parameters, respectively. The scale parameter controls the compactness or extension of a signal (frequency), and the shift parameter defines the position of the mobile window in the time domain. $S(t)$ is a signal time history. Equation (5.5) is referred to as the wavelet, which is the second portion of Eq. (5.4). The basic window function without scale and shift parameters is referred to as the mother wavelet.

$$\Psi_{a,b}(t) = |a|^{-0.5_\psi} \left(\frac{t-b}{a} \right)$$ (5.5)

The mother wavelet used in this study is the Gabor wavelet, which is based on a Gaussian function (Suzuki et al., 1996b). The results of wavelet analysis are the wavelet coefficients for the different combinations of time and frequency which are presented in a contour diagram or spectrogram. The 3D wavelet spectrum can also be presented as a 2D picture and used as input for an image processing algorithm.

3.3.3 Shannon's Entropy

Shannon's or information entropy was introduced by Shannon (Shannon, 1948). This parameter quantifies the randomness of a random variable. If the probability of an event is high, the event occurrence is expected and is not shocking; therefore, it delivers very little information. This concept can be used in signal processing and acoustic emission (Chai et al., 2018; Kahirdeh, 2014; Kahirdeh et al., 2017; Sauerbrunn, 2016; Soltangharaei et al., 2021a). Different methods are recommended to calculate signal entropy in the literature (Chai et al., 2018; Kahirdeh, 2014; Kahirdeh et al., 2016; Sauerbrunn, 2016). The methods are: voltage amplitude entropy, feature entropy, and fast Fourier transform (FFT) entropy.

In voltage entropy, the distribution of amplitude voltages of AE signals is estimated in different bins. The bin size is recommended to be as close as possible to the resolution of AE data acquisition (Chai et al., 2018). The distribution indicates the relative frequency of the currents within different voltage intervals, and the entropy for the voltages is calculated using the following equation (Shannon entropy equation):

$$Entropy = -\Sigma_{i=1}^{n} P(x_i) * \log(P(x_i)) \tag{5.6}$$

In the equation, n is the number of bins in each signal. The bins defined in each signal are represented as x_i and $P(x_i)$ is the relative frequency of each bin calculated according to the signal histogram. In this method, it is assumed that the voltage alteration is constant and independent between samples. Entropies of hits can be calculated either independently or cumulatively. The former method is referred to as discrete voltage entropy (DVE), and the later method is referred to as global voltage entropy (GVE). The two procedures are illustrated in Fig. 5.6.

The signals in Fig. 5.6 have a sampling rate of 5 million samples per second. In Fig. 5.6a, histograms are individually calculated for each hit, while in Fig. 5.6b, histograms are calculated by considering a hit and the hits which occurred prior to that.

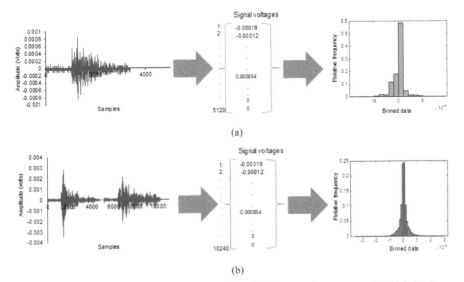

Fig. 5.6: Voltage entropy calculated in two procedures. (a) Discrete voltage entropy; (b) Global voltage entropy (Soltangharaei et al., 2021a).

Counts entropy is a feature entropy, which can be calculated using two procedures. Counts in AE is defined as the number of time that signal voltage exceeds the threshold set in the system. The first method is very straightforward, and the probability is estimated by dividing the counts for each hit in a specific time by the cumulative counts of hits occurring before that time. The entropy is calculated using Eq. (5.6). This method is referred to as CE in this document. The second procedure is referred to as CE_CDF since the cumulative distribution function (CDF) is estimated using the empirical CDF method, and the corresponding probability distribution function (PDF) is extracted from CDF (Kahirdehet al., 2016). The entropy is also estimated using Eq. (5.6) in this method.

FFT entropy is calculated using normalized FFT spectra of signals. The normalized spectra are assumed to be probability distributions, and Shannon entropies are estimated using the FFT spectra and Eq. (5.6) (Kahirdeh et al., 2016). Figure 5.7 shows how to calculate the FFT entropy.

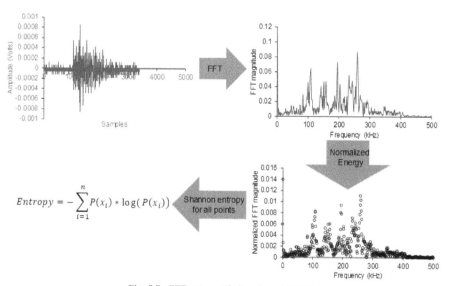

Fig. 5.7: FFT entropy (Soltangharaei, 2020).

3.4 Machine Learning and Clustering Methods

Machine learning and clustering methods can be categorized under data-driven methods, which are referred to as algorithms to recognize patterns and clusters in a data set and to predict and make decisions from the new data. In the following, some common algorithms are described.

3.4.1 k-means

One of the simplest and earliest unsupervised classification methods is k-means clustering. The algorithm starts with selecting k centroids (number of clusters), which is a user-defined parameter. Then each data set is assigned to the nearest centroid, forming new clusters. The centroids are recalculated based on the new clusters. This

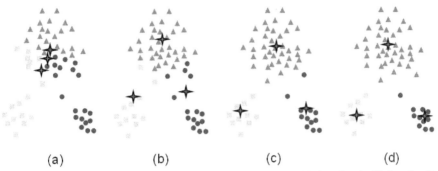

Fig. 5.8: Schematic figure of k-means algorithm for three clusters. (a) Iteration 1; (b) Iteration 2; (c) Iteration 3; (d) Iteration 4 (Tan et al., 2013).

process is repeated, and the centroid points are updated until the difference between the new centroid points and the previous ones become negligible (Tan et al., 2013). Figure 5.8 shows how the algorithm works in a 2D space.

The initial guess for the centroids of clusters is influential on the results of classification. The common method to start the algorithm is randomly selecting the centroid, which sometimes results in a poor classification. There are some methods recommended to improve the initial guess. One of the methods is randomly selecting the initial centroids several times and choosing the best arrangement of centroids with the lowest average cluster error. The other method is taking a sample of data and clustering them using hierarchical grouping. This method performs well if the data set is not large (a few hundreds to few thousands) and the number of desired clusters is small in comparison to the data set (Tan et al., 2013). Another common method is selecting the farthest point from the initially selected centroids. The latter method might not be effective when the data set has outliers (Abdelrahman et al., 2018). After selecting initial centroids, the distance between the data and centroids is calculated using a proximity measure and assigning each data point to the closest centroid point. The centroids in the clusters are calculated using the following equation:

$$c_i = \frac{1}{m_i} \Sigma_{x \in C_i} x \qquad (5.7)$$

where, m_i is the number of data point in the ith cluster (C_i) and x is the data point in a defined space for clustering, c_i is centroid for the ith cluster (C_i). The objective function for optimization in Euclidian space might be the sum of square error (SSE) within the clusters. A lower error for each classification indicates the higher performance of clustering and less scattering in the resulted groups (Tan et al., 2013).

3.4.2 Agglomerative Hierarchical Clustering

The agglomerative hierarchical algorithm is an unsupervised pattern recognition approach, which includes three main steps: distance calculation, grouping data into clusters, and determining the number of groups.

The result of the hierarchical clustering is usually presented in a tree-like graph called a dendrogram. The most important part of the algorithm is calculating the proximity between two clusters. Different methods are used to calculate the closeness between the clusters such as: a single link, which calculates the distance between the closest two points in two groups; a complete link, which calculates the nearness between the farthest points in two clusters and; a group average, which calculates the average distance between the points in two groups. In addition to the methods mentioned, Ward's procedure can also be utilized to calculate the proximity. In this process, clusters are presented by their centroids, and the criterion for merging two clusters is to minimize the sum of the squared distances of points in the groups (Murtagh and Legendre, 2014). The number of clusters is determined based on the height of each link with respect to the average height of the connections underneath the data dendrogram (Bouguettaya et al., 2015). The procedure for the agglomerative hierarchical clustering is shown in a flowchart in Fig. 5.9.

Fig. 5.9: Agglomerative hierarchical clustering.

3.4.3 Principal Component Analysis

Principal component analysis (PCA) is a method to reduce the dimensionality of a data set. Many features can be extracted from AE signals, such as duration, counts, amplitude, peak frequency, energy, etc. However, working with all features and finding the relation between them is difficult. PCA can reduce the dimensionality of a data set by projecting the data on new coordinates. Input for a PCA is a matrix where columns are features (variables) and rows are observations (hits). PCA initially calculates a covariance of the input matrix. Then, eigenvalue analysis is conducted on the covariance matrix, resulting in eigenvalues and eigenvectors. The number of eigenvalues and eigenvectors is the same as the number of features in the input matrix. The eigenvectors have components equal to the number of features. The eigenvalues and corresponding eigenvectors are sorted from the largest to smallest values. Then, the original input matrix is transferred to the new space by multiplying a matrix, which contains all eigenvectors. According to the eigenvalues, the least important principal components can be deleted without losing a significant amount of information.

3.4.4 Artificial Neural Network

Artificial neural network (ANN) can be considered as a supervised pattern recognition approach, which is trained by using labeled data to classify new unlabeled data or to estimate values according to new input figures.

Multilayer ANNs have three main layers: input, hidden, and output layers. The intermediary layers between input and output layers is referred to as hidden layers, where ANN learns the relationship between input and target data (Solomatine et al., 2009). The algorithm aims to solve an optimization problem to make an

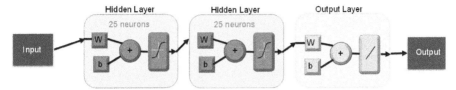

Fig. 5.10: Schematic artificial neural network with two hidden layers (Soltangharaei et al., 2020b).

output of ANN as close as possible to the target values. In this procedure, weight factors for each neuron are frequently updated to reach the target values. A schematic architecture of a multilayer ANN is shown in Fig. 5.10.

The input and target data are given to an ANN model. The ANN algorithm starts iteration by initial weight values (randomly selected) and computes output values, and compares it with target ones given to the model. Errors are calculated, and weight vectors are updated accordingly. This procedure continues until the output values approach the target ones as closely as possible. Nonlinear activation functions such as sigmoid or hyperbolic tangent functions can be employed to produce a nonlinear relationship between target and input values (Tan et al., 2018).

The input data is usually separated into three data sets: training, validation, and test data. The training data set is utilized for training the network. The validation data is selected from the input data to evaluate the network during the training process. This data is utilized to help a network to be generalized for a new dataset. The testing data is an independent dataset to test the performance of the trained network (Gavin, 2011).

3.4.5 Stacked Autoencoder Neural Network

This algorithm is a deep learning method. Deep learning networks have the ability to extract features from raw data rather than using prepared features. Deep learning methods can be used in computer vision, audio processing, signal processing, and language processing.

Each autoencoder has three layers similar to ANN: input, hidden, and output layers. Autoencoders are an unsupervised learning network, which discovers the embedded pattern in a data set by condensing the original input data and reconstructing the data set from the condensed data (Shin et al., 2012). A schematic structure of an autoencoder layer which is a bottleneck-shape in the hidden layer, is shown in Fig. 5.11.

If the input data has n dimensions $\{x^1, x^2, x^3 \dots x^n\}$ and the hidden layer has m dimensions, where $m < n$, the input data will be condensed using a nonlinear encoding function. The result of encoding is going through a decoding function and reconstructed in the output layer ($\{x'^1, x'^2, x'^3 \dots x'^m\}$). The encoding ($E$) and decoding ($D$) functions can be presented in the following equations:

$$E(x) = S_\theta (wx + b) \tag{5.8}$$

$$D(h) = S_{\theta'} (w'h + b') \tag{5.9}$$

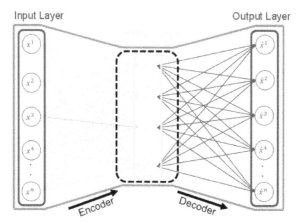

Fig. 5.11: Autoencoder structure.

In the above equations, w and w' are weights of encoder and decoder functions, and each is a $m \times n$ matrix whereas b and b' are bias vectors for encoder and decoder functions. The activation functions can be (i) sigmoid, (ii) hyperbolic tangent, or (iii) customized. An autoencoder's objective is to minimize the error between the original and the reconstructed version of the data.

A stacked autoencoder network can be built by piling two or several autoencoders (Wang et al., 2019). The first autoencoder derives the features from the input data, which are the input for the next autoencoder. The new features are then extracted from the second autoencoder. This procedure will be continued up to the last autoencoder. The resultant features from the last autoencoder layer are used as an input for either the classification, using a softmax layer or prediction using a regression layer. This part of the stacked autoencoder is supervised, and the network is trained for the assigned labels or values. The network can be fine-tuned by backpropagation to improve its performance (Wang et al., 2019). A schematic structure of a stacked autoencoder with two autoencoders is shown in Fig. 5.12 which shows how AE waveforms or FFT spectra of AE can be used as an input data set for a stacked autoencoder.

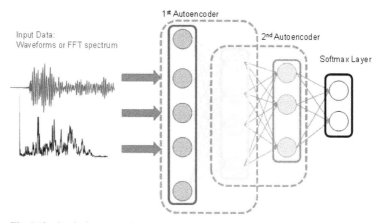

Fig. 5.12: Stacked autoencoder network with two autoencoders and classification layer.

3.4.6 Convolutional Neural Network

A convolutional neural network (CNN) is a deep learning method mostly used for image and signal recognition (Krizhevsky et al., 2012a). The main components of a CNN are an input layer, feature extraction layers, and a fully connected layer. A schematic CNN structure is illustrated in Fig. 5.13.

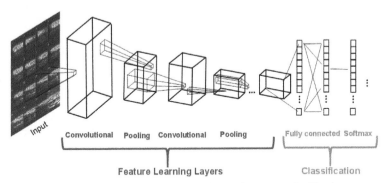

Fig. 5.13: A simplified CNN structure for wavelet contour classification.

The input data for picture classification is a tensor with (number of images) × (image height) × (image width) × (input channels) and is usually divided into training and testing data sets. Feature extraction layers include convolutional and pooling layers. In the feature extraction layers, the algorithm extracts features and learns to differentiate them. AE data can be presented as images using a wavelet transform. The resulting images are inserted as an input set in a CNN. Each picture goes through several convolutional filters, which highlight specific characteristics of the picture.

In a convolutional layer, several kernels (filters) are utilized to filter the input data and extract features. For instance, the output of the jth feature map in the nth convolutional layer is calculated by the following equation:

$$x_j^n = f(\sum_{i=1}^{M} x_i^{n-1} * k_{ij}^n + b_j^n) \tag{5.10}$$

where, $f(\,)$ is an activation function, "*" indicates convolution between the data and kernels, k_{ij}^n is the kernel of the nth filter, b_j^n is the bias parameter for the nth filter, x_i^{n-1} is the input feature map from the previous convolutional layer $(n-1)^{th}$.

The pooling layer takes samples of feature maps from the previous convolutional layer and reduces the size of features (Sun et al., 2020). If the down-sampling is not conducted from layer to layer, the computation process becomes expensive, and the model is prone to overfitting. Therefore, a pooling layer is required after each convolutional layer. The down-sampling can be conducted either using maximum pooling or average pooling. The general function for the pooling layer is as follows:

$$x_j^n = f(\beta_j^n s_{down} (x_j^{n-1}) + b_j^n) \tag{5.11}$$

In the equation, $f(\,)$ is the activation function, β_j^n and b_j^n refers to the multiplicative and additive bias, S_{down} is the down-sampling function, x_j^{n-1} is the input feature map, x_j^n is the output after down-sampling.

The fully connected layer is usually placed at the end of the network and transfers all features from the last pooling layer to a feature vector. The function for the fully connected layer has w^j weight and b^j bias as indicated below:

$$x^j = f(x^{j-1} w^j + b^j) \tag{5.12}$$

where:
f is the activation function for the fully connected layer, x^{j-1} and x^j are input and output of the layer.

CNNs have several parameters and complex structures. Designing and training a CNN from the beginning is very challenging and time-consuming and requires a large amount of data. Therefore, a concept of transfer learning was introduced to overcome the challenges with a new CNN (Pan and Yang, 2009). The transfer learning is based on a CNN that was already been trained by a large amount of data. The pre-trained network is fed by the new data (pictures) through the deeper layer of CNN. Learning rates for the initial layers (transferred layers) should be extremely decreased, and the learning rate factors for the fully connected layer should be increased. Some famous pre-trained CNNs for transfer learning are GoogLeNet (Szegedy et al., 2015), AlexNet (Krizhevsky et al., 2012b), ResNet (He et al., 2016), and Inception-v3 (Szegedy et al., 2016).

4. Case Studies

In this section, some examples using data-driven methods in AE data are discussed.

4.1 Stress Corrosion Cracking

The application of AE in monitoring the damage caused by SCC in stainless steel has been presented by several researchers (Shaikh et al., 2007; Alvarez et al., 2008; Du et al., 2011; Xu et al., 2012; Kovač et al., 2015). Some researchers focused on the fracture modes and signal differences between intergranular and transgranular cracking mechanisms (Alvarez et al., 2008). Recently, a research study was conducted regarding the capability of AE to recognize and localize SCC on stainless steel plates, which resembled a DCSS structure material at the University of South Carolina. Figure 5.14 shows large- and small-scale specimens, used for SCC tests. The large specimen is 5029 mm × 1524 mm × 16 mm and has four notches. The small-scale specimen is 305 mm × 311 mm × 16 mm with a single notch.

A potassium tetrathionate (K2S4O6) solution was used as an electrolyte to provide a corrosive environment at the notch locations. A plastic tube was placed on the notches and the solution was poured inside. The pH of the solution was decreased by adding sulfuric acid, which expedited the corrosion reaction. The top surface of the steel plates at the location of each notch was exposed to tensile stress by applying out-of-plane bending. The steel plates were heat-treated before the test for sensitization purposes, as shown in Fig. 5.1. The first visible crack was observed on the 9th day of the experiment, as shown in Fig. 5.15.

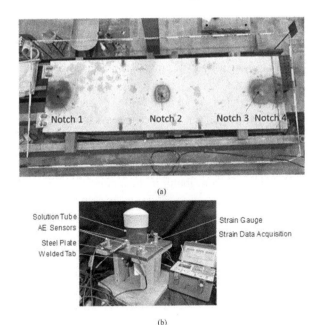

(a)

Solution Tube
AE Sensors
Steel Plate
Welded Tab

Strain Gauge
Strain Data Acquisition

(b)

Fig. 5.14: (a) Large-scale stainless-steel plate; (b) Small-scale stainless-steel plate (Ai et al., 2021a; Soltangharaei et al., 2020c).

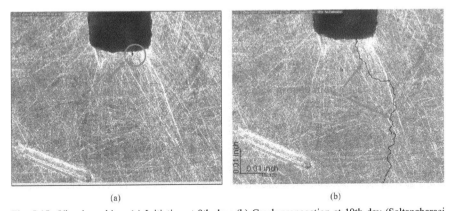

(a)   (b)

Fig. 5.15: Visual cracking: (a) Initiation at 9th day; (b) Crack propagation at 19th day (Soltangharaei et al., 2020c).

The signals were transferred to the frequency domain using FFT. The areas under the FFT spectra were divided in terms of frequency intervals. Energies enclosed in the frequency intervals were estimated. For example, the Nyquist frequency, which was 500 kHz, was divided into ten equal intervals. Therefore, ten energy values were calculated and used as the signal features. PCA was conducted on the data to reduce the dimension of the data set. The first four principal components, which accounted for 93% of the cumulative variance, were selected for pattern recognition.

An agglomerative hierarchical clustering method was used to classify the data based on the new features derived from PCA, and Ward's was used to estimate the similarity and link the objects (Murtagh and Legendre, 2014). The classified data was illustrated in terms of the first three principal components in 3D space in Fig. 5.16.

Average energy distributions in terms of frequency and cumulative signal strength in terms of experimental time for the three clusters are shown in Fig. 5.17.

The major jump in CSS occurred on the 8th day, which was close to the day (9th day) that the crack initiation was visually observed (Fig. 5.17b). Furthermore, Cluster 3 has the largest energy accumulation in frequencies higher than 200 kHz (Fig. 5.17a), and is the largest contributor in the major CSS jump (Fig. 5.17b).

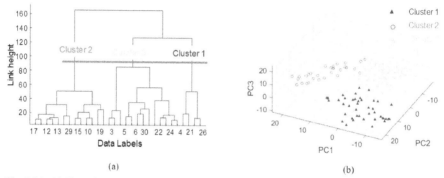

Fig. 5.16: (a) Clustering dendrogram; (b) Data in principal component space (Soltangharaei et al., 2020c).

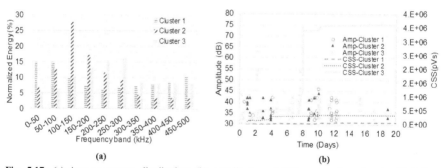

Fig. 5.17: (a) Average energy distribution; (b) Amplitude and CSS versus time for classified data (Soltangharaei et al., 2020c).

Coefficients of determination calculated from the global b-values in some instances during the test are presented in Fig. 5.18. A temporal trend is observed, where the increase in the coefficients of determination was followed by a sharp decrease after the first visible crack.

ANN and stacked autoencoder can be utilized for zonal localization in large-scale structures using minimal sensor numbers (Ai et al., 2021a). The inputs for the ANN model were parametric features such as duration, amplitude, rise time, average

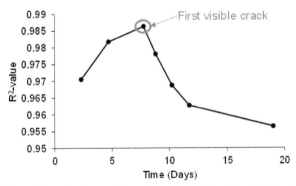

Fig. 5.18: Coefficient of determination evolution during the SCC test for global b-value (Soltangharaei et al., 2020c).

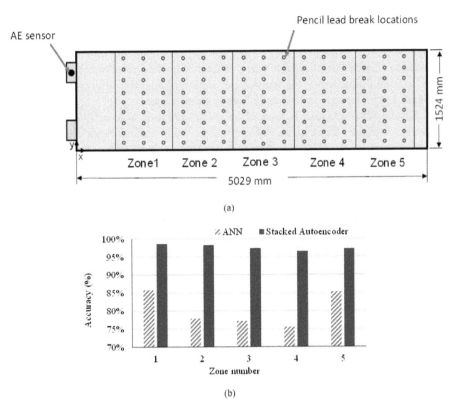

Fig. 5.19: (a) Zones on large-scale plate; (b) Zonal localization accuracy for ANN and stacked autoencoder (Ai et al., 2021a).

frequency, root mean square, energy, and peak frequency. However, the input dataset for the stacked autoencoder model was AE waveforms (time series). The large-scale plate was divided into five zones, as shown in Fig. 5.19a. The source location results indicated a significant improvement using the stacked autoencoder in Fig. 5.19b.

4.2 Alkali-silica Reaction

Several recent bodies of research have focused on using acoustic emission to monitor the ASR degradation of concrete structures (Abdelrahman et al., 2015; Farnam et al., 2015; Lokajíček et al., 2017; Pour-Ghaz et al., 2012; Soltangharaei et al., 2021a; Soltangharaei et al., 2020a; Soltangharaei et al., 2018). The ability of AE to record stress waves emitted during ASR was initially discussed by Pour-Ghaz et al. (2012), and a correlation between cumulative signal strength and ASR expansion was later discussed by Abdelrahman et al. (2015). Farnam et al. (2015), using feature analysis, discussed the frequency content of AE signals emitted during ASR. They compared the frequency content of AE during ASR with the AE data acquired from fracture tests. Lokajíček et al. (2017) discussed the relationship between the reactivity of aggregates and AE energy.

Several investigations were conducted by the University of South Carolina regarding ASR in reinforced and unreinforced concrete and large-scale specimens (Soltangharaei et al., 2021a; Soltangharaei et al., 2020a; Soltangharaei et al., 2018). Agglomerative hierarchical pattern recognition in combination with PCA were utilized to classify the AE data in terms of frequency-energy features extracted from FFT spectra. Furthermore, the temporal evolution of data for different clusters was evaluated in both large-scale and medium-scale specimens.

Spacial statistical distributions of AE events are used to identify the difference between damage mechanisms of structures with and without confinements (Soltangharaei et al., 2020a). Figure 5.20 presents the contour tomographies of AE

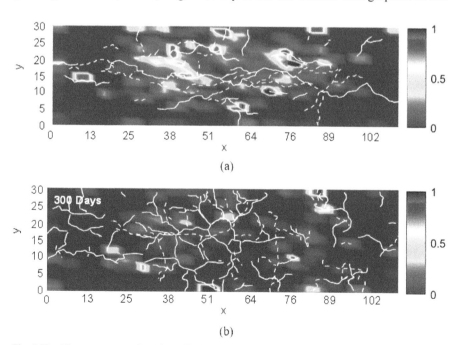

(a)

(b)

Fig. 5.20: AE event tomography: (a) confined specimen; (b) unconfined specimen (Soltangharaei et al., 2020a).

events for the confined and unconfined specimens. According to Figs. 5.20 and 5.5, the pattern of event distribution for the confined specimen is different from the unconfined specimen. The events for the confined specimen mostly occurred in the mid-third portion of the specimen width (Y direction) compared to the unconfined specimen.

The probabilities of event occurrence in the mid-third region of the specimen width were calculated at different instances during ASR for both specimens and shown in Fig. 5.21.

Shannon entropy can be employed to discriminate different phases during ASR, as mentioned in Section 3.3.3. The results of counts entropy using CDF (CE_CDF) for the confined and unconfined specimens are shown in Fig. 5.22. The entropy is initially increased and followed by a gradual decrease. In other words, the randomness of the data is rising during the first stage, which can be attributed to the microcrack formation in the concrete specimen. As ASR proceeds, the microcracks coalesce and form macrocracks in specific locations, decreasing the randomness of the data.

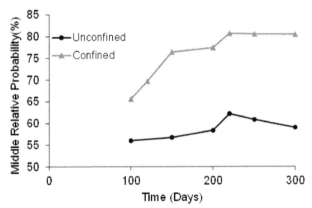

Fig. 5.21: Temporal evolution of middle relative probability of events for confined and unconfined specimens.

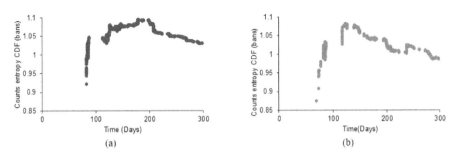

Fig. 5.22: Counts entropy for: (a) confined; (b) unconfined specimens (Soltangharaei et al., 2021a).

4.3 Impact on an Airplane Elevator

Fiber carbon composite materials are used in aerospace structures such as airplanes due to their high strength and low density. However, this material may be susceptible

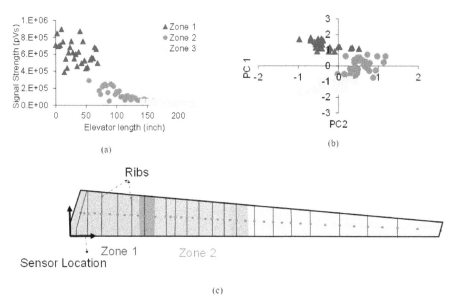

Fig. 5.23: K-means result: (a) signal strength variation of AE data along elevator; (b) PCA; (c) zone boundaries (Soltangharaei et al., 2019b).

to the impacts of debris and hail. Therefore, a structural health monitoring system, which can localize and determine impact levels, is essential for safe operation.

A zonal localization model based on data-driven methods was developed using a pattern recognition method and ANN (Soltangharaei et al., 2019b). The data was collected through a single broadband AE sensor (attached to the elevator spar) by impacting the surface of the structure using the steel impactors shown in Fig. 5.3. The impact locations are shown in Fig. 5.23c. K-means and PCA were used to classify the data along the elevator. Parametric features such as duration, rise time, signal strength, average frequency, amplitude, counts, and counts to peak were utilized for classification. The features evolved as the impact source moved farther from the sensor. PCA was conducted on the parametric features to reduce the data dimensions (Fig. 5.23b). The data is shown in Fig. 5.23b in a 2D principal component space. The resulting principal components were utilized as data features in the k-means method. The algorithm classified the data into three clusters, indicating the boundary of zones for the zonal source location model (Fig. 5.23a).

The data was labeled based on the k-means results (zone 1, zone 2, zone 3). The labeled data was then fed into an ANN model that was developed with two hidden layers (25 neurons in each layer) and a regression output layer. The data set was divided into three subsets: training, validation, and testing. The ANN was trained several times to get an optimum result with the lowest testing errors. The trained ANN can be used to estimate the zonal location of impacts on the elevator. The features from any of the impacts can be used as input for the model. The model result will be a zone number. The mislocalization error is reduced by considering overlaps in the zone boundaries (Fig. 5.23c) as shown in Fig. 5.24.

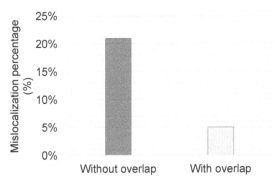

Fig. 5.24: ANN error for zonal source localization of elevator.

Deep learning methods such as autoencoder networks can be used to develop a zonal source location model for the elevator using raw data (signals in time or frequency domains) and are expected to have more precise results for a model with a larger zone number, as discussed in (Ai et al., 2021a), for a large-scale steel plate and elevator (Ai et al., 2021b). However, computation and training time for deep learning is much larger than ANN models. In addition, larger data storage is required for the input data of deep learning methods. These two drawbacks bring limitations for on-flight data acquisition and decision-making systems, making ANN models more suitable for this specific application at this point in time.

5. Conclusions and Recommendations

AE is a nondestructive method to detect damage in different structures due to different mechanisms. One of the advantages of this method compared to other techniques is the high sensitivity of AE sensors, which result in obtaining more data from micro- to macro-scale damages. This advantage brings challenges in terms of data analysis and management. The challenges can be attributed to the collection of extraneous data during acquisition or may be related to data analysis and interpretation. The former is resolved by providing different filtering approaches and protecting electrical connections, while the latter can be handled by different data-driven methods, some of which are presented in this chapter. The case studies illustrate the applications of some analysis methods for the AE data acquired from different loading or stress conditions applied to steel, concrete, and fiber reinforced composite structures.

Implementing more complex algorithms and data-driven models to analyze AE data becomes more feasible as computational hardware and systems advance. In addition, these methods can be a step forward in the development of an integrated autonomous system for data acquisition, data analysis, data interpretation, structural condition estimation, and decision making. Deep learning algorithms such as CNN, autoencoder, and recurrent neural networks such as long short-term memory (LSTM) networks are recommended to be employed for analyzing complex AE data and estimating dependencies of data in a time domain.

References

Abdelrahman, M., ElBatanouny, M.K. and Ziehl, P.H. 2014. Acoustic emission based damage assessment method for prestressed concrete structures: Modified index of damage. *Engineering Structures*, 60: 258–264.

Abdelrahman, M., ElBatanouny, M.K., Ziehl, P., Fasl, J., Larosche, C.J. and Fraczek, J. 2015. Classification of alkali–silica reaction damage using acoustic emission: A proof-of-concept study. *Construction Building Materials*, 95: 406–413.

Abdelrahman, M., ElBatanouny, M., Dixon, K., Serrato, M. and Ziehl, P. 2018. Remote monitoring and evaluation of damage at a decommissioned nuclear facility using acoustic emission. *Applied Sciences*, 8(9): 16–63.

Abdelrahman, M.A., ElBatanouny, M.K., Rose, J.R. and Ziehl, P.H. 2019. Signal processing techniques for filtering acoustic emission data in prestressed concrete. *Research in Nondestructive Evaluation*, 30(3): 127–148.

Ai, L., Soltangharaei, V., Bayat, M., Greer, B. and Ziehl, P. 2021a. Source localization on large-scale canisters for used nuclear fuel storage using optimal number of acoustic emission sensors. *Nuclear Engineering and Design*, 375: 111097.

Ai, L., Soltangharaei, V., Bayat, M., van Tooren, M. and Ziehl, P. 2021b. Detection of impact on aircraft composite structure using machine learning techniques. *Measurement Science Technology*, 32(8): 084013.

Allard, A., Bilodeau, S., Pissot, F., Fournier, B., Bastien, J. and Bissonnette, B. 2018. Expansive behavior of thick concrete slabs affected by alkali-silica reaction (ASR). *Construction Building Materials*, 171: 421–436.

Alvarez, M., Lapitz, P. and Ruzzante, J. 2008. AE response of type 304 stainless steel during stress corrosion crack propagation. *Corrosion Science*, 50(12): 3382–3388.

An, D., Kim, N.H. and Choi, J.-H. 2015. Practical options for selecting data-driven or physics-based prognostics algorithms with reviews. *Reliability Engineering & System Safety*, 133: 223–236.

Anay, R., Soltangharaei, V., Assi, L., DeVol, T. and Ziehl, P. 2018. Identification of damage mechanisms in cement paste based on acoustic emission. *Construction Building Materials*, 164: 286–296.

Anay, R., Lane, A., Jáuregui, D.V., Weldon, B.D., Soltangharaei, V. and Ziehl, P. 2020. On-site acoustic-emission monitoring for a prestressed concrete BT-54 AASHTO girder bridge. *Journal of Performance of Constructed Facilities*, 34(3): 04020034.

ASTM, E. 2006. *Standard Terminology for Nondestructive Examinations*.

Barbosa, R.A., Hansen, S.G., Hansen, K.K., Hoang, L.C. and Grelk, B. 2018. Influence of alkali-silica reaction and crack orientation on the uniaxial compressive strength of concrete cores from slab bridges. *Construction Building Materials*, 176: 440–451.

Bažant, Z.P. and Steffens, A. 2000. Mathematical model for kinetics of alkali–silica reaction in concrete. *Cement Concrete Research*, 30(3): 419–428.

Bishop, C.M. 2006. *Pattern Recognition and Machine Learning*. Springer.

Bouguettaya, A., Yu, Q., Liu, X., Zhou, X. and Song, A. 2015. Efficient agglomerative hierarchical clustering. *Expert Systems with Applications*, 42(5): 2785–2797.

Chai, M., Zhang, Z. and Duan, Q. 2018. A new qualitative acoustic emission parameter based on Shannon's entropy for damage monitoring. *Mechanical Systems Signal Processing*, 100: 617–629.

Colombo, I.S., Main, I. and Forde, M. 2003. Assessing damage of reinforced concrete beam using "b-value" analysis of acoustic emission signals. *Journal of Materials in Civil Engineering*, 15(3): 280–286.

Dron, R. and Brivot, F. 1992. Thermodynamic and kinetic approach to the alkali-silica reaction. Part 1: Concepts. *Cement Concrete Research*, 22(5): 941–948.

Du, G., Li, J., Wang, W., Jiang, C. and Song, S. 2011. Detection and characterization of stress-corrosion cracking on 304 stainless steel by electrochemical noise and acoustic emission techniques. *Corrosion Sci.*, 53(9): 2918–2926.

ElBatanouny, M.K., Ziehl, P.H., Larosche, A., Mangual, J., Matta, F. and Nanni, A. 2014. Acoustic emission monitoring for assessment of prestressed concrete beams. *Construction Building Materials*, 58: 46–53.

Farnam, Y., Geiker, M.R., Bentz, D. and Weiss, J. 2015. Acoustic emission waveform characterization of crack origin and mode in fractured and ASR damaged concrete. *Cement and Concrete Composites*, 60: 135–145.

Fowler, T.J., Blessing, J.A., Conlisk, P.J. and Swanson, T.L. 1989. The MONPAC system. *Journal of Acoustic Emission*, 8(3): 1–8.

Gavin, H. 2011. *The Levenberg-Marquardt Method for Nonlinear Least Squares Curve-fitting Problems.* Duke University.

Hanson, W. 1944. Studies relating to the mechanism by which the alkali-aggregate reaction produces expansion in concrete. *Proc., Journal Proceedings*, pp. 213–228.

Hawkins, D.M. 1980. *Identification of Outliers.* Springer.

Hayashi, C. 1998. What is data science? Fundamental concepts and a heuristic example. *Data Science, Classification, and Related Methods*, Springer, pp. 40–51.

He, K., Zhang, X., Ren, S. and Sun, J. 2016. Deep residual learning for image recognition. *Proc., Proceedings of the IEEE Conference on Computer Vision and Pattern Recognition*, pp. 770–778.

Jun, S.S. and Jin, C.S. 2010. ASR products on the content of reactive aggregate. *KSCE Journal of Civil Engineering*, 14(4): 539–545.

Jung, D.Y., Mizutani, Y., Todoroki, A. and Suzuki, Y. 2017. Frequency dependence of the b-value used for acoustic emission analysis of glass fiber reinforced plastics. *Open Journal of Composite Materials*, 7(03): 117.

Kahirdeh, A. 2014. *Energy Dissipation and Entropy Generation During the Fatigue Degradation: Application to Health Monitoring of Composites*. PhD. Philosophy, Louisiana State University.

Kahirdeh, A., Sauerbrunn, C. and Modarres, M. 2016. Acoustic emission entropy as a measure of damage in materials. *Proc., AIP Conference Proceedings*, AIP Publishing, p. 060007.

Kahirdeh, A., Sauerbrunn, C., Yun, H. and Modarres, M. 2017. A parametric approach to acoustic entropy estimation for assessment of fatigue damage. *International Journal of Fatigue*, 100: 229–237.

Karthik, M.M., Mander, J.B. and Hurlebaus, S. 2016. Deterioration data of a large-scale reinforced concrete specimen with severe ASR/DEF deterioration. *Construction Building Materials*, 124: 20–30.

Kovač, J., Legat, A., Zajec, B., Kosec, T. and Govekar, E. 2015. Detection and characterization of stainless steel SCC by the analysis of crack related acoustic emission. *Ultrasonics*, 62: 312–322.

Krizhevsky, A., Sutskever, I. and Hinton, G.E. 2012a. Imagenet classification with deep convolutional neural networks. *Proc., Advances in Neural Information Processing Systems*, pp. 1097–1105.

Krizhevsky, A., Sutskever, I. and Hinton, G.E. 2012b. Imagenet classification with deep convolutional neural networks. *Advances in Neural Information Processing Systems*, 25: 1097–1105.

Lokajíček, T., Přikryl, R., Šachlová, Š. and Kuchařová, A. 2017. Acoustic emission monitoring of crack formation during alkali silica reactivity accelerated mortar bar test. *Engineering Geology*, 220: 175–182.

MISTRASGroup. 2011. *PKWDI Sensor Wideband Low Power Integral Preamplifier Resonant Sensor*. P.A. Corporation, (Ed.), Copyright © 2011 MISTRAS Group, Inc. All Rights Reserved.

Murtagh, F. and Legendre, P. 2014. Ward's hierarchical agglomerative clustering method: Which algorithms implement Ward's criterion? *Journal of Classification*, 31(3): 274–295.

Ohtsu, M. 1995. Acoustic emission theory for moment tensor analysis. *Research in Nondestructive Evaluation*, 6(3): 169–184.

Ott, R.L. and Longnecker, M.T. 2015. *An Introduction to Statistical Methods and Data Analysis.* Nelson Education.

Pan, S.J. and Yang, Q. 2009. A survey on transfer learning. *IEEE Transactions on Knowledge Data Engineering*, 22(10): 1345–1359.

Ponce, J. and Batic, O.R. 2006. Different manifestations of the alkali-silica reaction in concrete according to the reaction kinetics of the reactive aggregate. *Cement Concrete Research*, 36(6): 1148–1156.

Pour-Ghaz, M., Spragg, R., Castro, J. and Weiss, J. 2012. Can acoustic emission be used to detect alkali silica reaction earlier than length change tests. *Proc., Proceeding of the 14th International Conference on Alkali-Aggregate Reaction in Concrete*.

Rajabipour, F., Giannini, E., Dunant, C., Ideker, J.H. and Thomas, M.D. 2015. Alkali–silica reaction: current understanding of the reaction mechanisms and the knowledge gaps. *Cement and Concrete Research*, 76: 130–146.

Rao, M. and Lakshmi, K.P. 2005. Analysis of b-value and improved b-value of acoustic emissions accompanying rock fracture. *Current Science*, 1577–1582.

Sagasta, F., Benavent-Climent, A., Fernández-Quirante, T. and Gallego, A. 2014. Modified Gutenberg–Richter coefficient for damage evaluation in reinforced concrete structures subjected to seismic simulations on a shaking table. *Journal of Nondestructive Evaluation*, 33(4): 616–631.

Sagasta, F.A., Torné, J.L., Sánchez-Parejo, A. and Gallego, A. 2013. Discrimination of acoustic emission signals for damage assessment in a reinforced concrete slab subjected to seismic simulations. *Archives of Acoustics*, 303–310.

Saouma, V. and Perotti, L. 2006. Constitutive model for alkali-aggregate reactions. *ACI Materials Journal*, 103(3): 194.

Saouma, V.E. and Hariri-Ardebili, M.A. 2014. A proposed aging management program for alkali silica reactions in a nuclear power plant. *Nuclear Engineering Design*, 277: 248–264.

Sauerbrunn, C.M. 2016. *Evaluation of Information Entropy from Acoustic Emission Waveforms as a Fatigue Damage Metric for Al7075-T6.* Master of Science, University of Maryland.

Shaikh, H., Amirthalingam, R., Anita, T., Sivaibharasi, N., Jaykumar, T., Manohar, P. and Khatak, H. 2007. Evaluation of stress corrosion cracking phenomenon in an AISI type 316LN stainless steel using acoustic emission technique. *Corrosion Sci.*, 49(2): 740–765.

Shannon, C.E. 1948. A mathematical theory of communication. *The Bell System Technical Journal*, 27(3): 379–423.

Shin, H.-C., Orton, M.R., Collins, D.J., Doran, S.J. and Leach, M.O. 2012. Stacked autoencoders for unsupervised feature learning and multiple organ detection in a pilot study using 4D patient data. *IEEE Transactions on Pattern Analysis Machine Intelligence*, 35(8): 1930–1943.

Shiotani, T. 1994. Evaluation of progressive failure using AE sources and improved b-value on slope model tests. *Progress in Acoustic Emission VII*, 529–534.

Shiotani, T., Nishida, T., Nakayama, H., Asaue, H., Chang, K.-C., Miyagawa, T. and Kobayashi, Y. 2017. Fatigue failure evaluation of RC bridge deck in wheel loading test by AE tomography. *Advances in Acoustic Emission Technology.* Springer, pp. 251–264.

Solomatine, D., See, L.M. and Abrahart, R. 2009. Data-driven modelling: Concepts, approaches and experiences. *Practical Hydroinformatics*, Springer, pp. 17–30.

Soltangharaei, V., Anay, R., Hayes, N., Assi, L., Le Pape, Y., Ma, Z. and Ziehl, P. 2018. Damage mechanism evaluation of large-scale concrete structures affected by alkali-silica reaction using acoustic emission. *Applied Sciences*, 8(11): 2148.

Soltangharaei, V., Anay, R., Begrajka, D., Bijman, M., ElBatanouny, M., Ziehl, P. and van Tooren, M.J. 2019a. An impact event detection system for composite box structures. *COMADEM 2019* Huddersfield, United Kingdom.

Soltangharaei, V., Anay, R., Begrajka, D., Bijman, M., ElBatanouny, M.K., Ziehl, P. and van Tooren, M.J. 2019b. A minimally invasive impact event detection system for aircraft movables. *Proc., AIAA Scitech 2019 Forum*, 1268.

Soltangharaei, V. 2020. *Evaluation of Temporal Damage Progression in Concrete Structures Affected by ASR Using Data-driven Methods.* Doctoral dissertation, University of South Carolina.

Soltangharaei, V., Anay, R., Ai, L., Giannini, E.R., Zhu, J. and Ziehl, P. 2020a. Temporal evaluation of ASR cracking in concrete specimens using acoustic emission. *Journal of Materials in Civil Engineering*, 32(10): 04020285.

Soltangharaei, V., Anay, R., Begrajka, D., Bijman, M., ElBatanouny, M.K., Ziehl, P. and van Tooren, M.J. 2020b. An impact event detection system for composite box structures. *Advances in Asset Management and Condition Monitoring*, Springer, pp. 1063–1073.

Soltangharaei, V., Hill, J., Ai, L., Anay, R., Greer, B., Bayat, M. and Ziehl, P. 2020c. Acoustic emission technique to identify stress corrosion cracking damage. *Structural Engineering Mechanics*, 75(6): 723–736.

Soltangharaei, V., Ai, L., Anay, R., Bayat, M. and Ziehl, P. 2021a. Implementation of information entropy, b-value, and regression analyses for temporal evaluation of acoustic emission data recorded during ASR cracking. *Practice Periodical on Structural Design Construction*, 26(1): 04020065.

Soltangharaei, V., Anay, R., Assi, L., Bayat, M., Rose, J. and Ziehl, P. 2021b. Analyzing acoustic emission data to identify cracking modes in cement paste using an artificial neural network. *Construction Building Materials*, 267: 121047.

Sun, Y., Zhang, H., Zhao, T., Zou, Z., Shen, B. and Yang, L. 2020. A new convolutional neural network with random forest method for hydrogen sensor fault diagnosis. *IEEE Access*, 8: 85421–85430.

Suzuki, H., Kinjo, T., Hayashi, Y., Takemoto, M., Ono, K. and Hayashi, Y. 1996a. Wavelet transform of acoustic emission signals. *Journal of Acoustic Emission*, 14: 69–84.

Suzuki, H., Kinjo, T., Hayashi, Y., Takemoto, M., Ono, K. and Hayashi, Y. 1996b. Wavelet transform of acoustic emission signals. *Journal of Acoustic Emission*, 14: 69–84.

Szegedy, C., Liu, W., Jia, Y., Sermanet, P., Reed, S., Anguelov, D., Erhan, D., Vanhoucke, V. and Rabinovich, A. 2015. Going deeper with convolutions. *Proc., Proceedings of the IEEE Conference on Computer Vision and Pattern Recognition*, pp. 1–9.

Szegedy, C., Vanhoucke, V., Ioffe, S., Shlens, J. and Wojna, Z. 2016. Rethinking the inception architecture for computer vision. *Proc., Proceedings of the IEEE Conference on Computer Vision and Pattern Recognition*, pp. 2818–2826.

Tan, P.N., Steinbach, M. and Kumar, V. 2018. Data mining cluster analysis: Basic concepts and algorithms. *Introduction to Data Mining*. New York, NY, USA: Pearson Education.

Villeneuve, V., Fournier, B. and Duchesne, J. 2012. Determination of the damage in concrete affected by ASR—The damage rating index (DRI). *Proc., 14th International Conference on Alkali-Aggregate Reaction (ICAAR). Austin, Texas (USA)*.

Wang, J.-G., Wang, Y., Yao, Y., Yang, B.-H. and Ma, S.-W. 2019. Stacked autoencoder for operation prediction of coke dry quenching process. *Control Engineering Practice*, 88: 110–118.

Ziehl, P. and ElBatanouny, M. 2015. Low-level acoustic emission in the long-term monitoring of concrete. *Acoustic Emission and Related Non-Destructive Evaluation Techniques in the Fracture Mechanics of Concrete*, Elsevier, pp. 205–224.

Chapter 6

Automatic Detection of Surface Thermal Cracks in Structural Concrete with Numerical Correlation Analysis

Diana Andrushia A.,[1] *Anand, N.,*[2,*] *Richard Walls,*[3] *Daniel Paul T.*[4]
and *Prince Arulraj*[5]

1. Introduction

Structural concrete is a versatile building material widely used for infrastructure. Concrete has a low conductivity and members are typically large and resistant to fire. During a destructive post-flashover fire, the temperature in buildings and structures may exceed 800–1000°C with heat fluxes well over 100 kW/m². However, structural concrete is good at resisting higher temperature exposure, as compared to other building materials. It is currently being used in various applications (e.g., firewalls, passive protection, load bearing elements in high-rise buildings) due to its inherent quality of fire resistance. Concrete cover creates an effective barrier in protecting reinforcing and structural steel in reinforced and composite construction. Even though concrete elements have excellent thermal properties, beyond around 500°C concrete, as a material, loses most of its compressive strength. Also, the temperature effects between 500°C and 800°C can be critical, as during this stage spalling occurs and thermal cracks typically develop. A rapid mass loss occurs in the concrete at temperatures above 800°C and strength loss is found to be directly proportional to the moisture loss that occurs. Mechanical and thermal properties of concrete also vary with increasing temperature (Akca and Zihnioğlu, 2013; Khaliq and Waheed, 2017; Peng et al., 2006; Sideris and Manita, 2013).

[1] Assistant Professor, Department of ECE, Karunya University, Coimbatore - 641114, India.
[2] Associate Professor, Department of Civil Engineering, Karunya University, Coimbatore - 641114, India.
[3] Professor, Structural Engineering Division, Stellenbosch University, Matieland 7602, Stellenbosch, South Africa.
[4] Research Scholar, Department of Civil Engineering, Karunya University, Coimbatore - 641114, India.
[5] Professor, Department of Civil Engineering, Karunya University, Coimbatore - 641114, India.
* Corresponding author: davids1612@gmail.com

Fig. 6.1: Examples of extensive fire damage to a reinforced concrete building showing (a) spalling, (b) buckling of a column, and (c) extensive cracking of the soffit along lines of reinforcing.

Concrete undergoes severe physical, structural, and microstructural changes because of higher temperature exposure. This effect may cause instability in the structural elements and could lead to a progressive collapse scenario. Figure 6.1 illustrates several of the phenomena discussed above, showing extensive structural damage to a reinforced concrete building.

Structural concrete is a preferred building material due to its ease of production, durability, and non-combustibility properties. Concrete has a high specific heat capacity and low thermal conductivity (Kodur, 2014). Concrete has an excellent ability to withstand the compressive stresses that arise during the application of loads, although it has limited tensile capacity. Due to this, concrete sections are sensitive to tensile loads and such forces cause cracks.

At elevated temperature, tensile stresses are developed inside the concrete microstructure because of pore pressure and induced stresses due to retrained expansion. When exposed to elevated temperature, the compressive strength of concrete varies with respect to the type of aggregate and its size, water-cement ratio, type of admixtures, and type of cement. Rate of temperature exposure also has an impact on residual compressive strength (Behnood and Ghandehari, 2009; Poon et al., 2004). Factors influencing development of induced tensile stresses from restraint include cross-sectional geometry, heating regime, and thermal properties. Micro-cracking and spalling occur in concrete once tensile stresses exceed permissible tensile strengths. Due to this reason, structural concrete is more prone to crack under tensile and transverse loads.

Structural elements are designed to satisfy strength and serviceability criteria. At high temperature, concrete loses its compressive strength and steel loses its tensile strength, thereby the elastic modulus and capacity of a cross-section decreases significantly. In addition, structural elements subjected to loading undergo deformations which can further exacerbate cracking. When concrete is exposed to elevated temperature, it affects the compressive strength, flexural strength, bond strength, shear strength, impact strength, stress-strain behaviour, and modulus of elasticity. This reduction may be due to the influence of constituent materials in the concrete and the rate of rise, and magnitude of temperature exposure (Youssef and Moftah, 2007). During exposure to elevated temperature, the breaking of interfacial bonds occurs within the microstructure of concrete.

The bending strength of concrete is considerably lesser than the compressive strength of concrete because the micro cracks developed in concrete can spread and widen under bending loads. When exposed to fire, bending strength of concrete is reduced due to the spalling that occurs within the concrete structure (Ding et al., 2012; Ding et al., 2009; Li et al., 2004). The decline in bond strength between steel and concrete can result. With an increase in temperature, a rapid reduction in tensile strength of concrete occurs and thus bonding between concrete and steel is reduced. Rate of temperature rise, and the duration of exposure are the parameters which influence the reduction in compressive and bond strength the most, and deformation/slips increases (Cülfik and Özturan, 2002; Özbay et al., 2015).

Cracking is the outcome of localised material failure which occurs because of elevated temperature in structural concrete. Evaluating the crack pattern, density, width, and area provide insight to assess the level of damage experienced. Assessment of crack development and detection can be useful to quantify the extent of damage experienced. Past studies have shown that surface thermal crack density and porosity is found to increase as the exposure temperature and duration increases. Crack detection and quantification are emerging areas in structural health monitoring. Research work and associated investigations on crack detection and quantification may also be helpful for repair and rehabilitation works. Several computational techniques have been employed in this area as a tool for the detection of different types of cracks (Andrushia et al., 2020, 2021a, 2021b). Therefore, in the present study the focus is on the detection of cracks in structural concrete exposed to elevated temperature.

To assess the residual load-bearing capacity in a scientific way, information is needed regarding the temperature distribution inside the concrete element and the residual material properties of both concrete and steel. The objective of this study is to detect and estimate the damage of fire affected concrete elements using Artificial Intelligence (AI) techniques. A machine learning-based technique is used to detect and quantify the damage to fire-affected concrete. Concrete specimens having a compressive strength of 20 to 50 MPa were exposed to high temperatures as per the standard time-temperature curve (ISO 834 curve). After fire tests, images showing the cracks at the surface of concrete specimens were captured using digital cameras which are processed to quantify damage and estimate fire exposure. A good correlation is obtained between the residual compressive strength and temperature at the core of the fire-damaged concrete, which was measured during experiments. The results show that the cracks detected from the developed technique are in good agreement with the visual observations.

2. Backgrounds

This section provides information about the damage detection methodologies developed. The techniques are related to basic image analysis coupled to advanced machine learning (ML)/deep learning (DL) techniques, as presented in detail below.

2.1 Image Processing Techniques

The primary advantage of image-based crack detection is that I produces accurate detection results. Conventional methods are time-consuming and subject to the knowledge of the inspector. Image-based crack detection and quantification is a promising field in computer vision. Several researchers have used computer vision-based methods to capture and analyse cracks efficiently (Patel et al., 2016; Lin et al., 2017; Andrushia et al., 2019). Vision-based techniques have been classified into four broad categories which are: filtering-based techniques, threshold-based techniques, morphological operator-based techniques, and transform-based techniques.

The basic image-based crack detection techniques have four fundamental processes which are: image acquisition, pre-processing, feature extraction, and crack detection. These techniques differ with respect to image type. Infra-red images, ultrasonic images, and laser images are some of the various sources that can be used. Noise removal and enhancement techniques are part of pre-processing techniques. Noise usually occurs because of the image acquisition setup. Illumination variations, stains, and shades all can be considered noise. This noise is removed by applying suitable noise elimination techniques. Image enhancements are done after the noise removal process. Enhancement techniques are used to enhance the lines, edges, points, and contours of the input images.

Iyer and Sinha (2005) investigated crack detection methods using high contrast images. Mathematical morphological techniques and curvature evaluation are used to find the crack patterns. Linear filtering is used to differentiate the cracks from noisy backgrounds. Yiyang (2014) applied an image smoothing process. The smoothened image is applied to the crack segmentation process. The cracks are quantified through crack area, crack perimeter, and roundness index. Fujita and Hamamoto (2011) investigated an automatic crack detection method for noisy crack images. To remove the noise from the input image, median filtering is applied, and a hessian matrix based multi-line filter is used to emphasise the cracks. The fine cracks are identified by using adaptive thresholds.

Wolf et al. (2015) proposed a system that detects cracks within concrete structures. An electronic ultrasonic sensor is attached with which the system helps to identify crack propagation. The accuracy of the detection methods has been compared with the non-destructive testing methods. Nguyen et al. (2014) presented a concrete crack detection method based on edge detectors. The cracks are detected as tree-like structures. Threshold filtering and morphological operations are used to detect the cracks in relation to backgrounds. Edge points are linked together to identify continuous cracks. Salman et al. (2013) have used the Gabor filter to find multidirectional cracks. This filter is effectively used to find cracks which are located at different orientations. The detection precision of this method achieved is 95%.

A method for crack detection in concrete walls has been proposed by Hoang et al. (2018). A combination of filters and edge detectors are used to identify damage. Median filters are used to remove noise on the input images and the edge detectors of Canny are used to find the cracks in the walls. Such basic filtering-based methods mostly adopt hybrid techniques to perform crack detection effectively, typically combined with morphological processing, filtering, and threshold techniques.

Transform based methods are also used to find damage in the concrete images. The images are accurately analysed in the frequency domain. The Fourier Transform, Wavelet Transform, Curvelet Transform, Shearlet Transform, and Ripplet Transform have been used by the researchers to identify damage. Abdel-Qader et al. (2003) investigated a transform domain-based crack detection techniques. The Fast Fourier Transform and Fast Haar Transform were used initially to identify cracks in concrete bridges. Edge detector techniques of Sobel and Canny are also used to enhance results. The Wavelet Transform is used by Patel et al. (2016) to detect damage in concrete structures. Wavelet coefficients are used to obtain crack properties. The Wavelet Transform is used by many researchers to find the cracks efficiently. The directional features of inputs are not captured well by conventional wavelet transforms. Hence, multiscale, and multidirectional wavelet transforms are leveraged to detect damage.

Li et al. (2014) have proposed a crack detection method using frequency domain analysis. The Fast Curvelet Transform is used to detect the cracks. Texture analysis also incorporated in the study to find the cracks accurately. The input image is decomposed initially to analyse input features in the frequency domain. An inverse transform is applied to reconstruct the input image and thereby the cracks are well identified. Cracks in fire affected concrete structures have been identified by the Ripplet Transform (Andrushia et al., 2020; Andrushia et al., 2021a). Major challenges linked to noise elimination, enhancement, and crack detection are presented in these techniques. The input images are decomposed into different sub-bands. The images' features are well analysed in the Ripplet domain, with the major and minor thermal cracks being accurately detected. Crack properties are quantified through crack length and crack width.

Zalama et al. (2014) used Gabor filters to find the major and minor cracks of a concrete structure. Specially designed feature extraction steps are used to determine the damage. The method also identifies thin micro cracks. As fire affected concrete structures have small micro cracks which may be detected as noise. Such cracks are missed by many computer-vision models because they require specialised feature extraction steps.

2.2 Machine Learning (ML)/Deep Learning (DL) Techniques

Following computer vision techniques, ML-based techniques are also used to find cracks in concrete structures. Traditional ML techniques usually have five stages, namely: image collection, pre-processing, feature extraction, damage detection, and classification. The initial pre-processing techniques are basic detection techniques coupled with ML methods. The feature extraction and damage detection phases are used with different optimization and neural network algorithms.

To process large input datasets, ML-based methods are adopted to classify the cracks in concrete (Dai et al., 2019). Support Vector Machine (SVM) (Li et al., 2017) and Extreme Learning Machine (ELM) (Dai et al., 2019) techniques have been applied to detect cracks from inputted images. Cha et al. (2016) discussed the structural damage detection based on SVM and Hough transform. Feature extraction processes are carried out with these methods as a primary step to learn the input images. If more characteristics need to be extracted, then feature reduction techniques are applied to find the optimal attributes. Principle Component Analysis

(PCA) is employed to perform feature selection, and thereby yield an informative feature set. The optimal attributes are given to classifiers to detect the cracks in an accurate manner. In addition, K nearest neighbour (KNN), Artificial Neural Network (ANN), fuzzy logic with SVM, genetic algorithms with SVM (Choudhary et al., 2012) are different types of classifiers used to detect cracks from input.

Traditional pattern recognition methods are involved with pre-processing, feature extraction, segmentation, classification, and recognition. Many algorithms are introduced in each step. However, due to the complex structure of inputs, it is hard to use the traditional process flow. The feature extraction step is difficult because the high-level characteristics that are related to the damage location are given by the user. As it is specified by the end-user, it will reduce the overall accuracy of the detection process. This is overcome by the DL techniques which can perform automatic feature extraction steps and deal with large numbers of input.

DL is an emerging tool under AI which has allowed for significant advancement in the pattern recognition field recently. The performances of these techniques are good in comparison with ML techniques. The methods are based on DL and have been applied in various research fields such as object detection, medical image segmentation, damage detection, land cover classification, and disease detection in precision agriculture (Malar et al., 2021). Irrespective of the fields, the DL techniques yield better results in the automated process. Recently, researchers in the remote sensing and geology fields have been using the DL procedures effectively.

Feng et al. (2020) explored deep convolution neural network-based surface crack detection in dams. The automatic pixel level crack identification technique is experimented with a detection accuracy of 80.31%. Tunnel crack diagnosis has been explored by Makantasis et al. (2015) where a multilayer perceptron and deep convolution neural network approach are used to identify cracks. Cha et al. (2017) used deep convolution neural network to detect damages of concrete surfaces. Andrushia and Lubloy (2021b) have used Unet with residual connection architecture for damage detection. Pixel wise detection is explored to detect the damages in an effective manner. Ali et al. (2021) compared multiple pre-trained models to detect the cracks in concrete structures.

Fan et al. (2020) used an ensemble network and encoder-decoder architecture for pavement crack detection. U-HDN is the U-Hierarchical Dilated Network that enables multiple dilation and end-to-end crack detection. It is used to highlight pixel-wise cracks. The efficiency of these models has been compared with different DL methods and achieved the precision values of 92.1%. Lin et al. (2017) investigated automatic crack detection using a convolutional neural network (CNN) model. The accuracy of this method is higher than other automatic methods.

The software system CrackNet has been proposed by Zhang et al. (2017) which consists of CNN architecture with five layers. A pixel-wise crack detection method is performed for asphalt surfaces. This architecture efficiently detects cracks, but the computational time is high for the detection. Hairline-type cracks are not well detected by this method. To overcome this challenge, CrackNet II, Zhang et al. (2018) have developed to reduce the computational time and noises associated with asphalt surfaces. The training time is reduced by fine tuning the model parameters of the architecture.

Even though multiple ML and DL methods are available for damage detection of concrete, techniques associated with fire affected concrete has received little attention. A primary reason for this is that significantly more time and effort is needed to measure and identify fire-induced damage.

Short et al. (2001) have investigated a colour image analysis of fire-affected concrete. The changes of colour in the images are considered with respect to the cracking and spalling of concrete subjected to elevated temperature. Li et al. (2018b) developed a method for crack detection of fire-affected concrete. The Local Binary Pattern (LBP) algorithm was used to analyse the damage. Cracks are quantified through calculating lengths and widths.

Based on the details above, the present study proposes a DL-based automatic crack detection methodology for fire-affected concrete. The customized CNN model has been used to detect the damage by quantifying crack formation. Additionally, a numerical correlation analysis on the fire-affected concrete is provided in depth. The automatic crack detection method is proposed and checked with crack parameters which are obtained from optical observations. Compressive strength tests are also done to measure the maximum strength of the fire-affected concrete and quantify the capacity loss.

2.3 Concrete Samples

In this work, 150 × 150 × 150 mm concrete samples were prepared and tested in an electric furnace as discussed below. For each temperature level (S-60, S-120, S-180, and S-240) 20 samples are considered for testing. These samples provide the data upon which the ML/DL techniques are developed.

3. Materials

Ordinary Portland Cement (OPC) of grade 53 conforming to IS: 12269-2013 is used in the present research work. It was purchased from a single source to maintain the quality being utilised for casting samples. The standard consistency of the cement was evaluated according to IS: 4031 (Part 4)—1988. Compressive strength of the samples made from the cement was determined as per IS: 4031 (Part 6)—1988 to confirm the strength grade. The specific gravity of the cement is 3.15.

3.1 Coarse Aggregate

In this study, crushed granite stone was used as normal weight coarse aggregates. The coarse aggregate used is mechanically broken granite drawn from the city of Coimbatore, India. The specific gravity of coarse aggregate is 2.96 and it conforms to IS 383-2016.

3.2 Fine Aggregate

The fine aggregate used during this investigation was natural river sand brought from the Cauvery Riverbed, Karur complying with IS: 383-2016. The specific gravity of the sand was determined as per IS: 2386(PIII)-1963. The specific gravity of was 2.70. Sieve analysis was carried out to confirm the grading as being Zone II. The bulk density of fine aggregate used was found to be 1620 kg/m³.

The gradation of coarse-and-fine aggregates plays an important role in the workability and development of paste. Particle size grading is important for characterisation of aggregates, as the effect of particle packing results in a reduction in voids. This, in turn, influences the water demand and requirement of the powder content in concrete.

Water is the key ingredient used for hydration of the cement. Potable drinking water was used for the concrete mix. The water used for the experiments satisfied the limits as per IS 456:2000.

4. Methodology

4.1 Testing of Specimens under Elevated Temperature

The concrete samples were kept in the curing tank for 28 days to achieve maximum strength. Dry reference specimens were tested in a laboratory before heating. The remaining specimens were heated in a computer controlled electrical furnace following the standard fire curve, with exposure times ranging from 60 to 240 minutes.

Concrete specimens were dried out for 24 hrs in an oven at 60°C before heating in the furnace at different exposure times according to the ISO 834-1:1999 standard fire time-temperature curve. The specimens are placed in the furnace through a boogie and the temperatures are set for the required time. The furnace consists of coils on four sides which heat the specimens through radiation when the target temperature is set. The furnace is 700 × 400 × 400 mm with a power capacity of 100 kW and is shown in Fig. 6.4. The furnace has two displays: (1) the set value display in which the gas temperature to heat to that will be attained at a specific time, and (2) a program value display which displays the temperature that is attained in the coil at the given time. Once the temperature is set, the coil heats the specimens till the target time. After the target temperature for the given duration is achieved, the furnace will stop. The specimens are taken out of the furnace and placed in a closed area for natural air-cooling. The maximum operating temperature of the furnace is 1200°C. The furnace is equipped with a micro controller which can be programmed to control the temperature as per the standard temperature curve given in Eq. 6.1

$$T - To = 345 \log 10 \ (8t + 1) \tag{6.1}$$

where: T—Furnace temperature at time t (degrees Celsius), To—Initial furnace temperature (degrees Celsius), t—Time (minutes). Figure 6.2 displays the ISO 834 Standard Time-Temperature Fire curve. Figure 6.3 gives the heating cooling cycles of natural air cooling used for the investigation, as obtained from experimental

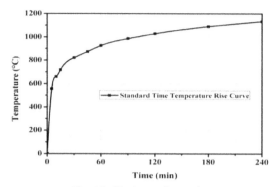

Fig. 6.2: Heating cooling cycle.

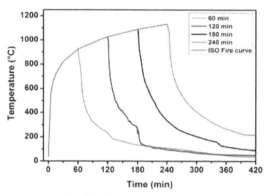

Fig. 6.3: Time temperature curve.

Fig. 6.4: Elevated temperature test.

samples. This typical ISO fire curve was selected since the rate of heating is quite high which is critical for concrete and several researchers have adopted this curve for their research (Ayudhya, 2016; Arioz, 2007; Xiaoyong and Fanjie, 2011; Peng et al., 2006; Hoff et al., 2000; Ali et al., 2011; Chung et al., 2016). The temperature of specimens was monitored by thermocouples installed at various points. The thermocouples connected to an automatic data acquisition were used. During the experiment the temperature of the furnace and surface of the specimens were recorded. Figure 6.4 shows the positions of the thermocouples used to measure the temperature of concrete specimens.

4.2 Compressive Strength

A digital compression testing machine of 200-ton capacity and capable of applying the load without shock, was used to apply the loading continuously at a rate of approximately 140 kg/cm² /min until failure. Cube specimens of size 150 × 150 mm were used for testing compressive strength of concrete as shown in Fig. 6.5. As introduced above, concrete specimens of different strength grades were cast and tested for different durations of heating as per the standard fire curve. A minimum of three samples were considered for each grade and duration of heating to confirm the test results.

Fig. 6.5: Compressive strength test on a 150 × 150 × 150 concrete cubes.

4.3 Compressive Strength

Figure 6.6 shows the residual compressive strength of the M50 grade concrete mix exposed to different durations of heating. After 60 mins exposure (max. 925°C), the residual compressive strength was 73.5% of the original strength for M50 grade mix, with a residual strength of 35.5% at 120 mins. At an exposure time of 180 mins, the specimens showed severe loss of strength and had a residual strength of 22% of the original compressive strength. At 240 mins (1133°C) duration of heating, the residual strength was found to be 11.3%. A severe deterioration in strength occurs

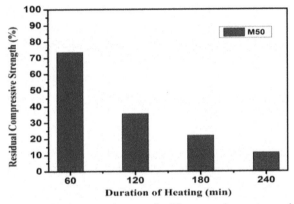

Fig. 6.6: Residual compressive strength of fire-exposed concrete samples.

because of the breakdown of CSH gel in the microstructure of the paste. Also, crystal structure transformation of aggregate occurs. Internal forces generated by non-uniform heating and the expansion of steam may also have generated significant stresses that increased the damage.

5. Methodology for Crack Detection

5.1 *Image Data Acquisition and Preprocessing*

After the fire test experiments, images of the fire damaged specimens were captured by a high-resolution digital camera. A camera with a stand was fixed permanently close to the heating furnace (Fig. 6.7). Specimens were cooled by natural air before capturing the images. As crack propagation increases during the cooling phase, the images were taken after the specimens reached room temperature. Cracks developed because of the effect of elevated temperatures and so are referred to as thermal cracks. However, the properties of the cracks may change based on the temperature, duration of heating, and type of concrete material, as discussed above.

Fig. 6.7: Image acquisition setup.

The image acquisition setup consists of a Canon SX510 HS camera for taking photographs with an optical image stabilizer of intelligent IS for sharp photos. It has a 30x optical zoom with a 24 mm wide lens in a mini body. It has a 60x Zoom Plus HS System with 12.1 Megapixel CMOS having DIGIC 4 facility and can produce quality images even in low light. Zoom framing assists the device to keep track of distant subjects with good clarity. Several images were captured repeatedly with a static position of concrete specimens, and the best images were used for the analysis. Figure 6.8 shows the proposed framework for the detection of thermal cracks in a structural concrete.

5.2 *Customized CNN Method*

In the proposed automatic crack detection method, customized CNN architecture is constructed from scratch by adjusting the different hyper-parameters which are the number of convolution layers, fully connected layers, strides, types of pooling, pooling locations, number of filters, and filter sizes. The hyper-parameters are

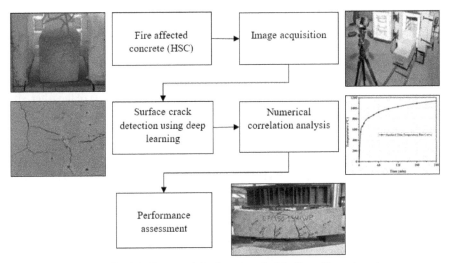

Fig. 6.8: Framework for the automatic detection of thermal cracks.

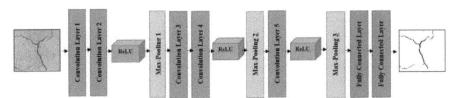

Fig. 6.9: Customized CNN for crack detection.

chosen by a trial-and-error method only. There is no exact mathematical calculation method to set the hyper-parameters for a particular dataset. Figure 6.9 shows the customized CNN architecture for the proposed work. It consists of 5 convolution layers, 3 pooling layers, 3 activations, 1 soft-max layer, and 2 fully connected layers. All these layers are used to increase the performance of the network by performing feature extraction, dimensionality reduction, and incorporating non-linearity.

The convolution layers are used to enhance the spatial invariance property for extracting all important features from the input image. The CNN architecture depends significantly on spatial features of the inputs. If the input image is predominantly sparse, then the learning rate of the network is significantly decreased. In this model, the Adam optimizer is used as an adaptive learning rate, to handle the sparse data. It handles the vanishing gradient problem in comparison with other optimizers such as RMSprop and Adadelta (Luo et al., 2020).

The input image is resized into 224 × 224 and passed onto the first layer. The input image is represented as a × b × c. It will pass into convolution layers and pooling layers to reduce the spatial size. 'a', 'b', and 'c' represent input image height, weight, and number of colour channels. After passing it into different layers, the final feature vector is obtained via a fully connected layer. Final crack detection is obtained from the fully attached layer. Initially, the convolution layer is used to find the features by identifying possible connections from the input data samples which are taken from the input layer through using the following relationship Eq. (6.2):

$$X = \sum (Input_{d \times d} + Weight_{e \times e}) + F \tag{6.2}$$

where X is the feature vector. $Input_{d \times d}$ represents the input receptive field where the convolution operations are performed. *Weight e* and *F* are the weights of the filter, size of the kernel, and filter size of the convolution layer. The feature vector is calculated by performing convolutions over image pixels and adding the pixel values together. With respect to the convolution kernels, the feature maps get varied. The final feature vector is given into the ReLU activation layer, which makes the non-negative values into zero, by executing element wise operations. It is done for the output values from the convolution layer. The ReLU activation function is primarily used in the majority of DL architectures because of its higher computation capabilities. Sigmoid, tanh, and swish are the different available activation functions. This layer assures better usage of feature maps that are taken from convolution layers by incorporating non-linearity. The Eq. (6.3) belongs to the activation layer.

$$\vartheta(J) = \max (0, J) \tag{6.3}$$

where '*J*' denotes the input vector elements

The pooling layer divides the feature map data into non-overlapping kernels. The maximum value of each kernel is forwarded into the next layer. Two sets of operations are performed by the max-pooling layer. Down-sampling incoming data and reducing the dimensionality is the first set of operations. This reduces model parameters and thereby reduces the computational time and is the second set of operations that are performed by pooling layers. The fully connected layer is the output layer that performs logical inferences. In this work, the fully connected layers convert the three-dimensional vector into a single-dimensional vector and performs a full convolution operation. The output of the fully connected layer is given by Eq. (6.4).

$$K_{n_o \times 1} = W_{n_o \times n_i} J_{n_i \times 1} \cdot B_{n_o \times 1} \tag{6.4}$$

where n_o and n_i represent size of input and output vector, K represents the fully connected layer output, W and B are the weights and bias matrix. Figure 6.10 shows the example output of activation and pooling layers.

The loss function is used to evaluate the efficiency of the proposed model. It is used to evaluate the deviation between the actual and the predicted value. For the large input, the network makes decisions based upon the input. At the time of

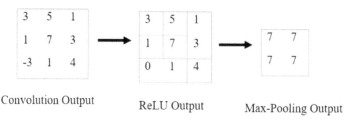

Convolution Output   ReLU Output   Max-Pooling Output

Fig. 6.10: The example outputs of activation and pooling layers.

the learning process, the loss function is continuously changed until the best fit occurs with respect to the error minimization. The softmax function or cross entropy function is used to find the difference between the actual and predicted values as shown by Eq. (6.5)

$$LF_{SMax} = N^{-1} \sum_{i=1}^{N} \log\left(\frac{e^{a_n}}{\sum e^{a_k}}\right) a$$

(6.5)

which represents the total number of neurons in the final output layer and a_n is the input data.

The hyper-parameters or internal parameters of the network play a vital role in performing the detection task accurately. The selection of a suitable optimizer is important to find the end detection results. The optimization procedures are used to minimize the loss function and update the hyper-parameters.

5.3 *Training and Testing*

The proposed customized model is implemented using MATLAB®, a 2.90 GHZ processor with 12 GB GPU card on a NVIDIA GTX1050. The performance of the proposed automatic crack detection method is ranked based on size of the input, network hyper-parameters, and rate of convergence. Six thousand images are used from experiments. The training and testing images are taken in the ratio of 80:20. The training and testing images are not overlapped.

6. Results and Discussion

This section discusses the experimental results of the proposed method and validations with respect to the optical observations. Figure 6.11 shows the damage detection results of the proposed customized CNN method. The images in the first row represent the original crack images investigated under different standard fire-time exposure. The second row provides the results of the proposed CNN model. The damage of the fire-affected concrete such as surface cracks, pores, and isolated edges are represented in white lines. It clearly shows the micro and macro damages are detected accurately by the proposed method.

The proposed results of the customized CNN model are validated by performing manual observations and measurements through AutoCAD. The sample results of the AutoCAD analysis are given in Fig. 6.12. It clearly reveals that the proposed method results agree with the observation results. The visual comparison highlights that the DL method is more accurate in comparison with the AutoCAD results. The sample images are taken from different heating durations such as 1 hour, 2 hours, 3 hours, and 4 hours. The total surface damage is given in terms of pixels in the proposed deep learning method. From the manual observations done, the total damage is given in millimetres of crack length. The crack length properties are detailed in Table 6.1.

Table 6.1 shows the quantitative results of the proposed DL method and optical observation method through AutoCAD. The damage is quantified in terms of crack

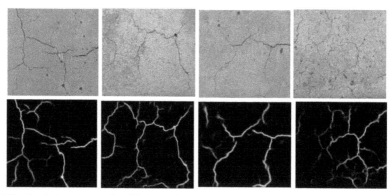

Fig. 6.11: Crack detection results with the top images showing concrete samples and the bottom images results from the CNN model.

Fig. 6.12: Crack detection results from AutoCAD.

Table 6.1: Quantitative results of crack properties.

Specimen ID	(x) Total Number of Pixels Identified as Damages by Proposed CNN (pixels)	(y) Total Crack Length Identified by AutoCAD (mm)	Ratio (x/y)
S-60	128705	1548	83.14×10^2
S-120	346341	4052	85.47×10^2
S-180	688438	7944	86.66×10^2
S-240	1346451	15142	88.92×10^2

length. The total number of pixels which are associated with the cracks are given as 'x'. The total crack length measured by optical observation is given in terms of 'y' and measured in millimetres (mm). Both quantitative results are not comparable directly. The crack widths are varying from 0.13 mm to 1.12 mm. Due to this variation, the manual measurement of crack widths is a time-consuming process. It is also difficult to determine an average crack width. Hence, the ratio calculation is provided. The ratio between the total number of crack pixels, which are obtained from the DL model, and the optical observation is performed. The ratio seems to be consistent between the samples. It is interesting to note that, under each category of specimens, the range in value is low with deviations being small.

This analysis highlights that the proposed DL method can detect the surface cracks on the fire-affected concrete specimens. The total number of pixels identified for the S-60 specimen is 128,705 and for S-240 is 1,346,451. The total number of crack pixels increases with respect to the heating duration of specimens. The specimen ID represents the heating durations such as 60 mins, 120 mins, 180 mins, and 240 mins. By considering the increase in the total number of pixels, samples with an increasing fire exposure time have escalating cracking and damage, as would be expected and as consistent with the post-fire strength tests. Hence, the DL method accurately detects the damages from the input image.

6.1 Correlation Analysis

This section discusses the thermal-structural behaviour of concrete specimens which are subjected to a standard fire exposure. The correlation analysis of crack detection can be used as one of the damage evaluation indicators for assessing fire damage to concrete structures. The primary reason is that cracks are evidence of high stresses and strains, and reduced material properties, for fire-affected concrete. As the temperature increases, it causes cracks in the structure due to the heat propagation and induced thermal induced. A correlation analysis between the cracks and duration of heating are given in Figs. 6.13 and 6.14. The crack lengths that are measured by the proposed DL method and through optical observations are highlighted.

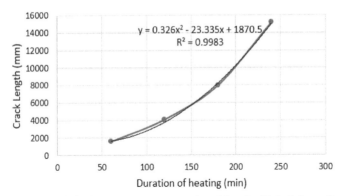

Fig. 6.13: Correlation between temperatures and crack lengths (Optical observation).

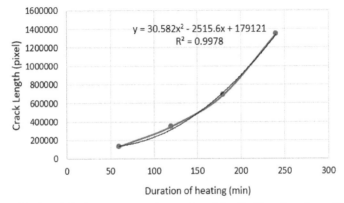

Fig. 6.14: Correlation between temperatures and crack lengths (Deep Learning method).

As the temperature and fire exposure time increases, the crack length also increases. The length measurement through optical observation and the DL method clearly highlight the change in crack length with standard fire time, and this appears to follow a parabolic trend. The total number of pixels that are detected as damage by the proposed DL method shows good correlation with the temperature exposure level.

7. Conclusion

This study investigates the automatic detection of thermal cracks in fire affected concrete structures. The customized CNN DL architecture is proposed to detect the cracks. The automatic detection of thermal cracks is validated with respect to the optical observations. The quantitative results of the proposed customized DL method show good agreement with the results of optical observation measurements. The numerical correlation analysis also depicts a similar acceptable agreement. From the study it is concluded that the damage propagation fully depends on the fire exposure time, as would be expected. As the temperature increases, the damage increases due to material degradation and induced stresses from temperature gradients. A major limitation of this work is that it is not able to determine the depth of cracking. The proposed methodology will be useful for automating the quantification of damage due to fires, either in experimental work or in real buildings. Further work is required to train the algorithms to be adaptable to a variety of scenarios and concrete finishes. When there are soot deposits on concrete samples, it is likely that the method will be less accurate.

References

Abdel-Qader, I., Abudayyeh, O. and Kelly, M.E. 2003. Analysis of edge-detection techniques for crack identification in bridges. *Journal of Computing in Civil Engineering*, 17(4): 255–263.

Akca, A.H. and Zihnioğlu, N.Ö. 2013. High performance concrete under elevated temperatures. *Construction and Building Materials*, 44: 317–328.

Ali, F., Nadjai, A. and Abu-Tair, A. 2011. Explosive spalling of normal strength concrete slabs subjected to severe fire. *Materials and Structures*, 44(5): 943–956.

Ali, L., Alnajjar, F., Jassmi, H.A., Gochoo, M., Khan, W. and Serhani, M.A. 2021. Performance evaluation of deep CNN-based crack detection and localization techniques for concrete structures. *Sensors*, 21(5): 1688.

Andrushia, D., Anand, N. and Arulraj, P.G. 2019. Anisotropic diffusion based denoising on concrete images and surface crack segmentation. *International Journal of Structural Integrity*, 11(3): 395–409.

Andrushia, D.A., Anand, N. and Arulraj, P.G. 2020. A novel approach for thermal crack detection and quantification in structural concrete using ripplet transform. *Structural Control & Health Monitoring*, 27(11).

Andrushia, A.D., Anand, N. and Arulraj, G.P. 2021a. Evaluation of thermal cracks on fire exposed concrete structures using Ripplet transform. *Mathematics and Computers in Simulation*, 180: 93–113.

Andrushia, A.D. and Lubloy, E. 2021b. Deep learning based thermal crack detection on structural concrete exposed to elevated temperature. *Advances in Structural Engineering*, 1369433220986637.

Arioz, O. 2007. Effects of elevated temperatures on properties of concrete. *Fire Safety Journal*, 42(8): 516–522.

Ayudhya, B.I.N. 2016. Comparison of compressive and splitting tensile strength of autoclaved aerated concrete (AAC) containing water hyacinth and polypropylene fibre subjected to elevated temperatures. *Materials and Structures*, 49(4): 1455–1468.

Behnood, A. and Ghandehari, M. 2009. Comparison of compressive and splitting tensile strength of high-strength concrete with and without polypropylene fibers heated to high temperatures. *Fire Safety Journal*, 44(8): 1015–1022.

Cha, Y.J., Choi, W. and Büyüköztürk, O. 2017. Deep learning-based crack damage detection using convolutional neural networks. *Computer-Aided Civil and Infrastructure Engineering*, 32(5): 361–378.

Cha, Y.J., You, K. and Choi, W. 2016. Vision-based detection of loosened bolts using the Hough transform and support vector machines. *Automation in Construction*, 71: 181–188.

Choudhary, G.K. and Dey, S. 2012. October. Crack detection in concrete surfaces using image processing, fuzzy logic, and neural networks. *In*: 2012 *IEEE Fifth International Conference on Advanced Computational Intelligence (ICACI)*. IEEE, 404–411.

Chung, C.H., Lee, J. and Choi, S.H. 2016. Temperature distribution within polypropylene fiber-mixed reinforced concrete slabs exposed to an ISO 834 standard fire. KSCE *Journal of Civil Engineering*, 20(5): 1878–1886.

Cülfik, M.S. and Özturan, T. 2002. Effect of elevated temperatures on the residual mechanical properties of high-performance mortar. *Cement and Concrete Research*, 32(5): 809–816.

Dai, B., Gu, C., Zhao, E., Zhu, K., Cao, W. and Qin, X. 2019. Improved online sequential extreme learning machine for identifying crack behavior in concrete dam. *Advances in Structural Engineering*, 22(2): 402–412.

Ding, Y., Azevedo, C., Aguiar, J.B. and Jalali, S. 2012. Study on residual behaviour and flexural toughness of fibre cocktail reinforced self-compacting high-performance concrete after exposure to high temperature. *Construction and Building Materials*, 26(1): 21–31.

Ding, Y., Zhang, Y. and Thomas, A. 2009. The investigation on strength and flexural toughness of fibre cocktail reinforced self-compacting high-performance concrete. *Construction and Building Materials*, 23(1): 448–452.

Fan, Z., Li, C., Chen, Y., Wei, J., Loprencipe, G., Chen, X. and Di Mascio, P. 2020. Automatic crack detection on road pavements using encoder-decoder architecture. *Materials*, 13(13): 2960.

Feng, C., Zhang, H., Wang, H., Wang, S. and Li, Y. 2020. Automatic pixel-level crack detection on dam surface using deep convolutional network. *Sensors*, 20(7): 2069.

Fujita, Y. and Hamamoto, Y. 2011. A robust automatic crack detection method from noisy concrete surfaces. *Machine Vision and Applications*, 22(2): 245–254.

Hoang, N.D. and Nguyen, Q.L. 2018. Metaheuristic optimized edge detection for recognition of concrete wall cracks: A comparative study on the performances of roberts, prewitt, canny, and sobel algorithms. *Advances in Civil Engineering*, 2018.

Hoff, G.C., Bilodeau, A. and Malhotra, V.M. 2000. Elevated temperature effects on HSC residual strength. *Concrete International*, 22(4): 41–48.

IS 2386-4 (1963) (Reaffirmed 2002). 2002. Methods of test for Aggregates for Concrete, Part 4: Mechanical properties [CED 2: Cement and Concrete]. *Bureau of Indian Standards*, Ministry of Consumer Affairs, Food and Public Distribution, New Delhi, India.

IS 383. 2016. Coarse and fine aggregate for concrete specification. *Bureau of Indian Standards*, Ministry of Consumer Affairs, Food and Public Distribution, New Delhi, India.

IS 4031-Part IV. 1988. Methods of physical tests for hydraulic cement. Part IV-Determination of consistency of standard cement paste. *Bureau of Indian Standards*, Ministry of Consumer Affairs, Food and Public Distribution New Delhi, India.

IS 4031-Part 6. 1988. Methods of physical tests for hydraulic cement. *Bureau of Indian Standards*, Ministry of Consumer Affairs, Food and Public Distribution, New Delhi, India.

IS 456:2000. 2000. Indian Standard Plain and Reinforced Concrete: Code of Practice. *Bureau of Indian Standards*, Ministry of Consumer Affairs, Food and Public Distribution, New Delhi, India.

IS:12269-2013. 2013. Ordinary portland cement, 53 grade specification (First Revision). *Bureau of Indian Standards*, Ministry of Consumer Affairs, Food and Public Distribution, New Delhi, India.

ISO 834-1. 1999. Fire-resistance tests: Elements of building construction—Part 1: General requirements. ISO Standard.

Iyer, S. and Sinha, S.K. 2005. A robust approach for automatic detection and segmentation of cracks in underground pipeline images. *Image and Vision Computing*, 23(10): 921–933.

Khaliq, W. and Waheed, F. 2017. Mechanical response and spalling sensitivity of air entrained high-strength concrete at elevated temperatures. *Construction and Building Materials*, 150: 747–757.

Kodur, V. 2014. Properties of concrete at elevated temperatures. *International Scholarly Research Notices*, 2014.

Li, G., Zhao, X., Du, K., Ru, F. and Zhang, Y. 2017. Recognition and evaluation of bridge cracks with modified active contour model and greedy search-based support vector machine. *Automation in Construction*, 78: 51–61.

Li, L., Wang, Q., Zhang, G., Shi, L., Dong, J. and Jia, P. 2018b. A method of detecting the cracks of concrete undergo high temperature. *Construction and Building Materials*, 162: 345–358.

Li, M., Qian, C. and Sun, W. 2004. Mechanical properties of high-strength concrete after fire. *Cement and Concrete Research*, 34(6): 1001–1005.

Li, X., Jiang, H. and Yin, G. 2014. Detection of surface crack defects on ferrite magnetic tile. *NDT & E International*, 62: 6–13.

Lin, Y.Z., Nie, Z.H. and Ma, H.W. 2017. Structural damage detection with automatic feature-extraction through deep learning. *Computer-aided Civil and Infrastructure Engineering*, 32(12): 1025–1046.

Luo, X., Wu, H., Yuan, H. and Zhou, M. 2019. Temporal pattern aware QoS prediction via biased non-negative latent factorization of tensors. *IEEE Transactions on Cybernetics*, 50(5): 1798–1809.

Makantasis, K., Protopapadakis, E., Doulamis, A., Doulamis, N. and Loupos, C. 2015. September. Deep convolutional neural networks for efficient vision-based tunnel inspection. *In: 2015 IEEE International Conference on Intelligent Computer Communication and Processing (ICCP)*. IEEE, 335–342.

Malar, B.A., Andrushia, A.D. and Neebha, T.M. 2021. May. Deep Learning-based disease detection in tomatoes. *In: 2021 3rd International Conference on Signal Processing and Communication (ICPSC)*. IEEE, 388–392.

Nguyen, H.N., Kam, T.Y. and Cheng, P.Y. 2014. An automatic approach for accurate edge detection of concrete crack utilizing 2D geometric features of crack. *Journal of Signal Processing Systems*, 77(3): 221–240.

Özbay, E., Türker, H.T., Balçıkanlı, M. and Lachemi, M. 2015. Effect of fiber types and elevated temperatures on the bond characteristic of fiber reinforced concretes. *International Journal of Civil and Environmental Engineering*, 9(5): 549–553.

Patel, S.S., Chourasia, A.P., Panigrahi, S.K., Parashar, J., Parvez, N. and Kumar, M. 2016. Damage identification of RC structures using wavelet transformation. *Procedia Engineering*, 144: 336–342.

Peng, G.F., Yang, W.W., Zhao, J., Liu, Y.F., Bian, S.H. and Zhao, L.H. 2006. Explosive spalling and residual mechanical properties of fiber-toughened high-performance concrete subjected to high temperatures. *Cement and Concrete Research*, 36(4): 723–727.

Poon, C.S., Shui, Z.H. and Lam, L. 2004. Compressive behavior of fiber reinforced high-performance concrete subjected to elevated temperatures. *Cement and Concrete Research*, 34(12): 2215–2222.

Salman, M., Mathavan, S., Kamal, K. and Rahman, M. 2013. October. Pavement crack detection using the Gabor filter. *In: 16th International IEEE Conference on Intelligent Transportation Sytems (ITSC 2013)*. IEEE, 2039–2044.

Short, N.R., Purkiss, J.A. and Guise, S.E. 2001. Assessment of fire damaged concrete using colour image analysis. *Construction and Building Materials*, 15(1): 9–15.

Sideris, K.K. and Manita, P. 2013. Residual mechanical characteristics and spalling resistance of fiber reinforced self-compacting concretes exposed to elevated temperatures. *Construction and Building Materials*, 41: 296–302.

Wolf, J., Pirskawetz, S. and Zang, A. 2015. Detection of crack propagation in concrete with embedded ultrasonic sensors. *Engineering Fracture Mechanics*, 146: 161–171.

Xiaoyong, L. and Fanjie, B. 2011, September. Residual strength for concrete after exposure to high temperatures. *In: International Conference on Information and Management Engineering*, Berlin, Heidelberg: Springer, pp. 382–390.

Yiyang, Z. 2014, December. The design of glass crack detection system based on image pre-processing technology. *In: 2014 IEEE 7th Joint International Information Technology and Artificial Intelligence Conference*. IEEE, pp. 39–42.

Youssef, M.A. and Moftah, M. 2007. General stress–strain relationship for concrete at elevated temperatures. *Engineering Structures*, 29(10): 2618–2634.

Zalama, E., Gómez-García-Bermejo, J., Medina, R. and Llamas, J. 2014. Road crack detection using visual features extracted by Gabor filters. *Computer-Aided Civil and Infrastructure Engineering*, 29(5): 342–358.

Zhang, A., Wang, K.C., Li, B., Yang, E., Dai, X., Peng, Y., Fei, Y., Liu, Y., Li, J.Q. and Chen, C. 2017. Automated pixel-level pavement crack detection on 3D asphalt surfaces using a deep-learning network. *Computer-aided Civil and Infrastructure Engineering*, 32(10): 805–819.

Zhang, A., Wang, K.C., Fei, Y., Liu, Y., Tao, S., Chen, C., Li, J.Q. and Li, B. 2018. Deep learning-based fully automated pavement crack detection on 3D asphalt surfaces with an improved CrackNet. *Journal of Computing in Civil Engineering*, 32(5): 04018041.

Chapter 7

State-of-the-Art Research in the Area of Artificial Intelligence with Specific Consideration to Civil Infrastructure, Construction Engineering and Management, and Safety

Islam H. El-adaway[1,*] and *Rayan H. Assaad*[2]

1. Introduction

With the prevalence of large amounts of data in different fields, digitization has emerged as a process associated with transforming information into a digital format that is often readable by computers. Digitization has led to a growing interest in data science and data analytics. Data science is a multidisciplinary field of study that relies on tools, algorithms, and knowledge of statistics and mathematics to find insights from the raw data (Volkova et al., 2019). On the other hand, data analytics is one of the subfields of data science and is defined as "the process of inspecting, cleaning, transforming, and modeling Big Data to discover and communicate useful information and patterns, suggest conclusions, and support decision making" (Cao et al., 2015).

Artificial intelligence (AI) has offered unprecedented opportunities to retrieve and reveal remarkable patterns, trends, relationships, and knowledge from big data

[1] Hurst-McCarthy Professor of Construction Engineering and Management, Professor of Civil Engineering, and Founding Director of the Missouri Consortium of Construction Innovation, Department of Civil, Architectural, and Environmental Engineering/Department of Engineering Management and Systems Engineering, Missouri University of Science and Technology, Butler-Carlton Hall, 1401 N. Pine St., Rolla, MO 65409.

[2] Assistant Professor of Construction and Civil Infrastructure, John A. Reif, Jr. Department of Civil and Environmental Engineering, New Jersey Institute of Technology, Newark, NJ 07102.
Email: rayan.hassane.assaad@njit.edu

* Corresponding author: eladaway@mst.edu

sets. Artificial intelligence is defined as "the study and implementation of techniques that allow actions requiring intelligence on the part of a human, to be performed on computational devices" (Plant, 2011). For many decades, AI has been applied to address challenges and provide solutions in different research areas and application domains. In relation to that, the goal of this chapter is to offer an introduction to the fundamental and essential methods, principles, and applications of AI. This chapter also provides definitions related to AI and presents the associated most widely used algorithms. Finally, this chapter focuses on the state-of-the-art research AI with specific consideration to the following three application domains: civil infrastructure, construction engineering and management, and safety.

2. Background

Artificial intelligence includes multiple fields; one of which is machine learning (ML), which is a "subfield of artificial intelligence that studies how computers can learn. It relies heavily on techniques and theory from statistics, optimization, [and] algorithms" (Kelchtermans et al., 2014). In fact, AI and ML techniques are one of the key catalyzers behind the revolution of computational science. In addition, a subset of ML is deep learning (DL) which is defined as "a type of representation learning method in which a complex multilayer neural network architecture learns representations of data automatically by transforming the input information into multiple levels of abstractions" (Chan et al., 2020). Figure 7.1 provides an overview and summary of the differences between AI, ML, and DL and the associated date or period that each one of these techniques has gained considerable attention and use.

Machine learning techniques or algorithms could be divided into three main categories: supervised learning, unsupervised learning, and reinforcement learning. In simple words, supervised learning is "a machine learning mode of studying a task that maps an input data into an output data based on a sample input-output pair" (Dike et al., 2018). On the other hand, unsupervised learning is defined as "the

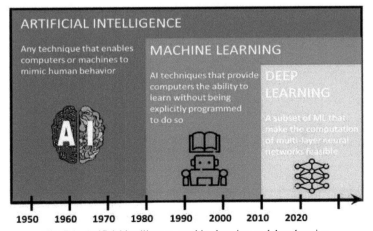

Fig. 7.1: Artificial intelligence, machine learning, and deep learning.

training of an artificial intelligence (AI) algorithm using information that is neither classified nor labeled and allowing the algorithm to act on that information without guidance" (El-Mashharawi et al., 2020). As for reinforcement learning, it could be defined as "a branch of machine learning concerned with using experience gained through interacting with the world and evaluative feedback to improve a system's ability to make behavioral decisions" (Littman, 2015). It is worth mentioning that different variants of the learning techniques have evolved, such as semi-supervised learning, among others. In relation to that, semi-supervised learning combines both supervised and unsupervised learning, and it is defined as "a learning paradigm concerned with the study of how computers and natural systems such as humans learn in the presence of both labeled and unlabeled data" (Zhu and Goldberg, 2009).

Furthermore, supervised learning could be used either for classification applications (i.e., where the output variable(s) is/are categorical or classes) or for regression purposes (i.e., where the output variable(s) is/are numerical). Since it is impossible to list all algorithms or techniques used in supervised learning as well as in unsupervised learning, this chapter lists the commonly used algorithms. In relation to that, according to ProjectPro (2021), the following are common traditional ML algorithms: artificial neural networks, naïve Bayes classifier, k-means clustering, support vector machine, Apriori algorithm, linear regression, logistic regression, decision tree, random forest, and nearest neighbors. Nevertheless, it is worth mentioning that this is not intended to be a comprehensive list of all AI algorithms since this chapter does not aim to provide a conclusive list of all possible AI techniques and algorithms or to explain how each algorithm works. Rather, there are many exceptional books that the readers could refer to acquire a more comprehensive list of AI algorithms and to learn the science, mathematics, and statistics behind each technique.

This chapter reviews the latest research in AI with specific consideration to three domains: civil infrastructure, construction engineering and management, and safety. Under each one of these domains, a review of three main applications is provided as shown in Fig. 7.2.

It is worth mentioning that this chapter focuses on providing a review of the current and most up-to-date (i.e., 2010–2021) research and knowledge on the use of AI techniques and algorithms in the three previously mentioned domains.

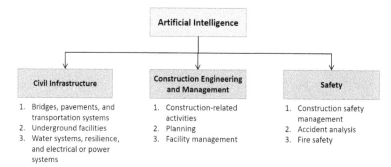

Fig. 7.2: Chapter focus.

3. State-of-the-Art Research in the Area of Artificial Intelligence

This section provides the advanced research in AI with specific consideration to the following three application domains: civil infrastructure systems, construction engineering and management, and safety.

3.1 Applications in Civil Infrastructure Systems

AI has been used in multiple civil infrastructure applications. These mainly include (1) bridges, pavements, and transportation systems; (2) underground facilities; and (3) water systems, resilience, and electrical or power systems.

3.1.1 Bridges, Pavements, and Transportation Systems

This subsection provides a summary of the state-of-the-art research in different infrastructure applications related to bridges, pavements, and transportation systems. In relation to that, a review of the current and most recent knowledge and use of the different AI techniques is presented.

Bridges, pavements, and transportation are the most important infrastructure systems since they establish the essential needs and services to maintain a modern society and to provide advantage within the global economy (Assaad et al., 2020a). Multiple applications of different AI techniques were conducted to study different characteristics of bridges, pavements, and transportation.

Starting with bridges, DL (deep feedforward neural networks and the TensorFlow framework) was used to assess the flood vulnerability of bridges by proposing a simulation-based probabilistic method to establish time-variant fragility surfaces of bridges under flood hazards (Khandel and Soliman, 2021). In addition, computer vision was employed for bridge risk assessment by obtaining topographical information that was used to simulation complex flow situations (Hackl et al., 2018). Also, bridge deterioration conditions were assessed using ML techniques (e.g., artificial neural networks and k-nearest neighbors) by developing a bridge infrastructure asset management system to evaluate and predict the deck conditions (Assaad and El-adaway, 2020a). Another bridge application is by using tensor factorization to assess and predict the network-level performance of bridges (Adarkwa et al., 2017). Moreover, genetic algorithms were used to propose a method/ model for cost-leveling models of bridge decks by generating multiple maintenance, repair, and rehabilitation solutions (Yoon et al., 2017). Furthermore, Elman neural networks were employed to predict the long-term deterioration of bridge elements and to model their performance (Lee et al., 2014). As far as failure of bridge main components is concerned, multiple ML techniques (mainly quadratic discriminant analysis, K-nearest neighbors, decision trees, random forests, naïve Bayes, and artificial neural networks) were used to predict the failure mode of circular concrete bridge columns (Mangalathu and Jeon, 2019). As for the safety of bridges, support vector machines were used for bridge scour depth and safety by proposing models for computing the natural frequency of bridges and implementing a series of simulation to examine the factors affecting scouring depth (Feng et al., 2016). Similarly, a new method was presented for safety screening and load capacity evaluation of large bridge populations using decision tees and random forest (Alipour et al., 2017). Additional bridge applications include the use of support vector regression to develop a stochastic fatigue truck load model to study the fatigue reliability of welded steel girder/deck bridges (Lu et al., 2017). Furthermore, research was conducted to provide a method for modeling and assessing the stochastic response of cross-sea bridges under correlated wing and waves using support vector regression and Latin hypercube sampling (Fang et al., 2020).

On the other hand, many AI techniques were used in different applications related to pavements and transportation systems, such as roads, highways, etc. In relation to that, support vector classifier and recurrent neural networks were used, and a model was proposed that predicts and classified the performance of pavement infrastructure systems (Tabatabaee et al., 2013). In addition, pavement deterioration conditions were assessed using decision tress and gradient-boosted trees by predicting the deterioration in the pavement condition index (PCI) over multiple year ranges (Piryonesi and El-Diraby, 2020). Moreover, ML (TensorFlow), transfer learning, and deep neural networks (Inception-v3) were employed to detect, evaluate, and classify the amounts of dust on gravel roads (Albatayneh et al., 2020). Also, the road condition or roughness was measured and predicted using different ML techniques; mainly, multi-variate linear regression, decision regression trees, artificial neural networks, and random forests (Wang et al., 2020). Additionally, unsupervised learning methods (such as self-organizing map and Kohonen map) were used to detect road cracks from pavement images with a focus on highly textured images (Mathavan et al., 2015). As far as flexible pavements are concerns, their performance was characterized for different interstates using fixed-parameters regression, random-parameters regression, and artificial neural networks (Yamany et al., 2020). Similarly, a method was proposed for flexible pavement deterioration in relation to climate change by predicting the alteration of infrastructure service life using fuzzy logic (Jeong et al., 2017). Different AI-based optimization techniques (mainly, genetic algorithms, simulated annealing, and Tabu search optimization) have proved their effectiveness in prioritizing interrelated road projects while minimizing the total system cost over a planning horizon (Shayanfar et al., 2016). Genetic algorithms were also used to optimize the protection planning of transportation-infrastructure systems (such as highway networks) that are subject to disasters given a limited budget (Chu et al., 2016). Another optimization application includes the development of a model to select the optimal sections of unpaved roads for dust suppressing chemical treatments within a limited budged using evolutionary algorithms (Saha et al., 2020). Research was also conducted to detect potholes and roadway pavement anomalies using artificial neural networks (Kyriakou et al., 2019). Artificial neural networks were also used for the condition assessment of highway culverts (Tatari et al., 2013) as well as for offering a cost-estimating system for rubberized asphalt road rehabilitation projects (Shehab and Meisami-Fard, 2013). Also, Bayesian networks were employed to propose an approach for more accurate stochastic degradation modeling for road pavements and for proposing inspection, maintenance, and rehabilitation strategies (Faddoul et al., 2013). As for the evaluation of pavement roughness based on the vehicles' characteristics, research works have been conducted to estimate the weighted longitudinal profile (WLP) of pavements based on vehicle response information using multilayer perceptron, support vector machines, and random forests (Nitsche et al., 2014) as well as to develop a method of processing connected vehicle responses capable of evaluating the road roughness condition using feature extraction and selection and artificial neural networks (Zhang et al., 2018).

3.1.2 Underground Facilities

This subsection provides a summary of the latest research in different infrastructure applications related to underground facilities. In relation to that, a review of the current and most up to date knowledge and use of the different AI techniques is presented.

Computer vision and DL, namely cascade region convolutional neural network (R-CNN), were used to propose a tunnel rapid detection equipment based on area-scan charge-coupled device (CCD) cameras (Huang et al., 2021). Another application is related to municipal asset condition where fuzzy logic was used for examining rehabilitation options to quantify the condition improvement required in municipal assets (Khan et al., 2015). As far as pipelines (e.g., sewer) or water pipes are concerned, Bayesian geoadditive regression was used to examine the criticality of components in large wastewater systems (Balekelayi and Tesfamariam, 2021). In addition, a method for automated anomaly detection and localization in the internal condition of sewers, using closed circuit television (CCTV) inspection videos, was proposed using scale invariant feature transform (SIFT), support vector machines, and end-to-end text detection and recognition (using maximally stable extremal regions (MSER) algorithm and a deep convolutional neural network) (Moradi et al., 2020). Moreover, computer vision and image processing were used for automated crack detection in sewer inspection based on CCTV images (Halfawy and Hengmeechai, 2014). Also, the prediction of sewer pipe deterioration scores was conducted using Bayesian geoadditive regression model from a set of predictors categorized as physical, maintenance, and environmental data (Balekelayi and Tesfamariam, 2019). Negative binomial regression with a Bayesian tuner was used to present a statistical method to model sewer blockage occurrences or the failure probability of sewer infrastructure (Erskine et al., 2014). Furthermore, genetic algorithms were employed to propose a method to examine the critical pipes for proactive seismic rehabilitation to enhance a network's post-earthquake serviceability (Pudasaini and Shahandashti, 2018).

3.1.3 Water Systems, Resilience, and Electrical/Power Systems

This sub-section provides a summary of the advanced research in different infrastructure applications related to water systems, resilience, and electrical/power systems. In relation to that, a review of the current and most modern knowledge and use of the different AI techniques is presented.

Starting with water systems, fuzzy analytical network process and minimum cut-set analysis were used for reliability assessment of water distribution networks by introducing a framework that assesses the mechanical and hydraulic reliabilities of water networks (Elshaboury et al., 2021). Another application of the AI techniques for water systems includes leak detection and geolocation where supervised learning and classification (namely, artificial neural networks and support vector machines) were used for early detection of high likelihood leaks, their geolocation, and the detection accuracy assessment in a water distribution system (Cantos et al., 2020). Moreover, research was conducted to predict the captured runoff volume from watershed area by permeable pavements using artificial neural networks (Radfar and Rockaway, 2016). Continuing with water systems, random forests were utilized in the topological representation of geospatial colocation to characterize the connectivity of water distribution–transportation interface networks and identify the nodal attributes that are most predictive of a given connectivity profile (Abdel-Mottaleb and Zhang, 2020). Other examples for the use of AI techniques in water-related systems include dams. In relation to that, while considering spillway gate maintenance activity, estimating dam overtopping probability was possible using Bayesian networks (Wang and Zhang, 2017). In addition, artificial neural networks and K-nearest neighbors were used for dam infrastructures hazard assessment by evaluating and predicting the hazard potential level of dams (Assaad and El-adaway, 2020b). Also, Gaussian process regression has been employed in dam safety monitoring by presenting a displacement model for health monitoring of concrete gravity dams, which can model the temperature effect by using long-term air temperature data (Kang and Li, 2020).

On the other contrary, different AI techniques and algorithms were used for multiple applications related to resilience and electrical/power systems. In regard to that, DL, machine vision, and deep CNNs were used to examine how to enhance resiliency within the electricity section by detecting threats to critical infrastructure before failures occur (Dick et al., 2019). Additionally, Bayesian networks were employed to develop an approach for modeling and assessing the reliability of a power distribution network (Tien and Der Kiureghian, 2017). Moreover, linear regression, support vector regression, and artificial neural networks have been applied for purposes related to costing power transmission projects by presenting data-driven cost estimation models that leverage a big data architecture to manage the required large and diverse data for power transmission projects (Davila Delgado et al., 2020). As for optimization-related applications, hybrid genetic algorithms have showed great value in post-disaster resilience of highway–bridge networks by being used to determine the optimal inspection routes and restoration schedules for damaged highway–bridge networks based on maximizing the networks' travel time (Zhang and Wei, 2021). Additional, research has been conducted for evacuation zone modeling by developing a grid cell-based data-driven method to predict future evacuation zones under climate change to help in the evaluation of the resilience of transportation systems in the presence of natural disasters (Xie et al., 2017). Finally, artificial neural networks—namely feedforward, Elman, and nonlinear autoregressive exogenous (NARX)—were applied to find and predict the performance and efficiency of flat-plate collectors (Handan et al., 2016).

3.2 Applications in Construction Engineering and Management

AI has been used in multiple construction engineering and management applications. These mainly include: (1) construction-related activities; (2) planning; and (3) facility management.

3.2.1 Construction-related Activities' Applications

This section aims to provide a summary of the state-of-the-art research in applications associated with construction-related activities. In relation to that, a review of the current and most recent knowledge and use of the different AI techniques is presented.

The automatic recognition of labor actions was possible due to the exceptional capabilities of ML techniques (namely k-nearest neighbors, multilayer perceptron, decision trees, and support vector machines) (Ryu et al., 2019). K-nearest neighbors were also used with case-based reasoning for the planning of deep foundation construction (Zhang et al., 2017). Studying the impact of dynamic workforce and workplace variables on the productivity of the construction industry was performed by relying on time series econometric analysis based on vector autoregression models (Assaad and El-adaway, 2021b), Productivity Indicator. In addition, association rule mining and Bayesian networks were employed to identify the relationships between defects and their occurrence probabilities (Fan, 2020). Furthermore, a method for detecting concrete structural components in color images was proposed based on image processing, Gaussian mixture model, artificial neural networks, and support vector machines (Son et al., 2012). Decision trees and naïve Bayesian classification were used for facilitating accurate project delay risk analysis and prediction (Gondia et al., 2020) to help in addressing schedule overruns that are highly present in the construction industry (Assaad et al., 2020c). Moreover, the integration of field submittals in project scheduling has been performed using random forests by mining project data to forecast delay during the project and to forecast the likelihood of acceptance of submittal requests (Awada et al., 2021). Additionally, clustering (expectation maximization) was utilized to propose an approach that can gather valuable knowledge from previously unanalyzed data to significantly improve resource and labor management practices (Hammad et al., 2014). Reinforcement learning was used to examine the impact of learning in bidding decision-making processes to help in having optimal outcomes in the long run (Assaad et al., 2021). Another application of AI-techniques in construction-related activities is the estimation of schedule to completion through neural network–long short-term memory (feedforward neural networks for nonsequential issue and recurrent neural networks for sequential issue) (Cheng et al., 2019). Long short-term memory architectures were also used with natural language processing to develop an automated specification reviewing model and an information extraction framework that is used for construction specification review (Moon et al., 2021). Chow–Liu tree and k-means clustering were used to develop a framework that measures the amount of uncertainty and information sharing among constraint discussions in construction planning meetings (Javanmardi et al., 2020). In addition, a method was proposed to offer a cascaded estimation approach for accurate forecasts of construction costs at the conceptual phases of building projects using linear regression and artificial neural networks (Dursun and Stoy, 2016).

In other ways, many AI-based optimization techniques and algorithms were used in different applications related to construction activities. In relation to that, genetic algorithms, along with k-nearest neighbors, showed the ability to estimate management reserve based on the cost and schedule performance ratios of international construction projects (Lee et al., 2017). Also, particle swarm optimization and genetic algorithms, combined with fuzzy set theory, were used for time-cost-resource optimization in construction project planning (Ashuri and Tavakolan, 2012). Moreover, research was conducted—based on genetic algorithms—to propose a simulation-based model to find an optimal lift plan for irregularly shaped high-rise building constructions and to minimize the total cost while satisfying the constraints of the "total lifting time" and "average journey time" (Kim et al., 2020). Genetic algorithms were also employed for evaluating the proactive material placement policy and presenting an integrated framework to determine the optimum layout for placing materials resulting in minimum material haulage time (Alanjari et al., 2015). Finally, constraint programming has showed its applicability in optimizing schedules both at the early planning stage and immediately before construction for projects with continuously evolving requirements and constraints (Abuwarda and Hegazy, 2016).

3.2.2 Planning Applications

This section aims to provide a summary of the state-of-the-art research in different planning applications. It is worth mentioning that this section does not focus on planning in construction management related aspects (which was touched upon in the previous section) but rather on planning applications related to general management considerations in the civil engineering domain. In reference to that, a review of the current and most up-to-date knowledge and use of the different AI techniques is presented.

Support vector regression was employed in strategic planning for drought mitigation by presenting a decision-support framework to assess the roles of strategic and tactical measures in drought preparedness and mitigation (Cai et al., 2015). In addition, the inspection planning of jacket platforms was made easier through Bayesian networks by evaluating their condition under the effect of damage, deterioration, and the applied loads and by computing the failure probabilities of structural elements (Gholami et al., 2020). The Bayes-adaptive partially observable Markov decision processes (BA-POMDP) were used to help in the optimal planning and learning for the management of wind farms by developing an approach that can learn in uncertain environments to reduce the total management cost of wind farms (Memarzadeh et al., 2015). Also, the operation planning of a multioutlet water reservoirs was possible by using batch-mode reinforcement learning (i.e., fitted Q-iteration) to account for both quantity and quality targets for a better management of water resources systems (Castelletti et al., 2014).

On the other hand, many AI-based optimization techniques and algorithms were used in different planning applications. In relation to that, evolutionary algorithms (namely, genetic algorithms, particle swarm optimization, and shuffled frog leaping algorithm) were utilized for developing optimal rules for multicrop irrigation areas associated with reservoir operation policies in reservoir-irrigation systems (Fallah-Mehdipour et al., 2013). In addition, scenario-based optimization has showed its effectiveness in multiperiod planning and design of water supply infrastructure (Kang and Lansey, 2014). Finally, genetic algorithms were used: (1) to plan emergency response services in road safety and traffic accidents (Kepaptsoglou et al., 2012); (2) Wang et al. (2016) proposed a method for anti-disaster differentiated planning of power distribution systems, which also relied on minimum spanning tree method; (3) to plan the capacity expansion of ground water supply systems with land subsidence constraints by relying also on constrained differential dynamic programming (Chu and Chang, 2010); (4) to plan optimal agricultural groundwater development for irrigation and to ensure optimal zonal crop patterns subject to the constraints on the maximum water-table depth and the stream-aquifer interflow at the dynamic equilibrium (Ghosh and Kashyap, 2012); and (5) to plan the optimal investment policy for infrastructure assets by evaluating the trade-off between total expenditure and the prevention of future asset failures (serviceability) (Gholami et al., 2020).

3.2.3 Facility Management Applications

This subsection provides a summary of the contemporary research in different facility management applications. Related to that, a review of the current and most up-to-date knowledge and use of the different AI techniques is presented.

Natural language processing and text mining (namely, naïve Bayes, logistic regression, and support vector machine) were used for work plan prediction for facility management through automatically forecasting the reaction type (regular or urgent) and the necessary crews based on textual descriptions of service requests (Mo et al., 2017). Many challenges have been present for decades in the construction industry (Assaad et al., 2020d), one of which is related to deferred maintenance in facility management. In relation to that, mitigation strategies for deferred maintenance were devised based on fault tree analysis and Bayesian network analysis to provide total-package-prioritization strategies for the evaluation and diagnosis of the deferred maintenance requirements of buildings (Yoon et al., 2021). Moreover, Baek et al. (2019) employed DL to develop a system for facility management which estimates the user's indoor position and orientation by comparing the user's perspective to building information modeling (BIM) which is experiencing substantial development and use in the construction industry (Assaad et al., 2020b). Also, research was conducted to predict the service life of external paint finishes on rendered facades and to identify the factors that have a substantial influence on the degradation rate of the paint finishes using ordinary linear regression models (Sousa et al., 2020). Furthermore, fuzzy multiple-criteria decision-making was used for facility risk assessment and maintenance systems to help service managers in selecting the maintenance policy for a single piece of equipment and in determining the maintenance priorities of equipment components in the initial stage of operation and maintenance (Wang and Piao, 2019).

On the other hand, many AI-based optimization techniques and algorithms were used in different facility management applications. In relation to that, a modified Dijkstra algorithm and the A* search algorithm showed exceptional capabilities in scheduling facility maintenance work orders (Chen et al., 2018). Finally, genetic algorithms were used for: (1) planning maintenance of hospital facilities by developing an optimal reliability-centered sustentation approach that takes into account the maintenance tasks and the expected downtime caused by failures of hospital systems (i.e., medical gases, primary heating, ventilation and air conditioning (HVAC), secondary HVAC, and elevators) (Salah et al., 2018); (2) examining building maintenance, repair, and renovation (MR&R) and by identifying and selecting multiyear building MR&R activities such that facility performance is maximized and life cycle costs are minimized (Grussing and Liu, 2014); and (3) minimizing the maintenance, repair, and rehabilitation cost profile of buildings by adjusting the timing of some activities to earlier points than the initial maintenance, repair, and rehabilitation schedule (Kim et al., 2016a).

3.3 Applications in Safety

AI has been used in multiple safety applications. These mainly include (1) construction safety management; (2) accident analysis; and (3) fire safety.

3.3.1 Construction Safety Management

This subsection summarizes the state-of-the-art research in various construction safety management applications. In relation to that, a review of the current and most recent knowledge and use of the different AI techniques is presented.

Support vector machines were used for scaffolding safety to enable real-time monitoring of large and complex scaffolds at construction sites to prevent construction workers from tragic scaffolding collapses (Sakhakarmi et al., 2019). Support vector machines were also used with dimension reduction for modeling and classifying unsafe actions in construction (Han et al., 2014). In addition, the estimation of unknown relationships between monitoring values/data and safety risks of deep foundation pits was possible with the implementation of random forests (Zhou et al., 2019). Research efforts also proposed a model that makes use of existing safety-related measures available within an organization to reliably evaluate safety performance by relying on case-based reasoning (Pereira et al., 2018). Natural language processing was applied to develop a rule-based approach for extracting domain knowledge elements from text documents in the domain of construction safety management (Xu et al., 2021). Genetic algorithms have been employed to optimize camera placements for site coverage while considering the occlusion dynamics caused by sidewalls and supports during excavation in metro station projects (Zhang et al., 2019). Moreover, research was conducted to fuse and integrate data from several sources with analysis and simulation components in a cost-, labor-, and time-efficient manner based on artificial neural networks (Pereira et al., 2020).

Additionally, the equivalence class transformation (EclaT) algorithm and association rules were used to develop a prediction method to improve workplace hazard identification (Wang et al., 2018). Latent class analysis and multiple correspondence analysis were conducted to examine construction-fatality reports for effective risk management and better control over the whole sociotechnical system (Zhao et al., 2016b).

Many AI-based image processing techniques and algorithms were used in different construction safety management applications. In relation to that, computer vision and fuzzy inference were employed to present an on-site safety-assessment system for monitoring struck-by accidents with moving entities (Kim et al., 2016b). The prediction of upcoming worker and equipment motion trajectories on construction sites was possible by relying on computer vision, long short-term memory networks, and mixture density networks (Tang et al., 2020). Machine vision and fuzzy neural networks have showed their ability in developing a real-time intelligence evaluation system to prevent collisions between workers and machines in excavation site construction (Hu et al., 2020). Image parameterization has provided great value in developing a skeleton-based real-time identification method by combining image-based technologies, construction safety knowledge, and ergonomic theory to recognize unsafe behaviors (Guo et al., 2018). A real-time warning system was developed using visual data acquired from cameras that are readily available in the heavy equipment to protect the workers from potentially dangerous situations involving equipment operations through the reliance on DL (computer vision and faster R-CNN) (Son et al., 2019).

3.3.2 Accident Analysis

This subsection provides a summary of the state-of-the-art research in different accident analysis applications. Regarding that, a review of the current and most advanced knowledge and use of the different AI techniques is presented.

Support vector machines were used to determine the safety risks that can materialize during the construction of deep pit foundations (Zhou et al., 2017). In addition, the exploration of the cognitive factors influencing unsafe behavior when working at a height was performed using multiple ML techniques, namely, multiple linear regression, artificial neural networks, and decision trees (Goh and Binte Sa'adon, 2015). Also, the drivers' unsafe actions at signal-controlled intersections were studied by extracting information from crash reports' narratives to better understand accident scenarios using natural language processing, boosted classification trees, and support vector machines (Kwayu et al., 2020). Similarly, classification trees and natural language processing (e.g., text mining) was used to interpret the mechanisms of an electrical/electrocution accident as a chain of decision mistakes throughout the entire task process (Zhao et al., 2016a). Natural language processing was also employed to propose a knowledge management system for construction accident cases to identify dangerous conditions and prevent accidents by controlling risks on-site (Kim and Chi, 2019). Furthermore, research was conducted to study highway–rail grade crossing accidents and collision risks using artificial neural networks (Zheng et al., 2019).

However, many AI-based clustering techniques and algorithms were used in different accident analysis applications. In relation to that, spectral clustering, and data mining techniques (e.g., Apriori algorithm and association rules) were used to determining the critical combinations of safety fatality causes on construction sites (Assaad and El-adaway, 2021a). Research was conducted to propose a model that predicts the outcomes of construction incidents to determine necessary preventative actions using artificial neural networks and latent class clustering analysis (LCCA) (Ayhan and Tokdemir, 2020). Similarly, latent class analysis was applied to classify the root causes of fatal fall-from-height accidents and to analyze their characteristics (Wong et al., 2016). Principal component analysis (PCA) and cluster analysis were used to perform an analysis of the fatal accidents in terms of the time of their occurrence and how they occur in construction trades (Chiang et al., 2018). Similarly, factor analysis has provided capabilities in identifying rework causes leading to increased safety risks and in determining the major underlying causal dimensions of rework and safety incidents (Yap et al., 2020). In fact, the construction industry is a complicated business that incorporates various uncertainties, risks, and ever-changing conditions (Assaad and El-adaway, 2020c). In addition, the investigation of shield machine operation errors and accidents based on the Technique for the Retrospective and Predictive Analysis of Cognitive Errors was performed using data mining and cluster analysis (k-means algorithm and association rules) (Li et al., 2020). Association rules were also used to study the characteristics of expressway traffic accidents and to analyze the influencing factors and causes of injuries and fatalities leading to such traffic crashes (Chen et al., 2020). Finally, the block clustering method was applied to investigate the heterogeneity in large truck crash datasets and to provide additional insights to develop potential countermeasures and strategies (Rahimi et al., 2019).

3.3.3 Fire Safety

This subsection provides a summary of the state-of-the-art research in different fire safety applications. In relation to that, a review of the current and most contemporary knowledge and use of the different AI techniques is presented.

Multiple ML techniques (namely Naive Bayes, generalized linear models, logistic regression, fast large margin, DL, decision trees, random forests, gradient boosted trees, and support vector machines) were used to examine the key features that govern the tendency of fire-induced spalling in reinforced concrete columns and to develop tools for instantaneous prediction of spalling (Naser, 2021). In addition, the fire safety management at building construction sites was improved by establishing a fire-recognition model and by developing a real-time construction fire detection using CNNs (Su et al., 2021). Moreover, regression analysis showed its value in exploring the effect of fire visibility, fire alarm voices, and individual characteristics on individuals' fire evacuation pre-movement time in underground commercial buildings (Wang et al., 2021). Furthermore, human behavior in fire pre-evacuation was studied to investigate the factors affecting building occupants' decision-making during a fire event using random forests (Zhao et al., 2020). Similarly, people's pre-evacuation behavior under fire was examined using support vector machines (Liu and Lo, 2011).

On the other hand, many AI-based optimization techniques and algorithms were used in different fire safety applications. In relation to that, genetic programming, along with computer vision, were used to introduce fire-based evaluation tools to comprehend structural behavior under fire conditions (Naser, 2020). In addition, particle swarm optimization and differential evolution reflect their ability in providing an integrated scheme for solving the stringent budget allocation problem for protecting historic buildings from fires (Naziris et al., 2016). Finally, artificial neural networks optimized by genetic algorithms were to predict the residual flexural capacity of reinforced concrete beams after fires (Cai et al., 2020).

4. Conclusion

The reliance on AI techniques have substantially increased in recent years, especially with advancements in technology and computational capabilities. This chapter provided the state-of-the-art research in AI with specific consideration to the following three application domains: civil infrastructure systems, construction engineering and management, and safety. In relation to that, scholars and research efforts have played an exceptional role in leveraging AI techniques in the domains, especially through proposing more sophisticated and advanced techniques and algorithms. Current and emerging trends and applications of AI techniques and how different tools and algorithms are being employed to help in different decisions and to address multiple problems and emerging challenges have been provided. This chapter has offered comprehensive and most contemporary knowledge of the current uses of AI techniques and shall be of great value to students, researchers, and practitioners since it has consolidated the existing and the latest information, and literature and has categorized them into multiple application fields, which offers a clearer understanding of the various AI techniques within the different domains.

References

Abdel-Mottaleb, N. and Zhang, Q. 2020. Water distribution–transportation interface connectivity responding to urban geospatial morphology. *Journal of Infrastructure Systems*, 26(3): 04020025.

Abuwarda, Z. and Hegazy, T. 2016. Work-package planning and schedule optimization for projects with evolving constraints. *Journal of Computing in Civil Engineering*, 30(6): 04016022.

Adarkwa, O., Attoh-Okine, N. and Schumacher, T. 2017. Using tensor factorization to predict network-level performance of bridges. *Journal of Infrastructure Systems*, 23(3): 04016044.

Alanjari, P., RazaviAlavi, S. and AbouRizk, S. 2015. Hybrid genetic algorithm-simulation optimization method for proactively planning layout of material yard laydown. *Journal of Construction Engineering and Management*, 141(10): 06015001.

Albatayneh, O., Forslöf, L. and Ksaibati, K. 2020. Image retraining using TensorFlow implementation of the pretrained inception-v3 model for evaluating gravel road dust. *Journal of Infrastructure Systems*, 26(2): 04020014.

Alipour, M., Harris, D.K., Barnes, L.E., Ozbulut, O.E. and Carroll, J. 2017. Load-capacity rating of bridge populations through machine learning: Application of decision trees and random forests. *Journal of Bridge Engineering*, 22(10): 04017076.

Ashuri, B. and Tavakolan, M. 2012. Fuzzy enabled hybrid genetic algorithm–particle swarm optimization approach to solve TCRO problems in construction project planning. *Journal of Construction Engineering and Management*, 138(9): 1065–1074.

Assaad, R. and El-adaway, I.H. 2020a. Bridge infrastructure asset management system: Comparative computational machine learning approach for evaluating and predicting deck deterioration conditions. *Journal of Infrastructure Systems*, 26(3): 04020032.

Assaad, R. and El-adaway, I.H. 2020b. Evaluation and prediction of the hazard potential level of dam infrastructures using computational artificial intelligence algorithms. *Journal of Management in Engineering*, 36(5): 04020051.

Assaad, R. and El-adaway, I.H. 2020c. Enhancing the knowledge of construction business failure: A social network analysis approach. *Journal of Construction Engineering and Management*, 146(6): 04020052.

Assaad, R., Dagli, C. and El-adaway, I.H. 2020a. A system-of-systems model to simulate the complex emergent behavior of vehicle traffic on an urban transportation infrastructure network. *Procedia Comput. Sci.* 168(Jan): 139–146.

Assaad, R., El-adaway, I.H., El Hakea, A.H., Parker, M.J., Henderson, T.I., Salvo, C.R. and Ahmed, M.O. 2020b. Contractual perspective for BIM utilization in US construction projects. *Journal of Construction Engineering and Management*, 146(12): 04020128.

Assaad, R., El-adaway, I.H. and Abotaleb, I.S. 2020c. Predicting project performance in the construction industry. *Journal of Construction Engineering and Management*, 146(5): 04020030.

Assaad, R., El-adaway, I.H., Hastak, M. and Needy, K.L. 2020a. Commercial and legal considerations of offsite construction projects and their hybrid transactions. *Journal of Construction Engineering and Management*, 146(12): 05020019.

Assaad, R., Ahmed, M.O., El-adaway, I.H., Elsayegh, A. and Siddhardh Nadendla, V.S. 2021. Comparing the impact of learning in bidding decision-making processes using algorithmic game theory. *Journal of Management in Engineering*, 37(1): 04020099.

Assaad, R. and El-adaway, I.H. 2021a. Determining critical combinations of safety fatality causes using spectral clustering and computational data mining algorithms. *Journal of Construction Engineering and Management*, 147(5): 04021035.

Assaad, R. and El-adaway, I.H. 2021b. Impact of dynamic workforce and workplace variables on the productivity of the construction industry: New gross construction productivity indicator. *Journal of Management in Engineering*, 37(1): 04020092.

Awada, M., Srour, F.J. and Srour, I.M. 2021. Data-driven machine learning approach to integrate field submittals in project scheduling. *Journal of Management in Engineering*, 37(1): 04020104.

Ayhan, B.U. and Tokdemir, O.B. 2020. Accident analysis for construction safety using latent class clustering and artificial neural networks. *Journal of Construction Engineering and Management*, 146(3): 04019114.

Badmos, O., Kopp, A., Bernthaler, T. and Schneider, G. 2020. Image-based defect detection in lithium-ion battery electrode using convolutional neural networks. *Journal of Intelligent Manufacturing*, 31(4): 885–897.

Baek, F., Ha, I. and Kim, H. 2019. Augmented reality system for facility management using image-based indoor localization. *Automation in Construction*, 99: 18–26.

Balekelayi, N. and Tesfamariam, S. 2019. Statistical inference of sewer pipe deterioration using Bayesian geoadditive regression model. *Journal of Infrastructure Systems*, 25(3): 04019021.

Balekelayi, N. and Tesfamariam, S. 2021. Operational risk-based decision making for wastewater pipe management. *Journal of Infrastructure Systems*, 27(1): 04020042.

Cai, B., Pan, G.L. and Fu, F. 2020. Prediction of the postfire flexural capacity of RC beam using GA-BPNN machine learning. *Journal of Performance of Constructed Facilities*, 34(6): 04020105.

Cai, X., Zeng, R., Kang, W.H., Song, J. and Valocchi, A.J. 2015. Strategic planning for drought mitigation under climate change. *Journal of Water Resources Planning and Management*, 141(9): 04015004.

Cantos, W.P., Juran, I. and Tinelli, S. 2020. Machine-learning–based risk assessment method for leak detection and geolocation in a water distribution system. *Journal of Infrastructure Systems*, 26(1): 04019039.

Cao, M., Chychyla, R. and Stewart, T. 2015. Big Data analytics in financial statement audits. *Accounting Horizons*, 29(2): 423–429.

Castelletti, A., Yajima, H., Giuliani, M., Soncini-Sessa, R. and Weber, E. 2014. Planning the optimal operation of a multioutlet water reservoir with water quality and quantity targets. *Journal of Water Resources Planning and Management*, 140(4): 496–510.

Chan, H.P., Samala, R.K., Hadjiiski, L.M. and Zhou, C. 2020. Deep learning in medical image analysis. *Deep Learning in Medical Image Analysis*, 1213: 3–21. doi:10.1007/978-3-030-33128-3_1.

Chen, L., Huang, S., Yang, C. and Chen, Q. 2020. Analyzing factors that influence expressway traffic crashes based on association rules: Using the Shaoyang–Xinhuang section of the Shanghai–Kunming expressway as an example. *Journal of Transportation Engineering, Part A: Systems*, 146(9): 05020007.

Chen, W., Chen, K., Cheng, J.C., Wang, Q. and Gan, V.J. 2018. BIM-based framework for automatic scheduling of facility maintenance work orders. *Automation in Construction*, 91: 15–30.

Cheng, M.Y., Chang, Y.H. and Korir, D. 2019. Novel approach to estimating schedule to completion in construction projects using sequence and non-sequence learning. *Journal of Construction Engineering and Management*, 145(11): 04019072.

Chiang, Y.H., Wong, F.K.W. and Liang, S. 2018. Fatal construction accidents in Hong Kong. *Journal of Construction Engineering and Management*, 144(3): 04017121.

Chu, H.J. and Chang, L.C. 2010. Optimizing capacity-expansion planning of groundwater supply system between cost and subsidence. *Journal of Hydrologic Engineering*, 15(8): 632–641.

Chu, J.C. and Chen, S.C. 2016. Optimization of transportation-infrastructure-system protection considering weighted connectivity reliability. *Journal of Infrastructure Systems*, 22(1): 04015008.

Davila Delgado, J.M., Oyedele, L., Bilal, M., Ajayi, A., Akanbi, L. and Akinade, O. 2020. Big data analytics system for costing power transmission projects. *Journal of Construction Engineering and Management*, 146(1): 05019017.

Dick, K., Russell, L., Souley Dosso, Y., Kwamena, F. and Green, J.R. 2019. Deep learning for critical infrastructure resilience. *Journal of Infrastructure Systems*, 25(2): 05019003.

Dike, H.U., Zhou, Y., Deveerasetty, K.K. and Wu, Q. 2018. Unsupervised learning based on artificial neural network: A review. *In: 2018 IEEE International Conference on Cyborg and Bionic Systems (CBS)*, pp. 322–327. IEEE.

Dursun, O. and Stoy, C. 2016. Conceptual estimation of construction costs using the multistep ahead approach. *Journal of Construction Engineering and Management*, 142(9): 04016038.

El-Mashharawi, H.Q., Abu-Naser, S.S., Alshawwa, I.A. and Elkahlout, M. 2020. Grape type classification using deep learning. *International Journal of Academic Engineering*, 3(12): 41–45.

Elshaboury, N., Attia, T. and Marzouk, M. 2021. Reliability assessment of water distribution networks using minimum cut set analysis. *Journal of Infrastructure Systems*, 27(1): 04020048.

Erskine, A., Watson, T., O'Hagan, A., Ledgar, S. and Redfearn, D. 2014. Using a negative binomial regression model with a Bayesian tuner to estimate failure probability for sewerage infrastructure. *Journal of Infrastructure Systems*, 20(1): 04013005.

Faddoul, R., Raphael, W., Soubra, A.H. and Chateauneuf, A. 2013. Incorporating Bayesian networks in Markov decision processes. *Journal of Infrastructure Systems*, 19(4): 415–424.

Fallah-Mehdipour, E., Bozorg Haddad, O. and Mariño, M.A. 2013. Extraction of multicrop planning rules in a reservoir system: Application of evolutionary algorithms. *Journal of Irrigation and Drainage Engineering*, 139(6): 490–498.

Fan, C.L. 2020. Defect risk assessment using a hybrid machine learning method. *Journal of Construction Engineering and Management*, 146(9): 04020102.

Fang, C., Tang, H. and Li, Y. 2020. Stochastic response assessment of cross-sea bridges under correlated wind and waves via machine learning. *Journal of Bridge Engineering*, 25(6): 04020025.

Feng, C.W., Ju, S.H. and Huang, H.Y. 2016. Using a simple soil spring model and support vector machine to determine bridge scour depth and bridge safety. *Journal of Performance of Constructed Facilities*, 30(4): 04015088.

Gholami, H., Asgarian, B. and Asil Gharebaghi, S. 2020. Practical approach for reliability-based inspection planning of jacket platforms using Bayesian networks. *ASCE-ASME Journal of Risk and Uncertainty in Engineering Systems, Part A: Civil Engineering*, 6(3): 04020029.

Ghosh, S. and Kashyap, D. 2012. Kernel function model for planning of agricultural groundwater development. *Journal of Water Resources Planning and Management*, 138(3): 277–286.

Goh, Y.M. and Binte Sa'adon, N.F. 2015. Cognitive factors influencing safety behavior at height: A multimethod exploratory study. *Journal of Construction Engineering and Management*, 141(6): 04015003.

Gondia, A., Siam, A., El-Dakhakhni, W. and Nassar, A.H. 2020. Machine learning algorithms for construction projects delay risk prediction. *Journal of Construction Engineering and Management*, 146(1): 04019085.

Grussing, M.N. and Liu, L.Y. 2014. Knowledge-based optimization of building maintenance, repair, and renovation activities to improve facility life cycle investments. *Journal of Performance of Constructed Facilities*, 28(3): 539–548.

Guo, H., Yu, Y., Ding, Q. and Skitmore, M. 2018. Image-and-skeleton-based parameterized approach to real-time identification of construction workers' unsafe behaviors. *Journal of Construction Engineering and Management*, 144(6): 04018042.

Hackl, J., Adey, B.T., Woźniak, M. and Schümperlin, O. 2018. Use of unmanned aerial vehicle photogrammetry to obtain topographical information to improve bridge risk assessment. *Journal of Infrastructure Systems*, 24(1): 04017041.

Halfawy, M.R. and Hengmeechai, J. 2014. Efficient algorithm for crack detection in sewer images from closed-circuit television inspections. *Journal of Infrastructure Systems*, 20(2): 04013014.

Hamdan, M.A., Badran, A.A., Abdelhafez, E.A. and Hamdan, A.M. 2016. Comparison of neural network models in the estimation of the performance of solar collectors. *Journal of Infrastructure Systems*, 22(4): A4014003.

Hammad, A., AbouRizk, S. and Mohamed, Y. 2014. Application of KDD techniques to extract useful knowledge from labor resources data in industrial construction projects. *Journal of Management in Engineering*, 30(6): 05014011.

Han, S., Lee, S. and Peña-Mora, F. 2014. Comparative study of motion features for similarity-based modeling and classification of unsafe actions in construction. *Journal of Computing in Civil Engineering*, 28(5): A4014005.

Hu, Q., Bai, Y., He, L., Cai, Q., Tang, S., Ma, G., Tan, J. and Liang, B. 2020. Intelligent framework for worker-machine safety assessment. *Journal of Construction Engineering and Management*, 146(5): 04020045.

Huang, Z., Fu, H.L., Fan, X.D., Meng, J.H., Chen, W., Zheng, X.J., Wang, F. and Zhang, J.B. 2021. Rapid surface damage detection equipment for subway tunnels based on machine vision system. *Journal of Infrastructure Systems*, 27(1): 04020047.

Javanmardi, A., Abbasian-Hosseini, S.A., Liu, M. and Hsiang, S.M. 2020. Improving effectiveness of constraints removal in construction planning meetings: Information-theoretic approach. *Journal of Construction Engineering and Management*, 146(4): 04020015.

Jeong, H., Kim, H., Kim, K. and Kim, H. 2017. Prediction of flexible pavement deterioration in relation to climate change using fuzzy logic. *Journal of Infrastructure Systems*, 23(4): 04017008.

Kang, D. and Lansey, K. 2014. Multiperiod planning of water supply infrastructure based on scenario analysis. *Journal of Water Resources Planning and Management*, 140(1): 40–54.

Kang, F. and Li, J. 2020. Displacement model for concrete dam safety monitoring via Gaussian process regression considering extreme air temperature. *Journal of Structural Engineering*, 146(1): 05019001.

Kelchtermans, P., Bittremieux, W., De Grave, K., Degroeve, S., Ramon, J., Laukens, K., Valkenborg, D., Barsnes, H. and Martens, L. 2014. Machine learning applications in proteomics research: How the past can boost the future. *Proteomics*, 14(4-5): 353–366.

Kepaptsoglou, K., Karlaftis, M.G. and Mintsis, G. 2012. Model for planning emergency response services in road safety. *Journal of Urban Planning and Development*, 138(1): 18–25.

Khan, Z., Moselhi, O. and Zayed, T. 2015. Identifying rehabilitation options for optimum improvement in municipal asset condition. *Journal of Infrastructure Systems*, 21(2): 04014037.

Khandel, O. and Soliman, M. 2021. Integrated framework for assessment of time-variant flood fragility of bridges using deep learning neural networks. *Journal of Infrastructure Systems*, 27(1): 04020045.

Kim, H., Kim, K. and Kim, H. 2016b. Vision-based object-centric safety assessment using fuzzy inference: Monitoring struck-by accidents with moving objects. *Journal of Computing in Civil Engineering*, 30(4): 04015075.

Kim, J., Han, S. and Hyun, C. 2016a. Minimizing fluctuation of the maintenance, repair, and rehabilitation cost profile of a building. *Journal of Performance of Constructed Facilities*, 30(3): 04015034.

Kim, T. and Chi, S. 2019. Accident case retrieval and analyses: Using natural language processing in the construction industry. *Journal of Construction Engineering and Management*, 145(3): 04019004.

Kim, T., Lee, D., Cha, M., Lim, H., Lee, M., Cho, H. and Kang, K.I. 2020. Simulation-based lift planning model for the lift transfer operation system. *Journal of Construction Engineering and Management*, 146(9), 04020098.

Kwayu, K.M., Kwigizile, V., Zhang, J. and Oh, J.S. 2020. Semantic N-gram feature analysis and machine learning–based classification of drivers' hazardous actions at signal-controlled intersections. *Journal of Computing in Civil Engineering*, 34(4): 04020015.

Kyriakou, C., Christodoulou, S.E. and Dimitriou, L. 2019. Smartphone-based pothole detection utilizing artificial neural networks. *Journal of Infrastructure Systems*, 25(3): 04019019.

Lee, J., Guan, H., Loo, Y.C. and Blumenstein, M. 2014. Development of a long-term bridge element performance model using Elman neural networks. *Journal of Infrastructure Systems*, 20(3): 04014013.

Lee, K.P., Lee, H.S., Park, M., Kim, D.Y. and Jung, M. 2017. Management-reserve estimation for international construction projects based on risk-informed k-NN. *Journal of Management in Engineering*, 33(4): 04017002.

Li, J., Wang, H., Xie, Y. and Zeng, W. 2020. Human error identification and analysis for shield machine operation using an adapted TRACEr Method. *Journal of Construction Engineering and Management*, 146(8): 04020095.

Littman, M.L. 2015. Reinforcement learning improves behaviour from evaluative feedback. *Nature*, 521(7553): 445–451.

Liu, M. and Lo, S.M. 2011. The quantitative investigation on people's pre-evacuation behavior under fire. *Automation in Construction*, 20(5): 620–628.

Lu, N., Noori, M. and Liu, Y. 2017. Fatigue reliability assessment of welded steel bridge decks under stochastic truck loads via machine learning. *Journal of Bridge Engineering*, 22(1): 04016105.

Mangalathu, S. and Jeon, J.S. 2019. Machine learning–based failure mode recognition of circular reinforced concrete bridge columns: Comparative study. *Journal of Structural Engineering*, 145(10): 04019104.

Mathavan, S., Rahman, M. and Kamal, K. 2015. Use of a self-organizing map for crack detection in highly textured pavement images. *Journal of Infrastructure Systems*, 21(3): 04014052.

Memarzadeh, M., Pozzi, M. and Zico Kolter, J. 2015. Optimal planning and learning in uncertain environments for the management of wind farms. *Journal of Computing in Civil Engineering*, 29(5): 04014076.

Mo, Y., Zhao, D., Syal, M. and Aziz, A. 2017. Construction work plan prediction for facility management using text mining. *Computing in Civil Engineering*, 2017: 92–100.

Moon, S., Lee, G., Chi, S. and Oh, H. 2021. Automated construction specification review with named entity recognition using natural language processing. *Journal of Construction Engineering and Management*, 147(1): 04020147.

Moradi, S., Zayed, T., Nasiri, F. and Golkhoo, F. 2020. Automated anomaly detection and localization in sewer inspection videos using proportional data modeling and deep learning–based text recognition. *Journal of Infrastructure Systems*, 26(3): 04020018.

Naser, M.Z. 2020. Autonomous fire resistance evaluation. *Journal of Structural Engineering*, 146(6): 04020103.

Naser, M.Z. 2021. Observational analysis of fire-induced spalling of concrete through ensemble machine learning and surrogate modeling. *Journal of Materials in Civil Engineering*, 33(1): 04020428.

Naziris, I.A., Lagaros, N.D. and Papaioannou, K. 2016. Selection and resource allocation model for upgrading fire safety of historic buildings. *Journal of Management in Engineering*, 32(4): 05016004.

Nitsche, P., Stütz, R., Kammer, M. and Maurer, P. 2014. Comparison of machine learning methods for evaluating pavement roughness based on vehicle response. *Journal of Computing in Civil Engineering*, 28(4): 04014015.

Pereira, E., Ali, M., Wu, L. and Abourizk, S. 2020. Distributed simulation–based analytics approach for enhancing safety management systems in industrial construction. *Journal of Construction Engineering and Management*, 146(1): 04019091.

Pereira, E., Hermann, U., Han, S. and AbouRizk, S. 2018. Case-based reasoning approach for assessing safety performance using safety-related measures. *Journal of Construction Engineering and Management*, 144(9): 04018088.

Piryonesi, S.M. and El-Diraby, T.E. 2020. Data analytics in asset management: Cost-effective prediction of the pavement condition index. *Journal of Infrastructure Systems*, 26(1): 04019036.

Plant, R. 2011. An introduction to artificial intelligence. *In: 32nd Aerospace Sciences Meeting and Exhibit*, American Institute of Aeronautics and Astronautics, p. 294. doi:10.2514/6.1994-294.

ProjectPro 2021. Top 10 Machine Learning Algorithms. https://www.dezyre.com/article/top-10-machine-learning-algorithms/202. Accessed 24 January 2021.

Pudasaini, B. and Shahandashti, S.M. 2018. Identification of critical pipes for proactive resource-constrained seismic rehabilitation of water pipe networks. *Journal of Infrastructure Systems*, 24(4): 04018024.

Radfar, A. and Rockaway, T.D. 2016. Captured runoff prediction model by permeable pavements using artificial neural networks. *Journal of Infrastructure Systems*, 22(3): 04016007.

Rahimi, A., Azimi, G., Asgari, H. and Jin, X. 2019. Clustering approach toward large truck crash analysis. *Transportation Research Record*, 2673(8): 73–85.

Ryu, J., Seo, J., Jebelli, H. and Lee, S. 2019. Automated action recognition using an accelerometer-embedded wristband-type activity tracker. *Journal of Construction Engineering and Management*, 145(1): 04018114.

Saha, P., Greer, N.A. and Ksaibati, K. 2020. Numerical model to optimize selection of unpaved roads for dust suppressing chemical treatments: Case Study. *Journal of Infrastructure Systems*, 26(1): 04019038.

Sakhakarmi, S., Park, J. and Cho, C. 2019. Enhanced machine learning classification accuracy for scaffolding safety using increased features. *Journal of Construction Engineering and Management*, 145(2): 04018133.

Salah, M., Osman, H. and Hosny, O. 2018. Performance-based reliability-centered maintenance planning for hospital facilities. *Journal of Performance of Constructed Facilities*, 32(1): 04017113.

Shayanfar, E., Abianeh, A.S., Schonfeld, P. and Zhang, L. 2016. Prioritizing interrelated road projects using metaheuristics. *Journal of Infrastructure Systems*, 22(2): 04016004.

Shehab, T. and Meisami-Fard, I. 2013. Cost-estimating model for rubberized asphalt pavement rehabilitation projects. *Journal of Infrastructure Systems*, 19(4): 496–502.

Son, H., Kim, C. and Kim, C. 2012. Automated color model-based concrete detection in construction-site images by using machine learning algorithms. *Journal of Computing in Civil Engineering*, 26(3): 421–433.

Son, H., Seong, H., Choi, H. and Kim, C. 2019. Real-time vision-based warning system for prevention of collisions between workers and heavy equipment. *Journal of Computing in Civil Engineering*, 33(5): p.04019029.

Sousa, V., Meireles, I. and Silva, A. 2020. Optimizing service life prediction models of external paint finishes. *Journal of Performance of Constructed Facilities*, 34(2): 04020014.

Su, Y., Mao, C., Jiang, R., Liu, G. and Wang, J. 2021. Data-driven fire safety management at building construction sites: Leveraging CNN. *Journal of Management in Engineering*, 37(2): 04020108.

Tabatabaee, N., Ziyadi, M. and Shafahi, Y. 2013. Two-stage support vector classifier and recurrent neural network predictor for pavement performance modeling. *Journal of Infrastructure Systems*, 19(3): 266–274.

Tang, S., Golparvar-Fard, M., Naphade, M. and Gopalakrishna, M.M. 2020. Video-based motion trajectory forecasting method for proactive construction safety monitoring systems. *Journal of Computing in Civil Engineering*, 34(6): 04020041.

Tatari, O., Sargand, S.M., Masada, T. and Tarawneh, B. 2013. Neural network approach to condition assessment of highway culverts: Case study in Ohio. *Journal of Infrastructure Systems*, 19(4): 409–414.

Tien, I. and Der Kiureghian, A. 2017. Reliability assessment of critical infrastructure using Bayesian networks. *Journal of Infrastructure Systems*, 23(4): 04017025.

Volkova, N.P., Rizun, N.O. and Nehrey, M.V. 2019. Data science: Opportunities to transform education. *In: Proceedings of the 6th Workshop on Cloud Technologies in Education (CTE 2018), Kryvyi Rih, Ukraine, 21 December 2018* (No. 2433), pp. 48-73. CEUR Workshop Proceedings.

Wang, D., Zhou, T. and Li, X. 2021. Impacts of environment and individual factors on human premovement time in underground commercial buildings in China: A virtual reality–based study. *ASCE-ASME Journal of Risk and Uncertainty in Engineering Systems, Part A: Civil Engineering*, 7(1): 04020056.

Wang, F. and Zhang, Q.L. 2017. Systemic estimation of dam overtopping probability: Bayesian networks approach. *Journal of Infrastructure Systems*, 23(2): 04016037.

Wang, F., Li, L., Li, C., Cao, Y., Zhang, Y. and Fang, B. 2016. Procedure and model of anti-disaster differentiated planning for a power distribution system. *Journal of Energy Engineering*, 142(1): 04015007.

Wang, G., Burrow, M. and Ghataora, G. 2020. Study of the factors affecting road roughness measurement using smartphones. *Journal of Infrastructure Systems*, 26(3): 04020020.

Wang, T.K. and Piao, Y. 2019. Development of BIM-AR-based facility risk assessment and maintenance system. *Journal of Performance of Constructed Facilities*, 33(6): 04019068.

Wang, X., Huang, X., Luo, Y., Pei, J. and Xu, M. 2018. Improving workplace hazard identification performance using data mining. *Journal of Construction Engineering and Management*, 144(8): 04018068.

Wong, L., Wang, Y., Law, T. and Lo, C.T. 2016. Association of root causes in fatal fall-from-height construction accidents in Hong Kong. *Journal of Construction Engineering and Management*, 142(7): 04016018.

Xie, K., Ozbay, K., Zhu, Y. and Yang, H. 2017. Evacuation zone modeling under climate change: A data-driven method. *Journal of Infrastructure Systems*, 23(4): 04017013.

Xu, N., Ma, L., Wang, L., Deng, Y. and Ni, G. 2021. Extracting domain knowledge elements of construction safety management: Rule-based approach using Chinese natural language processing. *Journal of Management in Engineering*, 37(2): 04021001.

Yamany, M.S., Saeed, T.U., Volovski, M. and Ahmed, A. 2020. Characterizing the performance of interstate flexible pavements using artificial neural networks and random parameters regression. *Journal of Infrastructure Systems*, 26(2): 04020010.

Yap, J.B.H., Rou Chong, J., Skitmore, M. and Lee, W.P. 2020. Rework causation that undermines safety performance during production in construction. *Journal of Construction Engineering and Management*, 146(9): 04020106.

Yoon, S., Weidner, T. and Hastak, M. 2021. Total-package-prioritization mitigation strategy for deferred maintenance of a campus-sized institution. *Journal of Construction Engineering and Management*, 147(3): 04020185.

Yoon, Y., Hastak, M. and Cho, K. 2017. Method for generating multiple MRR solutions for application in cost-leveling models. *Journal of Infrastructure Systems*, 23(3): 04016045.

Zhang, Y., Ding, L. and Love, P.E. 2017. Planning of deep foundation construction technical specifications using improved case-based reasoning with weighted k-nearest neighbors. *Journal of Computing in Civil Engineering*, 31(5): 04017029.

Zhang, Y., Luo, H., Skitmore, M., Li, Q. and Zhong, B. 2019. Optimal camera placement for monitoring safety in metro station construction work. *Journal of Construction Engineering and Management*, 145(1): 04018118.

Zhang, Z. and Wei, H.H. 2021. Modeling interaction of emergency inspection routing and restoration scheduling for post-disaster resilience of highway–bridge networks. *Journal of Infrastructure Systems*, 27(1): 04020046.

Zhang, Z., Sun, C., Bridgelall, R. and Sun, M. 2018. Application of a machine learning method to evaluate road roughness from connected vehicles. *Journal of Transportation Engineering, Part B: Pavements*, 144(4): 04018043.

Zhao, D., McCoy, A.P., Kleiner, B.M., Du, J. and Smith-Jackson, T.L. 2016a. Decision-making chains in electrical safety for construction workers. *Journal of Construction Engineering and Management*, 142(1): 04015055.

Zhao, D., McCoy, A.P., Kleiner, B.M., Smith-Jackson, T.L. and Liu, G. 2016b. Sociotechnical systems of fatal electrical injuries in the construction industry. *Journal of Construction Engineering and Management*, 142(1): 04015056.

Zhao, X., Lovreglio, R. and Nilsson, D. 2020. Modelling and interpreting pre-evacuation decision-making using machine learning. *Automation in Construction*, 113: 103140.

Zheng, Z., Lu, P. and Pan, D. 2019. Predicting highway–rail grade crossing collision risk by neural network systems. *Journal of Transportation Engineering, Part A: Systems*, 145(8): 04019033.

Zhou, Y., Li, S., Zhou, C. and Luo, H. 2019. Intelligent approach based on random forest for safety risk prediction of deep foundation pit in subway stations. *Journal of Computing in Civil Engineering,* 33(1): 05018004.

Zhou, Y., Su, W., Ding, L., Luo, H. and Love, P.E. 2017. Predicting safety risks in deep foundation pits in subway infrastructure projects: Support vector machine approach. *Journal of Computing in Civil Engineering,* 31(5): 04017052.

Zhu, X. and Goldberg, A.B. 2009. Introduction to semi-supervised learning. *Synthesis Lectures on Artificial Intelligence and Machine Learning,* 3(1): 1–130.

Chapter 8

Artificial Intelligence in Concrete Materials
A Scientometric View

Zhanzhao Li[1],* and *Aleksandra Radlińska*[2]

1. Introduction

The world population has grown from 1.5 to 7.7 billion over the last hundred years (Leridon, 2020), with more than half of the humanity currently living in urban areas (Chen et al., 2019). As the world population continues to grow, global urbanization will expand at a fast pace. It is expected that by 2050, two in every three people will live in cities (Buhaug et al., 2013). Such a high rate of urban development requires enormous quantities of materials for the construction of residential housing, commercial buildings, sanitation facilities, and other parts of the infrastructure. Concrete is the principal building material for the construction and infrastructure industries (Mehta and Monteiro, 2014). Despite decades of research and analysis, some scientific and engineering questions on concrete materials regarding mixture design optimization and service life prediction still remain unanswered (DeRousseau et al., 2018; Dolado and Van Breugel, 2011).While most of the early concrete research relied on expert knowledge and intuition, trial-and-error experiments, or physical modeling, recent advances in data-driven techniques such as artificial intelligence (AI) can provide fresh perspectives to tackle the existing research questions (Asteris et al., 2021; Cai et al., 2020; Young et al., 2019).

During the past few years, AI has attracted worldwide attention as the "fourth paradigm of science" owing to the exponential growth in computing power, higher accessibility of data repositories, and more availability of data science tools (Agrawal and Choudhary, 2016). Compared with labor-intensive experiments or

[1] Department of Civil and Environmental Engineering, The Pennsylvania State University, 3127 Research Drive, Office #122 State College, PA 16801, USA.

[2] Department of Civil and Environmental Engineering, The Pennsylvania State University, 231D Sackett Building University Park, PA 16802, USA. Email: azr172@psu.edu

* Corresponding author: zzl244@psu.edu

computationally expensive simulations, AI techniques take advantage of existing data, automatically learn patterns from them, and perform tasks without explicit instructions (Batra et al., 2020). Such data-driven techniques have offered an alternative route to accelerate concrete mixture design and optimization in a more effective manner (Gunasekera et al., 2020; Young et al., 2019): large concrete mixture datasets are constructed and fed into AI models; the models then screen and generate new mixtures with desired properties; and the best mixture can be identified and validated by experimental and computational tests, with the outcomes appended to the collected datasets and iteratively calibrating the models. Nevertheless, the interactions between AI and concrete are still in the nascent stages and the full power of AI in concrete research is far from being realized. Thus, an overview of the current research progress and future opportunities will be helpful for the wider adoption of AI in the construction industry.

Existing AI review studies in the concrete domain (Behnood and Golafshani, 2021; Ben Chaabene et al., 2020; DeRousseau et al., 2018; Nunez et al., 2021; Rafiei et al., 2016) have made valuable contributions. Yet, they have some limitations. First, these studies frequently adopted a relatively narrow perspective, focusing on limited AI applications in concrete science. For example, Nunez et al. (2021) examined different AI models for accurate prediction of compressive strength of concrete; Ben Chaabene et al. (2020) further extended the scope to AI applications in concrete mechanical property prediction, while DeRousseau et al. (2018) discussed several computational methods used to optimize concrete mixture design. Second, previous review studies largely depended on the manual review and appraisal, which may lead to subjective interpretation (Martinez et al., 2019; Pan and Zhang, 2021). As such, these review studies do not offer a full picture of the current knowledge structure of AI research within the concrete domain. As the first attempt to fill this gap, this chapter is intended to present a quantitative evaluation of the knowledge patterns and capture the research interests and emerging trends of AI in concrete science.

To this end, the present chapter utilized scientometric techniques to conduct quantitative analysis on the literature regarding AI in concrete materials. Scientometrics is a branch of informatics that quantifies features and characteristics of science and scientific research, and unravels emerging research trends and knowledge structures in the research domain (Chen et al., 2012). While manual review provides an insightful overview of the research field, scientometric tools take scientific literature as an input and generate interactive visualization and mapping in a more effective and efficient manner (Börner et al., 2003). The objectives of this chapter are to: (1) quantitatively analyze existing literature within the topic by scientometric techniques (i.e., keyword co-occurrence analysis and documentation co-citation analysis); (2) highlight hot research topics and applications of AI in concrete materials; and (3) identify research gaps for further adoption of AI in the construction industry.

2. Literature Survey

This chapter analyzes bibliographic data retrieved from the Web of Science (WoS) Core Collection database, which is one of the most established bibliographic data sources and includes peer-reviewed, high-quality scholarly journals published worldwide (Visser et al., 2021). A combination with other bibliographic databases (such as Google Scholar and Scopus) was not considered due to difficulties in checking duplicates of publications and dealing with different data formats from various databases.

Keywords were selected based on related review studies (Darko et al., 2020; Halilaj et al., 2018; Martinez et al., 2019; Pan and Zhang, 2021) and iterative search in the WoS platform. As a result, a list of keywords for topics related to concrete science and AI was created, with the query string being: ("concrete" OR "cement" OR "cementitious") AND ("artificial intelligence" OR "machine learning" OR "deep learning" OR "neural network*" OR "support vector machine*" OR "random forest" OR "decision tree*" OR "k-nearest" OR "k nearest" OR "k-nn" OR "knn" OR "k-mean*" OR "k mean*" OR "gaussian process*"). Note that the wildcard character * was used to capture plural forms, e.g., "neural network*" matches "neural network" and "neural networks". The keyword search was performed on topic terms from the title, abstract, author keywords, and keywords plus (automatically generated from the titles of cited articles by WoS) within a record. The search period was set to include the last 30 years, from January 1990 to December 2020, which is sufficient based on our preliminary search to represent the development of AI within the concrete domain. The language of the publications was limited to English, and the "document type" was set to "article" such that the evolution of the field can be represented by high-quality and original research (Darko et al., 2020). We further restricted the "research area" by the Boolean operation ("Construction & Building Technology" AND "Materials Science" NOT "Computer Science"), as keywords like "concrete" are often used in other senses of the words within other subject areas, such as computer science.

As of October 2021, 440 documents were identified. A further manual refining process was conducted to evaluate the collected articles by reading the source title and abstract for inclusion/exclusion. The exclusion was mainly for publications that were: (1) review articles and book chapters; (2) unrelated to concrete science (e.g., focused on other construction materials like steel but cited concrete-related articles as references); and (3) unrelated to AI (e.g., applied traditional statistical models like linear regression models and only included AI as a recommendation for future research). This approach filtered down the number of publications from 440 to 389. Detailed information of the retrieved publications is available in Li et al. (2022).

Figure 8.1 presents the list of top journals where publications on AI for concrete materials have been published. All the 12 journals in Fig. 8.1 had impact factors larger than 1.7 in the year 2020, which lends credence to the representativeness of the collected database. The majority (over 50%) of the publications on AI in concrete have been published in *Construction and Building Materials*, which is found to be the most influential journal in this research domain according to the number of citations (6,993). Interestingly, although the number of publications from *Cement and Concrete Research* ranked fourth place, the citation record of this journal ranked second in its field. *Steel and Composite Structures* and *ACI Structural Journal* were found to be the main sources of references for citations, while they were not in the list of top 10 journals in terms of number of publications.

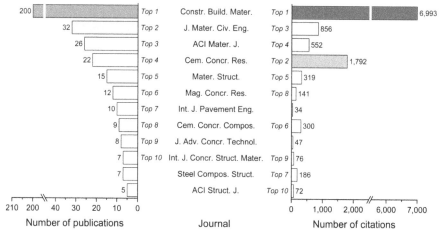

Fig. 8.1: Top 10 journals in terms of numbers of publications and citations in the area of AI in concrete science.

Note: Journal titles are presented using abbreviations based on ISO (International Standards Organization) standards.

The research field of AI dates back to the mid-20th century (Haenlein and Kaplan, 2019) and has been extended to the construction industry in the 1970s (Darko et al., 2020); however, the adoption of AI in concrete materials had not started until the early 1990s. The first study on AI in the concrete domain appears to be the work from Pratt and Sansalone (1992), published in the *ACI Materials Journal* in 1992, where AI was employed to automate signal interpretation for impact echo testing in the field. This was followed by the study from Mo and Lin (1994), where they used AI to model the behavior of reinforced concrete framed shear walls. Figure 8.2 shows the trend in research publications on AI in concrete science from 1990 to 2020. It reveals a relatively steady and gradual increase in research interest at the end of the 20th century. The number of publications did not reach double digits until 2009. Since then, exponential growth has been seen due to the ever-increasing computing power and availability of experimental data and computational tools. An exponential model was used to fit the data in Fig. 8.2: $N(t) = N_0(1 + p)^{(t-t_0)}$,

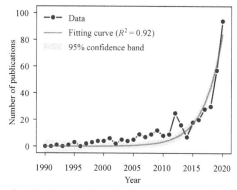

Fig. 8.2: Number of publications for AI applications in concrete science during 1990–2020.

where $N(t)$ and N_0 are the numbers of publications in the year of t and the initial year $t_0 = 1990$, respectively. Parameter p represents the annual percentage rate of increase. N_0 and p were determined as 0.006 and 0.376, respectively. This indicates that the estimated increase in the publication number over the period of a year is 37.6%, and that the numbers of relevant publications in 2021 and 2022 could be over 110 and 160, respectively.

3. Main Research Interests: Keyword Co-occurrence Analysis

Keywords are the words or phrases that deliver the most essential content of a published document (Martinez et al., 2019; Pan and Zhang, 2021). The analysis of keywords can offer opportunities to uncover the main research interests in any scientific field (Darko et al., 2020; van Eck and Waltman, 2014). In order to construct and map the knowledge domain between AI and concrete materials, a keyword co-occurrence network was developed using VOS viewer (van Eck and Waltman, 2013). The co-occurrence of two keywords can be defined as the situation when both keywords occur together in a publication (van Eck and Waltman, 2014). A typical co-occurrence network of keywords consists of nodes (i.e., keywords) and links (indicating relations between pairs of nodes). The strength of the relation between two keywords represents the connection of their respective knowledge domains. Visualization of the keyword co-occurrence network provides an understanding of existing research topics and how they are intellectually developed and connected (van Eck and Waltman, 2014).

Keywords can be extracted from the title and abstract of a publication by text mining approaches or from the list of author-supplied keywords (van Eck and Waltman, 2014). In this work, only author keywords were used to conduct the analysis and obtain reproducible and readable results. Identical terms (such as "neural network" and "neural networks"; "mix design" and "mixture design") were merged for the accuracy of the analysis. However, keywords like "concrete", "high performance concrete", "recycled aggregate concrete", "asphalt concrete", "self-consolidating concrete", "high strength concrete", and "reinforced concrete" remained in the database, as their presence and the absence of the terms regarding other types of concrete indicated hot topics and research interests for the specific types of concrete (Darko et al., 2020). Other examples include the keywords "strength", "compressive strength", "flexural strength", and "bond strength", which are mechanical performance of hardened concrete. The minimum number of occurrences of a keyword was set to five such that 45 of the 969 keywords were included in the co-occurrence network. This criterion was selected following previous related studies (Martinez et al., 2019) and based on multiple experiments to generate the optimal clusters in the network. Fractional counting methodology was adopted to fractionalize the weight of a link by considering the number of keywords in a publication (van Eck and Waltman, 2014). This method has been found to be preferable over a traditional full counting approach (Perianes-Rodriguez et al., 2016).

Figure 8.3 presents the resulting network of co-occurring keywords with 45 nodes and 536 links. In the network, each node represents a keyword term. The size

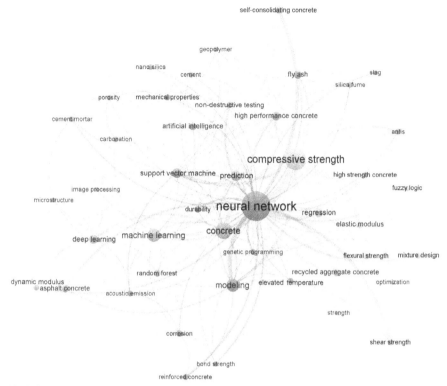

Fig. 8.3: Network of co-occurring keywords related to research on AI in concrete materials (cluster view).

of a node and a label denotes the number of publications that have the corresponding term in the keyword lists. The distance between two nodes indicates the relatedness of the nodes. In general, the smaller distance, the closer relationship between keywords. The width of a link represents the strength of the link, i.e., the number of publications where two keywords occur together. Nodes with the same color suggest a similar knowledge domain among the investigated publications (van Eck and Waltman, 2013). Table 8.1 summarizes the keyword occurrences and related network data.

As shown in Fig. 8.3 and Table 8.1, some research interests have received special attention. For example, keywords like "modeling", "prediction", "regression", and "optimization" have been frequently used in the publications. This could be explained by the fact that the main objectives of utilizing AI in concrete are to model concrete behavior, predict concrete properties, and optimize concrete mixture design for desirable performance.

Interestingly, "neural network" appeared most frequently among all the keywords, indicating that it has been the most commonly used model, followed by support vector machine and random forest as shown in Table 8.1. As the second most frequent keyword, "compressive strength" represented the strongest link with "neural network" (see Fig. 8.3). This shows that the most common AI application in concrete science has been the deployment of neural network models for compressive strength prediction of concrete, in agreement with other studies (Behnood and

Table 8.1: List of selected keywords and relevant network data (ranked by the number of occurrences).

Keyword	Occurrences	Average Publication Year	Links	Total Link Strength
Neural network	167	2014	42	130
Compressive strength	92	2015	32	85
Concrete	54	2014	25	45
Machine learning	34	2019	26	29
Modeling	32	2011	21	29
Prediction	22	2017	23	22
Support vector machine	19	2018	11	17
Deep learning	17	2019	13	13
Regression	17	2015	20	17
High performance concrete	15	2015	15	14
Fly ash	13	2014	14	12
Recycled aggregate concrete	13	2016	15	13
Elastic modulus	12	2014	15	11
Artificial intelligence	11	2016	15	9
Asphalt concrete	11	2018	6	9
Random forest	10	2019	7	7
Self-consolidating concrete	10	2013	6	8
Flexural strength	9	2018	11	9
Mixture design	9	2013	11	9
Corrosion	8	2017	9	6
Durability	8	2015	14	8
Fuzzy logic	8	2011	9	7
High strength concrete	8	2013	10	8
Mechanical properties	8	2018	11	8
Non-destructive testing	8	2015	9	8
Shear strength	8	2013	5	7
Acoustic emission	7	2017	5	6
ANFIS	7	2015	12	7
Cement mortar	7	2016	6	7
Dynamic modulus	7	2018	5	6
Genetic programming	7	2014	7	7
Reinforced concrete	7	2016	6	6
Carbonation	6	2014	10	6
Cement	6	2013	10	6
Elevated temperature	6	2017	7	5
Nano silica	6	2017	7	5
Optimization	6	2014	12	6

Table 8.1 contd. ...

...Table 8.1 contd.

Keyword	Occurrences	Average Publication Year	Links	Total Link Strength
Porosity	6	2015	7	6
Silica fume	6	2014	9	6
Strength	6	2011	9	4
Bond strength	5	2015	5	5
Geopolymer	5	2017	5	5
Image processing	5	2017	6	5
Microstructure	5	2015	6	5
Slag	5	2011	7	5

Note: ANFIS = Adaptive neuro-fuzzy inference system.

Golafshani, 2021; Ben Chaabene et al., 2020). Compressive strength is considered to be one of the most important design parameters in construction applications, since it determines the loading capacity of concrete structures and is correlated with several mechanical and durability properties of concrete, including tensile and flexural strength, elastic modulus, and impermeability (ACI Committee 211, 2002). The prediction of compressive strength of concrete as a function of mixture proportions has become a research focus to facilitate mixture design optimization during the design phase and assist project scheduling as well as quality control during the production phase. Although AI techniques are promising, Table 8.1 suggests that other models such as fuzzy logic, adaptive neuro-fuzzy inference system (ANFIS), and genetic programming have obtained far less attention in concrete research. Thus, future research in this area should explore different AI models (i.e., models other than neural networks) and full development and exploitation of various models could be a promising aid in accelerating adoption of AI in the construction industry.

Figure 8.3 and Table 8.1 also show that research interests have been focused on the prediction of hardened properties for concrete materials, including compressive strength, elastic modulus, flexural strength, shear strength, bond strength, dynamic modulus, porosity, and durability (e.g., corrosion and carbonation). However, modeling of the performance of concrete in fresh state has remained understudied by AI techniques. This must draw attention of concrete researchers, given that fresh properties, such as setting, bleeding, segregation, heat evolution, plastic shrinkage, and rheological properties, are also of significance for both workability in construction practice and microstructural development towards the hardened performance of concrete (e.g., strength and durability) (Kovler and Roussel, 2011). One major challenge that hinders practical applications of AI in fresh property prediction is the lack of large and universal datasets for concrete fresh properties. For example, absolute values (e.g., yield stress and plastic viscosity) calculated for a given concrete mixture are not identical and comparable among various rheometers (with different principles and geometries) (Brower and Ferraris, 2003). As such, data collection from different laboratories using different rheometers could become less meaningful. Moving forward, it would be promising to standardize reference materials for concrete rheometry in order to calibrate different concrete rheometers and normalize calculated values (Ferraris et al., 2014). With the development of large datasets for fresh properties and the availability of AI techniques, accurate prediction and modeling of concrete behavior at early ages could be very useful.

Based on the network, similar observations can be made for concrete types. Among various concrete types, high performance concrete, recycled aggregate concrete, asphalt concrete, self-consolidating concrete, high strength concrete, reinforced concrete, and geopolymer concrete have been extensively reported with AI methods (see Fig. 8.3 and Table 8.1). The absence of the other types of concrete in the network, including previous concrete, lightweight aggregate concrete, and 3D-printed concrete, indicates that these types of concrete have been overlooked in AI-related concrete research. AI technologies have demonstrated their vast potential to offer novel approaches to model and predict material behaviors of the majority of concrete. Current research could be extended to explore other concrete types with the use of AI approaches, which could facilitate the development of an intelligence ecosystem for various concrete materials.

As shown in Fig. 8.3, nodes with the same color suggest a similar topic among the publications, and seven distinct clusters of keywords were obtained in the network. From each cluster, main research interests and research gaps could be identified. For example, in the yellow cluster (Fig. 8.3, middle right), keywords such as "strength", "compressive strength", "flexural strength", "elastic modulus", were grouped together with "mixture design" and "optimization". This indicates that physical concrete performance measures have been the main objective for mixture design optimization (Ghafari et al., 2015; Han et al., 2020). However, concrete mixture design involves multiple competing criteria, such as minimizing both cost and environmental impacts (e.g., carbon emissions) while maximizing physical performance (DeRousseau et al., 2018; Huang et al., 2020; Young et al., 2019). From the viewpoint of engineering applications, more research efforts would be needed to explore the application of AI techniques in multi-objective problems towards cheaper, stronger, more workable, durable, and more environmentally sustainable concrete.

The green cluster was another cluster with significant size in Fig. 8.3. This cluster consisted of keywords such as "machine learning", "deep learning", and "image processing". It is worth pointing out that these topics are considered emerging, as indicated in Fig. 8.4. In contrast to the static representations of knowledge domains in Fig. 8.3, the network in Fig. 8.4 provides a timeline view of the knowledge map, where the evolution of AI applications in the concrete domain can be visualized according to the average publication year of each keyword. Notably, the average publication year of keyword "deep learning" was 2019, suggesting that research emphasis has been shifting towards the applications of deep learning techniques in concrete science, especially for image analysis and data recognition (corresponding to the keywords "image processing" and "acoustic emission", respectively). This could be attributed to the successful deep learning architectures developed in computer science, including convolutional neural networks for image classification (Krizhevsky et al., 2012) and long short-term memory networks for time-series modeling (Greff et al., 2017). Note that deep learning techniques have been mainly

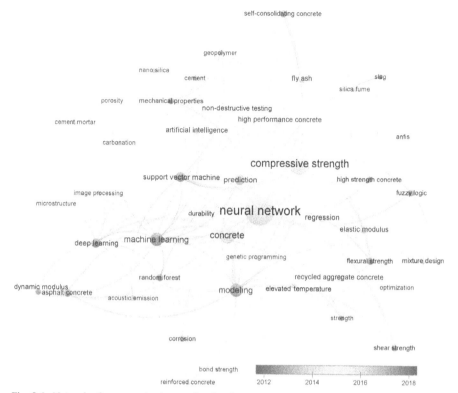

Fig. 8.4: Network of co-occurring keywords related to research on AI in concrete materials (timeline view).

Note: Timeline covers a period of time from 2012 to 2018 in terms of average publication year.

used for crack detection in field practice (van Eck and Waltman, 2014; Khallaf and Khallaf, 2021; Liu et al., 2019), whereas the integration of these techniques with concrete petrographic analyses (e.g., flatbed scanning (Song et al., 2020), scanning electron microscopy (Tong et al., 2019b, 2020), and X-ray computed tomography (Lorenzoni et al., 2020; Tong et al., 2019a)) could be promising yet under-researched.

Keywords such as "neural network", "compressive strength", and "concrete" were represented in the middle spectrum around 2015 (see Fig. 8.4 and Table 8.1). This is because these topics have been the research focuses during the whole period of time investigated. AI models including support vector machine (average publication year: 2018) and random forest (average publication year: 2019) have become prevalent in addition to deep learning approaches, which indicates that more research efforts have been made recently to explore the effectiveness of AI techniques (other than neural networks) in concrete research.

4. Citation Patterns: Document Co-citation Analysis

A publication typically includes a list of references, which delineates the knowledge base of the publication. Similar to co-occurrence, co-citation could be defined as the frequency with which two references are cited together by other publications (Small, 1973). Given that frequently cited references typically present key concepts, methods, or conclusions in a research field, co-citation patterns of references could offer valuable insights into the relationship between these key ideas. The analysis of document co-citation has demonstrated its potential to unravel the intellectual structures of scientific literature and thus to identify research interests and emerging trends in a research domain (Hou et al., 2018).

To further identify research trends in AI and concrete studies, a network of document co-citation was generated by CiteSpace (Chen, 2014, 2016) (see Fig. 8.5). Records of 389 citing publications and 11,842 cited references were retrieved from WoS, as described in Section 8.2. In Fig. 8.5, each node stands for a cited reference, and node size reflects the citation number of the reference. Links between nodes indicate co-citation. The colors of these links represent the first time when two references were co-cited. Links in more recent years are shown in red, whereas older co-citation links are colored in yellow. No network pruning (or link reduction) was performed to avoid reducing the characteristics of the bibliographic network. Figure 8.5 and Table 8.2 show the five major clusters identified in the network. These clusters were numbered and ranked in terms of their size, i.e., the number of references in a cluster. Each cluster was labeled using the log-likelihood ratio algorithm. This algorithm identifies a label based on the latent semantic analysis of publications that cite the references in each cluster. Although this approach has been found to provide the most accurate results in terms of unique labeling and sufficient coverage (Chen, 2014), additional manual examination of the publications was conducted to cross-validate the identified labels and determine the focus of each cluster.

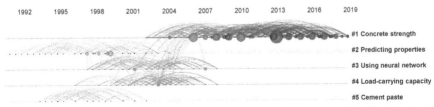

Fig. 8.5: Network of co-cited documents in the publications on AI in concrete materials (timeline view).

Table 8.2: Major clusters identified in the document co-citation analysis.

Cluster ID	Size	Silhouette Value	Average Publication Year	Cluster Label	Focus of the Cluster
1	169	0.913	2011	Concrete strength	Concrete compressive strength
2	65	0.957	1996	Predicting properties	Mixture design
3	47	0.979	2002	Using neural network	Neural network; durability
4	47	0.952	2002	Load-carrying capacity	Shear capacity; elastic modulus
5	46	0.976	1997	Cement paste	Cement paste; waste solidification

Two fundamental metrics, namely, the modularity Q index and the silhouette value, were calculated to examine the reliability of the network. The modularity Q index ranges from 0 to 1 and assesses the extent to which a network can be partitioned into multiple clusters; the silhouette value ranges between -1 and 1 and measures the quality of a clustering configuration. A higher modularity Q index and a higher silhouette value imply higher divisibility of a network and better homogeneity of a cluster, respectively (Chen, 2014; Lim and Aryadoust, 2021). The threshold of the silhouette value has been found to be 0.7, above which the clusters in a network could be considered highly isolated with clear boundaries and weak links between individual clusters (Chen, 2014). The network in Fig. 8.5 had a modularity Q index of 0.8867 and a weighted mean silhouette value of 0.9557. These values were relatively high, indicating that the network was highly divisible into clusters, each of which was, on average, highly homogeneous (Chen, 2014). In other words, studies in each identified cluster were consistent in addressing similar issues.

As per cluster sizes in Table 8.2, cluster #1 on concrete strength has been the largest cluster with 169 references. After examining the content of the 47 publications that cited these references, the focus of this cluster could be further specified as concrete compressive strength (Chou et al., 2014; Gholampour et al., 2017; Sobhani et al., 2010). This is in line with the earlier observation in Fig. 8.3 and Table 8.1, where "compressive strength" has been the second most frequently used keyword. The "average publication year" in Table 8.2 shows the average period within which the references in a given cluster were published. While most of the clusters showed activities over two decades ago (i.e., 1996–2002), studies in cluster #1 have been published more recently, around 2011 on average, and this cluster had the longest active duration of approximately 17 years from 2002 to 2019. These again confirm the significance of cluster #1 on concrete compressive strength in the field of AI in concrete science.

Detection of citation burstness was further conducted to identify references that were of great significance to the concrete domain. Citation burstness occurs when a reference received a considerable surge in terms of citations and attention within a specific duration (Chen et al., 2010) (e.g., large nodes in Fig. 8.5). Table 8.3 shows the top 10 references with the highest burst strengths (i.e., most co-cited and influential references). It is interesting to note that all these references listed in Table 8.3 were grouped in cluster #1. Most of them focused on the numerical prediction of compressive strength, and only two references (Behnood et al., 2015; Duan et al., 2013b) on that of elastic modulus (which is correlated to compressive strength). All these references utilized AI techniques for prediction. In agreement with Section 8.3, neural network was the most commonly used algorithm (Chithra et al., 2016; Dantas et al., 2013; Duan et al., 2013a,b; Öztaş et al., 2006; Pala et al., 2007; Sobhani et al., 2010; Topçu and Saridemir, 2008). The concrete type investigated in these studies varied from normal concrete (Behnood et al., 2017; Pala et al., 2007; Topçu and Saridemir, 2008) to high performance concrete (Behnood et al., 2017; Chithra et al., 2016) and recycled aggregate concrete (Behnood et al., 2015; Duan et al., 2013a,b). Note that most of these 10 cited references were included in the bibliographic database containing 389 citing publications, which cross-validates the completeness and representativeness of the retrieved database.

Table 8.3: Top 10 references regarding AI in concrete materials with the strongest document co-citation burstness.

No.	Reference	Burst Strength	Year	Begin	End	Span	Citation Counts	Cluster ID
1	Duan et al. (2013a)	11.27	2013	2015	2020	5	29	1
2	Öztaş et al. (2006)	10.15	2006	2008	2013	5	24	1
3	Sobhani et al. (2010)	7.84	2010	2012	2018	6	22	1
4	Behnood et al. (2017)	7.76	2017	2017	2020	3	15	1
5	Pala et al. (2007)	7.66	2007	2011	2016	5	16	1
6	Behnood et al. (2015)	7.59	2015	2018	2020	2	18	1
7	Topçu and Saridemir (2008)	7.38	2008	2016	2020	4	16	1
8	Chithra et al. (2016)	7.31	2016	2017	2020	3	16	1
9	Dantas et al. (2013)	7.07	2013	2008	2013	5	17	1
10	Duan et al. (2013b)	6.12	2013	2015	2020	5	15	1

Table 8.4: Top 10 most cited publications regarding AI in concrete materials.

No.	Reference	Year	Title	Total Citations
1	Yeh (1998b)	1998	Modeling of strength of high-performance concrete using artificial neural networks	548
2	Gopalakrishnan et al. (2017)	2017	Deep Convolutional Neural Networks with transfer learning for computer vision-based data-driven pavement distress detection	254
3	Ni and Wang (2000)	2000	Prediction of compressive strength of concrete by neural networks	243
4	Öztaş et al. (2006)	2006	Predicting the compressive strength and slump of high strength concrete using neural network	212
5	Yeh (2007b)	2007	Modeling slump flow of concrete using second-order regressions and artificial neural networks	163
6	Duan et al. (2013a)	2013	Prediction of compressive strength of recycled aggregate concrete using artificial neural networks	152
7	Sobhani et al. (2010)	2010	Prediction of the compressive strength of no-slump concrete: A comparative study of regression, neural network and ANFIS models	152
8	Dorafshan et al. (2018)	2018	Comparison of deep convolutional neural networks and edge detectors for image-based crack detection in concrete	150
9	Alshihri et al. (2009)	2009	Neural networks for predicting compressive strength of structural light weight concrete	138
10	Tam et al. (2007)	2007	Optimization on proportion for recycled aggregate in concrete using two-stage mixing approach	135

Table 8.4 provides the citation information of the 389 publications regarding AI in concrete. Three studies (Duan et al., 2013a; Öztaş et al., 2006; Sobhani et al., 2010) were present in both Table 8.3 and Table 8.4, suggesting their importance in the field of concrete research.

As shown in Table 8.2, cluster #2 with 65 cited references was identified as "predicting properties". There were eight citing publications in this cluster with a focus on mixture design (Dias and Pooliyadda, 2001; Yeh, 1998a,b, 2006, 2007a). Most studies (Dias and Pooliyadda, 2001; Yeh, 1998b, 2006) have emphasized sensitivity analyses for strength to investigate the effects of mixture constituents, such as cement, water, silica fume, and fly ash. Such sensitivity analyses could identify the relationship between strength and individual constituents of concrete mixtures. However, as mentioned earlier, there has been an ever-growing need for concrete mixtures to meet multiple objectives and satisfy many design criteria. Thus, developing appropriate AI models for multi-objective optimization (e.g., strength-cost-CO_2 optimization) will contribute to the concrete industry (Yeh, 1998a, 2007a; Young et al., 2019).

Cluster #3 was automatically labeled as "using neural network". Manual examination of the five citing publications (Chang and Yang, 2006; Kwon and Song, 2010; Öztaş et al., 2006; Pala et al., 2007; Song and Kwon, 2009) confirmed that neural network models were used in these studies. This suggests the key role of neural networks in concrete research, which is in entire agreement with the observation in Section 8.3. Another focus of this cluster was durability, such as carbonation (Kwon and Song, 2010), chloride penetration (Song and Kwon, 2009), later-age compressive strength (Pala et al., 2007), and residual strength after high temperature (Chang and Yang, 2006).

The fourth largest cluster with 47 cited references, cluster #4, was identified as "load-carrying capacity", which is not a clear labeling. We reviewed the content of the six publications in this cluster and identified shear capacity (Stegemann and Buenfeld, 2002) or elastic modulus (Demir, 2008; Gesoğlu et al., 2010; Yan and Shi, 2010) as the focus. Cluster #5 labeled as "cement paste" had 46 cited references but only two citing articles (Stegemann and Buenfeld, 2001, 2002). These two studies investigated the effects of pure inorganic or toxic metal compounds on compressive strength (Stegemann and Buenfeld, 2002) and setting time (Stegemann and Buenfeld, 2001) of cement paste systems by neural network models, providing financial benefits and environmental-friendly solutions for the treatment of industrial wastes by cement-based solidification. Hence, the focus of this cluster could be cement paste or waste solidification. Compared with other clusters, cluster #5 was the smallest cluster, which implies relatively little attention has been devoted to the research topic so far.

Note that all the high silhouette values in Table 8.2 were larger than the threshold 0.7, suggesting that each cluster is highly isolated and disjointed from each other with weak links in between (Chen, 2014) (see Fig. 8.5). This highlights the fact that researchers do not cite relevant work outside their clusters (Hicks, 1999), and thus fail to draw on a wide range of knowledge sources (Darko et al., 2020; Nerur et al., 2008). Consequently, the current body of research on AI in the concrete domain appears inward-looking, not benefiting from applicable theories or concepts from other research domains (Hosseini et al., 2018). As AI itself is a concept from the field of computer science, it is expected that interdisciplinary research could expand the frontier of knowledge in the concrete field.

5. Discussion and Future Trends

AI techniques have entered the concrete science toolbox in the early 1990s and an exponential increase in relevant publications has been witnessed from the last decade (Fig. 8.2). This trend confirms the growing research interest of AI in the field of concrete materials. Existing literature (period: 1990–2020) on AI in concrete materials was examined in the preceding sections by scientometric tools, including keyword co-occurrence analysis (Section 8.3) and documentation co-citation analysis (Section 8.4). The scientometric analysis extends earlier review studies of the field by providing a quantitative interpretation of the past and present knowledge structure and offering a complete picture and understanding of AI literature in the concrete domain. Most research efforts have been directed towards prediction of concrete properties, in particular compressive strength, and substantial successes have been achieved by the use of neural network models. For example, the highest cited work by Yeh (1998b) (Table 8.4) employed neural network models on a dataset of over 1,000 samples to predict the compressive strength of high performance concrete, and obtained a high coefficient of determination (R^2) of about 0.91 to 0.92. These neural network models with concrete mixture constituents as the input significantly outperformed the traditional regression analysis merely based on water-to-cement ratio and age ($R^2 \approx 0.57$). Since then, the neural network approach has received the most attention for predicting properties of different types of concrete (Ben Chaabene et al., 2020; Dias and Pooliyadda, 2001; Dingqiang et al., 2020; Duan et al., 2013a; Ramkumar et al., 2020; Young et al., 2019). In this section, we provide our perspectives on the key challenges and opportunities within the domain of AI in concrete science based on the results presented earlier.

5.1 Data

As discussed in Section 8.3, existing research efforts have focused on the prediction of hardened properties of concrete materials, while modeling behaviors of concrete in fresh state has received far less attention in current literature. This could be attributed to the difficulty in constructing datasets for concrete fresh properties given that measurement principles and results vary among testing methods or instruments (e.g., rheometers). Further examination of the retrieved 389 publications showed that the majority of studies have relied heavily on small datasets with less than 200 experimental data points (Ben Chaabene et al., 2020; Li et al., 2022), which could be insufficient for AI techniques, especially neural network models with a large number of parameters to be calibrated, and may lead to overfitting of AI models and poor generalization performance on new datasets.

The lack of large and universal datasets for concrete materials becomes a major challenge towards the wider adoption of AI in the construction industry. Many tens of thousands of concrete mixtures as well as their physical properties have been reported in the literature, but collecting these data could be a daunting data mining problem. Research articles are written in different formats by researchers in different laboratories, using different experimental parameters; methods and results are presented by different terms with inconsistent levels of completeness; thus, manual extraction of data from literature could be rather time-consuming and labor-intensive (Cao et al., 2018). In addition, null or negative results tend to be discarded and not published, although these results would be extremely useful in the context of AI. These lead to a call for standardization of data reporting criteria and sharing of research data, which are a big step towards open science (Gewin, 2016).

There are also other approaches worth exploring when dealing with data sparsity issues. One promising yet understudied topic in concrete research is transfer learning, where the knowledge learned from one material or property can be used to develop models for a similar material or property (Pan and Yang, 2010; Ravinder et al., 2021). As discussed in Section 8.3, existing literature has mainly focused on specific concrete types (e.g., high performance concrete) and concrete properties (e.g., compressive strength). Thus, it is reasonable to assume that more datasets related to these types of concrete and properties are available. By using transfer learning, reliable models could be developed for other types of concrete or other properties that have received far less attention with only sparse data available (Gopalakrishnan et al., 2017; Moussa and Owais, 2020; Tong et al., 2020; Yang et al., 2020).

Another solution to data sparsity is extracting composition-property datasets via natural language processing or text mining approaches (Batra et al., 2020). Data are typically presented in an unstructured form in a research article: researchers may report the compositions of concrete mixtures in one table, the testing conditions and experimental parameters in the body text of the methods section, and then the final properties in figures within the results section (Olivetti et al., 2020). Given the highly distributed nature of data, automated mining information from the literature could be more effective and reliable than manual collection, although there has been no work attempting to apply this approach in the concrete domain.

5.2 Models

Neural network models have been extensively used in current concrete literature (see Sections 8.3 and 8.4). The success of neural network models in concrete research provides a glimpse into possible future applications. An emerging trend is the development and deployment of convolutional neural networks, a popular class of deep neural networks or deep learning (Khallaf and Khallaf, 2021; Krizhevsky et al., 2012; Martinez et al., 2019). Deep networks herein refer to neural networks with deeper architectures and more processing layers. Table 8.4 presents two research articles that were listed among the top cited publications and highly related to the applications of convolutional neural networks (Dorafshan et al., 2018; Gopalakrishnan et al., 2017). These two studies have demonstrated the advantages of using convolutional neural networks for crack detection in field practice, e.g., in concrete bridges (Dorafshan et al., 2018) and pavements (Gopalakrishnan et al., 2017). This can be attributed to the high accuracy of convolutional neural networks to process data with grid patterns, e.g., images (Krizhevsky et al., 2012; Visser et al., 2021). While these techniques have been explored for monitoring concrete structures at macro scales, there have been limited studies on their applications in concrete microstructural analyses (e.g., scanning electron microscopy and X-ray computed tomography) (Dong et al., 2020; Lorenzoni et al., 2020; Song et al., 2020; Tong et al., 2020), where a large number of images could be produced and require extensive efforts from researchers to manage and interpret.

Many AI models have been treated as "black boxes" in concrete research, especially for neural networks that often provide accurate predictions at the cost of high model complexity. One challenge that hinders their systematic adoption in the construction industry is the lack of interpretability in AI models (Naser, 2021). Limited research efforts have been devoted to unravel how or why a model predicts the way it does (i.e., the cause-and-effect relationship). Besides building an accurate model, it is crucial to gain insights and knowledge from the model in the concrete domain. Testable hypotheses resulting from interpretable models can be cross-validated by proposed experiments. Interpretability has been an active area of research in computer science (Lipton, 2018; Molnar, 2019; Molnar et al., 2020; Montavon et al., 2017). Moving forward, AI models that offer scientific understanding will become increasingly desirable in the concrete research field.

A deep dive into the retrieved concrete literature reveals that most AI research has been based on commonly accessible AI models with a main emphasis on parameter tuning. As these models are designed for data mining purposes, they require large datasets and depend to a large degree on the information contained in the data without considering known scientific principles (Karpatne et al., 2017a). This raises challenges due to the inadequacy of the available data in concrete science (discussed in Section 5.1). Recent advances in physics-guided data science have illustrated the potential of integrating known physical laws or scientific knowledge (e.g., the inverse relationship between compressive strength and water-to-cement ratio in the context of concrete research) with AI models (Karpatne et al., 2017b; Willard et al., 2020). Such an approach offers novel solutions to increase model interpretability, enhance generalization performance, and reduce the amount of data needed for training.

6. Conclusion

AI has already been utilized in concrete research for almost three decades, where it has advanced the numerical modeling and prediction of concrete properties. This chapter has examined the existing literature using scientometric approaches such as keyword co-occurrence analysis and document co-citation analysis. Results revealed that current research interests have focused on the application of neural network models in concrete compressive strength prediction, while other AI techniques as well as other physical measures remain understudied. This implies that the existing research on AI in concrete science is still immature, with the power of their merging far from being fully realized. With continued advances in data (e.g., reporting standardization, data sharing, knowledge transferring by transfer learning, and automated data collection by natural language processing or text mining) and models (e.g., deeper architectures, higher interpretability, and more physics-aware), a veritable intelligence ecosystem is emerging in concrete science and will have a profound impact on this field.

Acknowledgements

The authors would like to thank Dr. Ismaila Dabo, Dr. Jinyoung Yoon, and Rui Zhang for their insightful thoughts and valuable discussions that form the basis of this work.

References

ACI Committee 211, 2002. 211.1-91 Standard Practice for Selecting Proportions for Normal, Heavyweight, and Mass Concrete (Reapproved 2009).

Agrawal, A. and Choudhary, A. 2016. Perspective: Materials informatics and big data: Realization of the "fourth paradigm" of science in materials science. *APL Materials*, 4. doi: 10.1063/1.4946894.

Alshihri, M.M., Azmy, A.M. and El-Bisy, M.S. 2009. Neural networks for predicting compressive strength of structural light weight concrete. *Construction and Building Materials*, 23: 2214–2219. doi: 10.1016/j.conbuildmat.2008.12.003.

Asteris, P.G., Skentou, A.D., Bardhan, A., Samui, P. and Pilakoutas, K. 2021. Predicting concrete compressive strength using hybrid ensembling of surrogate machine learning models. *Cement and Concrete Research*, 145, 106449. doi: 10.1016/j.cemconres.2021.106449.

Batra, R., Song, L. and Ramprasad, R. 2020. Emerging materials intelligence ecosystems propelled by machine learning. *Nature Reviews Materials*. doi: 10.1038/s41578-020-00255-y.

Behnood, A., Behnood, V., Modiri Gharehveran, M. and Alyamac, K.E. 2017. Prediction of the compressive strength of normal and high-performance concretes using M5P model tree algorithm. *Construction and Building Materials*, 142: 199–207. doi: 10.1016/j.conbuildmat.2017.03.061.

Behnood, A. and Golafshani, E.M. 2021. Artificial Intelligence to model the performance of concrete mixtures and elements: A review. *Archives of Computational Methods in Engineering*. doi: 10.1007/s11831-021-09644-0.

Behnood, A., Olek, J. and Glinicki, M.A. 2015. Predicting modulus elasticity of recycled aggregate concrete using M5' model tree algorithm. *Construction and Building Materials*, 94: 137–147. doi: 10.1016/j.conbuildmat.2015.06.055.

Ben Chaabene, W., Flah, M. and Nehdi, M.L. 2020. Machine learning prediction of mechanical properties of concrete: Critical review. *Construction and Building Materials*, 260: 119889. doi: 10.1016/j.conbuildmat.2020.119889.

Börner, K., Chen, C. and Boyack, K.W. 2003. Visualizing knowledge domains. *Annual Review of Information Science and Technology*, 37: 179–255. doi: 10.1002/aris.1440370106.

Brower, B.Y.L.E. and Ferraris, C.F. 2003. Comparison of concrete rheometers. *Concrete International*, 41–47.

Buhaug, H., Urdal, H. and Østby, G. 2013. Sustainable urbanization and human security. pp. 82–92. *In: A Changing Environment for Human Security*. London: Routledge. doi: 10.4324/9780203109885-15.

Cai, R., Han, T., Liao, W., Huang, J., Li, D., Kumar, A. and Ma, H. 2020. Prediction of surface chloride concentration of marine concrete using ensemble machine learning. *Cement and Concrete Research*, 136. doi: 10.1016/j.cemconres.2020.106164.

Cao, B., Adutwum, L.A., Oliynyk, A.O., Luber, E.J., Olsen, B.C., Mar, A. and Buriak, J.M. 2018. How to optimize materials and devices via design of experiments and machine learning: Demonstration using organic photovoltaics. *ACS Nano*, 12: 7434–7444. doi: 10.1021/acsnano.8b04726.

Chang, C.H. and Yang, C.C. 2006. Artificial neural networks in prediction of concrete strength reduction due to high temperature. *ACI Materials Journal*, 103: 68–69.

Chen, C. 2014. The CiteSpace manual. *College of Computing and Informatics*, 1: 1–84.

Chen, C. 2016. *CiteSpace: A Practical Guide for Mapping Scientific Literature*. Nova Science Publishers, Hauppauge, NY.

Chen, C., Hu, Z., Liu, S. and Tseng, H. 2012. Emerging trends in regenerative medicine: A scientometric analysis in CiteSpace. *Expert Opinion on Biological Therapy*, 12: 593–608. doi: 10.1517/14712598.2012.674507.

Chen, C., Ibekwe-SanJuan, F. and Hou, J. 2010. The structure and dynamics of cocitation clusters: A multiple-perspective cocitation analysis. *Journal of the American Society for Information Science and Technology*, 61: 1386–1409. doi: 10.1002/asi.21309.

Chen, M., Sui, Y., Liu, W., Liu, H. and Huang, Y. 2019. Urbanization patterns and poverty reduction: A new perspective to explore the countries along the Belt and Road. *Habitat International*, 84: 1–14. doi: 10.1016/j.habitatint.2018.12.001.

Chithra, S., Kumar, S.R., Chinnaraju, K. and Alfin Ashmita, F. 2016. A comparative study on the compressive strength prediction models for High Performance Concrete containing nano silica and copper slag using regression analysis and Artificial Neural Networks. *Construction and Building Materials*, 114: 528–535. doi: 10.1016/j.conbuildmat.2016.03.214.

Chou, J.S., Tsai, C.F., Pham, A.D. and Lu, Y.H. 2014. Machine learning in concrete strength simulations: Multi-nation data analytics. *Construction and Building Materials*, 73: 771–780. doi: 10.1016/j.conbuildmat.2014.09.054.

Dantas, A.T.A., Batista Leite, M. and De Jesus Nagahama, K. 2013. Prediction of compressive strength of concrete containing construction and demolition waste using artificial neural networks. *Construction and Building Materials*, 38: 717–722. doi: 10.1016/j.conbuildmat.2012.09.026.

Darko, A., Chan, A.P., Adabre, M.A., Edwards, D.J., Hosseini, M.R. and Ameyaw, E.E. 2020. Artificial intelligence in the AEC industry: Scientometric analysis and visualization of research activities. *Automation in Construction*, 112: 103081. doi: 10.1016/j.autcon.2020.103081.

Demir, F. 2008. Prediction of elastic modulus of normal and high strength concrete by artificial neural networks. *Construction and Building Materials*, 22: 1428–1435. doi: 10.1016/j.conbuildmat.2007.04.004.

DeRousseau, M.A., Kasprzyk, J.R. and Srubar, W.V. 2018. Computational design optimization of concrete mixtures: A review. *Cement and Concrete Research*, 109: 42–53. doi: 10.1016/j.cemconres.2018.04.007.

DeRousseau, M.A., Laftchiev, E., Kasprzyk, J.R., Rajagopalan, B. and Srubar, W.V. 2019. A comparison of machine learning methods for predicting the compressive strength of field-placed concrete. *Construction and Building Materials*, 228: 116661. doi: 10.1016/j.conbuildmat.2019.08.042.

Dias, W.P. and Pooliyadda, S.P. 2001. Neural networks for predicting properties of concretes with admixtures. *Construction and Building Materials*, 15: 371–379. doi: 10.1016/S0950-0618(01)00006-X.

Dingqiang, F., Rui, Y., Zhonghe, S., Chunfeng, W., Jinnan, W. and Qiqi, S. 2020. A novel approach for developing a green Ultra-High Performance Concrete (UHPC) with advanced particles packing meso-structure. *Construction and Building Materials*, 265: 120339. doi: 10.1016/j.conbuildmat.2020.120339.

Dolado, J.S. and Van Breugel, K. 2011. Recent advances in modeling for cementitious materials. *Cement and Concrete Research*, 41: 711–726. doi: 10.1016/j.cemconres.2011.03.014.

Dong, Y., Su, C., Qiao, P. and Sun, L. 2020. Microstructural crack segmentation of three-dimensional concrete images based on deep convolutional neural networks. *Construction and Building Materials*, 253: 119185. doi: 10.1016/j.conbuildmat.2020.119185.

Dorafshan, S., Thomas, R.J. and Maguire, M. 2018. Comparison of deep convolutional neural networks and edge detectors for image-based crack detection in concrete. *Construction and Building Materials*, 186: 1031–1045. doi: 10.1016/j.conbuildmat.2018.08.011.

Duan, Z.H., Kou, S.C. and Poon, C.S. 2013a. Prediction of compressive strength of recycled aggregate concrete using artificial neural networks. *Construction and Building Materials*, 40: 1200–1206. doi: 10.1016/j.conbuildmat.2012.04.063.

Duan, Z.H., Kou, S.C. and Poon, C.S. 2013b. Using artificial neural networks for predicting the elastic modulus of recycled aggregate concrete. *Construction and Building Materials*, 44: 524–532. doi: 10.1016/j.conbuildmat.2013.02.064.

Ferraris, C.F., Martys, N.S. and George, W.L. 2014. Development of standard reference materials for rheological measurements of cement-based materials. *Cement and Concrete Composites*, 54: 29–33. doi: 10.1016/j.cemconcomp.2014.01.008.

Flah, M., Suleiman, A.R. and Nehdi, M.L. 2020. Classification and quantification of cracks in concrete structures using deep learning image-based techniques. *Cement and Concrete Composites*, 114, 103781. doi: 10.1016/j.cemconcomp.2020.103781.

Gesoğlu, M., Güneyisi, E., Özturan, T. and Özbay, E. 2010. Modeling the mechanical properties of rubberized concretes by neural network and genetic programming. *Materials and Structures/Materiaux et Constructions*, 43: 31–45. doi: 10.1617/s11527-009-9468-0.

Gewin, V. 2016. Data sharing: An open mind on open data. *Nature*, 529: 117–119. doi: 10.1038/nj7584-117a.

Ghafari, E., Bandarabadi, M., Costa, H. and Júlio, E. 2015. Prediction of fresh and hardened state properties of UHPC: Comparative study of statistical mixture design and an artificial neural network model. *Journal of Materials in Civil Engineering*, 27: 1–11. doi: 10.1061/(ASCE)MT.1943-5533.0001270.

Gholampour, A., Gandomi, A.H. and Ozbakkaloglu, T. 2017. New formulations for mechanical properties of recycled aggregate concrete using gene expression programming. *Construction and Building Materials*, 130: 122–145. doi: 10.1016/j.conbuildmat.2016.10.114.

Gopalakrishnan, K., Khaitan, S.K., Choudhary, A. and Agrawal, A. 2017. Deep Convolutional Neural Networks with transfer learning for computer vision-based data-driven pavement distress detection. *Construction and Building Materials*, 157: 322–330. doi: 10.1016/j.conbuildmat.2017.09.110.

Greff, K., Srivastava, R.K., Koutnik, J., Steunebrink, B.R. and Schmidhuber, J. 2017. LSTM: A search space Odyssey. *IEEE Transactions on Neural Networks and Learning Systems*, 28: 2222–2232. doi: 10.1109/TNNLS.2016.2582924, arXiv:1503.04069.

Gunasekera, C., Lokuge, W., Keskic, M., Raj, N., Law, D.W. and Setunge, S. 2020. Design of alkali-activated slag-fly ash concrete mixtures using machine learning. *ACI Materials Journal*, 117: 263–278. doi: 10.14359/51727019.

Haenlein, M. and Kaplan, A. 2019. A brief history of artificial intelligence: On the past, present, and future of artificial intelligence. *California Management Review*, 61: 5–14. doi: 10.1177/0008125619864925.

Halilaj, E., Rajagopal, A., Fiterau, M., Hicks, J.L., Hastie, T.J. and Delp, S.L. 2018. Machine learning in human movement biomechanics: Best practices, common pitfalls, and new opportunities. *Journal of Biomechanics*, 81: 1–11. doi: 10.1016/j.jbiomech.2018.09.009.

Han, T., Siddique, A., Khayat, K., Huang, J. and Kumar, A. 2020. An ensemble machine learning approach for prediction and optimization of modulus of elasticity of recycled aggregate concrete. *Construction and Building Materials*, 244: 118271. doi: 10.1016/j.conbuildmat.2020.118271.

Hicks, D. 1999. The difficulty of achieving full coverage of international social science literature and the bibliometric consequences. *Scientometrics*, 44: 193–215. doi: 10.1007/BF02457380.

Hosseini, M.R., Martek, I., Zavadskas, E.K., Aibinu, A.A., Arashpour, M. and Chileshe, N. 2018. Critical evaluation of off-site construction research: A Scientometric analysis. *Automation in Construction*, 87: 235–247. doi: 10.1016/j.autcon.2017.12.002.

Hou, J., Yang, X. and Chen, C. 2018. Emerging trends and new developments in information science: A document co-citation analysis (2009–2016). *Scientometrics*, 115: 869–892. URL: https://doi.org/10.1007/s11192-018-2695-9, doi: 10.1007/s11192-018-2695-9.

Huang, Y., Zhang, J., Tze Ann, F. and Ma, G. 2020. Intelligent mixture design of steel fibre reinforced concrete using a support vector regression and firefly algorithm based multi-objective optimization model. *Construction and Building Materials*, 260: 120457. doi: 10.1016/j.conbuildmat.2020.120457.

Karpatne, A., Atluri, G., Faghmous, J.H., Steinbach, M., Banerjee, A., Ganguly, A., Shekhar, S., Samatova, N. and Kumar, V. 2017a. Theory-guided data science: A new paradigm for scientific discovery from data. *IEEE Transactions on Knowledge and Data Engineering*, 29: 2318–2331. doi: 10.1109/TKDE.2017.2720168, arXiv:1612.08544.

Karpatne, A., Watkins, W., Read, J. and Kumar, V. 2017b. Physics-guided neural networks (PGNN): An application in lake temperature modeling. *arXiv arXiv:1710.11431*.

Khallaf, R. and Khallaf, M. 2021. Classification and analysis of deep learning applications in construction: A systematic literature review. *Automation in Construction*, 129: 103760. doi: 10.1016/j.autcon.2021.103760.

Kovler, K. and Roussel, N. 2011. Properties of fresh and hardened concrete. *Cement and Concrete Research*, 41: 775–792. doi: 10.1016/j.cemconres.2011.03.009.

Krizhevsky, A., Sutskever, I. and Hinton, G.E. 2012. ImageNet classification with deep convolutional neural networks. *Advances in Neural Information Processing Systems*, 2: 1097–1105.

Kwon, S.J. and Song, H.W. 2010. Analysis of carbonation behavior in concrete using neural network algorithm and carbonation modeling. *Cement and Concrete Research*, 40: 119–127. doi: 10.1016/j.cemconres.2009.08.022.

Leridon, H. 2020. World population outlook: Explosion or implosion? *Population and Societies*, 573: 1–4. doi:10.3917/POPSOC.573.0001.

Li, Z., Yoon, J., Zhang, R., Rajabipour, F., Srubar III, W. V., Dabo, I. and Radlińska, A. 2022. Machine learning in concrete science: Applications, challenges, and best practices. Under Revision.

Lim, M.H. and Aryadoust, V. 2021. A scientometric review of research trends in computer-assisted language learning (1977–2020). *Computer Assisted Language Learning*, 0: 1–26. doi: 10.1080/09588221.2021.1892768.

Lipton, Z.C. 2018. The mythos of model interpretability: In machine learning, the concept of interpretability is both important and slippery. *Queue*, 16: 1–28. doi: 10.1145/3236386.3241340.

Liu, Z., Cao, Y., Wang, Y. and Wang, W. 2019. Computer vision-based concrete crack detection using U-net fully convolutional networks. *Automation in Construction*, 104: 129–139. doi: 10.1016/j.autcon.2019.04.005.

Lorenzoni, R., Curosu, I., Paciornik, S., Mechtcherine, V., Oppermann, M. and Silva, F. 2020. Semantic segmentation of the micro-structure of strainhardening cement-based composites (SHCC) by applying deep learning on micro-computed tomography scans. *Cement and Concrete Composites*, 108: 103551. doi: 10.1016/j.cemconcomp.2020.103551.

Martinez, P., Al-Hussein, M. and Ahmad, R. 2019. A scientometric analysis and critical review of computer vision applications for construction. *Automation in Construction*, 107. doi: 10.1016/j.autcon.2019.102947.

Mehta, P. Kumar, and Paulo J. M. Monteiro. 2014. *Concrete: Microstructure, Properties, and Materials*. 4th ed. New York: McGraw-Hill Education.

Mo, Y.L. and Lin, S.S. 1994. Investigation of framed shearwall behaviour with neural networks. *Magazine of Concrete Research*, 46: 289–299. doi: 10.1680/macr.1994.46.169.289.

Molnar, C. 2019. *Interpretable Machine Learning: A Guide for Making Black Box Models Explainable*. URL: https://christophm.github.io/interpretable-ml-book.

Molnar, C., König, G., Herbinger, J., Freiesleben, T., Dandl, S., Scholbeck, C.A., Casalicchio, G., Grosse-Wentrup, M. and Bischl, B. 2020. Pitfalls to avoid when interpreting machine learning models. *arXiv* arXiv:2007.04131.

Montavon, G., Lapuschkin, S., Binder, A., Samek, W. and Müller, K.R. 2017. Explaining nonlinear classification decisions with deep Taylor decomposition. *Pattern Recognition*, 65: 211–222. doi: 10.1016/j.patcog.2016.11.008, arXiv:1512.02479.

Moussa, G.S. and Owais, M. 2020. Pre-trained deep learning for hot-mix asphalt dynamic modulus prediction with laboratory effort reduction. *Construction and Building Materials*, 265: 120239. doi: 10.1016/j.conbuildmat.2020.120239.

Naser, M.Z. 2021. An engineer's guide to eXplainable artificial intelligence and interpretable machine learning: Navigating causality, forced goodness, and the false perception of inference. *Automation in Construction*, 129: 103821. doi: 10.1016/j.autcon.2021.103821.

Nerur, S.P., Rasheed, A.A. and Natarajan, V. 2008. The intellectual structure of the strategic management field: An author co-citation analysis. *Strategic Management Journal*, 29: 319–336. doi: 10.1002/smj.659.

Ni, H.G. and Wang, J.Z. 2000. Prediction of compressive strength of concrete by neural networks. *Cement and Concrete Research*, 30: 1245–1250. doi: 10.1016/S0008-8846(00)00345-8.

Nunez, I., Marani, A., Flah, M. and Nehdi, M.L. 2021. Estimating compressive strength of modern concrete mixtures using computational intelligence: A systematic review. *Construction and Building Materials*, 310: 125279. doi: 10.1016/j.conbuildmat.2021.125279.

Olivetti, E.A., Cole, J.M., Kim, E., Kononova, O., Ceder, G., Han, T.Y.J. and Hiszpanski, A.M. 2020. Data-driven materials research enabled by natural language processing and information extraction. *Applied Physics Reviews*, 7. doi: 10.1063/5.0021106.

Öztaş, A., Pala, M., Özbay, E., Kanca, E., Çağlar, N. and Bhatti, M.A. 2006. Predicting the compressive strength and slump of high strength concrete using neural network. *Construction and Building Materials*, 20: 769–775. doi: 10.1016/j.conbuildmat.2005.01.054.

Pala, M., Özbay, E., Öztaş, A. and Yuce, M.I. 2007. Appraisal of longterm effects of fly ash and silica fume on compressive strength of concrete by neural networks. *Construction and Building Materials*, 21: 384–394. doi: 10.1016/j.conbuildmat.2005.08.009.

Pan, S.J. and Yang, Q. 2010. A survey on transfer learning. *IEEE Transactions on Knowledge and Data Engineering*, 22: 1345–1359. doi: 10.1109/TKDE.2009.191.

Pan, Y. and Zhang, L. 2021. Roles of artificial intelligence in construction engineering and management: A critical review and future trends. *Automation in Construction*, 122, 103517. doi: 10.1016/j. autcon.2020.103517.

Perianes-Rodriguez, A., Waltman, L. and van Eck, N.J. 2016. Constructing bibliometric networks: A comparison between full and fractional counting. *Journal of Informetrics*, 10: 1178–1195. doi: 10.1016/j.joi.2016.10.006, arXiv:1607.02452.

Pratt, D. and Sansalone, M. 1992. Impact-echo signal interpretation using artificial intelligence. *ACI Materials Journal*, 89: 178–187. doi: 10.14359/2265.

Rafiei, M.H., Khushefati, W.H., Demirboga, R. and Adeli, H. 2016. Neural network, machine learning, and evolutionary approaches for concrete material characterization. *ACI Materials Journal*, 113: 781–789. doi: 10.14359/51689360.

Ramkumar, K.B., Kannan Rajkumar, P.R., Noor Ahmmad, S. and Jegan, M. 2020. A review on performance of self-compacting concrete—use of mineral admixtures and steel fibres with artificial neural network application. *Construction and Building Materials*, 261: 120215. doi: 10.1016/j. conbuildmat.2020.120215.

Ravinder, Venugopal, V., Bishnoi, S., Singh, S., Zaki, M., Grover, H.S., Bauchy, M., Agarwal, M. and Krishnan, N.M. 2021. Artificial intelligence and machine learning in glass science and technology: 21 challenges for the 21st century. *International Journal of Applied Glass Science*, 1–16. doi: 10.1111/ijag.15881.

Small, H. 1973. Co-citation in the scientific literature: A new measure of the relationship between two documents. *Journal of the American Society for Information Science*, 24: 265–269. doi: 10.1002/ asi.4630240406.

Sobhani, J., Najimi, M., Pourkhorshidi, A.R. and Parhizkar, T. 2010. Prediction of the compressive strength of no-slump concrete: A comparative study of regression, neural network and ANFIS models. *Construction and Building Materials*, 24: 709–718. doi: 10.1016/j.conbuildmat.2009.10.037.

Song, H.W. and Kwon, S.J. 2009. Evaluation of chloride penetration in high performance concrete using neural network algorithm and micro pore structure. *Cement and Concrete Research*, 39: 814–824. doi: 10.1016/j.cemconres.2009.05.013.

Song, Y., Huang, Z., Shen, C., Shi, H. and Lange, D.A. 2020. Deep learning-based automated image segmentation for concrete petrographic analysis. *Cement and Concrete Research*, 135: 106118. doi: 10.1016/j.cemconres.2020.106118.

Stegemann, J.A. and Buenfeld, N.R. 2001. Neural network modelling of the effects of inorganic impurities on calcium aluminate cement setting. *Advances in Cement Research*, 13: 101–114. doi: 10.1680/ adcr.13.3.101.39289.

Stegemann, J.A. and Buenfeld, N.R. 2002. Prediction of unconfined compressive strength of cement paste with pure metal compound additions. *Cement and Concrete Research*, 32: 903–913. doi: 10.1016/ S0008-8846(02)00722-6.

Tam, V.W., Tam, C.M. and Wang, Y. 2007. Optimization on proportion for recycled aggregate in concrete using two-stage mixing approach. *Construction and Building Materials*, 21: 1928–1939. doi: 10.1016/j.conbuildmat.2006.05.040.

Tong, Z., Gao, J., Wang, Z., Wei, Y. and Dou, H. 2019a. A new method for CF morphology distribution evaluation and CFRC property prediction using cascade deep learning. *Construction and Building Materials*, 222: 829–838. doi: 10.1016/j.conbuildmat.2019.06.160.

Tong, Z., Guo, H., Gao, J. and Wang, Z. 2019b. A novel method for multi-scale carbon fiber distribution characterization in cement-based composites. *Construction and Building Materials*, 218: 40–52. doi: 10.1016/j.conbuildmat.2019.05.115.

Tong, Z., Huo, J. and Wang, Z. 2020. High-throughput design of fiber reinforced cement-based composites using deep learning. *Cement and Concrete Composites*, 113: 103716. doi: 10.1016/j. cemconcomp.2020.103716.

Topçu, I.B. and Saridemir, M. 2008. Prediction of compressive strength of concrete containing fly ash using artificial neural networks and fuzzy logic. *Computational Materials Science*, 41: 305–311. doi: 10.1016/j.commatsci.2007.04.009.

van Eck, N.J. and Waltman, L. 2013. *VOSviewer Manual*. Technical Report July.

van Eck, N.J. and Waltman, L. 2014. *Visualizing Bibliometric Networks*. doi: 10.1007/978-3-319-10377-813.

Visser, M., van Eck, N.J. and Waltman, L. 2021. Large-scale comparison of bibliographic data sources: Scopus, Web of Science, Dimensions, Crossref, and Microsoft Academic. *Quantitative Science Studies*, 2: 20–41. doi: 10.1162/qssa00112.

Willard, J.D., Jia, X., Xu, S., Steinbach, M. and Kumar, V. 2020. Integrating physics-based modeling with machine learning: A survey. *arXiv arXiv:2003.04919*.

Yamashita, R., Nishio, M., Do, R.K.G. and Togashi, K. 2018. Convolutional neural networks: An overview and application in radiology. *Insights into Imaging*, 9: 611–629. doi: 10.1007/s13244-018-0639-9.

Yan, K. and Shi, C. 2010. Prediction of elastic modulus of normal and high strength concrete by support vector machine. *Construction and Building Materials*, 24: 1479–1485. doi: 10.1016/j.conbuildmat.2010.01.006.

Yang, Q., Shi, W., Chen, J. and Lin, W. 2020. Deep convolution neural network-based transfer learning method for civil infrastructure crack detection. *Automation in Construction*, 116: 103199. doi: 10.1016/j.autcon.2020.103199.

Yeh, I.C. 1998a. Modeling concrete strength with augment-neuron networks. *Journal of Materials in Civil Engineering*, 10: 263–268. doi: 10.1061/(asce)0899-1561(1998)10:4(263).

Yeh, I.C. 1998b. Modeling of strength of high-performance concrete using artificial neural networks. *Cement and Concrete Research*, 28: 1797–1808. doi: 10.1016/S0008-8846(98)00165-3.

Yeh, I.C. 2006. Analysis of strength of concrete using design of experiments and neural networks. *Journal of Materials in Civil Engineering*, 18: 597–604. doi: 10.1061/(asce)0899-1561(2006)18:4(597).

Yeh, I.C. 2007a. Computer-aided design for optimum concrete mixtures. *Cement and Concrete Composites*, 29: 193–202. doi: 10.1016/j.cemconcomp.2006.11.001.

Yeh, I.C. 2007b. Modeling slump flow of concrete using second-order regressions and artificial neural networks. *Cement and Concrete Composites*, 29: 474–480. doi: 10.1016/j.cemconcomp.2007.02.001.

Young, B.A., Hall, A., Pilon, L., Gupta, P. and Sant, G. 2019. Can the compressive strength of concrete be estimated from knowledge of the mixture proportions? New insights from statistical analysis and machine learning methods. *Cement and Concrete Research*, 115: 379–388. doi: 10.1016/j.cemconres.2018.09.006.

Chapter 9

Active Learning Kriging-Based Reliability for Assessing the Safety of Structures
Theory and Application

Koosha Khorramian[1] and *Fadi Oudah*[2,*]

1. Introduction

Structural reliability is concerned with the probabilistic evaluation of safety by considering the inherent randomness in structural response and applied loads. It has been historically used in calibrating structural design codes and standards by following the limit state design approach, where factored resistance should be equal to or greater than the factored load (Allen, 1975, 1991). One of the main reasons that hinder the widespread use of reliability analysis by structural engineers in the design and assessment of structures, as opposed to code calibration, is the lack of reliable computer tools designed to assess the reliability of rather complex structural interactions as is the case in real-life engineering problems. For complex structural systems, artificial intelligence (AI) can be used to supplement existing reliability methods to evaluate structural safety.

Kriging-based methods have been developed to help optimize the reliability calculation with the ultimate objective of inciting the use of structural reliability by practicing engineers to assess the safety of complex structural systems. Kriging, as detailed in this chapter, is a form of stochastic regression that can be used to develop surrogate models for reliability calculations (Kaymaz, 2005). It was originated in the geoscience mapping industry and found its way to structural reliability (Whitten,

[1] Postdoctoral Fellow, Department of Civil and Resource Engineering, Dalhousie University, 1360 Barrington Street, Halifax, NS, B3H 4R2 Canada. Email: Koosha.Khorramian@dal.ca
[2] Assistant Professor, Department of Civil and Resource Engineering, Dalhousie University, 1360 Barrington Street, Halifax, NS, B3H 4R2 Canada.
* Corresponding author: Fadi.Oudah@dal.ca

1997). It has been augmented with learning functions to further optimize the reliability calculation to form what is currently referred to as active learning Kriging (AK), a form of AI. The primary objective of using AK reliability is to reduce the computational cost for solving complex limit states functions that require calling advanced computational techniques to quantify the load part of the limit state, the resistance part, or both the load and resistance parts. Research related to AK in the geotechnical application is more active as compared with the use of AK in structural engineering applications (El Haj and Soubra, 2020; Buckley et al., 2021). Although the fundamental framework of conducting AK is insensitive to the application, the choice of key input functions is case-specific and should be determined based on a sensitivity analysis.

The impact of using AK in structural reliability analysis is presented in the following example. Let a structural engineer be tasked with quantifying the reliability of an existing industrial steel derrick tower subjected to lateral wind load. The existing derrick experiences localized damages in connections and reduced thickness due to corrosion. To quantify the reliability index of the structure, the engineer would need to establish the load model and the resistance model. The load model can be established relatively easily based on historical site-specific wind data or wind statistics used in the calibration of the respective code. Unlike the load model, building the resistance model is complex, especially for evaluating the resistance of existing structures with localized damage. Numerical analysis such as finite element (FE) simulation is typically used to quantify the global resistance of existing structures with localized deficiencies since a closed-form solution may not always be feasible. Random fields can also be used to account for the spatial variability in the material resistance and damage due to corrosion as part of the FE analysis. The engineer is now faced with a major challenge represented in the need to call the FE analysis for every random trial as part of the Monte Carlo (MC) simulation (which will be explained in this chapter) used to solve the reliability problem. The number of trails can be as high as 10^7 for highly reliable structures (i.e., structures with a low probability of failure). Our engineer appreciates that calling the FE analysis for every MC trail is not a feasible option due to time constraints—every FE analysis may take hours to run depending on the problem. The use of AK to conduct the MC simulation is an ideal solution to this problem. It can substantially reduce the number of required FE analyses to build the resistance model by learning the pattern of the analysis outcome derived from an initial small set of models. The objective of this chapter is to provide the engineer in our example with a single point of reference for conducting AK-based reliability analysis focused on structural engineering applications. The chapter starts with reviewing key concepts in structural reliability and Kriging, followed by presenting ordinary Kriging, and AK MC simulation. The chapter concludes with three practical examples of applying AK in structural engineering problems.

2. Review of Key Concepts

In this section, the key concepts of the reliability analysis using the active learning technique will be introduced. The basics of structural reliability will be elaborated, followed by providing key illustrations for the role of the surrogate models in the analysis. The concept of the active learning approach is explained building on the surrogate role. The main target is to find a way to increase the efficiency of the reliability analysis using nonlinear regression and optimization techniques.

Reliability-based analysis in structural engineering deals with the probability of failure of a structure under a prescribed loading condition. Once the effect of the applied loads [L(**X**)] exceeds the resistance of the structure [R(**X**)], failure occurs (Fig. 9.1(a)). In another form, the failure can be defined as a negative limit state function (LSF), also called the performance function [G(**X**)]. The performance function shown in Fig. 9.1, is defined as the resistance minus the load [G(**X**) = R(**X**) – L(**X**)]. It should be highlighted that this performance function is defined for the normal distributions of the load and resistance. For other types of distributions, different performance functions can be built by converting them to the normal distribution. The vector of random variables which is shown as **X** in this chapter is the input for load, resistance, and performance functions. Each vector **X** contains a set of random variables (i.e., X_1, X_2, ..., X_N), which can be selected based on the problem type. For example, in finding the reliability of a steel-reinforced concrete (RC) beam, concrete strength (f'_c), steel yielding stress (f_y), depth of steel bars in tension (*d*), the width of the beam (*b*), moment due to dead and live loads (M_D and M_L) can be considered as random variables, which form a vector of random variables as follows:

$$X = (X_1, X_2, X_3, X_4, X_5, X_6) = (f'_c, f_y, d, b, M_D, M_L) \tag{9.1}$$

In reliability problems, the performance function is related to many factors such as material behavior and properties, the geometry of the resisting structure, types of the applied loads, and the variability in the system. The accuracy of a reliability model can be improved by utilizing accurate and advanced methods in modeling the structural resistance, such as FE or finite difference (FD) models, and by

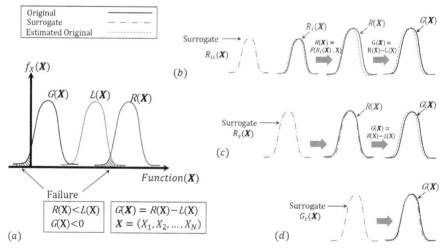

Fig. 9.1: Surrogate modeling in reliability analysis: (a) load and resistance distributions; (b) part of the resistance as the surrogate function; (c) surrogate resistance function; and (d) surrogate performance function.

considering as many details as possible. However, irrespective of the complexity and comprehensiveness of the model, there exist errors in the modeling due to uncertainty in the system and simplifying assumptions.

Rigorous resistance models lead to a high computational cost for reliability analysis purposes, where the number of required calls for the resistance model is high. For example, the number of calls required to determine the probability of failure of a certain structure can be in the order of 10^5 to 10^{10} using MCS, which depends on the type of the problem and the expected probability of failure.

To optimize the reliability analysis, the cost of the calculation can be reduced using the surrogate models instead of precise but time-consuming models, named as the original model in this chapter. The original model can be expressed at least in three different forms: (1) a part of the resistance model (Fig. 9.1(b)); (2) the resistance model (Fig. 9.1(c)); (3) the performance function (Fig. 9.1(d)).

For some resistance models, the model can be divided into computationally expensive and inexpensive sections. The original model can be set as the time-consuming part and be replaced with a surrogate model. For example, the section analysis for the moment capacity of an RC beam requires many iterations to achieve the force balance by varying the depth of the neutral axis. Alternatively, the calculation of the moment capacity can be separated into simple parts by knowing the strain profile and a surrogate model for finding the depth of the neutral axis. As illustrated in Fig. 9.1(b), the surrogate resistance of a portion of the resistance could be found first, followed by building the resistance model and the performance function. The original model can be considered as the whole resistance for analysis which is complex and cannot be separated into parts. For example, FE or FD modeling of a structural element such as a concrete column, a shear wall, or a steel frame can be considered as the surrogate model, as shown in Fig. 9.1(c). Also, the original model can be considered as the whole performance function. For example, a group of piles in the soil using both loads and resistance, or a full-bridge structure with dynamic loading (Fig. 9.1(d)). All these forms result in a surrogate performance function that estimates the original model with a lower cost of calculation.

To build a surrogate model, several realizations of the original model are required to train the surrogate model. The training process of the surrogate model is essentially beneficial in optimizing the calculation cost-efficiency in comparison to using the original model for the reliability analysis. For example, if a reliability analysis requires 1,000 different MCS, each of order of 10^6, a total of 10^9 calls for the original model is required while the surrogate model may need only 10^5 calls for the original model, which is 10,000 times less. Although surrogate models reduce the calculation cost, there exists a considerable cost to train the surrogate model which may require a high number of calls for the original model. Therefore, some methods can be applied to select the training database using a certain optimization criterion to reduce the cost of training, which are called active learning methods.

Active learning aids the surrogate in the sequential back-and-forth training and updating procedure using the original model and the surrogate model. An initial training set is built with a few numbers of realizations of the original model, called the design of experiment (DoE). A surrogate model is trained using this DoE, and a function is evaluated for several inputs using the surrogate model. The evaluated function is called the learning function, which determines the next required point to be selected for training from all evaluated inputs. The original model should be used to find a realization of the next point to update the DoE. The procedure is followed by building another surrogate model and updating the DoE up to reaching stopping criteria, which is set to obtain a desired level of accuracy. The procedure can significantly reduce the number of calls for the original model. For example, if for reliability analysis, a total of 10^9 calls for the original model is required, and surrogate training require 10^5 calls, the active learning procedure may only require 100 calls, which is 1,000 times less than the surrogate model without active learning and 10^7 times less than the reliability analysis with the original model.

One of the shortcomings in the active learning procedure is that the original model must be selected as the performance function. The idea behind the selectively chosen training dataset comes from optimizing a function to be accurate about a certain value. For reliability analysis, the failure corresponds to a certain value (i.e., zero) for the performance function. Therefore, using the active learning techniques, the value of the performance function can be found with great accuracy in the vicinity of zero, and the sign of the performance function can be determined using this method. The latter is sufficient to determine the failure for each trial using MCS for reliability analysis.

Figure 9.2 shows the MCS procedure and compares the use of active learning techniques, the surrogate model, and the original model. For MCS analysis with n trials, where each trial consists of N random variables, the trial evaluation is shown in Fig. 9.2(a). For each trial, the value of each random variable can be randomly selected based on the distribution type and first and second moment (i.e., mean and variance) of their corresponding distribution, to form a unique input vector (i.e., $\mathbf{X}$) for the performance function. As shown in Fig. 9.2(b), the performance function evaluation can be divided into two regions: (1) safety zone (i.e., $G(\mathbf{X}) \geq 0$); and (2) failure zone (i.e., $G(\mathbf{X}) < 0$). Once the performance function is evaluated for all n trials, the probability of failure can be obtained by dividing the number of failed trials (i.e., n_f) by the total number of trials (i.e., $P_f = n_f/n$). Figure 9.2(c) shows the distribution of the performance function for an MCS using the original model, a surrogate model, and active learning. The accuracy of these three simulations in the prediction of the probability of failure is almost the same, as the area of a negative performance function for all three methods (or the number of failed trials) is almost the same. The MCS using the original model and surrogate model share very similar distribution and statistical moments for the performance function. However, the MCS found by active learning cannot predict the distribution and statistical moments accurately.

Therefore, for the reliability analysis where the interest is only the probability of failure, active learning techniques offer great potential as they can predict the probability of failure with considerably lower calculation costs and a high degree of accuracy. However, for reliability analysis where the distribution of the performance function is required and its corresponding statistical moments, the surrogate models without active learning are reasonable. Finally, it should be mentioned that if the performance function is not complex and time-consuming, such as a linear function

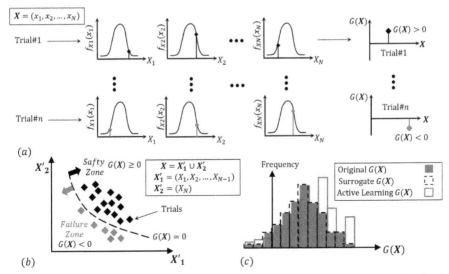

Fig. 9.2: Monte Carlo simulation (MCS): (a) generation of random vector trials; (b) reduced N-dimensional illustration of the trials; (c) active learning, surrogate, and original histograms of the performance function.

or a simple nonlinear function, the use of the original model is suggested. In other words, the surrogate models are recommended if a high degree of nonlinearity exists, or a complex time-consuming performance function is required to be evaluated multiple times.

3. Kriging

Kriging is a nonlinear stochastic regression determined by minimizing the variance of the response (Kaymaz, 2005). To predict a response, the model should be trained using a set of predefined inputs and outputs, where each input set is called a design site (S_i) and each output is called a response (Y_i). Each design site contains an *n*-dimensional vector of random variables, and each response is a *q*-dimensional output of the original model. For example, if the problem is finding the response of a steel RC beam under uniform dead and live loads, the design site can be a vector containing concrete strength (f'_c), steel yielding stress (f_y), depth of steel bars in tension (d), the width of the beam (b), moment due to dead and live loads (M_D and M_L), and the response can be a vector containing the ultimate moment capacity (M_Q) and deflection of the beam at the mid-span (Δ_u). For this example, Eqs. 9.2 and 9.3 show a single design site and its response, respectively.

$$S_i = [S_{i1}, S_{i2}, S_{i3}, S_{i4}, S_{i5}, S_{i6}] = [f'_c, f_y, d, b, M_D, M_L]_i \, ; n = 6 \qquad (9.2)$$

$$Y_i = [Y_{i1}, Y_{i2}] = [M_Q, \Delta_u]_i \, ; q = 2 \qquad (9.3)$$

A set of all design sites and their corresponding responses is called a DoE. An expansion of Eqs. 9.2 and 9.3 is presented for a general case with *m* design sites and *m* responses in Eqs. 9.4 and 9.5, respectively.

$$S = \begin{bmatrix} s_{11} & \cdots & s_{1n} \\ \vdots & \ddots & \vdots \\ s_{m1} & \cdots & s_{mn} \end{bmatrix} = \begin{bmatrix} S_1 \\ \vdots \\ S_m \end{bmatrix}; S_i = [s_{i1} \cdots s_{in}]; i = 1, ..., m \tag{9.4}$$

$$Y = \begin{bmatrix} y_{11} & \cdots & y_{1q} \\ \vdots & \ddots & \vdots \\ y_{m1} & \cdots & y_{mq} \end{bmatrix} = \begin{bmatrix} Y_1 \\ \vdots \\ Y_m \end{bmatrix}; Y_i = [y_{i1} \cdots y_{iq}]; i = 1, ..., m \tag{9.5}$$

where S and Y are design sites and response matrices containing m different, n-dimensional, and q-dimensional design sites and responses, respectively.

For a training set with a size of m, the responses to design sites in Eq. 9.4 can be obtained using the original model to build the response as Eq. 9.5. The design sites and responses can be used to train a Kriging predictor which can estimate the response of the system ($\hat{y}(X)$) for a new design site (X) whose response is unknown (Eq. 9.6), where the Kriging response ($\hat{y}(X)$) is a replacement for the actual response of the system ($y(X)$) that can be found using the original model.

$$\hat{y}(X): \mathbb{R}^n \rightarrow \mathbb{R}^q; \ \hat{y}(X) = [\hat{y}_1(X)...\hat{y}_q(X)]; \ X = [X_1 ... X_n] \tag{9.6}$$

Each estimated response (i.e., $\hat{y}_j(X), j = 1, ..., q$) can be determined separately, and the calculation for one response can be expanded to all of them.

The Kriging predictor for one response is presented in Eq. 9.7 that consists of a regression function and a stochastic process.

$$\hat{y}_j(X) = \mathcal{F}(\beta_j, X) + z_j(X); \ j = 1, ..., q \tag{9.7}$$

where β_j is a matrix of regression coefficients to be found by the predictor via optimization of mean squared error (MSE) of the predictor, $\mathcal{F}$ is the regression function, and z_j is the stochastic process for response j (out of q).

The regression function ($\mathcal{F}$) can be stated as a linear combination of p different functions with unknown coefficients, as presented in Eq. 11.8.

$$\mathcal{F}(\beta_j, X) = \beta_{1j}f_1(X) + \cdots + \beta_{pj}f_p(X) \tag{9.8}$$

where $f_i(X)$ is the regression function i (out of p), β_{ij} is the regression parameter for regression function i (out of p), and response j (out of q).

The matrix format of Eq. 118 is shown in Eq. 9.9.

$$\mathcal{F}(\beta_j, X) = [f_1(X) \cdots f_p(X)] \begin{bmatrix} \beta_{1j} \\ \vdots \\ \beta_{pj} \end{bmatrix} = f(X)^T \beta_j \tag{9.9}$$

where $f(X)$ is a vector of regression functions evaluated at a given input X.

By substituting Eq. 9.9 in Eq. 9.7, the predictor for one of the responses (i.e., ($\hat{y}_j(X)$) can be found and substituted in Eq. 11.6, to get the generalized matrix format of the predictor (i.e., Eq. 9.10).

$$\hat{y}(X) = \left[\hat{y}_1(X)...\hat{y}_q(X)\right] = \left[f_1(X)...f_p(X)\right]\begin{bmatrix} \beta_{11} & \cdots & \beta_{1q} \\ \vdots & \ddots & \vdots \\ \beta_{p1} & \cdots & \beta_{pq} \end{bmatrix} + \left[z_1(X)...z_q(X)\right] \quad (9.10)$$

The regression part of the predictor is a linear combination of functions and unknown coefficients. Three of the most popular regression functions are polynomials of order zero, one, and two which are called constant, linear, and quadratic regression functions, shown in Eqs. 9.11, 9.12, and 9.13, respectively.

$$f_1(X) = 1 \quad (9.11)$$

$$f_1(X) = 1;\ f_2(X) = x_1;...;\ f_{n+1}(X) = x_n \quad (9.12)$$

$$f_1(X) = 1;$$
$$f_2(X) = x_1;...;\ f_{n+1}(X) = x_n;$$
$$f_{n+2}(X) = x_1^2;...;\ f_{2n+1}(X) = x_1 x_n; \quad (9.13)$$
$$f_{2n+2}(X) = x_2^2;...;\ f_{3n}(X) = x_2 x_n;...;\ f_p(X) = x_n^2$$

where n is the number of design sites, and p is the number of polynomial functions that can be determined using Eq. 9.14.

$$p = \begin{cases} 1 & \text{for constant} \\ n+1 & \text{for linear} \\ \dfrac{1}{2}(n+1)(n+2) & \text{for quadratic} \end{cases} \quad (9.14)$$

The evaluation of the design sites using the regression function is shown in Eqs. 9.15 and 9.16, which are essential for the Kriging predictor.

$$F = \begin{bmatrix} f(S_1)^T \\ \vdots \\ f(S_m)^T \end{bmatrix} = \begin{bmatrix} f_1(S_1) & \cdots & f_p(S_1) \\ \vdots & \ddots & \vdots \\ f_1(S_m) & \cdots & f_p(S_m) \end{bmatrix} = \begin{bmatrix} F_{11} & \cdots & F_{1p} \\ \vdots & \ddots & \vdots \\ F_{m1} & \cdots & F_{mp} \end{bmatrix} \quad (9.15)$$

$$F_{ij} = f_j(S_i);\ i = 1,...,m;\ j = 1,...,p \quad (9.16)$$

where F is an m-by-p matrix that contains the evaluation of design sites using the regression function, F_{ij} is the evaluation of the regression function j for the design site S_i (like the example in Eq. 9.2).

The concept of regression is to minimize the error of the estimation from the real data for the training data (i.e., to minimize e). In other words, the vertical distance from each known design site to the regression function should be minimized as presented in Fig. 9.3. For the constant regression function, the prediction is the mean of the response for all the design sites. Therefore, if the predictor is used to

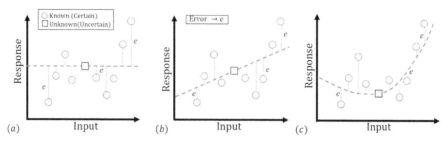

Fig. 9.3: Regression: (a) constant; (b) linear; (c) quadratic.

evaluate an unknown design site, the response is the mean value irrespective of the design site [Fig. 9.3(a)]. For the linear regression, a constant value is added to a set of linear functions in n-dimensional space which contains one line per design site (Fig. 9.3(b)). For the quadratic regression, a constant is added to a set of quadratic functions in n-dimensional space (Fig. 9.3(c)).

The difference between Kriging and conventional regression relates to the selection of the error function. For regression, the error is a deterministic value while for Kriging the error is a stochastic process with a mean of zero. The stochastic process vector for all q responses is shown in Eq. 9.17.

$$z(X): \mathbb{R}^n \rightarrow \mathbb{R}^q \; ; \; z(X) = \left[z_1(X)...z_q(X) \right]; \; X = [X_1...X_n] \tag{9.17}$$

It should be noted that the stochastic process can be also used separately to estimate the response of a system. For a stochastic process, the response for each design site and each desired unknown site is considered as a random variable. These random variables are related to each other with a correlation function (R). In other words, the response of every two design sites is two random variables correlated to each other by a correlation function. The n-dimensional correlation function can be formulated in Eq. 9.18, where n is the number of input variables vector X (similar to Eq. 9.2).

$$R(L, X_i, X_j) = \prod_{k=1}^{n} R_k(l_k, a_k, b_k); X_i = [a_1...a_n]; X_j = [b_1...b_n]; L = [l_1...l_n] \tag{9.18}$$

where R is the correlation function, R_k is component k for the correlation function, L is the vector containing correlation length of l_k (for $k = 1, ..., n$), X_i is the first n-dimensional site of a_k site values, and X_j the second n-dimensional site of b_k site values.

Two of the most popular correlation functions are Gaussian and Exponential correlation functions that are presented in Eqs. 9.19 and 9.20, respectively.

$$R_k(l_k, a_k, b_k) = \exp\left\{ -\frac{(a_k - b_k)^2}{l_k} \right\} = \exp\{-\theta_k(a_k - b_k)^2\} \tag{9.19}$$

$$R_k(l_k, a_k, b_k) = \exp\left\{ -\frac{|a_k - b_k|}{l_k} \right\} = \exp\{-\theta_k |a_k - b_k|\} \tag{9.20}$$

where θ_k is the correlation parameter for component R_k of the correlation function (correlation parameter is the inverse of the correlation length l_k).

Examples of one-, two-, and three-dimensional correlation functions are expanded in Eqs. 9.21 to 9.23.

$$1D \rightarrow R(\boldsymbol{L}, \boldsymbol{X}_i, \boldsymbol{X}_j) = \exp\left\{-\frac{(a_1 - b_1)^2}{l_1}\right\} \tag{9.21}$$

$$2D \rightarrow R(\boldsymbol{L}, \boldsymbol{X}_i, \boldsymbol{X}_j) = \exp\left\{-\frac{(a_1 - b_1)^2}{l_1} - \frac{(a_2 - b_2)^2}{l_2}\right\} \tag{9.22}$$

$$3D \rightarrow R(\boldsymbol{L}, \boldsymbol{X}_i, \boldsymbol{X}_j) = \exp\left\{-\frac{(a_1 - b_1)^2}{l_1} - \frac{(a_2 - b_2)^2}{l_2} - \frac{(a_3 - b_3)^2}{l_3}\right\} \tag{9.23}$$

Considering a one-dimensional problem, the estimation of an unknown site using the known sites is presented in Fig. 9.4. Regarding the correlation length, if the distance between two sites (i.e., $\Delta L = a_1 - b_1$) is zero (i.e., $\Delta L = 0$), the points are highly correlated, and the correlation is one (i.e., $\rho = 1$). As the distance between sites increases, the correlation between the responses decreases, as presented in Fig. 9.4(a). It should be clarified that different dimensions with different units are considered in an n-dimensional site, where the distance between two points in a dimension can be in units of length, time, force unit, or any other unit. For example, considering Gaussian and Exponential correlation functions for a one-dimensional problem, if the distance is almost equal to the correlation length (i.e., $\Delta L \approx 1$), the correlation would decrease to almost 0.36 (i.e., $\rho = 0.36$), and if the distance passes 5 times the correlation length (i.e., $\Delta L > 1$), the points will not be correlated (i.e., $\rho \approx 0$), as shown in Fig. 9.4(b).

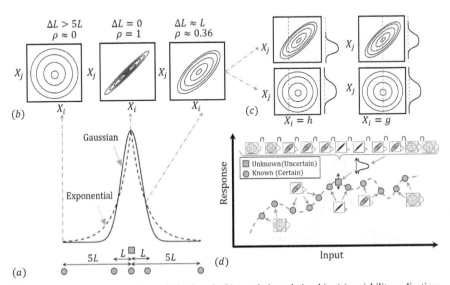

Fig. 9.4: Stochastic process: (a) correlation length; (b) correlation relationship; (c) variability realization; (c) stochastic interpolation.

For the evaluation of an unknown site, using only the stochastic process, each point is considered as a random variable that is correlated to the others. The unknown random variable would take a random posterior distribution using the certain values of the known sites, as presented in Fig. 9.4(c). The estimation of an unknown site using a set of known design sites with a stochastic process requires the union effect of the correlation of all known points with the unknown point, as presented in Fig. 9.4(d). As the distance between the points increases the effect of points on the prediction of the response would decrease (since they are less correlated), as shown in Fig. 9.4(d).

The correlation between each desired unknown and all known design sites can be expressed in a vector format, as presented in Eq. 9.24.

$$r = r(L, X, S) = \begin{bmatrix} r_1 \\ \vdots \\ r_m \end{bmatrix} = \begin{bmatrix} R(L, X, S_1) \\ \vdots \\ R(L, X, S_m) \end{bmatrix}; R(L, X, S_i) = \prod_{k=1}^{n} R_k (l_k, x_k, s_{ik}); i \quad (9.24)$$
$$= 1, ..., m$$

where x_k and s_{ik} are the components of X and S_i, respectively, and r is a vector of correlation whose components (i.e., r_i) are the correlation coefficients between the desired site (X) with an unknown response and each design site (S_i) out of m design sites.

Another important matrix is the correlation matrix between each two design sites, as expressed in Eq. 9.25.

$$R = \begin{bmatrix} R_{11} & \cdots & R_{1m} \\ \vdots & \ddots & \vdots \\ R_{m1} & \cdots & R_{mm} \end{bmatrix}; R_{ij} = R(L, S_i, S_j) = \prod_{k=1}^{n} R_k (l_k, s_{ik}, s_{jk}); i, j = 1, ..., m \quad (9.25)$$

where R is a symmetric correlation matrix, whose components are (i.e., R_{ij}) the correlation coefficients between each two design sites S_i and S_j.

The optimum correlation length for each component k from the inputs in the design sites (i.e., l_k) can be determined by Eq. 9.26 by considering the stochastic process as a Gaussian process.

$$L^* = \min_{L} \left\{ |R|^{\frac{1}{m}} \sigma_k^2 \right\}; k = 1, ..., n \quad (9.26)$$

where L^* is a vector of the optimized correlation length and σ_k is the standard deviation of component k in response (out of q responses), L is the correlation length vector, and $|R|$ is the determinant of the correlation matrix R. Determination of σ_k is explained later in this chapter.

From the correlation functions and standard deviation, the covariance between every two design sites can be expressed as the multiplication of their correlation coefficient by the variance (considering a constant variance for the whole process). Also, this covariance is the expected value of the product of the random process evaluated at these two sites. Equations 9.27 and 9.28 show the correlation between two different design sites and a design site with a site whose response is desired, respectively.

$$COV(L, S_i, S_j) = E\left(z_k(S_i)z_k(S_j)\right) = \sigma^2 R(L, S_i, S_j); i, j = 1,...,m \quad (9.27)$$

$$COV(L, X, S_i) = E\left(z_k(X)z_k(S_i)\right) = \sigma^2 R(L, X, S_i); i = 1,...,m \quad (9.28)$$

where COV is the covariance function, and $E(T)$ is the expected value of argument T.

The correlation matrices R and r are related to the covariance matrices V and v, which also can be found by expanding Eqs. 9.27 and 9.28 to form the matrix format, as expressed in Eqs. 9.29 and 9.30.

$$V = \begin{bmatrix} E\left(z_k(S_1), z_k(S_1)\right) & \cdots & E\left(z_k(S_1), z_k(S_m)\right) \\ \vdots & \ddots & \vdots \\ E\left(z_k(S_m), z_k(S_1)\right) & \cdots & E\left(z_k(S_m), z_k(S_m)\right) \end{bmatrix} = \sigma^2 R; \ k = 1, ..., q \quad (9.29)$$

$$v = \begin{bmatrix} E\left(z_k(X), z_k(S_1)\right) \\ \vdots \\ E\left(z_k(X), z_k(S_m)\right) \end{bmatrix} = \sigma^2 r; \ k = 1, ..., q \quad (9.30)$$

The correlation and covariance functions that represent the stochastic process and the regression function have been explained thus far. The predictor that combines both is the Kriging predictor, where the mean value of the predictor is determined via the regression functions, and the variance in the response is determined by the stochastic process as presented in Fig. 9.5. While for the regression, the predictor may or may not pass the responses at the design sites, for Kriging, the mean value always passes the responses corresponding to the design sites, because of the presence of the stochastic process. It should be mentioned that the type of regression function can change the predictor, as the stochastic process tends to oscillate about the regression function.

The predictor provides the mean value of the response. Other than the certain responses corresponding to the design site, all reactions corresponding to the desired site (unknown responses) are random, which means they are treated as random variables with a non-zero variance. In other words, the response to the design sites are

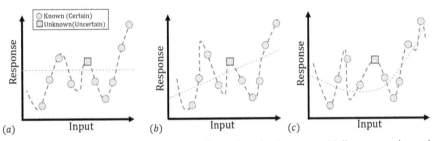

Fig. 9.5: Kriging predictor: (a) constant regression and stochastic process; (a) linear regression and stochastic process; (a) quadratic regression and stochastic process.

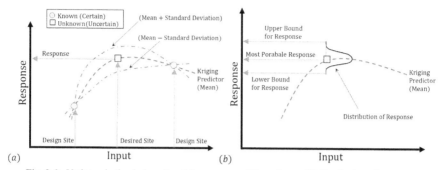

Fig. 9.6: Variance in the design sites: (a) response and its variance; (b) distribution of response.

random variables that take a certain value, but the response to the desired design sites is not certain, as shown in Fig. 9.6. Therefore, there is uncertainty in the prediction of the design site which can be shown by the standard deviation difference in Fig. 9.6(a), and by a distribution with upper and lower bounds corresponding to a proper confidence interval in Fig. 9.6(b). The variance is the measure of goodness of the curve, which means the predictor can be improved by reducing the variance of the predictor. Therefore, the idea behind finding the unknown coefficients in the Kriging predictor is minimizing the Kriging variance.

To find the unknown coefficients in the Kriging predictor, variance is required, and to build variance, an estimation of the real response is required. Consider the calculation of the response for a system with only one reaction. The evaluation of the Kriging predictor (i.e., Eq. 9.7) at each design site S_i can be shown in Eq. 9.31, and its expansion for all design sites can be expanded in Eq. 9.32.

$$y_{i1} = \beta_{11}f_1(S_i) + \cdots + \beta_{p1}f_p(S_i) + z_1(S_i) \tag{9.31}$$

$$Y = Y_1 = \begin{bmatrix} y_{11} \\ \vdots \\ y_{m1} \end{bmatrix} = \begin{bmatrix} F_{11} & \cdots & F_{1p} \\ \vdots & \ddots & \vdots \\ F_{m1} & \cdots & F_{mp} \end{bmatrix} \begin{bmatrix} \beta_{11} \\ \vdots \\ \beta_{p1} \end{bmatrix} + \begin{bmatrix} z_1(S_1) \\ \vdots \\ z_1(S_m) \end{bmatrix} = F\beta + Z \tag{9.32}$$

where Y is the matrix of the known responses for the design sites, Y_1 is the response vector for the first response set (out of q), y_{ij} is the response for the known design site S_i, and the j set of response (out of q), and Z is the vector of the stochastic process for design sites.

Assume that an estimated response ($\bar{y}(X)$) to a given site X with a desired unknown response is a linear combination of the responses for all design sites, which is presented in Eq. 9.33.

$$\bar{y}(X) = c(X)^T Y = [c_1(X) \ldots c_m(X)] \begin{bmatrix} y_{11} \\ \vdots \\ y_{m1} \end{bmatrix} = c^T (F\beta + Z) \tag{9.33}$$

where $\bar{y}(X)$ is an estimated response to X, $c(X)$ is a vector of the weight $c_i(X)$ for the response y_{i1}.

To build the variance, the estimation of the response is also required. For a Kriging predictor with only one response, the response to X is presented in Eq. 9.34.

$$\hat{y}(X) = \hat{y}_1(X) = [f_1(X)... f_p(X)]\begin{bmatrix} \beta_{11} \\ \vdots \\ \beta_{p1} \end{bmatrix} + z_1(X) = f(X)^T \beta + z \qquad (9.34)$$

where $\hat{y}(X)$ is the estimated response to site X using the Kriging equation (i.e., Eq. 9.10).

To have an unbiased estimator, it is required to have a mean of zero for the error, and the reduction in the variance is required to have the best possible estimator. Equation 9.35 shows the expected value of error, and by substituting Eqs. 9.33 and 9.34 in Eq. 9.35 and simplifying the results, the expected error or the mean of error can be shown in Eq. 9.36.

$$\mu_e = E[e] = E\left[(\bar{y}(X) - \hat{y}(X))\right] \qquad (9.35)$$

$$\begin{aligned} E\left[(\bar{y}(X) - \hat{y}(X))\right] &= E\left[\left(c^T (F\beta + Z) - (f(X)^T \beta + Z)\right)\right] \\ &= E[(c^T Z - z + (F^T c - f(X))^T \beta)] \qquad (9.36) \\ &= E[c^T Z] - E[z] + E\left[\left(F^T c - f(X)\right)^T \beta\right] \end{aligned}$$

where μ_e is the mean of the error and e is the error.

To satisfy the unbiased condition for the predictor, it is required to set the expected value of the error to zero. Thus, Eq. 9.36 should be zero. The expected values of the Z and z are zero because the mean of the stochastic process $z_i(X)$ is assumed zero for the Kriging predictor. Therefore, only the third term of the last sentence in Eq. 9.36 should be zero to satisfy the unbiased condition, as presented in Eq. 9.37.

$$E\left[(\bar{y}(X) - \hat{y}(X))\right] = E\left[\left(F^T c - f(X)\right)^T \beta\right] = 0 \rightarrow (F^T c - f(X))^T \beta = 0 \quad (9.37)$$

Since the regression coefficient matrix β cannot be zero (trivial answer), Eq. 9.37 leads to Eqs. 9.38 and 9.39 as the unbiased condition and the simplified error prediction, respectively.

$$F^T c = f(X) \qquad (9.38)$$

$$e = \bar{y}(X) - \hat{y}(X) = c^T Z - z \qquad (9.39)$$

The expected squared error is the variance of the error, also named as the mean squared error, as presented in Eq. 9.40.

$$\phi(X) = \sigma_e^2 = Var(e) = E\left[(e - \mu_e)^2\right] = E[e^2] = E\left[(\bar{y}(X) - \hat{y}(X))^2\right] \quad (9.40)$$

where σ_e^2 is the variance of the error which is a function of X and can be called $\varphi(X)$.

The use of an unbiased predictor leads to a zero mean for the error which is simplified in Eq. 9.40. By substituting Eq. 9.39 in Eq. 9.40, the variance of the error can be written as Eq. 9.41 for a response k (out of q) and can be expanded to form Eq. 9.42.

$$\varphi(X) = Var(e) = E\left[(c^T Z - z)^2\right] = E\left[z^2 + c^T Z Z^T c - 2c^T Z z\right] = E[z_k(X)z_k(X)] +$$

$$c^T E\left[\begin{bmatrix} z_k(S_1) \\ \vdots \\ z_k(S_m) \end{bmatrix}[z_k(S_1) \quad \cdots \quad z_k(S_m)]^T\right]c - 2c^T E\left[\begin{bmatrix} z_k(S_1) \\ \vdots \\ z_k(S_m) \end{bmatrix}z_k(X)\right] =$$

$$E[z_k(X)z_k(X)] + c^T \begin{bmatrix} E(z_k(S_1)z_k(S_1)) & \cdots & E(z_k(S_1)z_k(S_m)) \\ \vdots & \ddots & \vdots \\ E(z_k(S_m)z_k(S_1)) & \cdots & E(z_k(S_m)z_k(S_m)) \end{bmatrix}c -$$

$$2c^T \begin{bmatrix} E(z_k(X)z_k(S_1)) \\ \vdots \\ E(z_k(X)z_k(S_m)) \end{bmatrix} = \sigma^2 + c^T V c - 2c^T v; k = 1, ..., q \tag{9.41}$$

$$\varphi(X) = Var(e) = \sigma_k^2(1 + c^T R c - 2c^T r) \tag{9.42}$$

As discussed earlier, the unknown coefficients of the Kriging can be found by minimizing the unbiased predictor. Therefore, the conditional minimization shown in Eq. 9.43 leads to the solution.

$$\min_c \varphi(X) \ s.t. \ X \in \Omega = \{X \subseteq \mathbb{R}^n \mid F^T c(X) - f(X) = 0\} \tag{9.43}$$

Using a Lagrangian approach for the solution, Eq. 9.44 formulates the conditional optimization problem.

$$\mathcal{L}(c, \lambda) = \sigma_k^2(1 + c^T R c - 2c^T r) - \lambda^T \left(F^T c - f(X)\right) \tag{9.44}$$

where λ is the Lagrange undetermined multiplier, $\mathcal{L}$ is a Lagrangian expression, and (c, λ) is a stationary point (i.e., the point where the first partial derivatives of $\mathcal{L}$ are zero).

By setting the first derivative of Eq. 9.44, Eq. 9.45 can be written and expanded to Eqs. 9.46 and 9.47 and summarized in Eq. 9.48.

$$\frac{\partial \mathcal{L}(c, \lambda)}{\partial c} = 2\sigma_k^2(Rc - r) - F\lambda = 0 \tag{9.45}$$

$$2\sigma_k^2(Rc - r) - F\lambda = 0 \rightarrow Rc - r - F\frac{\lambda}{2\sigma_k^2} = 0 \rightarrow Rc + F\bar{\lambda} = r; \bar{\lambda} = -\frac{\lambda}{2\sigma_k^2} \tag{9.46}$$

$$R^{-1}Rc + R^{-1}F\bar{\lambda} = R^{-1}r \rightarrow c + R^{-1}F\bar{\lambda} = R^{-1}r \rightarrow c = R^{-1}r - R^{-1}F\bar{\lambda} \rightarrow c$$
$$= R^{-1}(r - F\bar{\lambda}) \tag{9.47}$$

$$c = R^{-1}(r - F\bar{\lambda}); \bar{\lambda} = -\frac{\lambda}{2\sigma_k^2} \tag{9.48}$$

By substituting the value of **c** from Eq. 9.48 into Eq. 9.38 which is the unbiased condition, Eq. 9.49, and its summarized version Eq. 9.50, determines the value of $\bar{\lambda}$ in terms of Kriging predictor parameters.

$$F^T c = f(X) \rightarrow F^T (R^{-1}(r - F\bar{\lambda})) = f(X) \rightarrow F^T R^{-1} r - F^T R^{-1} F\bar{\lambda} = f(X)$$
$$\rightarrow (F^T R^{-1} F)^{-1}(F^T R^{-1} F)\bar{\lambda} = (F^T R^{-1} F)^{-1}\left(F^T R^{-1} r - f(X)\right) \quad (9.49)$$

$$\bar{\lambda} = (F^T R^{-1} F)^{-1}\left(F^T R^{-1} r - f(X)\right) \quad (9.50)$$

By substituting the value of $\bar{\lambda}$ from Eq. 9.50 and **c** from Eq. 9.48 into the estimator $\bar{y}(X)$ from Eq. 9.33, the estimator can be presented as Eq. 9.51.

$$\bar{y}(X) = c(X)^T Y = \left(R^{-1}(r - F\bar{\lambda})\right)^T Y$$
$$= \left(r - F(F^T R^{-1} F)^{-1}\left(F^T R^{-1} r - f(X)\right)\right)^T R^{-1} Y \quad (9.51)$$
$$= r^T R^{-1} Y - \left(F^T R^{-1} r - f(X)\right)^T (F^T R^{-1} F)^{-1} F^T R^{-1} Y$$

It should be noted that **R** is a symmetric matrix, and R^{-1} is symmetric as a result. Also, since $F^T R^{-1} F$ is symmetric, $(F^T R^{-1} F)^{-1}$ is symmetric as well. Therefore, R^{-1^T} was replaced by R^{-1} and $(F^T R^{-1} F)^{-1^T}$ was replaced with $(F^T R^{-1} F)^{-1}$ in Eq. 9.51. By introducing a new parameter β^* in Eq. 9.52 and re-writing Eq. 9.51 in terms of β^*, the estimator can be written in form of Eq. 9.53.

$$\beta^* = (F^T R^{-1} F)^{-1} F^T R^{-1} Y \quad (9.52)$$

$$\bar{y}(X) = r^T R^{-1} Y - \left(F^T R^{-1} r - f(X)\right)^T \beta^*$$
$$= r^T R^{-1} Y + f(X)^T \beta^* - r^T R^{-1} F\beta^* \quad (9.53)$$
$$= f(X)^T \beta^* + r^T R^{-1}(Y - F\beta^*)$$

By defining a new parameter γ^* in Eq. 9.54, Eq. 9.53 can be written in form of Eq. 9.55 which is the Kriging predictor at its final shape.

$$\gamma^* = R^{-1}(Y - F\beta^*) \quad (9.54)$$

$$\bar{y}(X) = f(X)^T \beta^* + r(X)^T \gamma^* \quad (9.55)$$

By taking a close look at Eqs. 9.55, 9.54, and 9.52, the Kriging predictor is related to the design sites and their responses through β^* and γ^*, and it is related to the desired site only through $f(X)^T$ and $r(X)^T$, which are representative of the regression evaluation and stochastic process, respectively.

The MSE that satisfies the unbiased optimization condition (Eq. 9.43) for the Kriging predictor can be determined. By substituting **c** from Eq. 9.48 in Eq. 9.42, the optimized predictor condition is applied, and Eq. 9.56 can be derived.

$$\begin{aligned}
\varphi_k(X) &= \sigma_k^2(1 + c^T R c - 2c^T r) \\
&= \sigma_k^2 \left(1 + c^T(Rc - 2r)\right) \\
&= \sigma_k^2 \left(1 + \left(R^{-1}(r - F\bar{\lambda})\right)^T (RR^{-1}(r - F\bar{\lambda}) - 2r)\right) \\
&= \sigma_k^2 \left(1 + \left(F\bar{\lambda} - r\right)^T R^{-1}\left(r + F\bar{\lambda}\right)\right) \\
&= \sigma_k^2 \left(1 + \left(\bar{\lambda}^T F^T - r^T\right) R^{-1}\left(r + F\bar{\lambda}\right)\right) \\
&= \sigma_k^2 \left(1 + \bar{\lambda}^T F^T R^{-1} r + \bar{\lambda}^T F^T R^{-1} F\bar{\lambda} - r^T R^{-1} r - r^T R^{-1} F\bar{\lambda}\right)
\end{aligned}$$

$$(9.56)$$

where $\bar{\lambda}^T F^T R^{-1} r$ is a 1-by-1 matrix (i.e., by dimension check $(1 \times p) \times (p \times m) \times (m \times m) \times (m \times 1)$, a 1×1 matrix is the result), and it was set equal to its transpose $r^T(R^{-1})^T F\bar{\lambda}$.

Since R is symmetric, R^{-1} is a symmetric matrix, and its transpose can be replaced by itself so that $r^T(R^{-1})^T F\bar{\lambda}$ can be replaced by $r^T R^{-1} F\bar{\lambda}$ and Eq. 9.56 can be written as Eq. 9.57.

$$\varphi_k(X) = \sigma_k^2(1 + \bar{\lambda}^T F^T R^{-1} F\bar{\lambda} - r^T R^{-1} r) \tag{9.57}$$

By applying the unbiased condition and substituting $\bar{\lambda}$ from Eq. 11.50 into Eq. 9.57, Eqs. 9.58 and 9.59 can be determined.

$$\begin{aligned}
&\varphi_k(X) \\
&= \sigma_k^2 \left(1 + \left((F^T R^{-1} F)^{-1}(F^T R^{-1} r - f)\right)^T F^T R^{-1} F(F^T R^{-1} F)^{-1}(F^T R^{-1} r - f) - r^T R^{-1} r\right) \\
&= \sigma_k^2 \left(1 + (F^T R^{-1} r - f)^T \left((F^T R^{-1} F)^{-1}\right)^T F^T R^{-1} F(F^{-1} R F^{T^{-1}})(F^T R^{-1} r - f) - r^T R^{-1} r\right) \\
&= \sigma_k^2 \left(1 + (r^T R^{-1} F - f^T)\left(F^{-1} R F^{T^{-1}}\right) F^T R^{-1} F(F^{-1} R F^{T^{-1}})(F^T R^{-1} r - f) - r^T R^{-1} r\right) \\
&= \sigma_k^2 \left(1 + (r^T R^{-1} F - f^T)(F^{-1} R) R^{-1}(R F^{T^{-1}})(F^T R^{-1} r - f) - r^T R^{-1} r\right) \\
&= \sigma_k^2 \left(1 + (r^T R^{-1} F - f^T)(F^{-1} R F^{T^{-1}})(F^T R^{-1} r - f) - r^T R^{-1} r\right) \\
&= \sigma_k^2 \left(1 + (F^T R^{-1} r - f)^T (F^T R^{-1} F)^{-1}(F^T R^{-1} r - f) - r^T R^{-1} r\right)
\end{aligned}$$

$$(9.58)$$

$$\varphi_k(X) = \sigma_k^2 \left(1 + u^T (F^T R^{-1} F)^{-1} u - r^T R^{-1} r\right); \; u = F^T R^{-1} r - f \tag{9.59}$$

where k is the index that shows each response (out of q responses).

MSE or error variance can result in negative values due to the negative sign of f in Eq. 9.59. Therefore, for further use of this equation in analysis, it should be treated with care to avoid imaginary terms for the standard deviation of the error, in future calculations. To find the value of σ_k, generalized least squares fit approach can be applied (Lophaven et al., 2002a, 2002b) to form Eq. 9.60.

$$\sigma_k^2 = \frac{1}{m}(Y - F\beta^*)^T R^{-1}(Y - F\beta^*) \tag{9.60}$$

The executive summary of Kriging calculations is presented in Fig. 9.7, where the Kriging is used to find the response of any desired site (X) by starting with building the DoE, in which a set of input sites (S_i) with their corresponding responses (Y_i) are predetermined using the original model. A regression function can be chosen and the corresponding regression realization for design sites (F) can be built. A correlation function can be selected, and a correlation length vector (L) can be assumed to proceed with building the correlation matrix of the design sites (R). Then, an iterative process or a trial-and-error approach can be utilized to find the optimum correlation length (L^*) by building β^* and its corresponding σ_k^2 from

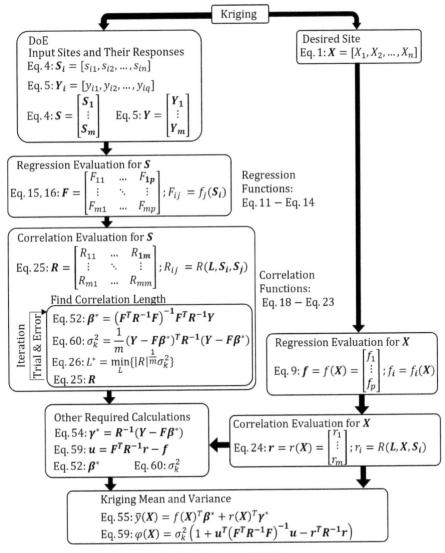

Fig. 9.7: Summary of Kriging.

Eqs. 9.52 and 9.60, respectively, and finding $|R|^{\frac{1}{m}}\sigma_k^2$ for a range of correlation length vectors (L) to find the one that satisfies Eq. 9.26. L is required to build R, and R is required to build β^* and σ_k^2, with which $|R|^{\frac{1}{m}}\sigma_k^2$ can be built. Once the optimum correlation length is found, the rest of the required parameters such as γ^*, u, r, and f can be built, with which the mean and variance of Kriging are determined, as presented in Fig. 9.7. For the calculations, the DACE toolbox for MATLAB® can be used (Lophaven et al., 2002a, 2002b), which uses advanced algebraic calculation techniques.

For reliability analysis algorithms, such as the first order reliability method (FORM) or other methods related to the local gradient search, the gradient of response and error is required, which can be found in Appendix 9A.1. As the scope of this chapter is limited to MCS, only Active learning Kriging Monte Carlo simulation (AK-MCS) will be discussed in the following section.

4. Active Learning Kriging Monte Carlo Simulation (AK-MCS)

The reliability of engineering structures can be determined using the Kriging predictor as a surrogate to the original model, and by combining it with a choice of a reliability method. For complex engineering problems that require heavy numerical calculations such as FE or FD analysis, the evaluation of the performance function is expensive and should be minimized to reach an optimized reliability analysis. Monte Carlo simulation (MCS) is the most accurate method if the required number of trials are considered. Therefore, with the advantage of having a cost-efficient surrogate model, MCS is chosen to be studied combined with Kriging.

There are gradient-based reliability methods such as first order second moment reliability method (FOSM), FORM, second order reliability method (SORM), or variance reduction techniques such as importance sampling (IS) or line sampling (LS) that can be compared with a combination of Kriging and reliability methods. For gradient-based reliability methods, an algorithmic search is required to find the most probable point (MPP) or the design point for the performance function and build a first order (for FORM) or second order (for SORM) approximation function to the performance function at MPP. In reliability analysis, MPP is defined as the point in standard normal space with the least distance from the origin. The gradient-based methods are efficient compared to MCS, but they may experience inaccuracies for highly nonlinear performance functions or multimodal problems. The algorithmic search for gradient-based methods may fail to converge for nonsmoothed stepwise gradients for highly nonlinear problems or may converge to a poor solution for problems with multiple local minima. For variance reduction techniques, a gradient-based search should be conducted to find the MPP at first, followed by an MCS about the MPP, which implies that the same inaccuracies would exist while using these techniques.

MCS, gradient-based methods, and variance reduction techniques can be used with the Kriging predictor, which has a smooth gradient and a low cost of calculation.

However, to build a reliable Kriging predictor for the analysis, it should be trained with enough design sites in the design space. The number of required sites depends on the degree of nonlinearity and the number of random variables for the performance function. The challenge with a combination of Kriging and other methods is that the Kriging predictor should have a good degree of accuracy for the whole domain of its input arguments. Although the training cost for Kriging is less than the cost of MCS calculation, training could be expensive for practical engineering reliability problems analyzed using standard computer devices. Therefore, to avoid the cost of training, the Kriging method can be further optimized to have enough accuracy around a certain value for the performance function instead of the whole domain. The latter can be achieved by active learning techniques that provide enough accuracy for the performance function about the desired value. This desired value corresponds to the failure surface or a performance function of zero for reliability analysis. The optimization is executed by minimizing the number of design sites by actively choosing the best candidates for training the Kriging to reach the desired accuracy. In this chapter, only the combination of MCS and optimized Kriging trained by active learning techniques is presented. The latter is also called active-learning Kriging Monte Carlo simulation or AK-MCS.

The algorithm for conducting AK-MCS is presented in Fig. 9.8. To start, m design sites can be selected randomly, and their corresponding responses can be determined using the original model, as shown in Fig. 9.8(a). This set of initial inputs and outputs is called the initial DoE. The selection of m random sites is adopted from a study conducted by Echard et al. (2011) for AK-MCS. The latter was selected by justifying the advantage of having fewer required initial points compared to an earlier study by Bichon et al. (2008), which recommended the Latin Hypercube Sampling (LHS) with a size of $n(n + 1)/2$ for the initial DoE (i.e., minimum points for the quadratic regression function in Eq. 9.14 where n is the number of input random variables). In this chapter, the initial DoE is selected only based on m randomly selected sites and their responses (Fig. 9.8(a)), where m can be set to be any value greater than the minimum required number of points based on the selected regression function (see Eq. 9.14).

Once the initial DoE is generated, using the Kriging algorithm explained in the previous section (see Fig. 9.7), the initial Kriging predictor can be built. The surrogate model is considered as a replacement for the original function at this stage. From the reliability side, many randomly generated input design sites (i.e., N_{MCS}), which form an input matrix of X_{MCS} is generated. The X_{MCS} is then evaluated using the initial Kriging predictor to have the mean and standard deviation of each MCS trial. It should be emphasized that by using the Kriging predictor, each predicted value is a random variable (Fig. 9.6) which is characterized as a normal distribution with its mean and standard deviation determined by Kriging (Fig. 9.7), as explained previously. While the mean value of the Kriging from Eq. 9.55 can be used directly, the value of Eq. 9.59, which is the variance, cannot be used directly. There is a chance for the variance to be negative using the Kriging formulation, whose square root is imaginary and should be avoided. Thus, the absolute value of Eq. 9.59 should be squared to be considered as the standard deviation.

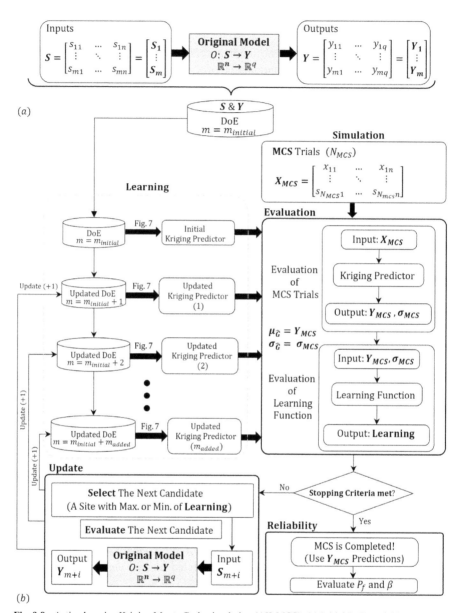

Fig. 9.8: Active learning Kriging Monte Carlo simulation (AK-MCS): (a) Initial DoE; and (b) procedure.

The matrices for the mean and the standard deviation of the predictor are called Y_{MCS} and σ_{MCS}, respectively, as presented in Fig. 9.8. These two are also equivalent to the mean (i.e., $\mu_{\hat{G}}$) and standard deviation (i.e., $\sigma_{\hat{G}}$) of the estimation to the performance function (i.e., $\hat{G}$) for each trial. The mean and standard deviation found by the initial Kriging predictor are the inputs for a learning function whose output is used to determine the level of the desired accuracy of the initial Kriging. The output of the learning function is compared to the stopping criteria for the selected learning function. The stopping criteria may consist of only one criterion or multiple conditions. If the stopping criteria are not met, the algorithm continues with updating the initial DoE with one more design site. The design site corresponding to the maximum or the minimum of the evaluated learning function for the MCS trials is selected as the next candidate for the update. The candidate site is then evaluated by the original model (based on the selected learning function). The candidate design site and its output will update the DoE. Using the updated DoE, an updated Kriging predictor can be built, as presented in Fig. 9.8.

The updated predictor will be used to evaluate the mean and standard deviation of the MCS trials again as well as the evaluation of their corresponding learning function. The candidates will be chosen sequentially and selectively, and the DoE will be updated so that the Kriging predictor is accurate enough for the reliability analysis. Once the desired Kriging predictor is trained and the stopping criteria are met, the reliability analysis can be conducted as shown in Fig. 9.8.

The final Kriging predictor will be used to evaluate all MCS trials and finding the number of failed trials (i.e., n_f). The probability of failure (i.e., P_f) can be found by dividing the number of failed trials by the total number of MCS trials (i.e., N_{MCS}) as presented in Eq. 9.61.

$$P_f = \frac{n_f}{N_{MCS}} \tag{9.61}$$

By considering a normal distribution for the performance function, the reliability index (i.e., β) can be calculated based on Eq. 9.62 or based on its summarized version in Eq. 9.63.

$$P_f = P[failure] = P[\hat{G} < 0] = P\left[\frac{\hat{G} - \mu_{\hat{G}}}{\sigma_{\hat{G}}} < -\beta\right] = 1 - \Phi(\beta) \tag{9.62}$$

$$\beta = \Phi^{-1}(1 - P_f) \tag{9.63}$$

where Φ and Φ^{-1} are the cumulative distribution function (CDF) and the inverse CDF of the standard normal distribution, respectively, while other parameters are defined previously.

The algorithm illustrated in Fig. 9.8 is general and can be applied to different learning functions and their corresponding stopping criteria. Many learning functions are available in the literature that can be used with the explained algorithm. However, in the following sections, only two commonly used learning functions are presented in detail, named EFF and U functions.

4.1 EFF Learning Function

The expected feasibility function (EFF) is an optimized learning function whose role is to find the best candidate for the AK-MCS based on the current Kriging predictor (Bichon et al., 2008). The concept of this function comes from efficient global optimization (EGO) developed for global optimization of a function with a gaussian process (GP) surrogate (Jones et al., 1998). The simplest form of optimization is to use the response surface method (RSM) to fit a surface (Rajashekhar and Ellingwood,

1993; Kaymaz and McMahon, 2005), find the minimum of the surface, and sample the best candidates iteratively, where samples are selected close to the current minimum. However, RSM optimization can easily converge to a local minimum and miss the global minimum. RSM does not acknowledge the uncertainty of the response surface, while EGO addresses this issue. In EGO, the next candidate is selected based on how much the expected improvement in the results of the current GP solution can be achieved. For EGO, the stopping criterion is reached once the expected improvement by adding the next candidate is acceptably small. This criterion addresses the need for an optimized number of added points for training the surrogate model. If too few samples are selected, the surrogate model suffers the lack of accuracy and if too many samples are chosen, the solution would be expensive without extra gain. The application of EGO in reliability is called efficient global reliability analysis (EGRA) which utilizes EFF function. It avoids too many samples for the reliability analysis by focusing on the regions where the performance function is close to zero or failure surface.

Derivation of the EFF learning function is conceptualized in Fig. 9.9. Let's consider that a set of training points (black points) are available (known) with which a surrogate function or a predictor ($\hat{G}$ shown in dashed black line) approximates the real performance function (solid black line), as presented in Fig. 9.9(a). To improve the predictor, the current minimum of the predictor is found first, and the next candidate is selected so that the choice of the next candidate improves the current minimum. For an arbitrary candidate (X_i shown as the red point), a set of possible values (blue points) for the real performance function are shown in Fig. 9.9(a). Since each predicted point is a random variable, there is a probability of occurrence associated with each of the possible values (blue point). The distribution

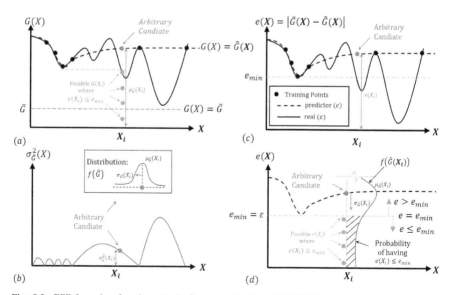

Fig. 9.9: EFF Learning function: (a) Performance function; (b) MSE functions (variance of error); (c) absolute error function; and (d) explanation of expected value of absolute error lower than prescribed minimum.

that shows this randomness is assumed to be a normal distribution with a mean value $(\mu_{\hat{G}}(X_i))$ equal to the prediction of the surrogate model at the studied point (X_i), as shown in Fig. 9.9(a). The variance of this distribution is set equal to the variance of the predictor $(\sigma_{\hat{G}}(X_i))$ at the studied point (X_i), as shown in Fig. 9.9(b). Having the distribution type, mean, and variance, the probability density function (PDF) of the predictor (i.e., $f(\hat{G}(X_i))$) for the studied point can be built, as shown in Fig. 9.9(b). With this probabilistic characteristic for the predictor, the probability of occurrence of each possible outcome or the expected value of the prediction at each point can be quantified.

As the goal is not global optimization, this expectation should be quantified for an error function $(e(X))$ which is the absolute value of the difference between the predicted performance function $(\hat{G}(X))$ and the desired value of the performance function $(\bar{G}(X))$, as presented in Fig. 9.9(c). For the error function, the minimum error (i.e., e_{min}) is shown in Fig. 9.9(d). For an arbitrary point (the red point), if the value of error (e) is more than the current minimum error (i.e., $e > e_{min}$), this point is not a good candidate as it cannot improve the current minimum error. And, for an arbitrary point, if the value of error is less than the current minimum error (i.e., $e < e_{min}$), the candidate can improve the minimum error, but it is uncertain that the candidate happens to have a value of less than the current minimum. Therefore, the expected value of the error function of less than the current minimum can give a great measurement for the chance of happening below the minimum error. This expectation is visualized as the shaded area under the predictor distribution (i.e., $f(\hat{G}(X_i))$) for error values of less or equal to the minimum error (i.e., $e \le e_{min}$) multiplied by all possible values less than the minimum, as shown in Fig. 9.9(d).

The EFF function can be expressed as the expected value of the error function for error values lower than the minimum error as presented in Eq. 9.64.

$$EFF(\hat{G}) = E[\max(e_{min} - e, 0)] \tag{9.64}$$

where E is the expectation of a function, and $\hat{G}$ is the short format for $\hat{G}(X)$, which is the predicted value of the performance function for the design site X, and e and e_{min} are the error function and minimum error defined in Eqs. 9.65 and 9.66, respectively.

$$e = e(X) = |\hat{G}(X) - \bar{G}(X)| = |\bar{G} - \hat{G}| \tag{9.65}$$

$$e_{min} - \varepsilon \tag{9.66}$$

where $\bar{G}$ is the desired value of the performance function and is zero for the reliability analysis (also called as a limit state), and ε defines the required level of error and is typically considered as twice the standard deviation at any studied point (i.e., $\varepsilon = 2\,\sigma_{\hat{G}} \ge 0$), in the literature (Bichon et al., 2008).

By substituting the value of e and e_{min} from Eq. 9.65 and Eq. 9.66 into Eq. 9.64, the EFF function can be defined as Eq. 9.67.

$$EFF(\hat{G}) = E[\max(\varepsilon - |\bar{G} - \hat{G}|, 0)] \tag{9.67}$$

To solve Eq. 9.67, first, the argument of the maximum value should be broken down, as Eq. 9.68 and Eq. 9.69 considered two scenarios of error greater than or equal to ε and less than ε, respectively.

$$\text{if } \varepsilon - |\,\overline{G} - \hat{G}\,| \le 0 \rightarrow |\,\overline{G} - \hat{G}\,| \ge \varepsilon \rightarrow \overline{G} - \hat{G} \ge \varepsilon \cup \overline{G} - \hat{G} \le -\varepsilon$$
$$\rightarrow \hat{G} \le \overline{G} - \varepsilon \cup \hat{G} \ge \overline{G} + \varepsilon \rightarrow \hat{G} \le G^- \cup \hat{G} \ge G^+ \quad (9.68)$$

$$\text{if } \varepsilon - |\,\overline{G} - \hat{G}\,| > 0 \rightarrow |\,\overline{G} - \hat{G}\,| < \varepsilon \rightarrow -\varepsilon < \overline{G} - \hat{G} < \varepsilon$$
$$\rightarrow \overline{G} - \varepsilon < \hat{G} < \overline{G} + \varepsilon \rightarrow G^- < \hat{G} < G^+ \quad (9.69)$$

where G^- and G^+ are the same as $\overline{G} - \varepsilon$ and $\overline{G} + \varepsilon$, that will be used for simplicity in the formulation.

Equations 9.68 and 9.69 can be used to write the maximum value in the form of a piecewise function as presented in Eq. 9.70.

$$\max(\varepsilon - |\,\overline{G} - \hat{G}\,|, 0) = \begin{cases} 0 & \text{if} & \hat{G} \le G^- \\ \varepsilon - |\,\overline{G} - \hat{G}\,| & \text{if} & G^- < \hat{G} < G^+ \\ 0 & \text{if} & G^+ \le \hat{G} \end{cases} \quad (9.70)$$

The piecewise function in Eq. 9.70 can be substituted in Eq. 9.67 and the expectation can be expanded as Eq. 9.71.

$$EFF(\hat{G}) = E[\max(\varepsilon - |\,\overline{G} - \hat{G}\,|, 0)]$$
$$= \int_{-\infty}^{\infty} \max(\varepsilon - |\,\overline{G} - \hat{G}\,|, 0) f(\hat{G}) d\hat{G} \quad (9.71)$$
$$= \int_{-\infty}^{G^-} (0) f(\hat{G}) d\hat{G} + \int_{G^-}^{G^+} (\varepsilon - |\,\overline{G} - \hat{G}\,|) f(\hat{G}) d\hat{G} + \int_{G^+}^{\infty} (0) f(\hat{G}) d\hat{G}$$

The solution to this integral can be followed by dividing Eq. 9.71 into two separate integrals (i.e., I_1 and I_2), as shown in Eq. 9.72.

$$EFF(\hat{G}) = \int_{G^-}^{G^+} (\varepsilon - |\,\overline{G} - \hat{G}\,|) f(\hat{G}) d\hat{G}$$
$$= \int_{G^-}^{\overline{G}} (\varepsilon - \overline{G} + \hat{G}) f(\hat{G}) d\hat{G} + \int_{\overline{G}}^{G^+} (\varepsilon - \hat{G} + \overline{G}) f(\hat{G}) d\hat{G} \quad (9.72)$$
$$= \int_{G^-}^{\overline{G}} (\hat{G} - G^-) f(\hat{G}) d\hat{G} + \int_{\overline{G}}^{G^+} (G^+ - \hat{G}) f(\hat{G}) d\hat{G} = I_1 + I_2$$

To solve I_1 and I_2 integrals, the PDF of the predictor which is a normal PDF can be expanded. As the solution is later written in terms of standard normal distributions, the PDF and CDF of normal and standard normal distributions are presented in Eq. 9.73 through Eq. 9.76.

$$\phi(z) = \frac{1}{\sqrt{2\pi}} \exp\left\{-\frac{z^2}{2}\right\} \quad (9.73)$$

$$\int_{-\infty}^{z} \frac{1}{\sqrt{2\pi}} \exp\left\{-\frac{x^2}{2}\right\} dx = \int_{-\infty}^{z} \phi(x) dx \rightarrow \int_{a}^{b} \phi(x) dx = \Phi(b) - \Phi(a) \quad (9.74)$$

$$f(\hat{G}) = \frac{1}{\sqrt{2\pi}\sigma} \exp\left\{-\frac{(\hat{G}-\mu_{\hat{G}})^2}{2\sigma_{\hat{G}}^2}\right\} = \frac{1}{\sigma}\phi\left(\frac{\hat{G}-\mu_{\hat{G}}}{\sigma_{\hat{G}}}\right) \tag{9.75}$$

$$F(\hat{G}) = \int_{-\infty}^{\hat{G}} f(\hat{G})d\hat{G} = \int_{-\infty}^{\hat{G}} \frac{1}{\sigma}\phi\left(\frac{\hat{G}-\mu_{\hat{G}}}{\sigma_{\hat{G}}}\right)d\hat{G} \tag{9.76}$$

where $\phi(z)$ and $\Phi(z)$ are PDF and CDF of standard normal distribution, respectively, $f(\hat{G})$ and $F(\hat{G})$ are the PDF and CDF of normal distribution for the $\hat{G}$ predictor, and z is a standard normal variable that can be defined as Eq. 9.77.

$$z = \frac{\hat{G}-\mu_{\hat{G}}}{\sigma_{\hat{G}}} \tag{9.77}$$

Using Eqs. 9.73 to 9.77, I_1 and I_2 integrals in Eq. 9.72 can be solved as presented in Eqs. 9.78 and 9.79, respectively.

$$
\begin{aligned}
I_1 &= \int_{G^-}^{\bar{G}} (\hat{G}-G^-)f(\hat{G})d\hat{G} \\[6pt]
&= \int_{G^-}^{\bar{G}} (\hat{G}-G^-+\mu_{\hat{G}}-\mu_{\hat{G}})\frac{1}{\sqrt{2\pi}\sigma_{\hat{G}}} \exp\left\{-\frac{(\hat{G}-\mu_{\hat{G}})^2}{2\sigma_{\hat{G}}^2}\right\}d\hat{G} \\[6pt]
&= \frac{\sigma_{\hat{G}}}{\sqrt{2\pi}} \int_{G^-}^{\bar{G}} \frac{(\hat{G}-\mu_{\hat{G}})}{\sigma_{\hat{G}}^2} \exp\left\{-\frac{(\hat{G}-\mu_{\hat{G}})^2}{2\sigma_{\hat{G}}^2}\right\}d\hat{G} \\[6pt]
&\quad + (\mu_{\hat{G}}-G^-)\int_{G^-}^{\bar{G}} \frac{1}{\sqrt{2\pi}} \exp\left\{-\frac{(\hat{G}-\mu_{\hat{G}})^2}{2\sigma_{\hat{G}}^2}\right\}\frac{d\hat{G}}{\sigma_{\hat{G}}} \\[6pt]
&= -\frac{\sigma_{\hat{G}}}{\sqrt{2\pi}} \int_{G^-}^{\bar{G}} \exp\left\{-\frac{(\hat{G}-\mu_{\hat{G}})^2}{2\sigma_{\hat{G}}^2}\right\}d\left(-\frac{(\hat{G}-\mu_{\hat{G}})^2}{2\sigma_{\hat{G}}^2}\right) \\[6pt]
&\quad + (\mu_{\hat{G}}-G^-)\int_{G^-}^{\bar{G}} \phi\left(\frac{\hat{G}-\mu_{\hat{G}}}{\sigma_{\hat{G}}}\right)d\left(\frac{\hat{G}-\mu_{\hat{G}}}{\sigma_{\hat{G}}}\right) \\[6pt]
&= -\sigma_{\hat{G}}\left(\frac{1}{\sqrt{2\pi}}\exp\left\{-\frac{(\hat{G}-\mu_{\hat{G}})^2}{2\sigma_{\hat{G}}^2}\right\}\right)\Bigg|_{G^-}^{\bar{G}} + (\mu_{\hat{G}}-G^-)\left(\Phi\left(\frac{\hat{G}-\mu_{\hat{G}}}{\sigma_{\hat{G}}}\right)\right)\Bigg|_{G^-}^{\bar{G}} \\[6pt]
&= -\sigma_{\hat{G}}\left[\phi\left(\frac{\bar{G}-\mu_{\hat{G}}}{\sigma_{\hat{G}}}\right)-\phi\left(\frac{G^--\mu_{\hat{G}}}{\sigma_{\hat{G}}}\right)\right] + (\mu_{\hat{G}}-G^-)\left[\Phi\left(\frac{\bar{G}-\mu_{\hat{G}}}{\sigma_{\hat{G}}}\right)-\Phi\left(\frac{G^--\mu_{\hat{G}}}{\sigma_{\hat{G}}}\right)\right]
\end{aligned}
\tag{9.78}
$$

$$I_2 = \int_{\bar{G}}^{G^+} (G^+ - \hat{G}) f(\hat{G}) d\hat{G}$$

$$= \int_{\bar{G}}^{G^+} (G^+ - \hat{G} + \mu_{\hat{G}} - \mu_{\hat{G}}) \frac{1}{\sqrt{2\pi}\sigma_{\hat{G}}} \exp\left\{-\frac{(\hat{G}-\mu_{\hat{G}})^2}{2\sigma_{\hat{G}}^2}\right\} d\hat{G}$$

$$= \frac{\sigma_{\hat{G}}}{\sqrt{2\pi}} \int_{\bar{G}}^{G^+} -\frac{(\hat{G}-\mu_{\hat{G}})}{\sigma_{\hat{G}}^2} \exp\left\{-\frac{(\hat{G}-\mu_{\hat{G}})^2}{2\sigma_{\hat{G}}^2}\right\} d\hat{G}$$

$$+ (G^+ - \mu_{\hat{G}}) \int_{\bar{G}}^{G^+} \frac{1}{\sqrt{2\pi}} \exp\left\{-\frac{(\hat{G}-\mu_{\hat{G}})^2}{2\sigma_{\hat{G}}^2}\right\} \frac{d\hat{G}}{\sigma_{\hat{G}}}$$

$$= \frac{\sigma_{\hat{G}}}{\sqrt{2\pi}} \int_{\bar{G}}^{G^+} \exp\left\{-\frac{(\hat{G}-\mu_{\hat{G}})^2}{2\sigma_{\hat{G}}^2}\right\} d\left(-\frac{(\hat{G}-\mu_{\hat{G}})^2}{2\sigma_{\hat{G}}^2}\right)$$ (9.79)

$$+ (G^+ - \mu_{\hat{G}}) \int_{\bar{G}}^{G^+} \phi\left(\frac{\hat{G}-\mu_{\hat{G}}}{\sigma_{\hat{G}}}\right) d\left(\frac{\hat{G}-\mu_{\hat{G}}}{\sigma_{\hat{G}}}\right)$$

$$= \sigma_{\hat{G}} \left(\frac{1}{\sqrt{2\pi}} \exp\left\{-\frac{(\hat{G}-\mu_{\hat{G}})^2}{2\sigma_{\hat{G}}^2}\right\}\right)\Big|_{\bar{G}}^{G^+} + (G^+ - \mu_{\hat{G}}) \left(\Phi\left(\frac{\hat{G}-\mu_{\hat{G}}}{\sigma_{\hat{G}}}\right)\right)\Big|_{\bar{G}}^{G^+}$$

$$= \sigma_{\hat{G}} \left[\phi\left(\frac{G^+ - \mu_{\hat{G}}}{\sigma_{\hat{G}}}\right) - \phi\left(\frac{\bar{G}-\mu_{\hat{G}}}{\sigma_{\hat{G}}}\right)\right] + (G^+ - \mu_{\hat{G}}) \left[\Phi\left(\frac{G^+ - \mu_{\hat{G}}}{\sigma_{\hat{G}}}\right) - \Phi\left(\frac{\bar{G}-\mu_{\hat{G}}}{\sigma_{\hat{G}}}\right)\right]$$

By substituting Eqs. 9.78 and 9.79 into Eq. 9.72, the EFF function can be written in Eq. 9.80 and its final form can be presented as Eq. 9.81.

$$EFF(\hat{G}) = I_1 + I_2$$

$$= -\sigma_{\hat{G}} \left[\phi\left(\frac{\bar{G}-\mu_{\hat{G}}}{\sigma_{\hat{G}}}\right) - \phi\left(\frac{G^- - \mu_{\hat{G}}}{\sigma_{\hat{G}}}\right)\right]$$

$$+ (\mu_{\hat{G}} - \bar{G} + \varepsilon) \left[\Phi\left(\frac{\bar{G}-\mu_{\hat{G}}}{\sigma_{\hat{G}}}\right) - \Phi\left(\frac{G^- - \mu_{\hat{G}}}{\sigma_{\hat{G}}}\right)\right]$$ (9.80)

$$+ \sigma_{\hat{G}} \left[\phi\left(\frac{G^+ - \mu_{\hat{G}}}{\sigma_{\hat{G}}}\right) - \phi\left(\frac{\bar{G}-\mu_{\hat{G}}}{\sigma_{\hat{G}}}\right)\right]$$

$$+ (\bar{G} + \varepsilon - \mu_{\hat{G}}) \left[\Phi\left(\frac{G^+ - \mu_{\hat{G}}}{\sigma_{\hat{G}}}\right) - \Phi\left(\frac{\bar{G}-\mu_{\hat{G}}}{\sigma_{\hat{G}}}\right)\right]$$

$$EFF(\hat{G}) = (\mu_{\hat{G}} - \bar{G})\left[2\Phi\left(\frac{\bar{G} - \mu_{\hat{G}}}{\sigma_{\hat{G}}}\right) - \Phi\left(\frac{G^+ - \mu_{\hat{G}}}{\sigma_{\hat{G}}}\right) - \Phi\left(\frac{G^- - \mu_{\hat{G}}}{\sigma_{\hat{G}}}\right)\right]$$

$$- \sigma_{\hat{G}}\left[2\Phi\left(\frac{\bar{G} - \mu_{\hat{G}}}{\sigma_{\hat{G}}}\right) - \Phi\left(\frac{G^+ - \mu_{\hat{G}}}{\sigma_{\hat{G}}}\right) - \Phi\left(\frac{G^- - \mu_{\hat{G}}}{\sigma_{\hat{G}}}\right)\right] \qquad (9.81)$$

$$+ \varepsilon\left[\Phi\left(\frac{G^+ - \mu_{\hat{G}}}{\sigma_{\hat{G}}}\right) - \Phi\left(\frac{G^- - \mu_{\hat{G}}}{\sigma_{\hat{G}}}\right)\right]$$

For AK-MCS, the procedure shown in Fig. 11.8 can be used with the EFF function and the stopping criteria are met once the maximum value for EFF function reaches feasibility of 0.001 (Bichon et al., 2008), as presented in Eq. 9.82.

$$\max\left(EFF(\hat{G}_i)\right) = 0.001; \ \hat{G}_i = \hat{G}(X_i); \ i = 1,...,N_{MCS} \qquad (9.82)$$

where $\hat{G}_i$ is the evaluation of the performance function at the MCS trial corresponding to site X_i.

As mentioned previously, the learning function is evaluated at each point by the Kriging outputs which are mean ($\mu_{\hat{G}_i}$) and standard deviation ($\sigma_{\hat{G}_i}$) of $\hat{G}_i$ for each MCS trial. The second learning function presented in the following section uses only these two values and characterizes the optimization.

4.1.1 U Learning Function

The importance of the sign of the performance function in the MCS led to the introduction of a learning function named as U learning function (Echard et al., 2011). The sign of the performance function can be changed close to the performance function of zero (also called the limit state). The U learning function (i.e., $U(X)$) is defined as a linear equation that determines the number of standard deviations (i.e., $\sigma_{\hat{G}}(X)$) between the distance of the predictor value (i.e., $\hat{G}(X)$) and the zero value (or limit state) for each design site X, as presented in Eq. 9.83.

$$|\hat{G}(X)| - U(X)\sigma_{\hat{G}}(X) = 0 \qquad (9.83)$$

where $|\hat{G}(X)|$ is the absolute value of the predictor.

U can be considered as the reliability index of making mistakes on the sign of the performance function. A value 2 for the minimum reliability of on making a mistake on the sign of the performance function per Eq. 9.83 is set as the stopping criteria for the U learning function (which is equal to a probability of $1 - \Phi(2)$ or 2.275% per Eq. 9.62). From Eq. 9.83, by replacing $|\hat{G}(X)|$ with the mean of the Kriging predictor, the value of the U learning function can be rewritten as presented in Eq. 9.84.

$$U(X) = \frac{\mu_{\hat{G}}(X)}{\sigma_{\hat{G}}(X)} \qquad (9.84)$$

The procedure of AK-MCS using U leaning function is presented in Fig. 9.10. The performance function and its absolute value determined by a Kriging predictor

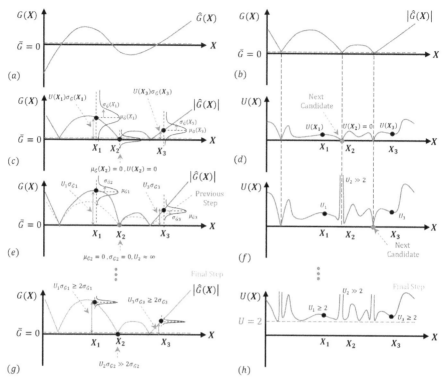

Fig. 9.10: U Learning function: (a) performance function; (b) absolute performance function; (c) distribution of the response at the beginning of AK-MCS; (d) U function at the beginning of AK-MCS; (e) updated predictor and distributions; (f) updated U function; (g) predictor at final step and distributions; (f) U function at the final step of AK-MCS.

are shown in Figs. 9.10(a) and 9.10(b), respectively, for the initial analysis stage using the initial DoE. In Fig. 9.3(c), the distribution of response at three design sites is shown (i.e., X_1, X_2, and X_3). To find the next candidate between these three, the one with the lowest U function (i.e., X_2) is selected as the next candidate for updating the Kriging function. The value of the U function corresponding to Fig. 9.10(c) is shown in Fig. 9.10(d). The value of U for X_2 is zero which means the point is on the limit state and should be added. The U function for X_1 is lower than X_3, although the mean value of X_1 is greater than X_3. The latter happened due to the sharper shaper of distribution at X_3 (i.e., smaller standard deviation), which means the distance between the point and the performance function of zero is defined by a larger number of standard deviations (as shown by red distance in Fig. 9.10(c)).

By updating the surrogate model with X_2, the standard deviations for the distributions of all input sites would decrease, and the standard deviation of the added point becomes zero, due to treating it as a certain point, as shown in Fig. 9.10(e). This change in the standard deviation of the distributions update the U function and since the standard deviation decreases at each step, the value of the

U function increases and the U function tends to infinity for certain points (for which the standard deviation is zero), as shown in Fig. 9.10(f). After enough design sites (i.e., X_i) were added to update the surrogate model, the uncertainty in the distributions decreases, and the stopping criterion of 2 for U function would be met, as shown in Figs. 9.10(g) and 9.10(h).

An executive summary of the discussed learning functions is presented in Table 9.1, for practical purposes. The explained AK-MCS algorithm in Fig. 9.8 and Table 9.1 can be used together to implement the procedure effectively.

It should be noted that this chapter is more focused on AK-MCS, with a discussion on the use of other reliability methods using the Kriging predictor. Different combinations of reliability methods and Kriging are available in the literature including Kriging-based importance sampling (IS) (Zhang et al., 2020; Yang et al., 2020), Kriging-based line sampling (LS) (Lv et al., 2015), Kriging-based subset simulation (SS) (Xiao et al., 2019). Furthermore, active learning techniques can be combined with artificial neural networks (ANN) in the form of modifying weights (Xiang et al., 2020), and reinforcement learning (RL) (Xiang et al., 2020) for reliability analysis. Also, only two learning functions were discussed in this chapter, while more learning functions can be found in the literature (Lv et al., 2015; Sun et al., 2017; Zhang et al., 2019; Shi et al., 2020). In the following section, three examples will be solved using AK-MCS. In these examples, except for the Kriging part, the algorithm was developed in MATLAB by the authors. For Kriging, the well-established DACE MATLAB toolbox was used (Lophaven et al., 2002a, 2002b).

Table 9.1: Learning functions.

No.	Learning Function	Stopping Criteria	Reference
1	$EFF(\hat{G}) = (\mu_{\hat{G}} - \bar{G})[2\bar{\Phi} - \Phi^+ - \Phi^-]$ $-\sigma_{\hat{G}}[2\bar{\phi} - \phi^+ - \phi^-] + \varepsilon[\Phi^+ - \Phi^-]$	$\max\left(EFF(\hat{G}_i)\right) \leq 0.001$	(Bichon et al., 2008)
2	$U(\hat{G}) = \dfrac{|\mu_{\hat{G}}|}{\sigma_{\hat{G}}}$	$\min\left(U(\hat{G}_i)\right) \geq 2$	(Echard et al., 2011)

Definitions:

$\hat{G} = \hat{G}(X); \ \sigma_{\hat{G}} = \sigma_{\hat{G}}(X); \ \mu_{\hat{G}} = \mu_{\hat{G}}(X)$

$\bar{G} = \bar{G}(X); \ G^+ = G^+(X) = \bar{G}(X) + \varepsilon; \ G^- = G^-(X) = \bar{G}(X) - \varepsilon;$

$\bar{Z} = \dfrac{\bar{G} - \mu_{\hat{G}}}{\sigma_{\hat{G}}}; \ Z^+ = \dfrac{G^+ - \mu_{\hat{G}}}{\sigma_{\hat{G}}}; \ Z^- = \dfrac{G^- - \mu_{\hat{G}}}{\sigma_{\hat{G}}};$

$\bar{\Phi} = \Phi(\bar{Z}); \ \Phi^+ = \Phi(Z^+); \ \Phi^- = \Phi(Z^-);$

$\bar{\phi} = \phi(\bar{Z}); \ \phi^+ = \phi(Z^+); \ \phi^- = \phi(Z^-).$

5. Examples

5.1 Example 1: Concrete Beam Reliability (Application of AK-MCS)

The reliability of a rectangular concrete beam with a width of 350 mm, a height of 400 mm, and four layers of steel rebar is desired. The steel rebars are in four layers with 600, 400, 400, and 600 mm² as the reinforcement area of each layer, and are located at 50, 150, 250, and 350 mm from the edge of the concrete. For this example, the concrete strength and yield stress of steel were considered as the only random variables. These random variables were considered normally distributed with a mean value of the concrete strength (f_c) and steel yield strength (f_y) as 34.0 and 465.1 MPa, respectively. The coefficient of variation (COV) of the concrete strength and the yield strength was considered as 0.145 and 0.04, respectively, which give standard deviations of 4.9 and 18.6 MPa for concrete strength and yield strength of steel, respectively. The performance function of the beam is presented in Eq. 9.85.

$$G(X) = M_r(X) - M_q \qquad (9.85)$$

where $G(X)$ is the performance function for the beam, $M_r(X)$ is the resistance of the beam, M_q is the factored applied moment, and X is the vector of random variables including concrete strength (f_c) and steel yield stress (f_y).

The value of M_q is considered as a certain value of 105 kN-m. The value of M_r is a function of f_c and f_y, which is calculated based on the Whitney stress block per CSA A23.3 (CSA 2019) with ultimate concrete strain of 0.0035 mm/mm (ε_{cu} = 0.0035 mm/mm), a concrete resistance factor of 0.65 (ϕ_c = 0.65), and a steel resistance factor of 0.85 (ϕ_s = 0.85). It should be noted that, for this example, the material resistance factors were considered inside the resistance function while for the calibration purposes, these values would be set to one for the resistance model. More information on finding the capacity of the steel-reinforced concrete (RC) beams can be found in the literature (Wight and MacGregor, 2012; CSA A23.3 2019). A summary of section properties and the modeling is presented in Fig. 9.11. It is desired to perform AK-MCS with EFF learning function to calculate the reliability

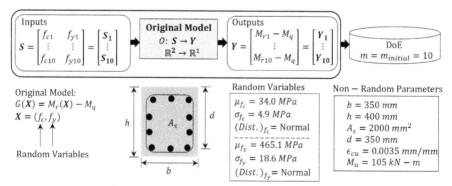

Fig. 9.11: Beam with two random variables (Example 1).

index of the beam in bending with the given characteristic and performance function. For the Kriging predictor, a constant regression function and Gaussian correlation function were selected. Also, the correlation length was optimized using the DACE MATLAB toolbox (Lophaven et al., 2002a, 2002b).

The original model for this problem is the performance function defined in Eq. 9.85, with two inputs (i.e., f_y and f_c) and one output (i.e., $M_r - M_Q$), as shown in Fig. 9.11. A total number of 10 randomly selected sites and corresponding responses are selected to build the initial DoE (i.e., $m = 10$). Therefore, the size of S and Y matrices in the DoE will be 10 by 2 and 10 by 1, respectively, whose rows are representative of the trial, as shown in Fig. 9.11. The algorithm shown in Fig. 9.7 was conducted using the DACE toolbox (Lophaven et al., 2002a, 2002b) to find the initial predictor. In this example, a total number of one million trials were considered. All trials were evaluated using the EFF learning function and the next candidate for the update was selected as the site whose response is corresponding to the maximum of EFF function. By adding the first candidate, the EFF learning function evaluation is presented in Fig. 9.12 whose range varies between 0 to 3. The algorithm of AK-MCS continues by adding the next candidates and the subsequent evaluation of the learning functions (see Table 9.1 and Fig. 9.8) up to reaching a maximum range of 0 to 0.0002 after adding 24 points. The latter meets the stopping criterion for the EFF function from Table 9.1 (i.e., a maximum of 0.001), as shown in Fig. 9.12. The value of the EFF function for each step of the analysis after updating the Kriging predictor with the active learning technique is presented in Fig. 9.12.

To assess the accuracy of the methodology, MCS with the original model was performed by considering 10^4, 10^5, 10^6, 10^7, and 10^8 trials, and the reliability index of each analysis is compared with that of the AK-MCS in Table 9.2.

The MCS with 10^8 trials was considered as the reference analysis with a reliability index of 10^8. Then, the reliability index of each analysis was compared to the reference analysis. The absolute error showed that AK-MCS with 10^6 trials has the lowest error (i.e., the closest reliability index to the 10^8 MCS), which shows the high degree of accuracy of the AK-MCS technique. It should be mentioned that for AK-MCS, the required calls for the performance function are only 24 while to reach the same level of accuracy with MCS, more than 10^7 trials are required. In other words, with AK-MCS, the cost of calls for the original model is as low as 0.00024% of the crude MCS. The substantial reduction in the number of MCS calls by using AK is most beneficial when the call for the original model is high, such as running the FE model for a bridge, a dam, or a complex structural system, as opposed to the simple section considered in this example.

The details of the AK-MCS as it relates to Example 1 are presented in Fig. 9.13. The evolution of the predictor from the first stage of the analysis (i.e., using the initial DoE) to the final stage of the analysis is shown in Fig. 9.13(a) where the black contour shows the initial Kriging predictor, the final Kriging predictor is presented in colors, the limit state is presented using a blue line, the initial points are shown in black, and the added points are shown as red points. As it can be observed, the initial contour is accurate about the initial DoE (i.e., black sites), while it cannot predict the limit state function accurately. By looking at the added points (i.e., the red points), most are located at the limit state line (i.e., $G(X) = 0$). The MSE of the Kriging (or

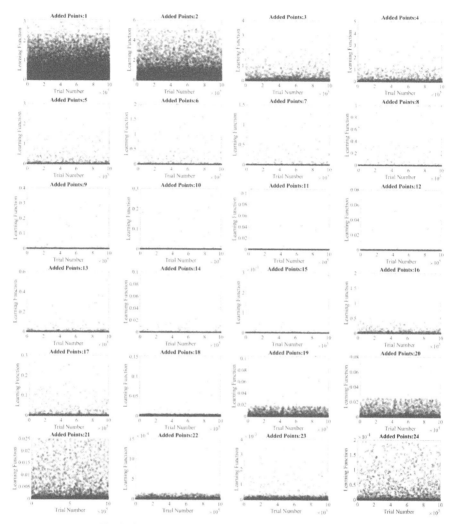

Fig. 9.12: Learning function evaluation steps for Example 1.

Table 9.2: MCS result comparison versus AK-MCS for Example 1.

Analysis	MCS (10^4)	MCS (10^5)	MCS (10^6)	MCS (10^7)	MCS (10^8)	AK-MCS (10^6)
Reliability Index (β)	∞	4.108	4.119	4.051	4.068	4.056
Error (%)	–	0.983	1.254	0.418	0	0.295

the standard deviation of the performance function) at the final stage of the analysis is presented in Fig. 9.13(b), where the values of the error are minimum at the areas of the limit state because of the concentration of the added points at its vicinity.

For comparing the limit state found by AK-MCS, the original model was used with a heavy grid varied between a wide range of random variables to give the surface

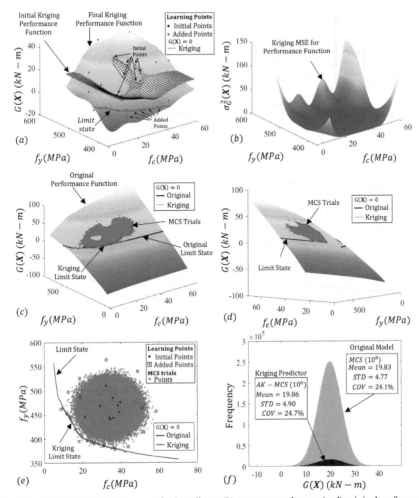

Fig. 9.13: AK-MCS: (a) Initial versus final predictor; (b) mean squared error; (c, d) original performance function contour and predicted limit state [G(X) = 0]; (e) real failure versus AK-MCS prediction; (f) histogram of performance function.

corresponding to the original performance function and its corresponding limit state, as presented in Figs. 9.13(c) and 9.13(d), respectively. The 10^6 MCS trials for AK-MCS are shown as red points in Figs. 9.13(c) and 9.13(d). The limit state found with the original model is shown as black and the one found with the Kriging predictor is shown in blue which are close to each other. To show the limit states and MCS trials, Fig. 9.13 illustrates the two-dimensional analysis results. It should be mentioned that the limit state found using AK-MCS (i.e., the blue line) is not extended as much as the black line since those extended regions were not required based on the available MCS trials (i.e., red points), as shown in Fig. 9.13(e).

The distribution of the performance function found by AK-MCS with 10^6 and the MCS with 10^8 trials are shown in Fig. 9.13(f). The comparison showed that

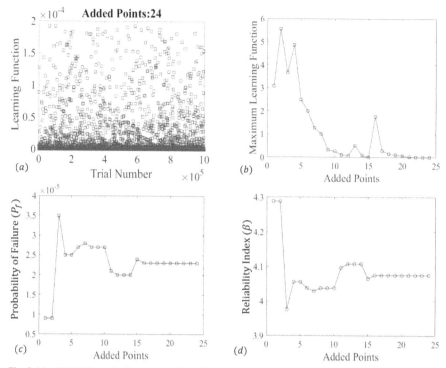

Fig. 9.14: AK-MCS results for example 1: (a) final evaluation of learning function with 24 added points; (b) maximum learning function versus added points; (c) probability of failure versus added points; and (d) reliability index versus added points.

distribution shape is normal for both analyses with very close characteristics. The mean, standard deviation, and coefficient of variation of the performance function using AK-MCS were 19.86 kN-m, 4.9 kN-m, and 24.7%, respectively, while that of the MCS with the original model were 19.83 kN-m, 4.77 kN-m, and 24.1%, respectively. Although the distribution shape and its characteristics obtained from AK-MCS is close to the one obtained by MCS, the fact that the distribution of AK-MCS is based on a surrogate model with high accuracy mostly near the limit state function should be emphasized, and the distribution found by AK-MCS should not be used directly without solid verification.

For further observations regarding the analysis, Fig. 9.14 was developed which shows the final stage of the learning function for all MCS trials (Fig. 9.14(a)), the maximum of the learning function (i.e., EFF function) for all MCS trials at each trial versus the number of the added points (Fig. 9.14(b)), the probability of failure calculated by Eq. 9.61 for the MCS trials in each step (Fig. 9.14(c)), and the reliability index of versus calculated by Eq. 9.63 (Fig. 9.14(c)).

The behavior of the maximum learning function versus added points showed an overall descending trend, which means that as the number of newly added points increases, the maximum EFF function decreases to reach the stopping criterion, as shown in Fig. 9.14(b). There are some local jumps in the maximum value which can be attributed to the predictor characteristics. The predictor at each step gives the mean and standard deviation of all MCS trials. For a certain trial, the mean and standard deviation change as the predictor changes, and it may result in a learning function evaluation for a specific predictor that could not be captured using other less accurate predictors. However, the accuracy of the model always increases as the new points are added.

The probability of failure and the reliability index versus added points are related (see Eq. 9.63). As the probability of failure increases, the reliability index decreases, and vice versa, as shown in Figs. 9.14(a) and 9.14(b). The reliability index and probability of the failure would finally converge to a value after which adding more points does not add to the accuracy. The required convergency can be assessed with the stopping criterion.

For AK-MCS analysis, the analysis included 10^6 trials. The number of required trials would be selected based on the expected probability of failure, where initially the probability of failure can be assumed subjectively. The accuracy of the assumption can be examined after the analysis, by evaluating Eq. 9.86 (Griffiths and Fenton, 2008).

$$COV_{P_f} = \sqrt{\frac{1-P_f}{P_f N_{MCS}}} \qquad (9.86)$$

where COV_{P_f} is the coefficient of variation of the probability of failure, P_f is the probability of failure, and N_{MCS} is the number of MCS trials.

If COV_{P_f} is a small value, the accuracy of the estimated P_f and its corresponding reliability index is acceptable. In this example, AK-MCS with 10^6 trials resulted in a probability of failure of 2.496×10^{-5} which has a COV_{P_f} of 20%. It should be noted that this value is higher than the COV_{P_f} of 5% recommended by Echard et al. (2011) for the required number of trials. However, a comparison of the AK-MCS and MCS results in Table 9.2 shows a good agreement between MCS and AK-MCS results.

5.2 Example 2: Concrete Beam Reliability: Different Learning Functions

This example is an expansion to Example 1 (Section 5.1), with the difference being that loads are considered as random variables. The problem considers the reliability of the beam in Example 1 under a given dead and live load with normal distribution. The mean and standard deviations of the dead load were 84 kN-m and 8.4 kN-m, respectively, and for live load, the mean and standard deviation were 21 kN-m and 3.8 kN-m, respectively, as shown in Fig. 9.15.

The input random variable vector (i.e., X) for the original model includes the concrete strength (f_c), the yield stress of steel (f_y), dead load (M_D), and live load (M_L). The performance function is defined in Eq. 9.87.

$$G(X) = M_r(X) - M_D(X) - M_L(X) \qquad (9.87)$$

The calculation of the resisting moment (i.e., M_r) is the same as the one explained in Example 1. To build the initial DoE, 20 design sites were randomly selected, and a total number of 10^5 MCS trials were considered for the AK-MCS procedure. In

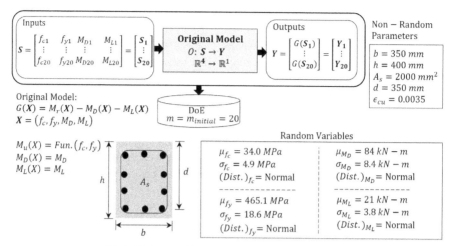

Fig. 9.15: Beam with four random variables (Example 2).

Table 9.3: Configurations for AK-MCS (Example 2).

No.	Configuration Number	Configuration ID	Regression Function	Correlation Function	Learning Function
1	Config1	U-G-0	Constant	Gaussian	U function
2	Config2	U-G-1	Linear	Gaussian	U function
3	Config3	U-G-2	Quadratic	Gaussian	U function
4	Config4	U-E-0	Constant	Exponential	U function
5	Config5	U-E-1	Linear	Exponential	U function
6	Config6	U-E-2	Quadratic	Exponential	U function
7	Config7	EFF-G-0	Constant	Gaussian	EFF function
8	Config8	EFF-G-1	Linear	Gaussian	EFF function
9	Config9	EFF-G-2	Quadratic	Gaussian	EFF function
10	Config10	EFF-E-0	Constant	Exponential	EFF function
11	Config11	EFF-E-1	Linear	Exponential	EFF function
12	Config12	EFF-E-2	Quadratic	Exponential	EFF function

this example, 12 different configurations for AK-MCS were examined to find the optimum configuration by considering three regression functions, two correlation functions, and two learning functions, as presented in Table 9.3.

There is an ID corresponding to each configuration, where the ID format is A-B-C. The letter "A" can be "U" or "EFF" corresponding to U and EFF learning functions, respectively. The letter "B" can be "G" or "E" corresponding to Gaussian and Exponential correlation functions, respectively, and the letter "C" can be "0", "1", or "2" corresponding to constant, linear, and quadratic regression functions.

The AK-MCS procedure detailed in Example 1 was implemented by accounting for a higher number of random variables (4 instead of 2) and a different performance function. The results of the analysis for Config1 to Config6 with U learning function

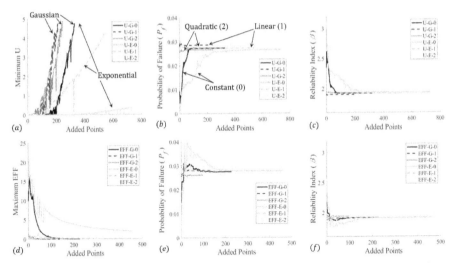

Fig. 9.16: Randomly selected initial DoE: (a) minimum learning function U versus added points; (b) probability of failure versus added points for Configs1 to Config6; (c) reliability index versus added points for Configs1 to Config6; (d) maximum learning function EFF versus added points; (e) probability of failure versus added points for Configs7 to Config12; (f) reliability index versus added points for Configs7 to Config12.

are presented in Figs. 9.16(a) to 9.16(c), respectively, while the results for Config7 to Config12 with EFF learning function are presented in Figs. 9.16(d) to 9.16(f), respectively.

The results revealed that a several added points are required to reach the stopping criteria of 0.001 for EFF function and 2 for the U function. To completely observe the behavior of the U function, the stopping criterion was set to 5, as shown in Fig. 9.5(a). It should be highlighted that for U-E-0, U-E-1, EFF-E-0, and EFF-E-1, the analysis was terminated manually because of the large number of added points and low chance of convergency in few steps.

The probability of failure and reliability index for all configurations converged way before reaching the stopping criterion. The latter suggests that the stopping criterion might be conservative and there is more room for optimizing the stopping criterion. Also, it should be mentioned that since the number of random variables increased from two random variables in Example 1 to four random variables in Example 2, the number of required added points increased.

For this example, the results also showed that using the Gaussian correlation function led to faster convergency for both U and EFF learning functions compared to the Exponential correlation function. Moreover, for the Gaussian correlation function, the linear regression function converged at a faster pace, while for the Exponential correlation function, the quadratic regression function provided a better convergence result, considering both U and EFF learning functions.

One of the possible reasons for having a larger number of added points in this example might be the way that the initial DoE is selected. Echard et al. (2011) suggest using a small number of randomly generated DoE and expand the DoE with the learning function while Bichon et al. (2008) suggest using a minimum number of required points for quadratic regression function which are selected with LHS. If the initial DoE is built using LHS, the initial Kriging predictor might be more accurate because of the selective choice of initial DoE, as a result, the number of required points might be lower, and the convergency might be faster. To examine the convergence and accuracy of the AK-MCS for which initial DoE is selected based on LHS, all configurations in Table 9.3 were studied.

LHS is a sampling technique in which the range of each input random variable is divided into several equally probable intervals and samples are selected so that at least a sample is selected from each division of each input random variable. In a one-dimensional problem, LHS suggests the division of cumulative density function (CDF) into the required number of divisions, and for two-dimensional problems, it suggests sampling from a square grid (named Latin square) if and only if one sample is selected in each row and column of the grid. For higher dimensions, the idea expands and can be named Latin Hypercube. For more details regarding LHS, referenced studies can be helpful (McKay et al., 2000; Shields and Zhang, 2016). In this example, the initial DoE was built using 20 points which means 20 samples out of 20^4 hypercubes were selected using LHS. The AK-MCS was performed using the initial DoE and the results are presented in Fig. 9.17.

Figure 9.17 revealed that the convergency of some cases has improved including U-G-2, while for many cases, no significant improvement was observed regarding the convergency. Also, the probability of failure and reliability index versus added

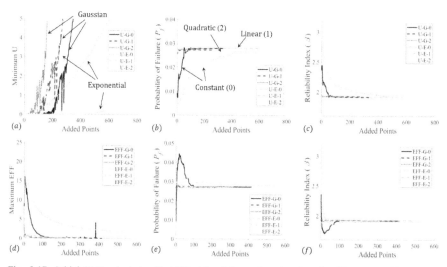

Fig. 9.17: Initial DoE selected with LHS: (a) minimum learning function U versus added points; (b) probability of failure versus added points for Configs1 to Config6; (c) reliability index versus added points for Configs1 to Config6; (d) maximum learning function EFF versus added points; (e) probability of failure versus added points for Configs7 to Config12; (f) reliability index versus added points for Configs7 to Config12.

points showed smoother curves which mean, by using LHS for initial DoE, the variation in the updated Kriging function in each step is smoother. In other words, the Kriging predictor between every two consecutive steps is less susceptible to extreme changes.

Other than the convergency, the accuracy of the methods was compared and the probability of failure and reliability index corresponding to each configuration presented in Table 9.3 are compared to the result of crude MCS with 10^6 trials, as presented in Fig. 9.18.

Overall, the range of reliability index and probability of failure found by AK-MCS were close to crude MCS. The results showed that the minimum number of required added points was 97 for Config9 which gave the highest error. The best configuration in terms of accuracy was Config8 with 0.08% error found by 228 points. The results do not show any improvement in the accuracy using LHS sampling for initial DoE except for some configurations.

It should be noted that AK-MCS was performed only using 10^5 trials and the expected reliability index was low which does not require further trials for AK-MCS. However, if the expected reliability index is high and the probability of failure is low, a larger number of trials may be required, which makes the calculation heavy if numerous added points are required. Therefore, to avoid the heavy cost of calculations, AK-MCS can be combined with other reliability methods as presented in the next example.

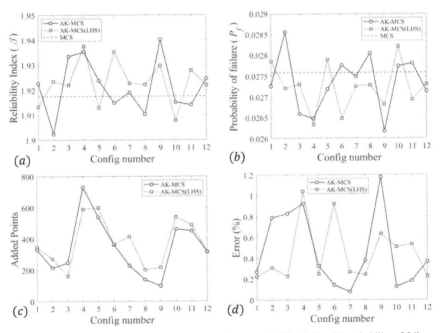

Fig. 9.18: Comparison of different configuration results: (a) reliability index; (b) probability of failure; (c) the number of added points; and (d) absolute error in percentage.

5.3 Example 3: Concrete Beam Reliability (Combined AK-MCS and Importance Sampling)

The reliability of the beam presented in Example 1 is required to be assessed using a two-step reliability approach where the mean of loads should be found based on the nominal resistance. A total of six random variables were considered including concrete strength (f_c), yield strength of steel (f_y), dead load (M_D), live load (M_L), transformation to live load (M_{LT}), and professional factor (ϕ_p). The statistical characteristics distributions, bias, and coefficient of variations of the random variables as well as their nominal values are presented in Table 9.4.

Table 9.4: Statistical characteristics and nominal values of random variables.

Random Variable	Nominal	Bias	COV	Distribution	Reference
Concrete strength (f_c)	27.6	1.235	0.145	Normal	(Nowak and Szerszen, 2003)
Yield strength of steel (f_y)	413.7	1.125	0.04	Normal	(Nowak and Szerszen, 2003)
Dead load (M_D)	–	1.05	0.1	Normal	(Bartlett et al., 2003)
Live load (M_L)	–	0.9	0.17	Gumbel	(Bartlett et al., 2003)
Transformation to live load (M_{LT})	1	1	0.206	Normal	(Bartlett et al., 2003)
Professional factor (ϕ_p)	1	1.02	0.06	Normal	(Nowak and Szerszen, 2003)

The mean value of dead load and live load can be found by setting the nominal capacity of the beam equal to the factored applied moment. The nominal capacity can be found using the resistance model with nominal values as input parameters. In this example, the factored moments per CSA A23.3 (CSA, 2019) were set equal to the factored capacity (i.e., by including the resistance factors), as presented in Eq. 9.88.

$$U_r M_n = 1.25 M_D + 1.5 M_L \tag{9.88}$$

where U_r is the utilization ratio defined as factored load over factored capacity, M_n is the nominal factored capacity, M_D is the nominal dead load, and M_L is the nominal live load.

For this example, the utilization ratio is considered 1.10 which represents a loading scenario where the section is overloaded by 10% of its nominal factored capacity. Also, the dead-to-live load ratio was assumed as 4.0 using the results of a recent survey for measurement of the dead-to-live load ratio (Oudah et al., 2019). The solution of Eq. 11.88 revealed 100.59 kN-m for M_n, 74.86 kN-m for M_D, and 18.71 kN-m for M_L.

The mean value for each distribution was calculated as the nominal value times its corresponding bias, and the standard deviation of each random variable was calculated as the mean value times the COV. The performance function for this example is presented in Eq. 9.89.

$$G(X) = \phi_p \times M_r(X) - M_D - M_L \times M_{LT} \tag{9.89}$$

where M_r is the unfactored resistance model, which means the concrete and steel resistance factors were set to 1.0 (i.e., $\phi_c = 1.0$ and $\phi_s = 1.0$).

The performance function should reflect the real resistance of the beams. Therefore, a random variable named the professional factor (ϕ_p), which can be built as the ratio of experimental to the theoretical model, is applied to the resistance model.

As observed in the previous example, considering four random variables, the number of required added points was high with 10^5 trials for the AK-MCS. For this example, a crude MCS with 10^7 trials was conducted as the reference, which showed a reliability index of 4.51 for the beam problem. The required number of trials for AK-MCS to give a reliability index of 4.5 for a normal performance function with a 20% COV is 7.36×10^7 using Eq. 9.86. Therefore, the cost of evaluation for the learning function, sampling from the distributions, and evaluation of the original model would be expensive. To reduce the cost of calculation, AK-MCS with 10^5 trials was used to find a Kriging predictor with which FORM and IS methods were implemented. The initial DoE was built using 28 design sites and their corresponding responses, where 28 is the minimum number of sites required for the quadratic regression function. The initial DoE is presented in Fig. 9.19.

All configurations in Table 9.3 were considered for Example 3. The AK-MCS with 10^5 trials and 28 randomly selected DoE was conducted for each configuration, three times to generate different Kriging predictors. Then FORM and IS were implemented for each configuration three times using the Kriging predictors found from AK-MCS, which are called AK-FORM and AK-IS, respectively, in this chapter.

FORM is a gradient-based search that examines the points on the limit state function (i.e., on $G(X) = 0$) with iteration to find the one with the minimum distance from the origin of the standard normal space for all random variables, by approximating the limit state with a line that passes the limit state at a point called the most probable point (MPP). The distance from the MPP to the origin of the standard normal random variable space is called the Hasofer-Lind reliability index (Hasofer and Lind, 1974; Nowak and Collins, 2000). It should be mentioned that for highly nonlinear problems, FORM may either face convergency problems or fail to find the global minimum. Kriging predictor provides a smooth surface and improves the convergency issues. However, because the Kriging was trained using only a limited number of points, the predictor surface is not reliable in the whole space and there would be inaccuracies in finding the reliability index using FORM and the Kriging predictor. It should be mentioned that to perform FORM, the gradient of Kriging is required which is available in Appendix 9A.1. More details on the theory and

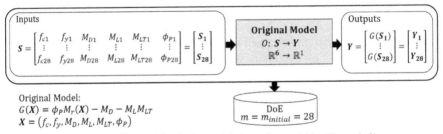

Fig. 9.19: Initial DoE for the beam with six random variables (Example 3).

implementation of FORM can be found in the referenced studies (Hasofer and Lind, 1974; Nowak and Collins, 2000).

To address this issue, IS was used in combination with AK-MCS to take an advantage of the predictor accuracy near the limit state. The MPP can be found using AK-FORM and a modified MCS can be conducted by sampling from new normal distributions (importance distributions) whose mean is MPP. The results, which are found this way, should be calibrated using proper weights for the IS method. This sampling helps to perform MCS with a lower number of trials and have more failed trials. MPP is on the limit state and the Kriging predictor is more accurate near the limit state since it was found by the AK-MCS procedure. Therefore, MCS sampled about this MPP (which is IS) may lead to reasonable results. More details on formulation, weight calculation, and implementation of IS can be found in the referenced studies (Owen et al., 2019; Melchers, 1989).

In this example, AK-MCS was conducted with 10^5 trials which failed to predict the reliability index because no failure was observed, and the number of trials was not adequate. However, the Kriging predictor found at the final stage of AK-MCS was used to perform AK-FORM for each reliability analysis case. The MPP found from AK-FORM was used to conduct AK-IS. The results include the reliability index found for each case based on AK-FORM and AK-IS, the number of added points for preliminary AK-MCS, and the error corresponding to each case, which is shown in Fig. 9.20.

The results indicate that AK-FORM can predict the reliability index with a maximum error of 10%, while AK-IS has a maximum error of 5% which is higher than the maximum of 1.2% error observed in AK-MCS in Example 2 (Section 5.2). However, the number of added points was only 5 for some configurations which reduces the computational cost substantially. It was observed that for the constant regression function, the Kriging predictor failed to converge for AK-FORM, while it was successfully used for linear and quadratic regression functions. It was also observed that for each configuration, the predictions of the reliability index are close to each other even when the initial DoE is randomly changed. It should be mentioned that the accuracy of the predictor about the performance function is a direct function of the number of AK-MCS trials. If the number of trials is not enough, there is a possibility that the added points are not close to the limit state, or the learning function cannot find a trial for the next step. As a result, the Kriging predictor might be poor if enough points are not selected. To study the effect of initial trials, the same study was conducted by increasing the number of trials from 10^5 to 5×10^5, and the results are presented in Fig. 9.21.

The results revealed that by increasing the number of AK-MCS trials, the reliability index prediction improved and the variance of the predictions for AK-IS decreased for analyses corresponding to the three different initial DoE. Also, Config6 showed an excellent prediction when the number of trials increased. It can be implied that to find the best configuration, there is no need for many trials. Overall, this example discussed alternative solutions which use AK-MCS and optimize it further to reduce the calculation costs.

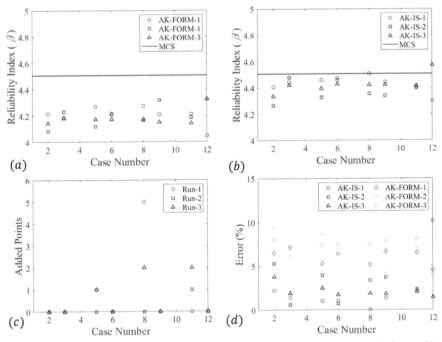

Fig. 9.20: Analysis results for Example 3: (a) AK-FORM; (b) AK-IS; (c) added points; and (d) error (%).

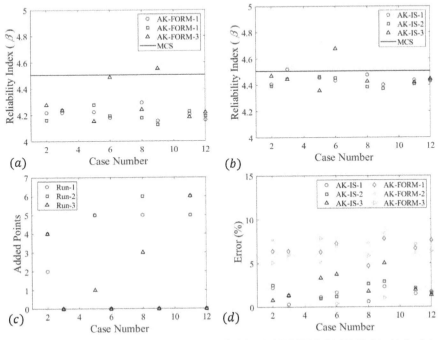

Fig. 9.21: Analysis results for Example 3 with 5×10^5 trials: (a) AK-FORM; (b) AK-IS; (c) added points; and (d) error (%).

References

Allen, D.E. 1975. Limit states design: A probabilistic study. *Canadian Journal of Civil Engineering*, 2(1): 36–49.

Allen, D.E. 1991. Limit states criteria for structural evaluation of existing buildings. *Canadian Journal of Civil Engineering*, 18(6): 995–1004.

Bartlett, F., Hong, H. and Zhou, W. 2003. Load factor calibration for the proposed 2005 edition of the National Building Code of Canada: Statistics of Loads and Load Effects. *Canadian Journal of Civil Engineering*, 30(2): 429–439.

Bichon, B.J., Eldred, M.S., Swiler, L.P., Mahadevan, S. and McFarland, J.M. 2008. Efficient global reliability analysis for nonlinear implicit performance functions. *AIAA Journal*, 46(10): 2459–2468.

Buckley, E., Khorramian, K. and Oudah, F. 2021 Application of adaptive kriging method in bridge girder reliability analysis. *CSCE Annual Conference*. Virtual: Canadian Society of Civil Engineering.

CSA A23.3-19. 2019. *Design of Concrete Structures*. Canadian Standard Association.

Echard, B., Gayton, N. and Lemaire, M. 2011. AK-MCS: An active learning reliability method combining Kriging and Monte Carlo simulation. *Structural Safety*, 33(2): 145–154.

El Haj, A.-K. and Soubra, A.-H. 2020. Efficient estimation of the failure probability of a monopile foundation using a Kriging-based approach with multi-point enrichment. *Computers and Geotechnics*, 121: 103451.

Griffiths, D.V. and Fenton, G.A. 2008. *Risk Assessment in Geotechnical Engineering*. Hoboken, New Jersey: John Wiley & Sons, Inc.

Hasofer, A.M. and Lind, N.C. 1974. Exact and invariant second-moment code format. *Journal of the Engineering Mechanics Division*, 100(1): 111–121.

Jones, D.R., Schonlau, M. and Welch, W.J. 1998. Efficient global optimization of expensive black-box functions. *Journal of Global Optimization*, 13(4): 455–492.

Kaymaz, I. 2005. Application of Kriging method to structural reliability problems. *Structural Safety*, 27(2): 133–151.

Kaymaz, I. and McMahon, C.A. 2005. A response surface method based on weighted regression for structural reliability analysis. *Probabilistic Engineering Mechanics*, 20(1): 11–17.

Lophaven, S.N., Nielsen, H.B. and Søndergaard, J. 2002a. *A Matlab Kriging Toolbox*. Technical University of Denmark, Kongens Lyngby, Technical Report No. IMM-TR-2002–12.

Lophaven, S.N., Nielsen, H.B. and Søndergaard, J. 2002b. *DACE: A Matlab Kriging Toolbox. Vol. 2. IMM Informatics and Mathematical Modelling*. The Technical University of Denmark, 1–34.

Lv, Z., Lu, Z. and Wang, P. 2015. A new learning function for Kriging and its applications to solve reliability problems in engineering. *Computers & Mathematics with Applications* 70(5): 1182–1197.

McKay, M.D., Beckman, R.J. and Conover, W.J. 2000. A comparison of three methods for selecting values of input variables in the analysis of output from a computer code. *Technometrics*, 42(1): 55–61.

Melchers, R.E. 1989. Importance sampling in structural systems. *Structural Safety*, 6(1): 3–10.

Nowak, A.S. and Collins, K.R. 2000. *Reliability of Structures*. McGraw-Hill.

Nowak, A.S. and Szerszen, M.M. 2003. Calibration of design code for buildings (ACI 318): Part 1— Statistical models for resistance. *ACI Structural Journal*, 100(3): 377–382.

Oudah, F., El Naggar, H.M. and Norlander, G. 2019. Unified system reliability approach for single and group pile foundations: Theory and resistance factor calibration. *Computers and Geotechnics*, 108: 173–182.

Owen, A.B., Maximov, Y. and Chertkov, M. 2019. Importance sampling the union of rare events with an application to power systems analysis. *Electronic Journal of Statistics*, 13(1): 231–254.

Rajashekhar, M.R. and Ellingwood, B.R. 1993. A new look at the response surface approach for reliability analysis. *Structural Safety*, 12(3): 205–220.

Shi, Y., Lu, Z., He, R., Zhou, Y. and Chen, S. 2020. A novel learning function based on Kriging for reliability analysis. *Reliability Engineering & System Safety*, 198: 106857.

Shields, M.D. and Zhang, J. 2016. The generalization of Latin hypercube sampling. *Reliability Engineering & System Safety*, 148: 96–108.

Sun, Z., Wang, J., Li, R. and Tong, C. 2017. LIF: A new Kriging based learning function and its application to structural reliability analysis. *Reliability Engineering & System Safety*, 157: 152–165.

Whitten, E.T. 1977. Stochastic models in geology. *The Journal of Geology*, 85(3): 321–330.

Wight, J.K. and MacGregor, J.G. 2012. *Reinforced Concrete: Mechanics and design (Vol. 6).* New Jersey: Pearson.

Xiang, Z., Bao, Y., Tang, Z. and Li, H. 2020. Deep reinforcement learning-based sampling method for structural reliability assessment. *Reliability Engineering & System Safety*, 199: 106901.

Xiang, Z., Chen, J., Bao, Y. and Li, H. 2020. An active learning method combining deep neural network and weighted sampling for structural reliability analysis. *Mechanical Systems and Signal Processing*, 140: 106684.

Xiao, M., Zhang, J., Gao, L., Lee, S. and Eshghi, A.T. 2019. An efficient Kriging-based subset simulation method for hybrid reliability analysis under random and interval variables with small failure probability. *Structural and Multidisciplinary Optimization*, 59(6): 2077–2092.

Yang, X., Cheng, X., Wang, T. and Mi, C. 2020. System reliability analysis with small failure probability based on active learning Kriging model and multimodal adaptive importance sampling. *Structural and Multidisciplinary Optimization*, 62: 581–596.

Zhang, X., Wang, L. and Sørensen, J.D. 2019. REIF: A novel active-learning function toward adaptive Kriging surrogate models for structural reliability analysis. *Reliability Engineering & System Safety*, 185: 440–454.

Zhang, X., Wang, L. and Sørensen, J.D. 2020. AKOIS: An adaptive Kriging oriented importance sampling method for structural system reliability analysis. *Structural Safety*, 82: 10187.

Appendix A9.1

The gradient of the Kriging predictor can be written down as Eq. A9.1.

$$
\frac{\partial \overline{y}(X)}{\partial X} = \begin{bmatrix} \dfrac{\partial \overline{y}(X)}{\partial x_1} \\ \vdots \\ \dfrac{\partial \overline{y}(X)}{\partial x_n} \end{bmatrix}
\tag{A9.1}
$$

By substituting Eq. 9.55 in Eq. A9.1, Eqs. A9.2 to A9.6 can be written.

$$
\begin{aligned}
\frac{\partial \overline{y}(X)}{\partial x_i} &= \frac{\partial}{\partial x_i}[f_1(X) \ldots f_p(X)]\boldsymbol{\beta}^* + \frac{\partial}{\partial x_i}[r_1(X) \ldots r_m(X)]\boldsymbol{\gamma}^* \\
&= \left[\frac{\partial f_1(X)}{\partial x_i} \ \ldots \ \frac{\partial f_p(X)}{\partial x_i}\right]\boldsymbol{\beta}^* + \left[\frac{\partial r_1(X)}{\partial x_i} \ \ldots \ \frac{\partial r_m(X)}{\partial x_i}\right]\boldsymbol{\gamma}^*
\end{aligned}
\tag{A9.2}
$$

$$
\begin{aligned}
\frac{\partial \overline{y}(X)}{\partial X} &= \begin{bmatrix} \dfrac{\partial \overline{y}(X)}{\partial x_1} \\ \vdots \\ \dfrac{\partial \overline{y}(X)}{\partial x_n} \end{bmatrix} = \begin{bmatrix} \left[\dfrac{\partial f_1(X)}{\partial x_1} \ \ldots \ \dfrac{\partial f_p(X)}{\partial x_1}\right]\boldsymbol{\beta}^* + \left[\dfrac{\partial r_1(X)}{\partial x_1} \ \ldots \ \dfrac{\partial r_m(X)}{\partial x_1}\right]\boldsymbol{\gamma}^* \\ \vdots \\ \left[\dfrac{\partial f_1(X)}{\partial x_n} \ \ldots \ \dfrac{\partial f_p(X)}{\partial x_n}\right]\boldsymbol{\beta}^* + \left[\dfrac{\partial r_1(X)}{\partial x_n} \ \ldots \ \dfrac{\partial r_m(X)}{\partial x_n}\right]\boldsymbol{\gamma}^* \end{bmatrix} \\[2ex]
&= \begin{bmatrix} \left[\dfrac{\partial f_1(X)}{\partial x_1} \ \ldots \ \dfrac{\partial f_p(X)}{\partial x_1}\right] & \left[\dfrac{\partial r_1(X)}{\partial x_1} \ \ldots \ \dfrac{\partial r_m(X)}{\partial x_1}\right] \\ \vdots & \vdots \\ \left[\dfrac{\partial f_1(X)}{\partial x_n} \ \ldots \ \dfrac{\partial f_p(X)}{\partial x_n}\right] & \left[\dfrac{\partial r_1(X)}{\partial x_n} \ \ldots \ \dfrac{\partial r_m(X)}{\partial x_n}\right] \end{bmatrix} \begin{bmatrix} \boldsymbol{\beta}^* \\ \boldsymbol{\gamma}^* \end{bmatrix} \\[2ex]
&= \begin{bmatrix} \dfrac{\partial f_1(X)}{\partial x_1} & \dfrac{\partial f_p(X)}{\partial x_1} & \dfrac{\partial r_1(X)}{\partial x_1} & \dfrac{\partial r_m(X)}{\partial x_1} \\ \vdots & \ddots & \vdots & \ddots \\ \dfrac{\partial f_1(X)}{\partial x_n} & \dfrac{\partial f_p(X)}{\partial x_n} & \dfrac{\partial r_1(X)}{\partial x_n} & \dfrac{\partial r_m(X)}{\partial x_n} \end{bmatrix} \begin{bmatrix} \boldsymbol{\beta}^* \\ \boldsymbol{\gamma}^* \end{bmatrix}
\end{aligned}
\tag{A9.3}
$$

$$
\boldsymbol{J}_f(X) = \begin{bmatrix} \dfrac{\partial f_1(X)}{\partial x_1} & \ldots & \dfrac{\partial f_1(X)}{\partial x_n} \\ \vdots & \ddots & \vdots \\ \dfrac{\partial f_p(X)}{\partial x_1} & \ldots & \dfrac{\partial f_p(X)}{\partial x_n} \end{bmatrix} ; \ \boldsymbol{J}_f(X)_{ij} = \frac{\partial f_i(X)}{\partial x_j} ; \ i = 1, \ldots, n ; \ j = 1, \ldots, p
\tag{A9.4}
$$

$$J_r(X) = \begin{bmatrix} \dfrac{\partial r_1(X)}{\partial x_1} & \cdots & \dfrac{\partial r_1(X)}{\partial x_n} \\ \vdots & \ddots & \vdots \\ \dfrac{\partial r_m(X)}{\partial x_1} & \cdots & \dfrac{\partial r_m(X)}{\partial x_n} \end{bmatrix}; \; J_r(X)_{ij} = \dfrac{\partial r_i(X)}{\partial x_j}; \; i = 1, ..., n \; ; j = 1, ..., m \quad \text{(A9.5)}$$

$$\frac{\partial \overline{y}(X)}{\partial X} = \frac{\partial f(X)^T}{\partial X} \beta^* + \frac{\partial r(X)^T}{\partial X} \gamma^* = J_f(X)^T \beta^* + J_r(X)^T \gamma^* \quad \text{(A9.6)}$$

where J_f and J_r are the Jacobian matrices for $f(X)$ and $r(X)$, respectively. Similarly, the gradient of the Kriging MSE can be found by taking the derivative of Eq. 9.59 with respect to vector X, as presented in Eqs. A9.7 and A9.8.

$$\frac{\partial \varphi_k(X)}{\partial X} = 2\sigma_k^2 \left((F^T R^{-1} F)^{-1} \frac{\partial u}{\partial X} - R^{-1} \frac{\partial r}{\partial X} \right)$$

$$= 2\sigma_k^2 \left((F^T R^{-1} F)^{-1} \frac{\partial}{\partial X} (F^T R^{-1} r - f) - R^{-1} \frac{\partial r}{\partial X} \right) \quad \text{(A9.7)}$$

$$= 2\sigma_k^2 \left((F^T R^{-1} F)^{-1} \left(F^T R^{-1} \frac{\partial r}{\partial X} - \frac{\partial f}{\partial X} \right) - R^{-1} \frac{\partial r}{\partial X} \right)$$

$$\frac{\partial \varphi_k(X)}{\partial X} = 2\sigma_k^2 \left((F^T R^{-1} F)^{-1} (F^T R^{-1} J_r(X) - J_f(X)) - R^{-1} J_r(X) \right) \quad \text{(A9.8)}$$

Chapter 10

A Bayesian Estimation Technique for Multilevel Damage Classification in DBHM

William Locke,[1,] Stefani Mokalled,[2] Omar Abuodeh,[1]*
Laura Redmond[3] and Christopher McMahan[2]

1. Introduction

Drive-by health monitoring (DBHM) is a relatively new structural health monitoring (SHM) strategy that was originally developed to address limitations associated with visual inspection practices and installing traditional SHM systems on highway bridges (Agdas, 2015; Huang and Beck, 2015; Lynch and Loh, 2006; Moore et al., 2001; Sohn and Law, 2000; Yang and Yang, 2018). By mounting a network of specialized sensors on a single or multiple vehicles, DBHM theoretically enables the collection of health data across a network of bridges in a more efficient, cost effective, and less labor-intensive manner than the traditional SHM (Yang and Yang, 2018). The purpose of this chapter is to present in detail a novel approach of conducting DBHM. In particular, this chapter reproduces and reviews (with permission from the authors) the methodology proposed in (Locke, 2021; Locke et al., 2021). This methodology provides three levels of damage classification within SHM; Level 1 (L1) identifies the presence of damage; Level 2 (L2) determines the geometric location of the damage; Level 3 (L3) quantifies the damage severity. To achieve these levels of damage classification, this DBHM methodology leverages a Bayesian estimation technique that can be used to analyze noisy vehicle data to

[1] Glenn Department of Civil Engineering, Clemson University, Clemson, SC-29631.
 Email: oabuode@g.clemson.edu
[2] School of Mathematical and Statistical Sciences, Clemson University, Clemson, SC-29631.
 Email: smokall@g.clemson.edu
[3] Glenn Department of Civil Engineering and Department of Mechanical Engineering, Clemson University, Clemson, SC-29631. Email: lmredmo@clemson.edu
* Corresponding author: wrlocke@g.clemson.edu

update a physics-based vehicle-bridge finite element model (FEM) for the purpose of detecting, locating, and quantifying crack damage. To determine if damage is present under this framework, a spike and slab prior is specified on the FEM's crack depth, which allows for the description of two separate populations; i.e., healthy and damaged. The vehicle-bridge FEM and the Bayesian model are outlined in the Methodology Section. To relate crack ratios identified in the simplified vehicle-bridge FEM to approximate levels of cracking on a physical bridge, a damage mapping methodology is also presented in the Methodology Section. The vehicle bridge FEM and the mapping methodology are validated in the Model and Damage Mapping Validation Section, while the functionality of the Bayesian model is provided in the Demonstration Section.

2. Methodology

The Bayesian estimation technique for damage classification employs a series of models including a simplified vehicle-bridge FEM, a damage model (an open crack model), a damage mapping methodology to relate identified crack ratios to physical levels of damage, and a Bayesian model updating routine (Locke, 2021). The simplified vehicle-bridge FEM is embedded within the Bayesian model updating routine, which makes use of spike and slab priors, to perform multilevel damage classification. Crack location (δ_1) and crack ratio (δ_2) estimates from the Bayesian model updating routine are leveraged by the damage mapping methodology to identify an equivalent crack ratio on a high-fidelity bridge FEM by relating the localized changes in flexibility of both FEMs. The crack ratio identified in the higher fidelity FEM is considered to be closely representative of the true crack conditions present on the physical structure. Figure 10.1 provides a flow diagram for the overall Bayesian estimation technique (Locke, 2021).

2.1 Simplified Vehicle-Bridge FEM

2.1.1 Vehicle-Bridge Interactions

Vehicle-bridge interactions use the procedure in the model of Yang et al. (Yang et al., 2004) which is based on the concept of vehicle-bridge interaction elements; the description for the subject modeling strategy is referenced from (Locke, 2021; Locke et al., 2020). In this procedure, the bridge is represented by a one-dimensional model that is discretized into a number of Euler-Bernoulli beam elements (N_{ele}). Note that N_{ele} depends on the crack damage condition being applied, please see the Open Crack Model Section for more details. Equation (10.1) shows the equation-of-motion (EoM) for a beam element occupied by a vehicle:

$$(\mathbf{M}_{b,i}\ddot{\mathbf{d}}_{b,i} + \mathbf{C}_{b,i}\dot{\mathbf{d}}_{b,i} + \mathbf{K}_{b,i}\mathbf{d}_{b,i})_{t+\Delta t} = (\mathbf{f}_{be,i} - \mathbf{f}_{bc,i})_{t+\Delta t}, \tag{10.1}$$

where $\mathbf{M}_{b,i}$, $\mathbf{C}_{b,i}$, and $\mathbf{K}_{b,i}$ are the ith element's mass, damping, and stiffness matrices, respectively. Vectors $\ddot{\mathbf{d}}_{b,i}$, $\dot{\mathbf{d}}_{b,i}$, and $\mathbf{d}_{b,i}$ represent the nodal acceleration, velocity, and displacement for a subject element, respectively. External forces introduced from

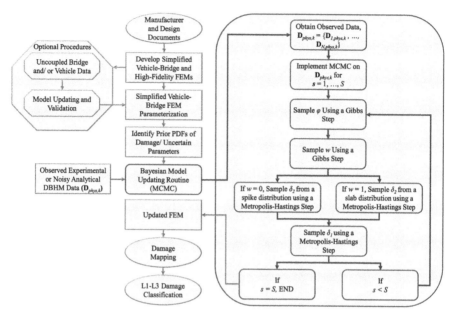

Fig. 10.1: Order of operations for the Bayesian estimation technique for damage classification, including optional initial model validation with an uncoupled bridge and/or vehicle data.

dead loads and environmental effects are represented by $\mathbf{f}_{be,i}$, while contact loads introduced by vehicles are represented by $\mathbf{f}_{bc,i}$. Subscript $t + \Delta t$ indicates the EoM is being solved for in-a-future time step.

Parameters $\mathbf{K}_{b,i}$ and $\mathbf{M}_{b,i}$ are modeled using the traditional four degree-of-freedom (DoF) stiffness matrix and continuous mass matrix (Chopra, 2012). In the example presented, axial DoFs are ignored, as thermal effects are presently not considered and vehicle velocities (V) are held constant; i.e., axial forces are not introduced from breaking or accelerating. Global mass $\mathbf{M}_b$ and stiffness $\mathbf{K}_b$ matrices, along with the global external load vector $\mathbf{f}_{be}$, are created by piecing together the local element matrices and vectors when the bridge is free of traffic. The global damping matrix $\mathbf{C}_b$ is calculated using the Rayleigh damping method outlined below in Eqs. 10.2 and 10.3:

$$\mathbf{C}_b = b_0\mathbf{M}_b + b_1\mathbf{K}_b, \tag{10.2}$$

$$b_0 = \xi\frac{2\omega_1\omega_2}{\omega_1 + \omega_2} \quad b_1 = \xi\frac{2}{\omega_1 + \omega_2}, \tag{10.3}$$

where parameters b_0 and b_1 are numerical constants calculated using the damping ratio ξ and the first two undamped modal frequencies w_1 and w_2 (Chopra, 2012).

To solve for the bridge's global contact forces vector $\mathbf{f}_{bc}$, the EoM for a vehicle is needed. In this example, vehicles are represented using the six DoF half-vehicle model outlined in Fig. 10.2 (Locke, 2021). The half-vehicle model provides more locations for measuring time history data than a quarter-vehicle model (Gillespie, 1992; Yang and Yang, 2018), which helps to improve the reliability of the Bayesian estimation technique that is discussed later. For a half-vehicle model, Eq. 10.4 represents the EoM for the vehicle in the future time step $t + \Delta t$:

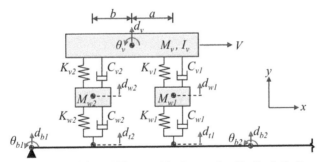

Fig. 10.2: Simplified vehicle-bridge FEM leveraged for damage classification in the Bayesian estimation technique. The vehicle is represented as a six DoF half-vehicle model, while bridge elements are represented as traditional four DoF Euler-Bernoulli beam elements.

$$\left(\begin{bmatrix} \mathbf{M}_{uu} & \mathbf{M}_{ul} \\ \mathbf{M}_{lu} & \mathbf{M}_{ll} \end{bmatrix} \begin{Bmatrix} \ddot{\mathbf{d}}_u \\ \ddot{\mathbf{d}}_l \end{Bmatrix} + \begin{bmatrix} \mathbf{C}_{uu} & \mathbf{C}_{ul} \\ \mathbf{C}_{lu} & \mathbf{C}_{ll} \end{bmatrix} \begin{Bmatrix} \dot{\mathbf{d}}_u \\ \dot{\mathbf{d}}_l \end{Bmatrix} + \begin{bmatrix} \mathbf{K}_{uu} & \mathbf{K}_{ul} \\ \mathbf{K}_{lu} & \mathbf{K}_{ll} \end{bmatrix} \begin{Bmatrix} \mathbf{d}_u \\ \mathbf{d}_l \end{Bmatrix} \right)_{t+\Delta t} =$$

$$\left(\begin{Bmatrix} \mathbf{f}_{ue} \\ \mathbf{f}_{le} \end{Bmatrix} + \begin{Bmatrix} 0 \\ \mathbf{f}_{lc} \end{Bmatrix} \right)_{t+\Delta t}, \tag{10.4}$$

where subscript u represents properties associated with the upper four DoFs on the sprung and unsprung masses of the vehicle in Fig. 10.2 (i.e., d_v, θ_v, d_{w1}, and d_{w2}), and subscript l represents the properties associated with the lower two DoFs where the vehicle makes contact with the bridge; i.e., d_{t1} and d_{t2} in Fig. 10.2. As can be seen, matrices for mass, damping, and stiffness are divided into sub-matrices; the structure of these sub-matrices can be observed in Appendix 10A or referenced from (Locke, 2021; Yang et al., 2004). In the same manner as Eq. 10.1, vectors $\ddot{\mathbf{d}}$, $\dot{\mathbf{d}}$, and $\mathbf{d}$, respectively, represent the vehicle's acceleration, velocity, and displacement, while the force vectors with subscripts e and c represent the vehicle's external and contact forces, respectively. Note that external forces are set equal to zero for the examples in this chapter and need only be included when vehicle acceleration/deceleration is considered, or an external excitation is applied to the monitoring vehicle to further excite the bridge system (Oshima et al., 2008; Zhang et al., 2012).

To calculate the vehicle's contact forces $\mathbf{f}_{lc}$, the future acceleration, velocity, and displacement of the upper vehicle must be analyzed first; Newmark Beta numerical integration is leveraged in Eqs. 10.5–10.7 to solve for these parameters in the future time step:

$$\ddot{\mathbf{d}}_{u,t+\Delta t} = a_0 \Delta \mathbf{d}_u - a_1 \dot{\mathbf{d}}_{u,t} - a_2 \ddot{\mathbf{d}}_{u,t}, \tag{10.5}$$

$$\dot{\mathbf{d}}_{u,t+\Delta t} = \dot{\mathbf{d}}_{u,t} + a_3 \ddot{\mathbf{d}}_{u,t} + a_4 \ddot{\mathbf{d}}_{u,t+\Delta t}, \tag{10.6}$$

$$\mathbf{d}_{u,t+\Delta t} = \mathbf{d}_{u,t} + \Delta \mathbf{d}_u, \tag{10.7}$$

where vector $\Delta \mathbf{d}_u$ represents the change in upper vehicle displacements between the present time step t and future time step $t + \Delta t$. Parameters a_0–a_4 are constants used to perform the numerical integration. These constants, along with a_5–a_7 employed later in the modeling procedures, are calculated as such (Eq. 10.8):

$$a_0 = \frac{1}{\beta \Delta t^2} \quad a_1 = \frac{1}{\beta \Delta t} \quad a_2 = \frac{1}{2\beta} - 1 \quad a_3 = (1-\gamma)\Delta t,$$

$$a_4 = \gamma \Delta t \quad a_5 = \frac{\gamma}{t\beta} \quad a_6 = \frac{\gamma}{\beta} - 1 \quad a_7 = \frac{\Delta t}{2}\left(\frac{\gamma}{\beta} - 2\right),$$

$$(10.8)$$

where β (1/4) and γ (1/2) represent the variation in acceleration during the incremental time step Δt and artificial damping introduced by discretization in the time domain, respectively (Chopra, 2012; Huebner et al., 2008); Δt is equal to a thousandth of a second ($\Delta t = 0.001$ s) for all simulations throughout this chapter. The equations for the future acceleration $\ddot{\mathbf{d}}_{u,t+\Delta t}$, velocity $\dot{\mathbf{d}}_{u,t+\Delta t}$, and displacement $\mathbf{d}_{u,t+\Delta t}$ can be entered into Eq. 10.4 to solve for the unknown $\Delta \mathbf{d}_u$. The calculated $\Delta \mathbf{d}_u$ can then be entered back into Eqs. 10.5–10.7 to solve for $\ddot{\mathbf{d}}_{u,t+\Delta t}$, $\dot{\mathbf{d}}_{u,t+\Delta t}$, and $\mathbf{d}_{u,t+\Delta t}$ directly.

Having related the future acceleration, velocity, and displacement of the upper vehicle EoM to be in terms of the present time step and $\Delta \mathbf{d}_u$, $\mathbf{f}_{lc}$ in Eq. 10.4 can be determined by employing the Newmark Beta numerical integration scheme to reformulate the EoM of the lower vehicle as seen in Eq. 10.9:

$$\mathbf{f}_{lc,t+\Delta t} = \left(\mathbf{M}_c \ddot{\mathbf{d}}_l + \mathbf{C}_c \dot{\mathbf{d}}_l + \mathbf{K}_c \mathbf{d}_l + \mathbf{p}_c\right)_{t+\Delta t} + \mathbf{q}_{c,t}, \qquad (10.9)$$

where $\mathbf{M}_c$, $\mathbf{C}_c$, and $\mathbf{K}_c$ are contact mass, damping, and stiffness matrices, respectively. Vectors $\mathbf{p}_c$ and $\mathbf{q}_c$ represent the effects of external forces on the vehicle in the future time step and the effects of the vehicle's displacement vector at the beginning of the present time step, respectively. Equations 10.10–10.14 indicate how these matrices and vectors are calculated:

$$\mathbf{M}_c = \mathbf{L}_l^{-1}(\mathbf{M}_{ll} - \mathbf{\Psi}_{lu}\mathbf{\Psi}_{uu}^{-1}\mathbf{M}_{ul}), \qquad (10.10)$$

$$\mathbf{C}_c = \mathbf{L}_l^{-1}(\mathbf{C}_{ll} - \mathbf{\Psi}_{lu}\mathbf{\Psi}_{uu}^{-1}\mathbf{C}_{ul}), \qquad (10.11)$$

$$\mathbf{K}_c = \mathbf{L}_l^{-1}(\mathbf{K}_{ll} - \mathbf{\Psi}_{lu}\mathbf{\Psi}_{uu}^{-1}\mathbf{K}_{ul}), \qquad (10.12)$$

$$\mathbf{p}_{c,t+\Delta t} = \mathbf{L}_l^{-1}(\mathbf{\Psi}_{lu}\mathbf{\Psi}_{uu}^{-1}\mathbf{f}_{ue,t+\Delta t} - \mathbf{f}_{le,t+\Delta t}), \qquad (10.13)$$

$$\mathbf{q}_{c,t} = \mathbf{L}_l^{-1}(\mathbf{\Psi}_{lu}\mathbf{\Psi}_{uu}^{-1}\mathbf{q}_{u,t} - \mathbf{q}_{l,t}), \qquad (10.14)$$

where $\mathbf{L}_l$ is an identity matrix whose dimensions are governed by the number of contact points; i.e., 2x2 for the six DoF vehicle model in Fig. 10.2. Parameters $\mathbf{\Psi}_{lu}$, $\mathbf{\Psi}_{uu}$, $\mathbf{q}_{u,t}$, and $\mathbf{q}_{l,t}$ are representative parameters used to simplify Eqs. 10.10–10.14, and are derived during the Newmark Beta integration of the upper and lower vehicle EoM; Eqs. 10.15–10.18 provide the full expression for these parameters:

$$\mathbf{\Psi}_{lu} = a_0 \mathbf{M}_{lu} + a_5 \mathbf{C}_{lu} + \mathbf{K}_{lu}, \qquad (10.15)$$

$$\mathbf{\Psi}_{uu} = a_0 \mathbf{M}_{uu} + a_5 \mathbf{C}_{uu} + \mathbf{K}_{uu}, \qquad (10.16)$$

$$\mathbf{q}_{u,t} = \left(\mathbf{M}_{uu}(a_1 \dot{\mathbf{d}}_u + a_2 \ddot{\mathbf{d}}_u) + \mathbf{C}_{uu}(a_6 \dot{\mathbf{d}}_u + a_7 \ddot{\mathbf{d}}_u) - \mathbf{K}_{uu}\mathbf{d}_u\right)_t, \qquad (10.17)$$

$$\mathbf{q}_{l,t} = \left(\mathbf{M}_{lu}(a_1 \dot{\mathbf{d}}_u + a_2 \ddot{\mathbf{d}}_u) + \mathbf{C}_{lu}(a_6 \dot{\mathbf{d}}_u + a_7 \ddot{\mathbf{d}}_u) - \mathbf{K}_{lu}\mathbf{d}_u\right)_t. \qquad (10.18)$$

To relate the vehicle's contact forces $\mathbf{f}_{lc}$ to the bridge contact forces $\mathbf{f}_{bc,i}$, $\ddot{\mathbf{d}}_l$, $\dot{\mathbf{d}}_l$, and $\mathbf{d}_l$ in Eq. 10.9 are reformulated to be in terms of an occupied bridge element's nodal accelerations, velocities, and displacements Eqs. 10.19–10.21:

$$\mathbf{d}_{l,t+\Delta t} = \mathbf{n}'_s \mathbf{d}_{b,i,t+\Delta t}, \tag{10.19}$$

$$\dot{\mathbf{d}}_{l,t+\Delta t} = \dot{\mathbf{n}}'_s \mathbf{d}_{b,i,t+\Delta t} + \mathbf{n}'_s \dot{\mathbf{d}}_{b,i,t+\Delta t}, \tag{10.20}$$

$$\ddot{\mathbf{d}}_{l,t+\Delta t} = \ddot{\mathbf{n}}'_s \mathbf{d}_{b,i,t+\Delta t} + \dot{\mathbf{n}}'_s \dot{\mathbf{d}}_{b,i,t+\Delta t} + \mathbf{n}'_s \ddot{\mathbf{d}}_{b,i,t+\Delta t}, \tag{10.21}$$

where $\mathbf{n}_s$ is a vector containing Cubic Hermitian polynomial shape functions, which are used to derive the Euler-Bernoulli elemental stiffness matrix (Chopra, 2012). Equation 10.22 indicates the vector of shape functions:

$$\mathbf{n}_s = \left\{ 1 - 3x_b^2 + 2x_b^3; \quad x_c(1 - 2x_b + x_b^2); \quad 3x_b^2 - 2x_b^3; \quad x_c(x_b^2 - x_b) \right\}, \tag{10.22}$$

where x_c is a vehicle's local position on an element, and x_b is the local coordinate on an element; i.e., $x_b = x_c / l_e$. Note l_e represents the length of an element and is calculated such that $l_e = L/N_{ele}$, where L is the length of the structure. Having reformulated $\ddot{\mathbf{d}}_l$, $\dot{\mathbf{d}}_l$, and $\mathbf{d}_l$ in Eq. 10.9 to be in terms of $\ddot{\mathbf{d}}_{b,i}$, $\dot{\mathbf{d}}_{b,i}$, and $\mathbf{d}_{b,i}$, the final EoM for an occupied bridge element can be derived by multiplying the now reformulated $\mathbf{f}_{lc}$ by $\mathbf{n}_s$ to obtain $\mathbf{f}_{bc,i}$. Equation 10.23 indicates the final EoM for an occupied bridge element:

$$(\mathbf{M}_{b,i} \ddot{\mathbf{d}}_{b,i} + \mathbf{C}_{b,i} \dot{\mathbf{d}}_{b,i} + \mathbf{K}_{b,i} \mathbf{d}_{b,i})_{t+\Delta t} =$$
$$(\mathbf{f}_{be,i} - \mathbf{M}_c^* \ddot{\mathbf{d}}_{b,i} - \mathbf{C}_c^* \dot{\mathbf{d}}_{b,i} - \mathbf{K}_c^* \mathbf{d}_{b,i} - \mathbf{p}_c^*)_{t+\Delta t} - \mathbf{q}_{c,t}^*, \tag{10.23}$$

where the asterisked matrices and vectors (e.g., $\mathbf{M}_c^*$) are the contact parameters in Eq. 10.9 that have been modified by the Cubic Hermitian shape functions (Yang et al., 2004).

Having obtained the finalized EoM for a vehicle occupied bridge element, the modified contact matrices are added into $\mathbf{M}_{b,i}$, $\mathbf{C}_{b,i}$, and $\mathbf{K}_{b,i}$ on the left side of Eq. 10.23; the local element matrices and vectors are then assembled into the previously assembled global matrices at their respective global coordinates. The Newmark Beta scheme is then utilized to solve for the Bridge's global $\ddot{\mathbf{d}}_{b,t+\Delta t}$, $\dot{\mathbf{d}}_{b,t+\Delta t}$, and $\mathbf{d}_{b,t+\Delta t}$. The nodal acceleration, velocity, and displacement values for the vehicle occupied bridge element(s) are then substituted into Eqs. 10.19–10.21 to calculate $\ddot{\mathbf{d}}_{l,t+\Delta}$, $\dot{\mathbf{d}}_{l,t+\Delta}$, and $\mathbf{d}_{l,t+\Delta}$; the resulting values are substituted into Eq. 10.4 to calculate $\ddot{\mathbf{d}}_{u,t+\Delta}$, $\dot{\mathbf{d}}_{u,t+\Delta}$, and $\mathbf{d}_{u,t+\Delta}$. Once the future accelerations, velocities, and displacements are known for both the lower and upper portions of the vehicle occupying the bridge, the global (x_g) and local x_c positions for the vehicle are updated for the next-time step, and the analysis is re-iterated. This iterative procedure is repeated until the rear wheel of a subject vehicle reaches the end of the bridge.

As per Fig. 10.1, the simplified vehicle-bridge FEM is solved for within the Bayesian model updating routine of the Bayesian estimation technique to perform L1–L3 damage classification. The Bayesian model updating routine leverages raw acceleration data from the upper vehicle's sprung and unsprung vertical DoFs (i.e., $\ddot{\mathbf{d}}_u([d_v, d_{w1}, d_{w2}], \mathbf{t}_{tot})$) to update unknown model input parameters; i.e., δ_1 and δ_2. Note that $\mathbf{t}_{tot}$ is a time vector ranging from $0 : \Delta t : T_f$, and T_f is the total length of a simulation in seconds, which is governed by vehicle velocity V and the total distance traveled (D_t). In this study, D_t is the sum of the bridge length L and the length of the approach slabs (L_a); therefore, $T_f = D/V$. Cracking is modeled in the simplified vehicle-bridge FEM using the procedures outlined in the following section, while the Bayesian model is discussed in more detail in the Bayesian Model Updating Routine Section.

2.1.2 Open Crack Model

Crack damage is modeled in the simplified vehicle-bridge FEM by employing the crack disturbance methodology proposed by Qian et al. (Qian et al., 1990); the description for the subject modeling strategy is referenced from (Locke, 2021). The proposed methodology uses fracture mechanics to provide empirical expressions of stress intensity factors that can be leveraged to approximate changes in element flexibility. Note that the methodology can be directly modified to introduce stiffness and damping nonlinearities caused by breathing effects (Nguyen, 2013; Nguyen and Tran, 2010).

The calculation of increasing stress field energy caused by cracking has been thoroughly studied in fracture mechanics, and the flexibility coefficient expressed by a stress intensity factor can be derived by means of Castigliano's theorem when operating in the linear-elastic range (Qian et al., 1990). Through this approach, the flexibility matrix $\mathbf{F}_L$ of a cracked element can be calculated as shown in Eq. (10.24):

$$\mathbf{F}_L = \frac{1}{6EI}\begin{bmatrix} 2l_e^3 & 3l_e^2 \\ 3l_e^2 & 6l_e \end{bmatrix} + \left(\frac{18\pi(1-v^2)}{EWH^2} \begin{bmatrix} l_e^2 & 2l_e \\ 2l_e & 6 \end{bmatrix} \right) \int_0^{\delta_2} (\psi F_I^2(\psi))d\psi, \qquad (10.24)$$

where E and v are material properties that represent a cracked element's modulus of elasticity and Poisson ratio, respectively (Bovsunovsky and Surace, 2005; Lee and Chung, 2000; Qian et al., 1990). Variables W, H, and I are geometric properties that represent a cracked element's width, height, and moment of inertia, respectively. Relative crack size is represented by ψ, and as can be seen is integrated from 0 to the crack ratio δ_2. Note, δ_2 is defined as the depth of cracking (d_c) divided by height; i.e., $\delta_2 = d_c/H$. Coefficient $F_I^2(\psi)$ is the stress intensity correction factor, which can be approximated as shown in Eq. (10.25):

$$F_I^2(\psi) = \left(\frac{0.923 + 0.199\left(1 - \sin(\frac{\pi\psi}{2})\right)^4}{\cos(\frac{\pi\psi}{2})} \right) \sqrt{\frac{\tan(\frac{\pi\psi}{2})}{\frac{\pi\psi}{2}}}. \qquad (10.25)$$

Using conditions of equilibrium, the cracked element's stiffness matrix $\mathbf{K}_{crack,i}$ in the free-free state can now be calculated as a function of the flexibility matrix (Eq. 10.26):

$$\mathbf{K}_{crack,i} = \mathbf{T}_r' \mathbf{F}_L^{-1} \mathbf{T}_r, \qquad (10.26)$$

where $\mathbf{T}_r$ is the transformation matrix calculated as follows (Eq. 10.27):

$$\mathbf{T}_r = \begin{bmatrix} -1 & -l_e & 1 & 0 \\ 0 & -1 & 0 & 1 \end{bmatrix} \qquad (10.27)$$

It is worth mentioning that to use this crack modeling strategy, cracks must be located at the center of a subject element; this makes it difficult to incrementally update δ_1 during the Bayesian model updating routine while continuously meshing the structure with the same number and size elements. To address this issue, a meshing scheme is adapted that continuously re-meshes the structure for each iteration of the Bayesian model updating routine. During the re-meshing procedure, the global coordinates of the cracked element's nodes are first assigned such that they are equal distance from the estimated δ_1; the global coordinates of the nodes are assigned such that $l_e = L/N_{ele}$. Once the cracked element's nodes have been assigned, the remaining unmeshed sections of the model are uniformly divided such that the average element length is approximately equal to l_e. As a precaution, lower and upper bounds of $0.5 l_e$ and $1.5 l_e$ are assigned to prevent any single element from being disproportionately small or large. Additionally, N_{ele}, which governs the average mesh size of all elements of the bridge, is selected to be sufficiently large (i.e., $N_{ele} = 30$) such that for every δ_1 estimate along the bridge, modeling δ_2 as zero (undamaged) results in no measurable change in bridge frequencies up to two decimal places. As a final precaution, cracks are prevented from being assigned within l_e of the supports; this is done to enforce the upper and lower bounds, and to allow the prior distribution for δ_1 to behave as a truncated normal distribution. Overall, the subject re-meshing scheme functions in a similar manner as the delete-and-fill, moving point, and other adaptive meshing schemes that are employed for simulating crack propagation in higher fidelity FEMs (Cornejo et al., 2019; Koenke et al., 1998; Verfürth, 2013).

2.2 Damage Mapping

Although the computational efficiency of the simplified vehicle-bridge FEM makes it ideal for the Bayesian estimation technique, there is a need to map estimated crack ratios δ_2 back to equivalent levels of physical damage in order to sufficiently describe the L3 damage classification. Therefore, a mapping methodology is provided that identifies an equivalent crack size in a high-fidelity Abaqus FEM that produces the same magnitude change in system flexibility as the estimated δ_2 from the simplified bridge FEM. The higher fidelity FEM is more representative of a physical bridge, and thus provides a better indication of the true crack conditions capable of being detected by the Bayesian estimation technique.

Damage-induced changes in system flexibility can be detected by comparing flexibility matrices obtained from healthy and damaged mode shapes, if the identified mode shapes are mass-normalized to unity (Ndambi et al., 2002; Pandey and Biswas, 1994). Equations 10.28 and 10.29 provide the relationships between the system flexibility matrix and the modal properties of the structure:

$$\mathbf{F}_{u,d} = \phi \Omega^{-1} \phi' = \sum_{m=1}^{N_{mod}} \frac{\phi_m \phi_m'}{\omega_m^2}, \tag{10.28}$$

$$\Delta \mathbf{F} = \mathbf{F}_u - \mathbf{F}_d, \tag{10.29}$$

where $\mathbf{F}_{u,d} \in R^{l \times l}$ represents the undamaged or damaged flexibility matrices and $\Delta \mathbf{F} \in R^{l \times l}$ is the damage indicator matrix; note that l corresponds to the number of measurement points along the length of the bridge. Parameter $\Omega \in R^{m \times m}$ is the modal stiffness matrix that is equal to $\text{diag}(1/\omega_m^2)$, ω_m is the mth frequency along the diagonal of $\boldsymbol{\Omega}$, ϕ_m is the mth mode shape, and N_{mod} is the number of modes being considered (Ndambi et al., 2002; Pandey and Biswas, 1994). In this chapter, only the first mode is considered for calculating the change in system flexibility.

Traditionally, experts have either leveraged the absolute maximum values of each column of $\Delta\mathbf{F}$ or its diagonals to identify and locate damage on a structure, while the work in this chapter does the reverse. After leveraging the Bayesian model updating routine to identify mean estimates for δ_1 and δ_2 on the simplified bridge FEM, $\Delta\mathbf{F}$ is calculated and the value associated with δ_1 is recorded. The mapping methodology is then employed to identify the equivalent crack depth that yields the same $\Delta\mathbf{F}$ value for the DoF associated with the given longitudinal location in the high-fidelity Abaqus FEM. To assure the measurement locations match between the simplified and high-fidelity FEMs, a mesh size in the simplified FEM is selected to match the longitudinal dimension of the elements in the high-fidelity FEM. Furthermore, because the simplified FEM is incapable of capturing transverse variations in cross-section displacement, the mode shapes identified for each girder in a high-fidelity FEM are averaged together to obtain a single averaged mode shape for the entire cross-section. The outlined approach is demonstrated in the Model and Damage Mapping Validation Section and subsequently leveraged to interpret the results of the DBHM study in the Demonstrations Section. Note that this damage mapping methodology can be applied to FEMs that use other open crack or complex breathing crack models.

2.3 Bayesian Model Updating Routine

The preceding sections have described established methods used to model coupled vehicle-bridge dynamics and crack damage, and a new damage mapping methodology to relate cracking in simplified FEMs to high-fidelity representations of bridge damage. Attention is now turned towards embedding the simplified vehicle-bridge FEMs into the Bayesian model updating routine to detect damage and obtain damage location δ_1 and crack ratio δ_2 estimates for the mapping methodology. The novelty of this model updating routine is the use of a spike-and-slab prior distribution as an approach to damage detection. It is worth noting that more traditional Bayesian based approaches employ Bayes' theorem for damage detection via model updating (Asadollahi, 2018). Spike and slabs are typically employed when sampling a significant (slab prior) versus insignificant (spike prior) coefficient in a model. By assuming a spike-and-slab prior distribution on δ_2 in this study, δ_2 either introduces insignificant changes to the bridge model and is sampled from the healthy (spike) population, or introduces a significant change to the bridge model and is sampled from the damaged (slab) population.

Once it is determined whether the bridge belongs to a healthy or damaged population (providing L1 damage classification), δ_1 and δ_2 can be estimated to provide L2 and L3 damage classifications, respectively. These parameters are initially treated as unknown inputs required for solving the simplified vehicle-bridge FEM. Note that δ_1 and δ_2 are constrained to a range of feasible quantities during model updating, with the lower and upper bounds for δ_1 being $l_1 = l_e$ and $u_1 = L - l_e$, and the lower and upper bounds for δ_2 being $l_2 = 0$ and $u_2 = 0.2$. The bounds for δ_1 were selected per the rules established for meshing in the Open Crack Section, while the upper bound for δ_2 was selected because it was assumed for the work in this chapter that cracks should be detected and repaired before reaching a crack-to-depth ratio of 0.2. For simplicity, let $\delta = (\delta_1, \delta_2)$ represent a vector of these unknown parameters being updated within the Bayesian model updating routine.

As mentioned, the raw acceleration data from the upper vehicle's sprung and unsprung vertical DoFs $\ddot{d}u([d_v, d_{w1}, d_{w2}], \mathbf{t}_{tot})$ are leveraged within the Bayesian model updating routine. Let $\mathbf{D}_{sim}(\delta,V) = \ddot{d}u([d_v, d_{w1}, d_{w2}], \mathbf{t}_{tot})$ such that $\mathbf{D}_{sim}(\delta,V)$ is denoted as a function of the unknown input parameters δ and the known velocity V. Let $\mathbf{D}_{sim,k}(\delta,V)$ denote the kth row of matrix $\mathbf{D}_{sim}(\delta,V)$. It is assumed (Eq. 10.30):

$$\mathbf{D}_{n,phys,k} = \mathbf{D}_{sim,k}(\delta,V) + \epsilon_{n,k}, \qquad (10.30)$$

where $\mathbf{D}_{n,phys,k}$ represents the kth row of the physical observations matrix for the nth vehicle run. Note that physical observations are obtained on N different vehicle runs (i.e., $n = 1, 2, \ldots, N$), which are assumed to be independent of one another. The matrix $\epsilon_{n,k}$ consists of measurement and model errors, which are assumed to be normally distributed with a mean of zero and covariance $\varphi^{-1}\mathbf{I}$ such that the errors are uncorrelated. By assuming $\epsilon_{n,k}$ is normally distributed, it is assumed that $\mathbf{D}_{n,phys,k}$ given the known and unknown inputs is also normally distributed with mean $\mathbf{D}_{sim,k}(\delta,V)$ and variance $\varphi^{-1}\mathbf{I}$ as shown in Eq. 10.31.

$$\mathbf{D}_{n,phys,k}|V, \delta, \varphi \sim \mathrm{MVN}(\mathbf{D}_{sim,k}(\delta,V)\,\varphi^{-1}\mathbf{I}), \qquad (10.31)$$

where $\mathrm{MVN}(\mu, \Sigma)$ is the multivariate normal distribution with mean vector μ and variance-covariance matrix Σ. In this chapter, it is assumed that $\mathbf{D}_{sim,k}(\delta,V)$ captures the true acceleration responses at the time steps corresponding to the physical measurements in $\mathbf{D}_{n,phys,k}$. The only discrepancy between $\mathbf{D}_{sim,k}(\delta,V)$ obtained from the simplified vehicle-bridge FEM and the observed $\mathbf{D}_{n,phys,k}$ is the measurement and model errors. This assumption is reasonable, as physical vehicle position data can be obtained using GPS tracking to identify the time of initial vehicle loading during physical tests, which could then be used to align time steps. Additionally, adaptations can be made on the variances $\varphi^{-1}\mathbf{I}$ to incorporate correlation in error terms via functional forms on the covariances (i.e., with nonzero off-diagonal elements) as in (Kennedy and O'Hagan, 2001).

To enable Bayesian estimation, the following priors are specified on the model parameters:

$$\varphi \sim \mathrm{gamma}(a_0, b_0)$$

$$\delta_1 \sim \mathrm{TN}\{\mu_1, \tau_1, (l_1, u_1)\}$$

$$\delta_2|w \sim \mathrm{TN}\{\mu_2, r(w)\tau_2, (l_2, u_2)\}$$

$$w|p \sim \mathrm{Bernoulli}(p)$$

where $\mathrm{TN}(\mu, \tau, (l, u))$ denotes a truncated normal distribution with mean μ, variance τ, and bounds (l, u). This formulation emits a stochastic search variable selection (SVSS) spike and slab prior for δ_2 (George and McCulloch, 1993). The SVSS is accomplished via the binary switch $r(w)$, which transitions the prior between the spike (when $w = 0$) and the slab (when $w = 1$) by taking values $r(0) = r$ and $r(1) = 1$, respectively. Note that the tuning parameter r is set relative to τ_2; meaning $r \ll \tau_2$, where τ_2 is taken to be comparatively large and is set relative to the desired δ_2 threshold for L1 damage classification. Specifically, $r(0)\tau_2$ is set such that $\sqrt[3]{r(0)\tau_2}$ (i.e., three standard deviations from μ_2) is equal to the δ_2 threshold. For the work in this chapter, the δ_2 threshold for L1 classification is selected based on the sensitivity of the vehicle responses in the embedded model to damage induced changes in the bridge response when considering a variety of δ_1 values, velocities, and moderate levels of noise. Note that a risk-based approach can alternatively be employed to identify a δ_2 threshold that enables early damage detection and limits the probability of false positives. As indicated, an enticing feature of this prior specification is that L1 damage classification can be directly obtained by computing the posterior probability of inclusion of δ_2. Details on this computation can be found at the end of this section.

To facilitate model fitting, a posterior sampling algorithm is developed. This algorithm draws realizations from the posterior distribution, which is given by (Eq. 10.32):

$$
p(\varphi, \delta_1, \delta_2, w, p \mid \mathbf{D}_{1,\dots,N,phys}) \alpha \, \varphi^{\frac{3NT_{step}}{2}}
$$
$$
\cdot \exp\left[\frac{-\varphi \sum_{n=1}^{N} \sum_{k=1}^{3} \{\mathbf{D}_{n,phys,k} - \mathbf{D}_{sim,k}(\delta, V)\}'\{\mathbf{D}_{n,phys,k} - \mathbf{D}_{sim,k}(\delta, V)\}}{2} \right]
$$
$$
\cdot \tau_1^{-\frac{1}{2}} \exp\left\{ \frac{-(\delta_1 - \mu_1)^2}{2\tau_1} \right\} \cdot \{r(w)\tau_2\}^{-\frac{1}{2}} \exp\left\{ \frac{-(\delta_2 - \mu_2)^2}{2r(w)\tau_2} \right\} p^w (1-p)^{1-w}
$$
$$
\cdot \varphi^{a_0-1} \exp\{-b_0\varphi\},
$$
(10.32)

where $\mathbf{D}_{1,\dots,N,phys} = \{\mathbf{D}_{1,phys}, \dots, \mathbf{D}_{N,phys}\}$ and T_{step} represents the total number of time steps in $\mathbf{t}_{tot}$. Since the posterior distribution is not of an amendable form (i.e., of a known distributional family), a Markov Chain Monte Carlo (MCMC) sampling algorithm consisting of both Gibbs and Metropolis-Hastings (MH) steps is considered. In particular, w and φ are sampled using Gibbs steps as it is easy to directly sample from their full conditional distributions, while δ_1 and δ_2 are sampled using MH steps since directly sampling from their full conditional distributions is difficult. A general outline of the following sampling steps is provided in Fig. 10.1. To elucidate these steps, the full conditional distribution of w is given in Eqs. 10.33 and 10.34:

$$
w|\tilde{p}, \delta_2 \sim \text{Bernoulli}(\tilde{p}), \tag{10.33}
$$

$$
\tilde{p} = \{p\pi_{\delta 2}(\delta_2; \mu_2, \tau_2)\}/\{p\pi_{\delta 2}(\delta_2; \mu_2, \tau_2) + (1-p)\pi_{\delta 2}(\delta_2; \mu_2, r\tau_2)\}. \tag{10.34}
$$

Note that $p = 0.5$. In the aforementioned expression, $\pi_{\delta 2}(\cdot; \mu, \tau)$ denotes the density of a truncated normal distributions with mean μ and variance τ. Similarly, the full conditional distribution of φ is given in Eq. 10.35:

$\varphi \mid \delta, \mathbf{B} \sim$

$$\text{gamma}\left(\frac{3NT_{step}}{2} + a_0, \frac{\sum_{n=1}^{N}\sum_{k=1}^{3}\{\mathbf{D}_{n,phys,k} - \mathbf{D}_{sim,k}(\delta,V)\}'\{\mathbf{D}_{n,phys,k} - \mathbf{D}_{sim,k}(\delta,V)\}}{2} + b_0\right),$$

(10.35)

where the inputs to the gamma distribution are shape and scale parameters, respectively. Equations 10.33–10.35 are shown in the steps of sampling φ and w using Gibbs steps in Fig. 10.1. Finally, the full conditional distributions of δ_1 and δ_2 are, respectively, given by Eqs. 10.36 and 10.37:

$$p(\delta_1 \mid \mathbf{D}_{1,...,N,phys}, \varphi, \delta_2) \alpha$$

$$\exp\left[\frac{-\varphi\sum_{n=1}^{N}\sum_{k=1}^{3}\{\mathbf{D}_{n,phys,k} - \mathbf{D}_{sim,k}(\delta,V)\}'\{\mathbf{D}_{n,phys,k} - \mathbf{D}_{sim,k}(\delta,V)\}}{2}\right]$$ (10.36)

$$\cdot \exp\left\{\frac{-(\delta_1 - \mu_1)^2}{2\tau_1}\right\},$$

$$p(\delta_2 \mid \mathbf{D}_{1,...,N,phys}, \varphi, \delta_1, w) \alpha$$

$$\exp\left[\frac{-\varphi\sum_{n=1}^{N}\sum_{k=1}^{3}\{\mathbf{D}_{n,phys,k} - \mathbf{D}_{sim,k}(\delta,V)\}'\{\mathbf{D}_{n,phys,k} - \mathbf{D}_{sim,k}(\delta,V)\}}{2}\right]$$ (10.37)

$$\cdot \exp\left\{\frac{-(\delta_2 - \mu_2)^2}{2r(w)\tau_2}\right\},$$

It is important to note that directly sampling from these distributions is difficult, and MH steps must be used to sample from these full conditionals. Equations 10.36–10.37 can be observed in Fig. 10.1 as the two MH steps; as a reminder, δ_2 in Fig. 10.1 is sampled from a healthy or damaged distribution depending on current value of w. For complete details on the implementation of the MH steps, as well as a step-by-step description of the entire MCMC posterior sampling routine, see Appendix 10B or reference (Locke, 2021). In implementing the MCMC algorithm, parameters are sampled for $s = 1,...,S$ iterations, where S is the total number of iterations, with a burn-in period such that the first s_{burn} iterations are removed and not included in estimation and inference. The burn-in period allows the algorithm to move from the point of initialization to a region of high probability. Thinning is also performed such that every s_{thin} iterate is retained; this is done to reduce the autocorrelation in the MCMC chains. Burn-in and thinning in this procedure are standard practice; for more details see (Hoff, 2009).

After implementing this algorithm, posterior estimation and inference proceed as usual; for further discussion see (Gelman et al., 2013). In particular, the mean of the resulting chains of sampled values for w, δ_1, and δ_2 are summarized as the estimates provided for the numerical studies in the Section. The MCMC chain of values for w includes sampled values of 0 or 1, where a value of 1 indicates the bridge is damaged. Hence, by computing the proportion of MCMC iterations in which $w = 1$ (i.e., the posterior probability of inclusion), the probability the bridge is damaged and, therefore, L1 damage classification is obtained. Computing the mean of the sampled values of δ_1 and δ_2 across the MCMC chains results in point estimates for the L2–L3 classifications, respectively.

3. Subject Vehicle-Bridge System

The half-vehicle model and bridge system employed for the numerical studies in this chapter are based on full-scale vehicle and bridge systems employed in previous research studies. The selected half-vehicle model closely resembles the H-20 AASHTO truck employed for bridge live-load testing, and was adopted from previous research focused on coupled vehicle-bridge systems and crack damage detection (Deng and Cai, 2009; Nguyen and Tran, 2010). Table 10.1 indicates the properties for the half-vehicle model; note that the length properties of the vehicle (i.e., a and b) are equal to 3 m. The bridge is a full-scale steel girder bridge designed for laboratory testing at the University of Nebraska-Lincoln (Kathol et al., 1995). Two analytical studies are conducted with the subject bridge to demonstrate the damage mapping methodology and Bayesian estimation technique for damage classification.

The bridge is composed of a single simply supported span with a length of 21.3 m and a width of 7.9 m, which is enough to accommodate two lanes of vehicle traffic. As demonstrated by the cross-sectional view in Fig. 10.3, the superstructure is built compositely with a reinforced concrete deck resting on top of three welded plate girders. The concrete deck has typical Nebraska Department of Road (NDOR) open concrete barrier rails. The welded plate girders are composed of a 22.86×1.905 cm top flange, a 137.2×0.953 cm web, a 35.56×3.175 cm bottom flange for the center girder, and a 35.56×1.905 cm bottom flange for the edge girders. K-Frame diaphragms are installed in 6.83 m increments starting from the beginning of the bridge, while web shear stiffener plates are installed at the locations indicated in the longitudinal view of Fig. 10.3. Please reference Kathol et al. (Kathol et al., 1995) for more details regarding bridge design and dimensions. For the simplified bridge FEM, the equivalent area method was employed to represent the entire cross-section in Fig. 10.3 as an equivalent cross-section of steel. The properties of the equivalent cross-section are: $E = 200$ GPa, $H = 1.9282$ m, $W = 0.1409$ m, $I = 0.0842$ m^4, $\rho = 5600$ kg/m, and $\xi = 3\%$. Lastly, an approach slab length of $L_a = 6.5$ m was included on both sides of the bridge in the simplified vehicle-bridge FEM. The value for L_a was selected such that the whole vehicle could start off of the bridge and the front of the vehicle could continue to travel forward while the rear still occupied the bridge. Note that the start position of the vehicle was such that $D_t = L + L_a$.

Table 10.1: Properties for half-vehicle model.

Mass Properties		Stiffness Properties		Damping Properties	
M_v (kg)	12,404	$K_{v,1}$ (kN/m)	1,969	$C_{v,1}$ (Ns/m)	7,182
I_v (kg/m^2)	172,160	$K_{v,2}$ (kN/m)	728	$C_{v,2}$ (Ns/m)	2,190
M_{w1} (kg)	725	$K_{w,1}$ (kN/m)	4,735	C_{w1} (Ns/m)	0
M_{w2} (kg)	725	$K_{w,2}$ (kN/m)	1,973	C_{w2} (Ns/m)	0

Note: Vehicle bounce, pitch, and unsprung natural frequencies are 1.3 Hz, 2.2 Hz, 9.7 Hz, and 15.4 Hz, respectively.

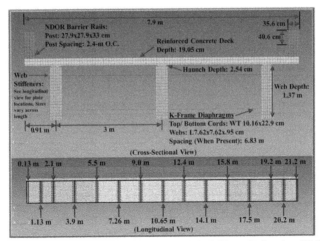

Fig. 10.3: Cross-sectional and longitudinal views of steel girder bridge at University of Nebraska-Lincoln campus, used as the basis of the simulation study.

4. Model and Damage Mapping Validation

A summary of the validation of the simplified and high fidelity bridge FEMs and the damage mapping methodology is presented here. For a more detailed presentation of these studies please see (Locke, 2021). The high-fidelity Abaqus FEM served as the physical representation of the structure employed to map crack ratios estimated for the simplified vehicle-bridge FEM back to equivalent levels of physical damage. The high-fidelity FEM represented cracking in a simplified manner using linear elastic material properties with disjoined nodes on the lower flanges and along the height of webs.

Static and dynamic data from previous studies were leveraged to validate the reliability of the bridge FEMs in this chapter (Abedin and Mehrabi, 2019; Kathol et al., 1995). Table 10.2 provides a comparison between the static deflections measured by Kathol et al. (Kathol et al., 1995) and the deflections measured for the subject FEMs in this chapter, while Table 10.3 provides a comparison between the healthy and damaged frequencies obtained by Abedin and Mehrabi (Abedin and Mehrabi, 2019) and the frequencies identified for the subject FEMs. As can be seen by Table 10.2, the static displacements of the Abaqus and simplified FEM are close to the deflections observed on the physical structure. Note that due to the one-dimensional representation of the simplified bridge FEM, it is impossible to transversely capture variations in vertical deflections; this was not considered an issue, as the static displacement of the simplified FEM falls within the bounds of deflections observed on the physical structure. As can be seen by Table 10.3, both the simplified and high-fidelity FEM reasonably capture the healthy and damaged dynamic behavior of the subject structure observed in previous studies.

To demonstrate the reliability of damage mapping, the methodology was first leveraged to identify equivalent levels of cracking on the high-fidelity Abaqus FEM when $\delta_2 = 0.10$ was applied at mid-span ($\delta_1 = 10.65$ m) on the simplified FEM. As cracks can occur on a single girder or simultaneously on multiple girders of a physical

Table 10.2: Comparing static deflections of the simplified and Abaqus bridge FEMs against physical measurements.

Physical Bridge Mid-span Deflections		
Diaphragm Present	**Center Girder (cm)**	**Edge Girders (cm)**
No	1.91	1.40
Yes	1.75	1.45
Abaqus FEM Mid-Span Deflections		
Diaphragm Present	**Center Girder (cm)**	**Edge Girders (cm)**
No	1.96	1.52
Yes	1.85	1.56
Simplified FEM Mid-Span Deflection		
Diaphragm Present	**Entire Cross-Section (cm)**	
N.A.	1.75	

Note: Static deflections are obtained from elastic load tests when 2.5xHS-20 trucks were used to load each lane of the bridge. Please reference Kathol et al. (Kathol et al., 1995) for more details on front and rear wheel load positions.

Table 10.3: Validation study comparing healthy and damaged frequencies of FEMs.

Abedin's and Mehrabi's FEM		
Damage State	**Frequency (Hz)**	**Percent Change (%)**
Healthy	6.00	–
Cracked	5.00	16.67
High-Fidelity Abaqus FEM		
Damage State	**Frequency (Hz)**	**Percent Change (%)**
Healthy	6.00	–
Cracked	4.87	18.83
Simplified FEM		
Damage State	**Frequency (Hz)**	**Percent Change (%)**
Healthy	6.00	–
Cracked	4.88	18.67

Note: The diaphragm is present in the high-fidelity Abaqus FEM.

structure, multiple scenarios needed to be mapped in the high-fidelity Abaqus FEM. In this chapter, three damage scenarios (DS_1–DS_3) were considered on the Abaqus FEM: DS_1 cracking occurring equally on all girders, DS_2 a single crack occurring on the interior girder, and DS_3 a single crack occurring on one of the exterior girders. During the mapping analysis, equivalent crack ratios of 0.0035, 0.0162, and 0.0180 were identified for DS_1–DS_3, respectively, to introduce the same peak change in flexibility (i.e., $\Delta F = 2.75e-10$) on the high-fidelity Abaqus FEM as the simplified FEM for the location associated with $\delta_1 = 10.65$ m. Additionally, each of the identified crack ratios captured both the damaged fundamental bridge frequency of 5.95 Hz and the general change in flexibility across the length of the structure. Note that when comparing the fundamental mode shapes of the simplified and Abaqus FEMs, it was determined that the healthy and damaged mode shapes were in excellent agreement, with a modal assurance criterion (MAC) value of 0.999 indicating almost perfect consistency between the FEMs. The agreement between the healthy and damaged mode shapes demonstrates that the method of calculating an average mode shape for the high-fidelity FEM during damage mapping is reasonable.

5. Demonstration of the Bayesian Estimation Technique for DBHM

This section provides a series of examples to demonstrate the reliability of the Bayesian estimation technique for damage classification. Simulation studies are conducted where 'physical' data $\mathbf{D}_{n,phys,k}$ is generated for a number of damage states and operating conditions. Additionally, L2 and L3 classification results from the simulation studies are leveraged in the damage mapping methodology to provide an indication of the physical crack ratios capable of being detected by the Bayesian estimation technique.

5.1 Scope

A base study was conducted to demonstrate the L1–L3 damage classification accuracy of the Bayesian estimation technique for a crack ratio equal to 0.05 at locations 2.1, 5.5, and 10.65 m away from the left support of the 21.3 m long bridge. These locations were selected based on the presence of welded connections for the shear stiffeners and diaphragms outlined in the longitudinal view of Fig. 10.3. Per the Bridge Inspectors Reference Manual, welded connections, especially those on tension faces, are prone to fatigue cracking due to a combination of fabrication flaws and high levels of stress (Ryan et al., 2012). Note that $\delta_2 = 0.05$ for all DBHM studies in this chapter.

Studies were also conducted to demonstrate the impact noise and the quantity of experimental data N have on L1–L3 classification accuracy when $\delta_1 = 2.1$, and 10.65 m. For the base study, N was set to one, V was set to a moderate 15.66 m/s, and a low level of noise was added to each response vector such that an average signal-to-noise ratio (SNR) of 30 dB was recorded across each DoF in $\mathbf{D}_{1,phys,k}$. SNR was calculated as shown in Eq. 10.38:

$$\text{SNR} = 20\text{Log}_{10}\left(\frac{A_s}{A_n}\right), \tag{10.38}$$

where A_s and A_n represent the root mean squares of a clean acceleration response and noise vector, respectively. Simulations with SNRs of 20 and 10 dB were also conducted to demonstrate the impact increasing levels of noise have on damage classification. The subject range of SNR values was selected, as they represent levels of additive noise that have typically been employed for SHM studies (Li et al., 2014; OBrien et al., 2015; O'Brien et al., 2014; Zhang et al., 2020). Equation (10.38) was reformulated to calculate the A_n required to achieve these noise ratios (Eq. 10.39):

$$A_n = \frac{A_s}{10^{\frac{\text{SNR}}{20}}}. \tag{10.39}$$

Hence, it was assumed that $\epsilon_{n,k} \sim \text{MVN}(\mathbf{0}, A_n\mathbf{I})$ such that $\varphi^{-1} = A_n$ in all simulation configurations. Figure 10.4 demonstrates how the acceleration response of the vehicle's sprung and unsprung DoFs change for SNRs of 10, 20, and 30 dB. To demonstrate the impact N has on L1–L3 damage classification, $\mathbf{D}_{n,phys,k}$ was obtained for N equal to 5, 10, 15, and 20 runs and a multi-run average was taken for each vehicle DoF. Tests were conducted using an ideal V determined by the velocity study for $\delta_1 = 2.1$ m and an SNR of 10 dB and 20 dB.

Under all simulation study configurations, the hyperparameters on the prior distribution of φ were assumed to be $a_0 = 1$ and $b_0 = 0$ such that the prior was weakly informative, allowing the $\mathbf{D}_{n,phys,k}$ to provide more information to the model. Since fatigue cracks are most likely to occur in higher flexural regions closer to the center of the bridge, δ_1 was assumed to have a prior distribution with $\mu_1 = 10.65$ m and bounds $l_1 = 0.72$ m and $u_1 = 20.58$ m. The variance was set to be $\tau_1 = 12.60$ m^2, allowing for the beginning and end of the bridge to be approximately three standard deviations from the mean. From a sensitivity analysis conducted on the simplified vehicle-bridge FEM, $\delta_2 = 0.015$ was the smallest crack ratio that caused any reasonably measurable change in vehicle response data. To establish $\delta_2 = 0.015$ as the threshold for L1 damage classification, τ_2 was set to 100 and $r(0)$ was set to 2.5e–7. In this study, $\delta_2 = 0.015$ lies three standard deviations from $\mu_2 = 0$; i.e., $3\sqrt{(2.5e-7)100} = 0.015$. See Fig. 10.5 for a visual representation of the spike and slab prior distributions employed.

To perform posterior estimation and inference, the posterior sampling algorithm was used to draw $S = 16{,}000$ MCMC samples with a burn in period of $s_{burn} = 1{,}000$ iterations. To reduce the effects of autocorrelation, thinning was performed such that every $s_{thin} = 15$th iterate was retained, leaving 1,000 samples for analysis. The number of iterations was chosen to ensure that after the burn in period and thinning, a sufficiently large posterior chain was attained for reliable point estimation and inference. Note that for these simulated data studies, noisy experimental data $\mathbf{D}_{n,phys,k}$ was generated for N equal to 1, 5, 10, 15, or 20 runs prior to running the MCMC algorithm, which is equivalent to obtaining experimental data from 1, 5, 10, 15, or 20 physical vehicle runs. Once the MCMC algorithm was initialized, sampling depended only on running the vehicle-bridge FEM. For each

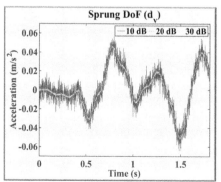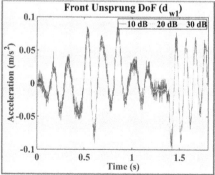

Fig. 10.4: Acceleration response of the vehicle's sprung DoF (d_v) and front unsprung DoF (d_{w1}) when considering SNRs of 10, 20, and 30 dB.

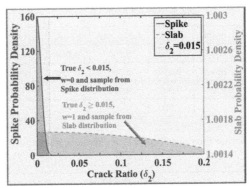

Fig. 10.5: Example demonstrating both a spike distribution with $\mu_2 = 0$ and standard deviation of 0.005 (i.e., $\sqrt{r(0)\tau_2} = \sqrt{(2.5e-7)100}$), and a slab distribution with $\mu_2 = 0$ and standard deviation of 10; i.e., $\sqrt{r(1)\tau_2} = \sqrt{(1)100}$. When $w = 0$, the bridge is considered healthy and δ_2 is sampled from a spike distribution. When $w = 1$, the bridge is considered cracked and δ_2 is sampled from a slab distribution.

damage case presented in this chapter, 100 data sets were simulated and analyzed; i.e., a 100 instances of the operations in Fig. 10.1 were executed. The 100 simulated data sets were generated for each damage case to validate the methodology; i.e., to ensure that the mean estimates well approximated the truth. Note that if working with a single data set obtained from physical DBHM experiments, the operations in Fig. 10.1 would only need to be performed once to update the simplified vehicle-bridge FEM and obtain L1–L3 damage classifications.

The results of the simulation studies are summarized in the following sections. In particular, the presented results are the average posterior mean estimates for the δ_1 and δ_2. To quantify uncertainty, the average estimate of the posterior standard deviation (SSD) are also provided. Finally, to assess the efficacy of the model with respect to L1 damage classification, the average estimated posterior probability of inclusion (PI) is also provided.

5.2 Base Study

As can be seen in Table 10.4, 100% L1damage classification was achieved at each damage location for the base case where data was collected from a single vehicle run at $V = 15.66$ m/s and the SNR was 30 dB. These results demonstrate that for low levels of noise, the Bayesian estimation technique was not only able to detect the subject crack ($\delta_2 = 0.05$) with absolute certainty, but also identify its location and magnitude with high accuracy and precision across the length of the structure.

For the base study, DS_1–DS_3 outlined in the Damage Mapping Section for the high fidelity Abaqus FEM were leveraged to map the δ_2 estimates for damage locations 2.1 and 10.65 m. For each damage scenario in Table 10.5, both the true δ_2 value and the δ_2 estimate from the base DBHM study are mapped to the equivalent crack ratios (defined consistently with section) on the high fidelity FEM. As can be seen by Table 10.5, the subtle discrepancies between the truth and the estimates on the simplified model when $\delta_2 = 0.05$ did not result in significant differences between the truth and the mapped estimates on the Abaqus FEM, indicating that accurate

Table 10.4: Results for L1-L3 damage classification when $N = 1$, $V = 15.66$ m/s, and the SNR was 30 dB.

	L2: δ_1 (m)	L3: δ_2	L2: δ_1 (m)	L3: δ_2	L2: δ_1 (m)	L3: δ_2
Truth	2.100	0.050	5.500	0.050	10.650	0.050
Mean	2.084	0.049	5.500	0.050	10.636	0.050
SSD	0.026	0.002	0.022	0.001	0.013	0.001
PI (L1)	–	1.000	–	1.000	–	1.000

Table 10.5: Mapping damage from simplified FEM to physical representations on high-fidelity Abaqus FEM.

Simplified FEM					
$\delta_{1, true}$ (m)	$\delta_{2, true}$	ΔF_{true} (m/N)	$\delta_{1, estimate}$ (m)	$\delta_{2, estimate}$	$\Delta F_{estimate}$ (m/N)
2.100	0.0500	4.369e-12	2.084	0.0490	5.423e-11
10.650	0.0500	7.150e-11	10.636	0.0500	7.150e-11
Abaqus FEM DS$_1$: All Girders Damaged					
$\delta_{1, true}$ (m)	$\delta_{2, true}$ Mapped	ΔF_{true} (m/N)	$\delta_{1, estimate}$ (m)	$\delta_{2, estimate}$ Mapped	$\Delta F_{estimate}$ (m/N)
2.100	0.0009	4.393e-12	2.084	0.0009	4.102e-12
10.650	0.0009	7.150e-11	10.636	0.0009	7.150e-11
Abaqus FEM DS$_2$: Interior Girder Damaged					
$\delta_{1, true}$ (m)	$\delta_{2, true}$ Mapped	ΔF_{true} (m/N)	$\delta_{1, estimate}$ (m)	$\delta_{2, estimate}$ Mapped	$\Delta F_{estimate}$ (m/N)
2.100	0.0029	4.393e-12	2.084	0.0027	4.102e-12
10.650	0.0029	7.150e-11	10.636	0.0029	7.150e-11
Abaqus FEM DS$_3$: Exterior Girder Damaged					
$\delta_{1, true}$ (m)	$\delta_{2, true}$ Mapped	ΔF_{true} (m/N)	$\delta_{1, estimate}$ (m)	$\delta_{2, estimate}$ Mapped	$\Delta F_{estimate}$ (m/N)
2.100	0.0031	4.393e-12	2.084	0.0029	4.102e-12
10.650	0.0031	7.150e-11	10.636	0.0031	7.150e-11

Note: Physical crack ratios are based on the crack to depth ratio of the girders only.

estimates for the true crack conditions on a physical structure can still be obtained if simplified model estimates are slightly inaccurate. To visually demonstrate that the mapped crack ratios cause the same change in flexibility as $\delta_2 = 0.05$ on the simplified model, Fig. 10.6 shows how well the estimated crack ratios capture the peak ΔF and the general change in flexibility across the length of the structure when compared to the simplified FEM.

5.3 Influence of Noise

It can be seen by Table 10.6 and Fig. 10.7 that as the SNR decreases (i.e., signal noise increases), the mean estimates for δ_1 move away from the truth and the standard deviations increase. Note that when L1 classification is below 50%, mean estimates for δ_1 move towards its prior mean specification $\mu_1 = 10.65$ m; similarly,

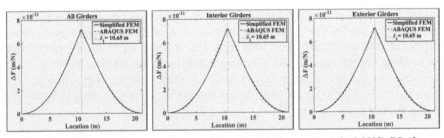

Fig. 10.6: Flexibility curves from mapping $\delta_1 = 10.65$ m for DS$_1$ ($\delta_{2,estimate}$ mapped $= 0.0009$), DS$_2$ ($\delta_{2,estimate}$ mapped $= 0.0029$), and DS$_3$ ($\delta_{2,estimate}$ mapped $= 0.0031$), respectively. The plots from all three scenarios demonstrate the Abaqus FEM's ability to accurately capture the simplified FEM's ΔF profile and peak ΔF at $\delta_1 = 10.65$ m.

Table 10.6: Results for L1–L3 damage classification when $N = 1$, $V = 15.66$ m/s, and the SNR varied between 10 and 20 dB.

	Cracking with 20 dB SNR			
	L2: δ_1 (m)	**L3: δ_2**	**L2: δ_1 (m)**	**L3: δ_2**
Truth	2.100	0.050	10.650	0.050
Mean	8.986	0.021	10.633	0.050
SSD	2.785	0.017	0.075	0.001
PI (L1)	–	0.459	–	1.000
	Cracking with 10 dB SNR			
	L2: δ_1 (m)	**L3: δ_2**	**L2: δ_1 (m)**	**L3: δ_2**
Truth	2.100	0.050	10.650	0.050
Mean	10.018	0.009	10.702	0.052
SSD	2.373	0.014	1.156	0.005
PI (L1)	–	0.099	–	1.000

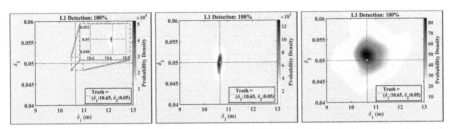

Fig. 10.7: Bivariate kernel density distributions developed using δ_1 and δ_2 estimates obtained from 100 data sets for a true $\delta_1 = 10.65$ m and $\delta_2 = 0.05$ when $N = 1$, $V = 15.66$ m/s and the SNR was equal to 30, 20, and 10 db, respectively. Visually demonstrate show L2–L3 classification accuracy and precision decrease with lower SNRs; i.e., higher levels of noise.

mean estimates for δ_2 move towards its prior mean specification $\mu_2 = 0$. Another observation from Table 10.6 is that decreasing SNR has more of an impact on L1–L3 classification accuracy for cracks occurring closer to the supports than at mid-span. The crack ratio of 0.05 was accurately detected at all locations for a SNR of 30 dB, but the Bayesian estimation technique could only reliably detect, locate, and quantify cracking at $\delta_1 = 10.65$ m for a SNR of 10 dB. This observation was further supported by the SSD for δ_1 estimates increasing as cracking occurred closer to the support during the 20 and 10 dB SNR tests. Lower SNRs are believed to have more of an impact on damage classification moving closer to the supports due to $\delta_2 = 0.05$ introducing smaller changes in system flexibility that are increasingly more difficult to detect under high levels of noise; this is supported by the $\Delta \mathbf{F}$ values in Table 10.5 being smaller at $\delta_1 = 2.1$ m.

5.4 Influence of Quantity of Experimental Data

To evaluate the impact N has on L1–L3 damage classification, $\mathbf{D}_{n, phys, k}$ was obtained for N equal to 5, 10, 15, and 20 runs, and a multi-run average was taken for each vehicle DoF. By taking a multi-run average, the root mean squares of noise A_n in the vehicle response decrease, effectively increasing the SNR to enable reliable damage classification. It can be seen in Table 10.7 that increasing the number of vehicle runs in a multi-run average does indeed result in improved L1–L3 classification. Particularly, when N was increased for the 10 dB SNR case, L1 classification increased significantly from the levels of detection observed in Table 10.6, reaching approximately 90% at $N = 20$ runs; L2–L3 classification accuracy and precision also continued to increase as N increased. Note that as δ_1 moves away from the supports, less experimental runs are required to obtain similar improvements in L1–L3 classification accuracy and precision (Locke, 2021). As damage- induced changes in the bridge's dynamic response became more pronounced, in the vehicle response away from the supports, less runs are required to reduce A_n to a level that enabled reliable L1–L3 damage classification; this conclusion is further supported by the fact that less runs are required to improve classification results at $\delta_1 = 2.1$ m for the 20 dB SNR beyond the improvements observed for the 10 dB SNR.

Table 10.7: Damage classification results from the quantity of experimental data study where N varied between 5–20 runs, $V = 17.88$ m/s, and the SNR varied between 10 and 20 dB.

10 dB SNR				
N Runs	15		20	
	L2: δ_1 (m)	L3: δ_2	L2: δ_1 (m)	L3: δ_2
Truth	2.100	0.050	2.100	0.050
Mean	6.084	0.034	4.688	0.041
SSD	3.226	0.016	3.204	0.013
PI (L1)	–	0.768	–	0.915
20 dB SNR				
N Runs	5		10	
	L2: δ_1 (m)	L3: δ_2	L2: δ_1 (m)	L3: δ_2
Truth	2.100	0.050	2.100	0.050
Mean	2.290	0.049	2.127	0.050
SSD	1.056	0.005	0.315	0.003
PI (L1)	–	0.998	–	1.000

5.5 Additional Things to Consider

Velocity studies should be conducted on a per bridge basis to evaluate the impact vehicle-on-bridge occupation times and the magnitude of dynamic impulses on L1–L3 damage classification for a given vehicle-bridge system. As an example, a velocity study was conducted during the development of the subject Bayesian estimation technique to gain a better understanding of how V impacted L1–L3 damage classification when high levels of noise were present at different damage locations across the length of the bridge (Locke, 2021). The results from the study demonstrated that longer vehicle-on-bridge occupations times (i.e., slightly lower velocities) were more beneficial for L1–L3 damage classification when damage occurred closer to mid-span, while larger magnitude impulses (i.e., slightly faster velocities) were more beneficial for L1–L3 damage classification when damage occurred closer to the supports. Overall, the findings from the velocity study suggest that there is an optimal trade-off between the magnitude of bridge excitation and vehicle-on-bridge occupation times for a given δ_1 and SNR; this conclusion is in line with other studies from the literature (Krishnanunni and Rao, 2021; Locke, 2021).

Unlike most model updating strategies that locate damage-induced changes in stiffness by minimizing an objective function based on differences in measured and analytical modal properties, the subject Bayesian estimation technique identifies damage by tuning the embedded vehicle-bridge FEM to minimize the difference between measured and predicted acceleration time histories (Asadollahi, 2018; Feng et al., 2006). The benefit of directly leveraging time history data is that more response features are available to improve tuning. However, by using time history data for model updating, model uncertainties from environmental and operational effects need to be identified to enable reliable damage classification (Asadollahi, 2018; Behmanesh and Moaveni 2016; Doebling et al., 1996; Feng et al., 2006). Additionally, the properties of the vehicle and bridge models must be accurate to correctly capture the dynamics of the physical systems. The Bayesian estimation technique can potentially identify uncertain bridge, vehicle, environmental, and operational parameters simultaneously with the unknown damage parameters via uncertainty quantification (Locke et al., 2021). Note, however, that uncertainty quantification has its own limitations that need to be considered to reliably and efficiently classify damage (Asadollahi, 2018; Doebling et al., 1996; Feng et al., 2006; Kennedy and O'Hagan, 2001; Lam et al., 2018; Yang and Lam, 2018).

When performing model updating with real system response data, a series of dynamic tests need to be conducted to capture time history responses under controlled but realistic operating conditions. Obtaining data under controlled conditions is important when performing updating directly with time-history data, as unknown sources of excitation, such as that introduced by random traffic, cannot easily be captured in an FEM and will introduce more model uncertainty (Feng et al., 2006; Khouri Chalouhi, 2016). When tuning a FEM with experimental data, it is imperative that ambient temperature effects, surface roughness, and nonlinear crack damage effects also be accurately represented in the model or mitigated during testing. Ambient temperature is known to cause significant linear and nonlinear fluctuations in bridge modal properties that can easily be mistaken as damage (Behmanesh and Moaveni, 2016; Farrar et al., 1996; Khouri Chalouhi, 2016). Temperature effects can be mitigated by conducting tests at the same time of day and/or around the same ambient temperatures (Khouri Chalouhi, 2016). Alternatively, a number of statistical and analytical tools are available to account for temperature effects within an FEM (Behmanesh and Moaveni, 2016; Khouri Chalouhi, 2016; Moser and Moaveni, 2011; Peeters and De Roeck, 2001; Xia et al., 2012; Yuen, 2010). Surface roughness is known to amplify the dynamic interaction between vehicles and bridges, and if not properly modeled can lead to numerical ill-conditioning. Physical road surface profiles can potentially be obtained for numerical models using specialized instruments (e.g., profilometers) to directly measure a profile during physical testing, or by calculating a profile from experimental vehicle acceleration data (Blum, 2015; Krishnanunni and Rao, 2021; González et al., 2008; O'Brien et al., 2014). Additionally, any discrepancies between profiles identified for the simplified FEM and the physical profiles can potentially be addressed through uncertainty quantification (Asadollahi, 2018; Feng et al., 2006; Kennedy and O'Hagan, 2001). Nonlinear crack effects are introduced when breathing is excited within a crack, causing continuous fluctuations in stiffness and damping that can make identifying damage difficult. Non-linear crack effects can be captured using rotational springs with bilinear stiffness or other fracture mechanics based methods (Dimarogonas, 1996; Friswell and Penny, 2002; Nguyen, 2013; Nguyen and Tran, 2010). The need to model nonlinear cracking can be mitigated if experimental data is measured for low levels of excitation where breathing is not excited (Dimarogonas, 1996; Friswell and Penny, 2002).

The inclusion of the above-mentioned uncertainties, in addition to any uncertainties in vehicle-bridge mass, stiffness, and damping parameters, can significantly increase the computational cost of the Bayesian estimation technique and result in higher run-times (Asadollahi, 2018; Jensen and Papadimitriou, 2019; Kennedy and O'Hagan, 2001; Khouri Chalouhi, 2016). One approach to reducing run times is to perform a sensitivity analysis to identify the impact parameter uncertainties have on fluctuations in response data. Through this approach, parameters demonstrating little to no influence on response data can be held constant and removed from the model updating space. An additional factor to consider when estimating uncertain parameters is the interdependencies between two parameters (e.g., the relationship of stiffness and mass to changes in frequency) as this can lead to a non-identifiability issue. Care must also be taken in specifying prior distributions to reflect the physical properties of each parameter. If knowledge does not exist on the distribution, a weakly informative prior can be specified such that the observed data provides more information to the model. Lastly, validation metrics (e.g., RMSE) should be levered to identify how well a model fits the true system response.

6. Conclusion

In this chapter, a novel Bayesian estimation technique for multilevel damage classification was presented that addresses limitations with other damage classification strategies in DBHM; i.e., the need to reference labeled data, required to travel at slow velocities, noise intolerance, or the inability to detect damage along an entire bridge length. Unlike most Bayesian techniques, the proposed methodology makes use of an embedded FEM, thus reducing the parameter space and simultaneously guiding the Bayesian model via physics-based principles. Furthermore, unlike most model updating strategies that locate damage-induced changes in stiffness by minimizing an objective function based on differences in measured and analytical modal properties, the subject Bayesian estimation technique identifies damage by tuning the embedded vehicle-bridge FEM to minimize the difference between measured and predicted acceleration time histories. The benefit of directly leveraging time-history data is that: more response features are available to improve tuning; nonlinearities are easier to identify; and time-history data is less sensitive to higher order modes with low levels of excitation that are difficult to accurately capture in simplified FEMs. When updating embedded FEMs with time-history data, it is important that model uncertainties from environmental and operational effects be identified and accounted for to reliably classify damage. Additionally, the properties of the vehicle and bridge models must be accurate to capture the dynamics of the physical system. Various computational tools can be employed under the subject Bayesian framework to update FEMs by tuning against experimental data, providing the benefit of being able to classify damage while simultaneously identifying uncertain model parameters (Asadollahi, 2018; Feng et al., 2006; Huang and Beck, 2015; Kennedy and O'Hagan, 2001; Locke, 2021; Locke et al., 2021). Being able to identify uncertain model parameters is especially beneficial when monitoring composite bridge structures, as the presence of material and/or geometric nonlinearities are difficult to accurately predict and capture (Locke, 2021; Locke et al., 2021). An additional benefit of the subject Bayesian estimation technique is the damage-mapping methodology used to relate damage identified on the embedded bridge FEM to more physically representative levels of damage on a higher-fidelity bridge FEM. Through this approach, simplified and computationally efficient vehicle-bridge FEMs are able to be employed for damage classification without having to sacrifice the physical interpretation of crack magnitudes. Ultimately, it has demonstrated that the Bayesian estimation technique is a very promising approach for damage classification and uncertainty quantification in DBHM, and has the potential to be employed to evaluate the effect of damage on structural performance, optimize maintenance strategies, and forecast a bridge's remaining service life.

References

Abedin, M. and Mehrabi, A.B. 2019. Novel approaches for fracture detection in steel girder bridges. *Infrastructures*, 4(3): 42.

Agdas, D., Rice, J.A., Martinez, J.R. and Lasa, I.R. 2015. Comparison of visual inspection and structural-health monitoring as bridge condition assessment methods. *Journal of Performance of Constructed Facilities*, 30(3): 04015049.

Asadollahi, P. 2018. *Bayesian-based Finite Element Model Updating, Damage Detection, and Uncertainty Quantification for Cable-stayed Bridges*. PhD thesis, University of Kansas.

Behmanesh, I. and Moaveni, B. 2016. Accounting for environmental variability, modeling errors, and parameter estimation uncertainties in structural identification. *Journal of Sound and Vibration*, 374: 92–110.

Blum, N.C. 2015. *System Identification of Vehicle Dynamics and Road Conditions using Wireless Sensors*. PhD Thesis, University of Maryland.

Bovsunovsky, A.P. and Surace, C. 2005. Considerations regarding superharmonic vibrations of a cracked beam and the variation in damping caused by the presence of the crack. *Journal of Sound and Vibration*, 288(4-5): 865–886.

Chopra, A.K. 2012. *Dynamics of Structures*. Prentice Hall, Upper Saddle River, NJ, pp. 174–196.

Cornejo, A., Mataix, V., Zárate, F. and Oñate, E. 2019. Combination of an adaptive remeshing technique with a coupled fem–dem approach for analysis of crack propagation problems. *Computational Particle Mechanics*, pp. 1–18.

Deng, L. and Cai, C. 2009. Identification of parameters of vehicles moving on bridges. *Engineering Structures*, 31(10): 2474–2485.

Dimarogonas, A.D. 1996. Vibration of cracked structures: A state of the art review. *Engineering Fracture Mechanics*, 55(5): 831–857.

Doebling, S.W., Farrar, C.R., Prime, M.B. and Shevitz, D.W. 1996. *Damage Identification and Health Monitoring of Structural and Mechanical Systems from changes in their Vibration Characteristics: A Literature Review*. Technical report, Los Alamos National Lab., NM (United States).

Farrar, C., Doebling, S., Cornwell, P. and Straser, E. 1996. *Variability of Modal Parameters Measured on the Alamosa Canyon Bridge*. Technical report, Los Alamos National Lab., NM (United States).

Feng, M., Fukuda, Y., Chen, Y., Soyoz, S. and Lee, S. 2006. Long-term structural performance monitoring of bridges: Phase II: Development of baseline model and methodology for health monitoring and damage assessment.

Friswell, M.I. and Penny, J.E. 2002. Crack modeling for structural health monitoring. *Structural Health Monitoring*, 1(2): 139–148.

Gelman, A., Carlin, J.B., Stern, H.S., Dunson, D.B., Vehtari, A. and Rubin, D.B. 2013. *Bayesian Data Analysis*. CRC Press. Taylor and Francis Group, Informa Business, 3rd Edition, First Published: 2013, eBook Published: 6 July 2015, Pub. Location: New York. Imprint: Chapman and Hall/CRC.doi: https://doi.org/10.1201/b16018, Pages: 675, eBook: ISBN9780429113079. Subjects: Behavioral Sciences, Mathematics & Statistics.

George, E.I. and McCulloch, R.E. 1993. Variable selection via gibbs sampling. *Journal of the American Statistical Association*, 88(423): 881–889.

Gillespie, T.D. 1992. *Fundamentals of Vehicle Dynamics*. Technical report, SAE Technical Paper.

González, A., O'brien, E.J., Li, Y.-Y. and Cashell, K. 2008. The use of vehicle acceleration measurements to estimate road roughness. *Vehicle System Dynamics*, 46(6): 483–499.

Hoff, P.D. 2009. *A First Course in Bayesian Statistical Methods*. Springer, New York, NY.

Huang, Y. and Beck, J.L. 2015. Hierarchical sparse bayesian learning for strucutral health monitoring with incomplete modal data. *International Journal for Uncertainty Quantification*, 5(2).

Huebner, K., Dewhirst, D., Smith, D. and Byrom, T. 2008. *The Finite Element Method for Engineers*. John Wiley & Sons, Inc., New York, NY.

Jensen, H. and Papadimitriou, C. 2019. Sub-structure coupling for dynamic analysis.

Kathol, S., Azizinamini, A. and Luedke, J. 1995. *Strength Capacity of Steel Girder Bridges*. Department of Civil Engineering, College of Engineering and Technology. URL: https://books.google.com/books?id=GZxVewAACAAJ.

Kennedy, M.C. and O'Hagan, A. 2001. Bayesian calibration of computer models. *Journal of the Royal Statistical Society: Series B (Statistical Methodology)*, 63(3): 425–464.

Khouri Chalouhi, E. 2016. Structural health monitoring of bridges using machine learning: The influence of temperature on the health prediction.

Koenke, C., Harte, R., Krätzig, W.B. and Rosenstein, O. 1998. On adaptive remeshing techniques for crack simulation problems. *Engineering Computations*, 15(1): 74–88.

Krishnanunni, C.G. and Rao, B.N. 2021. Indirect health monitoring of bridges using Tikhonov regularization scheme and signal averaging technique. *Structural Control and Health Monitoring*, 28(3): e2686.

Lam, H.-F., Yang, J.-H. and Au, S.-K. 2018. Markov chain monte carlo-based bayesian method for structural model updating and damage detection. *Structural Control and Health Monitoring*, 25(4): e2140.

Lee, Y.-S. and Chung, M.-J. 2000. A study on crack detection using eigenfrequency test data. *Computers & Structures*, 77(3): 327–342.

Li, W.-m., Jiang, Z.-h., Wang, T.-l. and Zhu, H.-p. 2014. Optimization method based on generalized pattern search algorithm to identify bridge parameters indirectly by a passing vehicle. *Journal of Sound and Vibration*, 333(2): 364–380.

Locke Jr, W.R. 2021. *Experimental Evaluation of System Identification Techniques and Development of a Bayesian Damage Detection Strategy for Drive-By Health Monitoring*. PhD thesis, Clemson University.

Locke, W., Sybrandt, J., Redmond, L., Safro, I. and Atamturktur, S. 2020. Using drive-by health monitoring to detect bridge damage considering environmental and operational effects. *Journal of Sound and Vibration*, 468: 115088.

Locke, W.R., Mokalled, S.C., Abuodeh, O.R., Redmond, L.M. and McMahan, C.S. 2021. An intelligently designed AI for structural health monitoring of a reinforced concrete bridge. *In 'SP-350—The Concrete Industry in the Era of Artificial Intelligence*. ACI, pp. 103–112.

Lynch, J.P. and Loh, K.J. 2006. A summary review of wireless sensors and sensor networks for structural health monitoring. *Shock and Vibration Digest*, 38(2): 91–130.

Moore, M., Phares, B., Graybeal, B., Rolander, D. and Washer, G. 2001. Reliability of visual inspection for highway bridges. *Federal Highway Administration*, 1(FHWA-RD-01-020). URL: https://www.fhwa.dot.gov/publications/research/nde/01020.cfm.

Moser, P. and Moaveni, B. 2011. Environmental effects on the identified natural frequencies of the dowling hall footbridge. *Mechanical Systems and Signal Processing*, 25(7): 2336–2357.

Ndambi, J.-M., Vantomme, J. and Harri, K. 2002. Damage assessment in reinforced concrete beams using eigenfrequencies and mode shape derivatives. *Engineering Structures*, 24(4): 501–515.

Nguyen, K.V. 2013. Comparison studies of open and breathing crack detections of a beam-like bridge subjected to a moving vehicle. *Engineering Structures*, 51: 306–314.

Nguyen, K.V. and Tran, H.T. 2010. Multi-cracks detection of a beam-like structure based on the on-vehicle vibration signal and wavelet analysis. *Journal of Sound and Vibration*, 329(21): 4455–4465.

O'Brien, E.J., McGetrick, P. and González, A. 2014. A drive-by inspection system via vehicle moving force identification. *Smart Structures and Systems*, 13(5): 821–848.

OBrien, E., Carey, C. and Keenahan, J. 2015. Bridge damage detection using ambient traffic and moving force identification. *Structural Control and Health Monitoring*, 22(12): 1396–1407.

Oshima, Y., Yamaguchi, T., Kobayashi, Y. and Sugiura, K. 2008. Eigenfrequency estimation for bridges using the response of a passing vehicle with excitation system. *In Proceedings of the Fourth International Conference on Bridge Maintenance, Safety and Management*, pp. 3030–3037.

Pandey, A. and Biswas, M. 1994. Damage detection in structures using changes in flexibility. *Journal of Sound and Vibration*, 169(1): 3–17.

Peeters, B. and De Roeck, G. 2001. One-year monitoring of the z24-bridge: Environmental effects versus damage events. *Earthquake Engineering & Structural Dynamics*, 30(2): 149–171.

Qian, G.-L., Gu, S.-N. and Jiang, J.-S. 1990. The dynamic behaviour and crack detection of a beam with a crack. *Journal of Sound and Vibration*, 138(2): 233–243.

Ryan, T., Mann, J., Chill, Z. and Ott, B. 2012. Bridge inspector's reference manual (birm).*Publication No. FHWA NHI*, pp. 12–049.

Sohn, H. and Law, K.H. 2000. Bayesian probabilistic damage detection of a reinforced-concrete bridge column. *Earthquake Engineering & Structural Dynamics*, 29(8): 1131–1152.

Verfürth, R. 2013. *A Posteriori Error Estimation Techniques for Finite Element Methods*. OUP Oxford.

Xia, Y., Chen, B., Weng, S., Ni, Y.-Q. and Xu, Y.-L. 2012. Temperature effect on vibration properties of civil structures: A literature review and case studies. *Journal of Civil Structural Health Monitoring*, 2(1): 29–46.

Yang, J.-H. and Lam, H.-F. 2018. An efficient adaptive sequential monte carlo method for Bayesian model updating and damage detection. *Structural Control and Health Monitoring*, 25(12): e2260.

Yang, Y. and Yang, J.P. 2018. State-of-the-art review on modal identification and damage detection of bridges by moving test vehicles. *International Journal of Structural Stability and Dynamics*, 18(02): 1850025.

Yang, Y.-B., Yau, J., Yao, Z. and Wu, Y. 2004. *Vehicle-bridge Interaction Dynamics: With Applications to High-speed Railways*. World Scientific.

Yuen, K.-V. 2010. *Bayesian Methods for Structural Dynamics and Civil Engineering*. John Wiley & Sons (Asia) Pte Ltd, 2 Clementi Loop, # 02-01, Singapore 129809.

Zhang, C., Gao, Y.-W., Huang, J.-P., Huang, J.-Z. and Song, G.-Q. 2020. Damage identification in bridge structures subject to moving vehicle based on extended kalman filter with l 1-norm regularization. *Inverse Problems in Science and Engineering*, 28(2): 144–174.

Zhang, Y., Wang, L. and Xiang, Z. 2012. Damage detection by mode shape squares extracted from a passing vehicle. *Journal of Sound and Vibration*, 331(2): 291–307.

Appendix 10A: Vehicle Matrices for Vehicle-Bridge FEM

Mass Matrices

$$\mathbf{M}_{uu} = \begin{bmatrix} M_v & 0 & 0 & 0 \\ 0 & I_v & 0 & 0 \\ 0 & 0 & M_{w1} & 0 \\ 0 & 0 & 0 & M_{w2} \end{bmatrix} \tag{10A.1}$$

$$\mathbf{M}_{lu} = \mathbf{M}'_{ul} = [0]_{2x4} \tag{10A.2}$$

$$\mathbf{M}_{ll} = [0]_{2x2} \tag{10A.3}$$

Stiffness Matrices

$$\mathbf{K}_{uu} = \begin{bmatrix} K_{v1} + K_{v2} & K_{v2} \cdot b - K_{v1} \cdot a & -K_{v1} & -K_{v2} \\ K_{v2} \cdot b - K_{v1} \cdot a & K_{v1} \cdot a^2 + K_{v2} \cdot b^2 & K_{v1} \cdot a & -K_{v2} \cdot b \\ -K_{v1} & K_{v1} \cdot a & K_{v1} + K_{w1} & 0 \\ -K_{v2} & -K_{v2} \cdot b & 0 & K_{v2} + K_{w2} \end{bmatrix} \tag{10A.4}$$

$$\mathbf{K}_{lu} = \mathbf{K}'_{ul} = \begin{bmatrix} 0 & 0 & -K_{w1} & 0 \\ 0 & 0 & 0 & -K_{w2} \end{bmatrix} \tag{10A.5}$$

$$\mathbf{K}_{ll} = \begin{bmatrix} K_{w1} & 0 \\ 0 & K_{w2} \end{bmatrix} \tag{10A.6}$$

Damping Matrices

$$\mathbf{C}_{uu} = \begin{bmatrix} C_{v1} + C_{v2} & C_{v2} \cdot b - C_{v1} \cdot a & -C_{v1} & -C_{v2} \\ C_{v2} \cdot b - C_{v1} \cdot a & C_{v1} \cdot a^2 + C_{v2} \cdot b^2 & C_{v1} \cdot a & -C_{v2} \cdot b \\ -C_{v1} & C_{v1} \cdot a & C_{v1} + C_{w1} & 0 \\ -C_{v2} & -C_{v2} \cdot b & 0 & C_{v2} + C_{w2} \end{bmatrix} \tag{10A.7}$$

$$\mathbf{C}_{lu} = \mathbf{C}'_{ul} = \begin{bmatrix} 0 & 0 & -C_{w1} & 0 \\ 0 & 0 & 0 & -C_{w2} \end{bmatrix} \tag{10A.8}$$

$$\mathbf{C}_{ll} = \begin{bmatrix} C_{w1} & 0 \\ 0 & C_{w2} \end{bmatrix} \tag{10A.9}$$

External Force Vector

$$\mathbf{f}_{le} = -g \left\{ \begin{array}{c} \frac{b}{a+b} \cdot M_v + M_{w1} + M_{w2} \\ \frac{a}{a+b} \cdot M_v + M_{w1} + M_{w2} \end{array} \right\} \tag{10A.10}$$

Appendix 10B: Metropolis-Hastings and MCMC Routine

Metropolis-Hastings Steps

The Metropolis-Hastings step for sampling δ_1 begins by sampling γ_1^* from

$$\gamma_1^* \sim N(\psi_1, c_1^2), \tag{10B.1}$$

where $\psi_1 = \log\{g(\delta_1^{(s-1)})\}/\{1 - g(\delta_1^{(s-1)})\}$, $g(z) = (z - l_1)/(u_1 - l_1)$, $\delta_1^{(s-1)}$ is the previous value of δ_1, and c_1 is a tuning parameter. The proposed value of δ_1^* is obtained by transforming γ_1^* by $\delta_1^* = (u_1 - l_1)\exp(\gamma_1^*)/\{\exp(\gamma_1^*) + 1\} + l_1$ and $\delta_1^* \in (l_1, u_1)$. This transformation yields the following proposal distribution of δ_1^*

$$\pi(\delta_1^* \mid \delta_1^{(s-1)}, c_1^2) =$$

$$c_1^{-1} \exp\left(\frac{-\left[\log\left\{\frac{g(\delta_1^*)}{1-g(\delta_1^*)}\right\} - \psi_1\right]^2}{2c_1^2} \right) \frac{1}{g(\delta_1^*)\{1 - g(\delta_1^*)\}} \frac{1}{u_1 - l_1}. \tag{10B.2}$$

Using this density, compute

$$\alpha_1 = \min\left\{1, \frac{p(\delta_1^* \mid \mathbf{D}_{1,\dots,N,phys}, \varphi, \delta_2)\pi(\delta_1^{(s-1)} \mid \delta_1^*, c_1^2)}{p(\delta_1^{(s-1)} \mid \mathbf{D}_{1,\dots,N,phys}, \varphi, \delta_2)\pi(\delta_1^* \mid \delta_1^{(s-1)}, c_1^2)}\right\}, \tag{10B.3}$$

where $p(\delta_1 \mid \mathbf{D}_{1,\dots,N,phys}, \varphi, \delta_2)$ is the density function of the full conditional distribution of δ_1, which is given by (19). Finally, set

$$\delta_1^{(s)} = \begin{cases} \delta_1^* & \text{with probability } \alpha_1 \\ \delta_1^{(s-1)} & \text{with probability } 1 - \alpha_1 \end{cases}. \tag{10B.4}$$

The Metropolis-Hastings step to sample δ_2 is practically identical and is therefore omitted for brevity.

MCMC Routine

(1) Initialize $\delta_1^{(0)}$ and $\delta_2^{(0)}$ using a sequence of values for each parameter and choosing the combination that maximizes the observed data likelihood function.

(2) For $s = 1, \dots, S$:

(a) Using a Gibbs step, sample $\varphi^{(s)}$ from:

$$\text{gamma}\left(\frac{3NT_{step}}{2} + a_0, \frac{\sum_{n=1}^{N}\sum_{k=1}^{3}\{\mathbf{D}_{n,phys,k} - \mathbf{D}_{sim,k}(\delta, V)\}'\{\mathbf{D}_{n,phys,k} - \mathbf{D}_{sim,k}(\delta, V)\}}{2} + b_0 \right). \tag{10B.5}$$

(b) Sample $\delta_1^{(s)}$ using a Metropolis-Hastings step.

   (i) Compute $\psi_1 = \log\{g(\delta_1^{(s-1)})\}/\{1 - g(\delta_1^{(s-1)})\}$.

   (ii) Sample γ_1^* from N (ψ_1, c_1^2).

(iii) Transform to get $\delta_1^* = (u_1 - l_1)\exp(\gamma_1^*)/\{\exp(\gamma_1^*) + 1\} + l_1$.

(iv) Compute

$$\alpha_1 = \min\left\{1, \frac{p(\delta_1^* \mid \mathbf{D}_{1,\ldots,N,phys}, \varphi^{(s)}, \delta_2^{(s-1)})\pi(\delta_1^{(s-1)} \mid \delta_1^*, c_1^2)}{p(\delta_1^{(s-1)} \mid \mathbf{D}_{1,\ldots,N,phys}, \varphi^{(s)}, \delta_2^{(s-1)})\pi(\delta_1^* \mid \delta_1^{(s-1)}, c_1^2)}\right\}. \quad (10B.6)$$

(v) Set

$$\delta_1^{(s)} = \begin{cases} \delta_1^* & \text{with probability } \alpha_1 \\ \delta_1^{(s-1)} & \text{with probability } 1 - \alpha_1 \end{cases}. \quad (10B.7)$$

(c) Sample $\delta_2^{(s)}$ using a Metropolis-Hastings step.

(i) Compute $\psi_2 = \log\{g(\delta_2^{(s-1)})\}/\{1 - g(\delta_2^{(s-1)})\}$.

(ii) Sample γ_2^* from $N(\psi_2, c_2^2)$.

(iii) Transform to get $\delta_2^* = (u_2 - l_2)\exp(\gamma_2^*)/\{\exp(\gamma_2^*) + 1\} + l_2$.

(iv) Compute

$$\alpha_2 = \min\left\{1, \frac{p(\delta_2^* \mid \mathbf{D}_{1,\ldots,N,phys}, \varphi^{(s)}, \delta_1^{(s-1)}, w^{(s)})\pi(\delta_2^{(s-1)} \mid \delta_2^*, c_2^2)}{p(\delta_2^{(s-1)} \mid \mathbf{D}_{1,\ldots,N,phys}, \varphi^{(s)}, \delta_1^{(s-1)}, w^{(s)})\pi(\delta_2^* \mid \delta_2^{(s-1)}, c_2^2)}\right\}. \quad (10B.6)$$

(v) Set

$$\delta_2^{(s)} = \begin{cases} \delta_2^* & \text{with probability } \alpha_2 \\ \delta_2^{(s-1)} & \text{with probability } 1 - \alpha_2 \end{cases}. \quad (10B.9)$$

(d) Increment s and return to Step (a).

Chapter 11

Machine Learning and IoT Data for Concrete Performance Testing and Analysis

Andrew Fahim,* Tahmid Mehdi, Ali Taheri, Pouria Ghods,
Aali Alizadeh and *Sarah De Carufel*

1. Introduction

Internet of Things (IoT) devices have gained widespread adoption in recent years for many industrial applications, including concrete monitoring. Such devices/ sensors allow the collection of valuable, insight-rich data from previously untapped sources. With sensor usage/adoption growing at a significant rate, and subsequently the significant growth in volume of generated data, a significant opportunity exists in developing algorithms that rely on this real-time data to generate insights to practitioners.

For the concrete industry, IoT sensors are typically used for several purposes among which temperature, relative humidity, and strength monitoring (using the maturity method) are currently the most common. Figure 11.1 shows an example of a sensor used by concrete practitioners to monitor concrete properties in real-time. The sensor is installed on the rebar and remains embedded in concrete, where it monitors and records concrete properties. The sensor is Bluetooth®-enabled and therefore practitioners can use any modern mobile phone to connect to the sensor and extract data in real-time. Alternatively, users can use a fixed gateway capable of autonomously connecting to such sensors and backing up the collected data. In both cases, the data is pushed to a cloud platform where it can be accessed by practitioners to view the collected data as well as analytics and insights generated. Algorithms and

Giatec Scientific Inc., 245 Menten Place, Ottawa, Ontario, Canada.
* Corresponding author: andrew@giatec.ca

Fig. 11.1: Concrete-embedded IoT sensor used for monitoring temperature and *in situ* strength.

concepts presented in this chapter have been trained using data collected from tens of thousands of similar sensors, used in over 7,500 projects representing geographical regions of over 45 countries and representing several thousand unique concrete mixtures.

Machine learning (ML)/artificial intelligence algorithms are currently used in a wide array of applications. Significant advancements have been made in the past decade in implementing these algorithms in fields such as finance, medicine, marketing, material science, process optimization, autonomous vehicles, object and speech recognition for software development and web search content filtering. With the ongoing advancements in this field, it is expected that this technology will soon make its way to every possible field and become a part of everyday lives in a very noticeable manner. Despite the significant advancements in the field of ML, little progress has been made (outside of academic research) in implementing these technologies at an industrial scale for the concrete industry.

Due to the influence of numerous factors on concrete production and performance, the concrete industry generates and tracks significant volumes of data. This data ranges from raw material characteristics (e.g., cement composition/chemistry, aggregate gradation and mineralogy, chemical admixture nature and dosages), ambient conditions (e.g., weather conditions at the time of pouring), transit conditions (e.g., time from plant to site), performance characteristics (e.g., strength, slump, air content, fresh concrete temperature, shrinkage), and specifications (e.g., minimum cementitious materials content, maximum supplementary cementitious materials, etc.). Considering the abundance of this data available to concrete practitioners and researchers, the concrete industry presents an excellent opportunity to utilize machine algorithms in a range of applications including concrete performance prediction, anomaly detection, mixture formulation or process optimization. This chapter presents on how concrete-embedded IoT sensors data is used to train artificial intelligence algorithms to perform several tasks that assist sensor end-users (concrete producers, general contractors, etc.) in day-to-day tasks.

2. Machine Learning Algorithms

The most utilized form of ML, which is used in this work, is termed supervised machine learning. In this form, algorithms are 'trained' to perform a specific task, such as classification or regression, utilizing a dataset where the inputs and outputs are accessible to the algorithm. During this process of training, the algorithm knows both the input features that are expected to correlate to/map the output, and it also knows the final desired output result. The outputs and inputs are correlated through nonlinear functions that use arbitrary adjustable parameters. During the process of algorithm training, the objective is to adjust these parameters to correlate the input features and outputs. During training, an objective function (e.g., mean absolute percent error, mean absolute error, etc.) is calculated which finds the error between the calculated output results and the desired result at every iteration. The algorithm modifies the adjustable parameters to reduce this error to acceptable limits in several iterations. In several ML systems, there may be millions of these adjustable parameters. To properly adjust the parameters, the learning algorithm computes a "gradient vector" that indicates by what amount the error would increase or decrease if the parameters were increased or decreased by a specific amount. Using the gradient vector, the parameters are then adjusted to refine them and obtain a better agreement between the expected result and the outputted result (LeCun et al., 2015). Many options exist for algorithms that utilize this concept. Such algorithms could include but are not limited to support vector machines (SVMs), linear or logistic regression, decision trees, k-nearest neighbor (KNN), neural networks (multilayer perceptrons), and recurrent neural networks (RNNs) or long short-term memory (LSTM) algorithms. An in-depth introduction to these concepts can be found in (Goodfellow et al., 2016).

The choice of algorithm is usually governed by the task in hand. For instance, algorithms such as RNNs and LSTMs are most frequently (but not exclusively) used in learning long-term dependencies for sequence prediction problems (e.g., temporal data or natural language data) (Hochreiter and Schmidhuber, 1997) while SVM or KNN models are used most frequently for classification problems (Cortes and Vapnik, 1995). Neural networks are applied for classification or regression problems while convolutional neural networks, a special example of a neural network, are used mostly for classification problems and are the backbone of most modern image detection and classification algorithms (LeCun et al., 1998). Nevertheless, these models could be engineered to perform many tasks other than the mentioned ones based on the required accuracy, complexity, inference speed, available resources, etc.

Regardless of the algorithms or methods chosen, the most significant portion of time is spent in three distinct (often linked) areas of development: data preparation/ feature selection, algorithm hyperparameter tuning, and algorithm evaluation. The chapter focuses on challenges faced in each of these areas when IoT data is involved. The data preparation step typically includes data collection pipelines, formatting, cleansing, anomaly detection, and feature selection. Although all these tasks have a significant importance, in most cases, feature selection is the most challenging aspect. The performance of the algorithms depends highly on how data preparation is performed, specifically the 'representation' of the data provided.

When algorithms are developed to perform a specific task, for instance the prediction of concrete strength, the algorithms use 'features' that may represent concrete characteristics. These are relevant pieces of information that a human expert or an algorithm can use to predict concrete strength. The algorithm learns (typically from historical data) how each of the provided features relate to concrete strength and uses this information to predict performance when faced with new examples. Any task to be automated through algorithms (such as predicting strength) can only be performed if the right features are chosen. For instance, it cannot be claimed that concrete strength can be determined using only water-to-cementitious materials ratio and cement content. It is well established that many other factors can affect the performance. On the other hand, for most practical applications, not every factor can be collected, or 'significant' historical data may not be available for every such factor. During the stage of feature selection, the goal is to determine the factors that would map the system to an acceptable degree of accuracy (determined based on the task in hand). For example, if 20 variables can be used to predict concrete strength to a mean absolute percent error (MAPE) of 15% and 200 variables to an MAPE of 7% when an algorithm is trained on the same volume of data, practitioners need to be interviewed to determine the acceptable error and the availability of such features in their day-to-day operations to determine the level of complexity required. It is evident that the errors/uncertainties in inputs grow as the number of input features increases. This may yield models with larger number of inputs (and a theoretically better error rate) having a lower overall accuracy in practice.

Hyperparameter tuning is essential in algorithm development although it has not been given significant attention in concrete literature; with some work presenting on the final chosen model architecture and not the hyperparameter ranges tested to reach such architecture. All the mentioned algorithms incorporate numerous hyperparameters. For example, a neural network/multilayer perceptron will have the following hyperparameters: number of layers, layer size/number of neurons, learning rate (and often rate decay), batch size, regularization method and rate, momentum, optimization method used/optimizer parameters, dropout use/dropout rate, activation function/loss function, etc. Regardless of the algorithm developer's expertise and knowledge, these hyperparameters must be determined through trial and error; whether random search is performed or other methods such as coordinate descent or grid search or other advanced methods like Bayesian optimization or genetic algorithms. All these methods rely on trying many combinations of hyperparameters and determining 'ideal' values for the problem in hand. Note that significant changes in input features or outputs variables or size of dataset often require a re-determination of hyperparameters. It is important that the hyperparameter search is exhaustive and solid methods are implemented to evaluate one set of hyperparameters versus another. Typically, during this stage, the dataset is divided into a train, test and validate datasets and the validation dataset is used to compare one set of hyperparameters to another. This is a portion of data that is held back from training the model and used solely for hyperparameter tuning (as it is not biased to the model fit on the training dataset). The hyperparameter choice is also often based on the inference speed required from the model and not only accuracy. For instance, in cases where a fast model inference/near-real-time prediction is required, a lightweight network may be more useful than one with significantly higher computation time/complexity that outperforms in accuracy.

Hyperparameter tuning and feature selection are not distinctly separate but are typically iterative. In many of the cases presented in this chapter, the input features grow as the project grows and more data is collected or as the algorithm developer's knowledge of the problem develops, this typically requires continuous model training and hyperparameter tuning.

Finally, the algorithm evaluation is not only done on the evaluation and test sets, but also in collaboration with algorithm end-users. This is done by demonstrating algorithm inputs and outputs to domain experts and practitioners to judge on the experience, accuracy, and reliability of the model. This is especially important for cases such as concrete performance analysis and optimization; due to the vast differences in operations, specifications, governing bodies, applications, and raw material characteristics depending on application and geographical location. For instance, if the train/validation/test datasets are obtained from North America and Europe and then used to predict performance in Asia, where the standards and operations are very different (let alone the material characteristics) the algorithm needs to be separately evaluated with practitioners to determine use cases, reliability, suitability, etc. In the cases developed in this work, it was found essential to develop mechanisms to collect users' feedback to evaluate the model performance and further develop it for the specific use-cases. For example, when algorithm suggestions are shown to the user, the user can input actions on whether they agree with the suggestion or are satisfied with the analytics. For instance, when a prediction of strength is shown, the algorithm subsequently collects the actual laboratory determination of compressive strength to evaluate itself and improve using such data. This data and the feedback loops have been instrumental in the evolution of these models.

2.1 Data and Algorithms

At the time of model development, IoT concrete sensors, developed by Giatec Scientific and used in this work, have been used in over 7,500 projects in 45 countries. This sensor data includes concrete temperature, maturity, strength, and humidity data for thousands of unique mixtures composed of a wide range of raw materials, from various sources, exposed to varying environmental conditions, used for different applications, and utilizing different construction practices. This data is collected from every sensor/project in a centralized cloud database where it is accessible to algorithm developers for the purpose of algorithm training, development, validation, and testing. As this database has been exponentially growing with the increased adoption of IoT concrete monitoring, Giatec has been developing algorithms, some of which are presented in this work, that make use of this data to provide the users with tools to analyze their mixtures, predict future performance, detect anomalies, optimize mixtures, etc.

The first algorithm presented in this work was developed to capture the concrete pouring time from temperature data captured by embedded IoT concrete sensors. The time of pouring is a critical input to any concrete maturity monitoring system. Maturity is a non-destructive approach to testing concrete that allows an estimation of the early-age compressive strength of in-place concrete in real-time. To estimate the *in situ* strength, a strength-maturity calibration is done through laboratory tests on the concrete mixture to be used. The temperature history of the concrete is recorded *in situ* from the time of concrete placement to the time when the strength estimation is desired. The temperature history is used to calculate the maturity index of the concrete using, for instance, the Nurse-Saul or equivalent age method. Using the maturity index and the strength-maturity relationship, the *in situ* strength of concrete can be estimated (Carino et al., 1983).

Since the maturity-index is calculated starting from the time of concrete placement, an accurate estimation of this time is essential for an precise approximation of the maturity index and subsequently the strength. Since the temperature-measuring sensors are small, embedded devices in large concrete elements, determining the time at which the sensor and the concrete came in contact becomes rather challenging for complex or large elements. Sensor users usually input approximate times such as the time of truck arrival, start of pour or end of pour. It is rather rare for users to note the exact time at which the sensor is covered with concrete. The goal of this work was to develop a model that can accurately infer the time of pouring solely from temperature data.

To perform this, historical data from over 20,000 sensors was collected from Giatec's database. The concrete pouring time data was labelled by the sensor users as it is a required input to the system. This data was reviewed by concrete practitioners/ subject-matter experts with the purpose of answering two questions: (1) whether the time of pouring can be found by domain experts solely from temperature data, and (2) the average number of data points, after pouring, required by the experts to detect concrete pouring. Figure 11.2 is an example of the data shown to experts.

It was found to be evident to concrete practitioners that the time of pouring can be inferred from the sensor temperature data through observing the concrete temperature rise attributed to the exothermic nature of hydration. The experts had a degree of accuracy of 95% in finding the time of pouring when presented with the overall temperature-time figures. This accuracy generally decreased as the experts were only presented with a few (less than 6) hours of data. Finding that

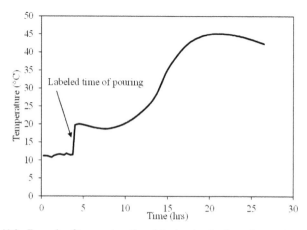

Fig. 11.2: Example of temperature-time data showing the time of concrete pouring.

domain experts were able to detect this pouring time with a great degree of accuracy generally indicated that ML algorithms can be used to autonomously find this time through training the algorithm on data where this time is correctly labelled.

The temperature data collected from the sensors was cleaned using manual data reviewing, as well as using certain criteria such as removing sensors where all the datapoints fall in ranges that are outside of those typical of concrete placements, e.g., if all the temperatures throughout the time were lower than 0°C. The sensor data was then divided into cases with different lengths ranging from 2 hours to 4 days of temperature measurements before algorithm training. This was done to expose the algorithm to data like that seen in practice where the length of data is unpredictable and is variable based on the time the user retrieves sensor data using their mobile application.

Data from the 20,000 sensors was divided into training, validation, and testing datasets with a split of 60/20/20, respectively. On an average, each sensor had 20 days of temperature data. Data from each sensor was truncated into many windows of varying lengths that fall in the range of 2 hours to 4 days, leading to over 500,000 windows of varying lengths (varying number of temperature data points) used for algorithm training. Several models were trained on the data including neural networks, convolutional neural networks, and LSTMs. The input for all these models was one-dimensional temperature data. Four metrics were used to evaluate and compare the models: the mean absolute error (in hours) calculated between the user-chosen pouring time and model-predicted pouring time; the percent success in finding the pouring time (to be compared to domain experts); the time-period (after pouring) required for the model to predict a pouring event; and finally, algorithm inference time/speed.

For each of the models, hyperparameter tuning was performed. This is essential in determining variables including number of neural network layers, layer size, learning rate, activation function, number of convolutional layers, number of filters, filter size, Batch Norm (Loffee and Szegedy, 2015), dropout rate (Srivastava et al., 2015), etc. Based on initial hyperparameter tuning work, a convolutional neural network was chosen. This type of network is very common for applications of image recognition, image classification, and financial time series since it is hypothesized to be able to successfully capture spatial and temporal dependencies through the application of convolutional filters. These networks are used frequently when translation invariance is desired, such as in problems akin to the pouring time; where the network is required to predict pouring time correctly regardless of the point in time at which this happens (i.e., whether it happens at the beginning of the temperature-time signal or following days of temperature monitoring). The network made use of stacked convolution layers that are followed by multilayer perceptron layers. The number and size of filters for the convolution portion and size and number of layers for the multilayer perceptron was determined through the results of the hyperparameter tuning. The network made use of dropout and batch normalization concepts. The output of the network was the point in time in which pouring occurred.

The developed model had a 90% success rate in finding the pouring time for sensors with a labelled time, which is rather close to the success of domain experts.

The mean absolute error was 3.2 hours. On average, the model was able to detect the correct pouring time after 6 hours of data points following a pouring event. This was generally attributed to the signature heat rise (due to hydration) occurring after such a time-period. Finally, the model had an inference time of 200–500 ms (depending on the length of the temperature–time data) when deployed. This was considered acceptable for the user experience considering that such models are being run in the background without interrupting the usage of the mobile application.

Some examples of predictions are shown in Figs. 11.3, 11.4, and 11.5. Blue vertical lines indicate the algorithm prediction, while red lines show the users' choices. Figure 11.3 shows cases where the user relies only on the algorithm to predict the time of pouring. Figure 11.4 shows examples of cases where the algorithm

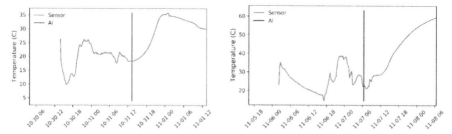

Fig. 11.3: Examples of detected time of pouring where the user relies on the algorithm suggestion (blue).

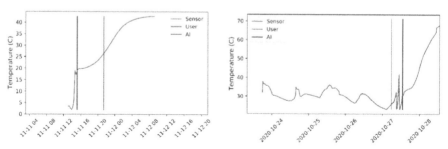

Fig. 11.4: Examples of detected time of pouring where the detection (blue) has a higher accuracy than the user's choice (red).

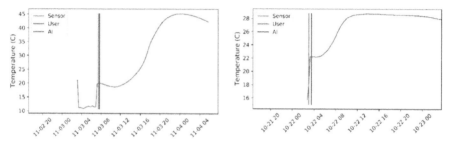

Fig. 11.5: Examples of detected time of pouring where the detection (blue) is within 2 hours of the user's choice (red).

corrects the user's wrong choice of time of pouring and Fig. 11.5 shows examples of cases where the user and the algorithm agree within less than 2 hours. In such cases of agreement, the user is not notified to change the time of pouring, considering the minor differences in the maturity index (strength correlates to the logarithm of the index).

To further support customers during the process of implementing maturity for *in situ* strength monitoring, it was desirable to develop an algorithm where inaccuracies in the maturity calibration can be detected and flagged to the user. Although maturity is a rather well-understood concept within the North American and European markets, with established ASTM (American Society for Testing and Materials), CSA (Safety Standards in Canada), and EN (European) standards, the concept is relatively new for other parts of the world. This requires a significant amount of support from maturity sensor suppliers in supporting customers during the development of the maturity calibration and its utilization in the field. Therefore, a model where such inaccuracies can be autonomously found can provide a significant value to practitioners using such a concept for the first time and provide monitoring system suppliers with an increased efficiency in customer support.

To develop the algorithm, mixture data was collected from Giatec's cloud database and from historical QA/QC data from 10 other producers around the U.S., Canada, and Europe. This data accumulated to over 7,500 unique mixtures and over 300,000 compressive strength datapoints from historical QA/QC records spanning 5–15 years of history dependent on the producer. The data represents a range of mixture proportions, used for various applications, utilizing varying raw materials (cementitious materials, supply chain managements (SCMs), various admixtures, aggregate characteristics, etc.) and exposed to a wide range of environmental conditions and construction practices. The performance data represented tickets from over 20,000 projects. The data included cementitious materials (Portland cement and SCMs) from 300 sources, over 200 commercially available chemical admixtures and included aggregates representing every gradation in ASTM C33 from over 40 different sources.

Many challenges are associated with the pre-processing and preparation of such data. The selection of input features and outlier detection were the most technically challenging tasks and the most instrumental in the model's success. The selection of input features was based on two criteria: (1) the availability of data for such a feature and (2) the importance of the feature in predicting concrete performance. The available features were divided into two categories, most available features (e.g., mixture proportions, cement type, aggregate gradation, ambient conditions, curing method, admixtures dosages, etc.) and least available features (e.g., cementitious materials and SCMs chemistry/mill cert information, aggregate shape, and mineralogy, mixing water temperature, transit time, etc.). These features were collected and presented to several domain experts to determine their relative importance. The features that had the most importance and most availability in the data were used for initial model training. The features selected for model training were extended gradually in many rounds while evaluating the influence of increasing input features on MAPE and mean absolute error (MAE). Other statistical feature selection methods were used including filter methods and wrapper methods. It was desirable to not only predict laboratory performance of concrete used in maturity calibration development, but also other performance criteria for field concrete (e.g., slump) for other modules developed within Giatec's platform. Many variables that may not have a strong correlation with strength were therefore kept in the dataset for such models.

To improve the quality of the data, an outlier detection model was developed. This model made use of unsupervised learning models. Such unsupervised outlier detectors are typically classification models. Two models were used and compared for this purpose: isolation forest (Liu et al., 2008) and local outlier factor (Breunig et al., 2000). The models were evaluated by comparing the outliers and inliers detected in the data versus those labelled as outliers and inliers by domain experts. An f1 score was calculated based on this and the local outlier factor model was chosen as it outperformed the isolation forest model. This model was not only used to improve the dataset quality but is further used as an anomaly detection module within Giatec's platform.

After outlier detection, several regression models were trained on this data with the goal of predicting the maturity calibration of specific mixtures using a knowledge of mixture proportions. These models include, but are not limited to neural networks, support vector regression, gaussian process regression, decision trees, logistics regression, ridge regression and lasso regression. After initial testing for these models, a neural network (multilayer perceptron) was chosen for this problem.

As with the pouring time detection problem, the dataset was divided into training, validation, and testing sets, with a split of 60/20/20. It was ensured that mixtures that were in the training set were not in other sets. For example, if a datapoint for a certain mixture tested at a certain age was in the training set, all other ages of testing or instances of testing for the same mix were added to the training set. This was done to prevent the network from gaining insights on all mixtures and raw materials. Hyperparameter tuning was performed to determine the network's number of layers, layer size, activation function, normalization method, learning rate, regularization method and parameters, dropout rate, batch normalization use, etc. The hyperparameters were chosen based on the minimization of the MAPE and MAE. The optimized neural network made use of concepts of batch normalization, dropout, L2 normalization and used the rectified linear unit (ReLU) activation function. The specific size of the network (in terms of number of layers and layer size) is not described here as many different network sizes were found to be adequate and provide a similar accuracy after hyperparameter tuning (e.g., learning rate, activation function, dropout rate, etc.) is performed. The deployed algorithm achieved an MAPE of 13.8% and an MAE of 720 PSI (pound per sq inch) on strength prediction when evaluated on the testing set (average of all data sources). These metrics generally change based on the concrete testing age, with the error being the lowest at 28 days. This is due to the higher availability of data at 28-days and the lower influence of curing conditions/ambient conditions on the 28-day strength when compared to a 1-day strength.

It is worth mentioning that after model development and release, feature selection/ extension work continues to be ongoing as the model develops, more outputs are requested by the end users and more trends are seen. For instance, the model initially did not incorporate the use of natural pozzolans or limestone fillers. However, the growing industry trends (such as the use of Portland limestone cement) necessitated the collection of data related to such materials. This requires consistent updates to the database, data pipeline, model characteristics, and subsequently hyperparameters.

To use the model to detect inaccuracies in maturity calibration, a module was developed on Giatec's cloud platform (Giatec360™) where the user-inputted mixture calibrations (strength development over time/maturity based on inputted mixture proportions) are compared to the model predictions and large deviations are flagged to the user as inaccuracies. Mixtures can be flagged as having an abnormally high or abnormally low strength. Other inaccuracies in datum temperature, activation energy, reference temperature, etc., can also be flagged using this method.

Furthermore, the reverse solution of the network is performed to determine the potential for cement reduction and mixture optimization. When a strength-maturity calibration is performed and inputted into the system, the calibration is verified using the mentioned method. Subsequently, for mixtures passing this verification, the specified/target strength, inputted by the user, is compared to the performance of the mixture, putting into consideration a user-chosen safety factor. This is shown schematically in Fig. 11.6. When the mixture performance exceeds the target strength, the cementitious materials' reduction capacity is determined by sweeping the cement and SCM input of the network (while maintaining yield) until the mixture strength meets the target strength. This leads to a suggestion of the cementitious materials reduction capacity. This feature is currently used by hundreds of users monthly with a mean suggested cement reduction potential of 10%.

As mentioned previously, the algorithm was trained on historical data of not only strength, but characteristics including slump, slump flow, air content, delivery temperature, *in situ* heat rise and maximum temperature, shrinkage, and diffusion coefficient/rapid chloride permeability test prediction. The full algorithm

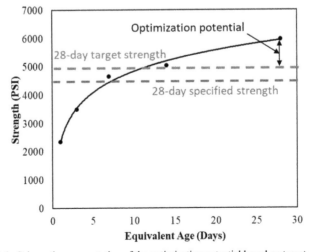

Fig. 11.6: Schematic representation of the optimization potential based on target strength.

with these capabilities has been deployed (Giatec Scientific Inc., 2021) for the users to be able to perform a full performance prediction of their mixtures depending on mixture proportions, raw material compositions and characteristics, and ambient conditions.

More importantly, the algorithm can be used for the purpose of mixture optimization/alteration, where the user can specify information on their performance-based specifications, such as specified strength, slump, or air content; prescriptive specifications, such as minimum cementitious materials content, and maximum SCM replacement; as well as ambient and environmental conditions, such as temperature, humidity, location of pour, etc. An optimization algorithm has been specifically designed to allow the determination of optimized mixture proportions to meet these performance criteria, while minimizing for either cost, carbon footprint, or a combined metric of cost and carbon footprint. This yields mixtures that are continuously changing based on the altering specifications or ambient conditions, ensuring that optimization is continuously achieved regardless of other influencing factors. Such algorithms also make use of continuously collected IoT sensor data to ensure that the optimization and the performance levels are achieved.

The algorithm has been used in a pilot test with one partner producer, where the goal was to reduce the carbon footprint of two commonly used mixtures, with a nominal design strength of 25 and 35 MPa. The mixtures had specifications required for slump and air content of 100–150 mm and an air content of 5–8% as well as prescriptive requirements for maximum slag replacement of 30%. The algorithm was capable to adjust these mixtures to minimize the cementitious materials content and increase the slag replacement and the use of high-range water reducing admixtures. Although the mixtures are not presented here as they are considered proprietary by the producer, the carbon footprint is shown to demonstrate the impact of algorithm usage.

To compare the suggested mixtures to the producer's business as usual, as well as compositions widely used in the literature, the method by Damineli et al. (2010) was used. The author proposed an index correlating the amount of binder required to develop one unit of concrete property (i.e., compressive strength). The so-called binder intensity index (bi) shown in the following Eq. (11.1) quantifies the eco-efficiency of concrete mixtures,

$$bi = \frac{b}{p} \tag{11.1}$$

where b is the total consumption of binder materials (kg/m³) and p is the performance requirement (i.e., compressive strength in MPa). Figure 11.7 shows the binder intensity of these mixtures compared to the ready-mix producer business as usual as well as thousands of those summarized in Damineli et al. (2010). It is evident that the algorithm-suggested mixtures have a significantly lower binder intensity for a comparable strength when compared to literature data or the producers' data. Note that the strength for all these mixtures is higher than their nominal strength values of 25 and 35 MPa.

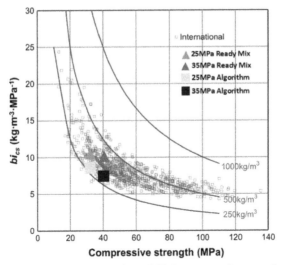

Fig. 11.7: Relationship between binder intensity and 28-day compressive strength from international records adapted from (Damineli et al., 2010).

3. Conclusions

As shown by the cases presented, the concrete industry presents an excellent opportunity for utilizing ML algorithms for analysing concrete performance, owing to the significant amount of readily available data and the growth in the use of IoT sensors. The cases presented in this work show how this data can be used to develop algorithms that can provide significant value to concrete practitioners. The first presented case shows how IoT data collected from embedded concrete sensors was used to develop an algorithm that can infer the time of concrete pouring using only temperature data. The model was found to have a detection accuracy of 90% and predicts pouring times that are on average within 3 hours of the user-selected time. The second case shows the development of an algorithm that can verify the strength-maturity calibration developed in concrete laboratories as well as suggest cement saving potential for concrete practitioners by comparing the current performance to the specified/target strength. This model had a MAPE score of 13.8% and has generally suggested a cement saving potential upwards of 10%. Finally, this algorithm has been deployed for the purpose of performance prediction and mixture optimization and piloted with a ready-mix producer showing the enhancements in carbon footprint realized with the model usage.

References

Breunig, M.M., Kriegel, H.P., Ng, R.T. and Sander, J. 2000. Lof: Identifying density-based local outliers. *Acm Sigmod Record*, 29(2): 93–104.

Carino, N.J., Lew, H.S. and Volz, C.K. 1983. Early age temperature effects on concrete strength prediction by the maturity method. *ACI Journal Proceedings*, 80(2): 93–101.

Cortes, C. and Vapnik, V. 1995. Support-vector networks. *Machine Learning*, 20(3): 273–297.

Damineli, B.L., Kemeid, F.M., Aguiar, P.S. and John, V.M. 2010. Measuring the eco-efficiency of cement use. *Cement and Concrete Composites*, 32(8): 555–562.

Giatec Scientific Inc. 2021. SmartMix™. www.giatecscientific.com/products/software/smartmix/.

Goodfellow, I., Bengio, Y. and Courville, A. 2016. *Deep Learning* (Ser. Adaptive computation and machine learning). USA: MIT Press.

Hochreiter, S. and Schmidhuber, J. 1997. Long short-term memory. *Neural Computation*, 9(8): 1735–80.

LeCun, Y., Bengio, Y. and Hinton, G. 2015. Deep learning. *Nature*, 521: 436–444.

Lecun, Y., Bottou, L., Bengio, Y. and Haffner, P. 1998. Gradient-based learning applied to document recognition. *Proceedings of the IEEE*, 86(11).

Liu, F.T., Ting, K.M. and Zhou, Z.-H. 2008. Isolation forest. *Proceedings—IEEE International Conference on Data Mining, (ICDM)*, pp. 413–422.

Loffe, S. and Szegedy, C. 2013. Batch normalization: Accelerating deep network training by reducing internal Covariate shift. *arXiv*:1502.03167

Srivastava, N., Hinton, G., Krizhevsky, A., Sutskever, I. and Salakhutdinov, R. 2015. Dropout: A simple way to prevent neural networks from overfitting. *Journal of Machine Learning Research (JMLR)*, 15(2): 1929–1958.

Chapter 12

Knowledge-enhanced Deep Learning for Efficient Response Estimation of Nonlinear Structures

Haifeng Wang[1] and Teng Wu[2],*

1. Introduction

The accurate estimation of nonlinear structural responses beyond the design load level is critical for ensuring the safety of civil structures and infrastructure (Chen, 2013, 2014; Di Matteo et al., 2014; Joyner and Sasani, 2018; Luo et al., 2014; Tabbuso et al., 2016; Wang and Mahin, 2018; Zeng et al., 2017; Zhang et al., 2017, 2018; Zhang and Xu, 2001). For the statistical analysis of structure performance under various hazards, the efficiency of structural response estimation is critical (Spence and Kareem, 2014; Chuang and Spence, 2017; Cui and Caracoglia, 2018; Suksuwan and Spence, 2018). While the high-fidelity finite element method can provide accurate nonlinear structural response estimations, it is computationally expensive and time-consuming, especially when several response samples are needed for the statistical analysis.

To increase the nonlinear response estimation efficiency, the physics-based reduced-order modeling has been widely implemented. For example, the reduced-order finite element model, where a simpler finite element model is used to capture the behaviors of the original complex model, has been successfully used for the efficient response estimation of nonlinear structures (Spottswood and Allemang, 2006; Wu et al., 2013; Xiong et al., 2016). The statistical linearization approach, where the nonlinear system is represented by an equivalent linear system, has also

[1] Postdoc Research Associate, Department of Civil and Environmental Engineering, ATLSS Engineering Research Center, Lehigh University, 117 ATLSS Drive, Bethlehem, PA 18015.
Email: haw621@lehigh.edu
[2] Associate Professor, Department of Civil, Structural and Environmental Engineering, University at Buffalo, Ketter Hall 226, Buffalo, NY 14260.
* Corresponding author: tengwu@buffalo.edu

been used for the efficient nonlinear response estimation (Di Matteo et al., 2014; Feng and Chen, 2018; Saitua et al., 2018; Zhao et al., 2019). While the physics-based reduced-order models have performed well on nonlinear response estimation tasks, it is generally a specialized method that needs careful design. Moreover, great attention needs to be paid to the identification of corresponding model parameters from the numerical and/or experimental data.

Compared with physics-based reduced-order modeling, the data-driven reduced-order modeling can be treated as a generalized method for increasing the nonlinear structural response estimation efficiency. With the fast growth of computing power, especially heterogeneous computing, the data-driven reduced-order modeling has shown great promise for the efficient simulation of complex systems (Bottou, 2012; Iungo et al., 2015; Mittal and Vetter, 2015; Peherstorfer and Willcox, 2015).

With the advantage of the wide spectrum of application, the artificial neural network (ANN) has become the most popular method for data-driven reduced-order modeling (Abiodun et al., 2018; Wu and Snaiki, 2022). Actually, the modeling approaches based on deep neural networks (DNNs) have been extensively used in various fields, such as natural language processing, speech recognition, image compression and restoration, and self-driving (Goodfellow et al., 2016). In feedforward ANNs, the input is sequentially processed by multiple layers of neurons to get the output. With the increase of layer numbers, the modeling ability of the ANNs is greatly improved (Bengio, 2012; Duda et al., 2012; Goodfellow et al., 2016). Thus, DNNs, i.e., ANNs with more than two hidden layers, have been widely utilized. Additionally, novel network architectures, such as convolutional neural network (CNN), deconvolutional network (DeConvNet), recurrent neural network (RNN), and residual neural network (ResNet) have also been proposed to greatly increase the feedforward ANN modeling ability for different tasks (Gers et al., 1999; Simonyan et al., 2013; He et al., 2016; Kim, 2017).

The ANNs have also been extensively employed for the safety analysis of civil structures and infrastructure (Elfadel et al., 2018; Facchini et al., 2014; Jiang and Adeli, 2005; Linse and Stengel, 1993; Mead and Ismail, 2012). For example, the feedforward network has been used for the response estimation and parameter identification of both single-degree-of-freedom (SDOF) and multi-degree-of-freedom (MDOF) systems (Pei et al., 2005; Vaidyanathan et al., 2005; Guarize et al., 2007; Derkevorkian et al., 2015). With the ability of automatically detecting important features, CNN has also been widely used for the response estimation of structural systems (Gholizadeh et al., 2009; Wu and Jahanshahi, 2019).

The successful applications of ANNs usually rely on a large amount of high-quality training data, especially for deep neural networks (Ajiboye et al., 2015). For example, the successful application of CNNs to image classification relies heavily on the development of large image datasets (Krizhevsky et al., 2012). However, the high-quality excitation-response datasets are usually unavailable or expensive to acquire for civil structures and infrastructure, limiting the application of deep learning (DL). To this end, Snaiki and Wu (2019) proposed the knowledge-enhanced deep learning (KEDL) methodology, which uses the prior knowledges to guide the training process. Specifically, the physical-based and empirical formulas are added to the loss function to take advantage of the gradient descent mechanism (Bottou, 2012). Actually, the idea of using prior knowledge to enhance the network training efficiency has been proposed for several decades (Psichogios and Ungar, 1992; Dissanayake and Phan-Thien, 1994). Raissi et al. (2017a, 2017b) integrated the physical governing equations into loss functions to efficiently train partial differential equation (PDE)-solving networks. Shin et al. (2020) explored the theoretical proof that physical governing equations can guide the neural network for solving PDEs. In addition to the equation-based 'explicit' domain knowledge (e.g., physics-based, or empirical formulas), the equation-free 'tacit' domain knowledge has recently been incorporated into a deep reinforcement learning-based shape optimizer for efficient aerodynamic mitigation of wind-sensitive structures (Li et al., 2021). In this study, the equation-based KEDL methodology is used for the data-efficient training of the response-estimation network.

The structural response at each time-step is essentially determined by the response of the previous time-step and the excitation of current time-step. Thus, the last-time-step response and the current-time-step excitation need to be considered in the response-estimation network. In this viewpoint, RNN, with the ability of recurrently passing the output of the last step to the current input, is a natural fit for the response estimation task. RNN has been successfully used for the response estimation of the nonlinear systems (e.g., Wu and Kareem, 2011; Chaves et al., 2015; Spiridonakos and Chatzi, 2015; Gonzalez and Yu, 2018). In this study, the long short-term memory (LSTM) neurons (a.k.a. fancy RNN) are utilized for the estimation of structural responses. Specifically, the stacked LSTM architecture is utilized to learn the excitation–response relationship. The wavelet-domain projection is introduced to increase the training efficiency by simplifying the network input–output relationship. The performance of the neural network-based response estimation is comprehensively examined with wind-induced nonlinear structural response.

2. Theoretical Background

2.1 Deep Neural Network

An ANN is a composition of simple arithmetic operations represented by the neurons. As the data flow sequentially through neurons in a network, the neurons can be categorized into different layers according to their positions in the network. In this viewpoint, an ANN is composed of the input layer, the hidden layers, and the output layer. Figure 12.1 depicts the structure of a 3-layer feedforward neural network. The input layer takes the data, the hidden layers process the data, and the output layer generates the results. Each neuron in the hidden layers and the output layer performs operation on the input x_i as:

$$y = \sigma\left(\sum_i w_i x_i + b\right) \tag{12.1}$$

where y is the output of the neuron; $\sigma(\cdot)$ is the nonlinear activation function; w_i is the weight corresponding to input x_i; and b is the bias of the neuron.

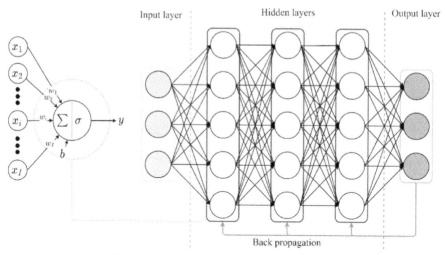

Fig. 12.1: Schematic of a 3-layer feedforward ANN.

In the viewpoint of input–output relationship, an ANN can be treated as a generalized function approximator. By adjusting the weight and bias of each neuron, the input–output relationship can be adjusted. The difference between the network output and the corresponding target is quantified by the loss function, which can be a metric of distance, probability distribution, and entropy. Given training data, the network training process essentially minimizes the loss function. The most successful training methodology is the backpropagation, where the neural network parameters of each layer are updated by gradient descent (Rumelhart et al., 1985; Bottou, 2012).

2.1.1 RNN

Compared with conventional feedforward neural network (Fig. 12.1), the neurons in RNNs have a hidden state to memorize the past data and recurrently pass the hidden state to the next-step input. Figure 12.2a shows the data flow of a single-neuron RNN unfolded in time steps $i - 1$ and i. At each time-step, the neurons receive data from both the input (feedforward edge) and the hidden state of the last time-step (recurrent edge). With the hidden state, the output of at time i is related to both the input at time i and the previous outputs. Thus, the RNN architecture is well suited for 'remembering' the previous outputs (LeCun et al., 2015). With the ability of remembering the previous outputs, the RNN has been extensively applied to sequence data-related tasks, including natural language processing, speech recognition, and audio classification (Graves et al., 2013; Mogren, 2016; Sezer and Ozbayoglu, 2018).

The architecture of RNN is particularly suitable for structural response estimation. Specifically, for the structural dynamics, the current responses (i.e., displacement u_i, velocity $\dot{u}_i$, and acceleration $\ddot{u}_i$) are determined by last-time-step structural status $(u_{i-1}, \dot{u}_{i-1}, \ddot{u}_{i-1})$ and the current-time-step excitation f_i. Accordingly, both the last-time-step structure status (output at time $i-1$) and the current excitation (input at time i) need to be considered to accurately estimate the structural response, which naturally fit the advantages of RNN (Fig. 12.2a).

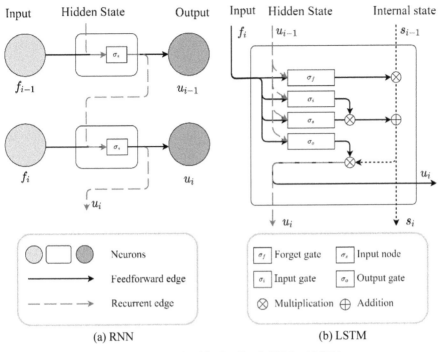

(a) RNN (b) LSTM

Fig. 12.2: Schematic of the data flow in RNN and LSTM.

2.1.2 LSTM

While the hidden state captures the previous outputs, RNN can only effectively pass information across very limited time-steps due to the problem of vanishing and exploding gradients (Bengio et al., 1994; Jozefowicz et al., 2015). The LSTM neuron shown in Fig. 12.2b is proposed to overcome this problem (Hochreiter and Schmidhuber, 1997). Compared with the RNN shown in Fig. 12.2a, the LSTM has an internal state to maintain information for a longer time and a set of gates to control how the input affects the internal state, the hidden state, and the output (Gers et al., 1999). Specifically, the forget gate learns to flush the contents of the internal state; the internal node and the input gate decide how the internal state is updated; the output gate and the internal state s_i jointly determine the output u_i. With these gates, the LSTM neurons can circumvent the problem of vanishing and exploding gradients for long-time series and effectively model nonlinear structural dynamics (Li et al., 2020).

2.2 KEDL Methodology

In conventional DL, a large amount of training data is critical for effectively training and validating an ANN. The difficulty of obtaining high-quality training data is one of the major challenges that limit the application of DL to engineering problems. For structural dynamics, it is typically very expensive to measure the structural dynamics responses. In addition, the measurement data are inevitably contaminated by noises due to the harsh environment. To this end, Snaiki and Wu (2019) proposed the KEDL methodology, where the prior knowledge of the target system is integrated into the loss function to better guide the training process and increase noise robustness. In KEDL, the loss function L calculated as:

$$L = (1 - a)L_D + aL_K \qquad (12.2)$$

where L_D is the data-based sub-loss function; L_K is the knowledge-based sub-loss function; a is the parameter that determines the importance of each sub-loss function to the training process. In the case that a is set to zero, the KEDL is equivalent to the conventional data-driven training process.

Figure 12.3 depicts the application of KEDL to a simple RNN. The neuron weight update is guided by two loss functions, i.e., the data loss L_D and the knowledge loss L_K. The data loss L_D is a metric that quantifies the difference between the neural network output y_{output} and the target output y_{target} provided by the training data. The knowledge loss is a metric that quantifies the difference between the neural network output y_{output} and the knowledge-based output $f_K(x_{input}, y_{output})$, where $f_K(\cdot)$ is the input-output relation described by the prior knowledge.

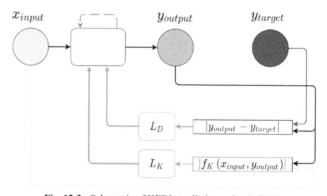

Fig. 12.3: Schematic of KEDL applied to a simple RNN.

In this study, the knowledge sub-loss function L_K is determined by both physics-based equations and empirical information as:

$$L_K = f_{L_K}(L_p, L_E) \qquad (12.3)$$

where $f_{L_K}(\cdot)$ is the combination function of L_p and L_E; L_p quantifies deviation of the network input-output relation from physics-based governing equations; and L_E quantifies the deviation from empirical information. In KEDL, the training dataset, the physics-based equations, and the empirical information are jointly used to guide the training process by minimizing the loss function. It should be noted the importance of L_p and L_E to the training process is adjusted by the combination function $f_{L_K}(\cdot)$.

2.3 Wavelet-domain Projection

The formats and features of input and output determine the ANN complexity, and hence the difficulty of training (Pascanu et al., 2013). Thus, manually extracted features have been extensively used as the network input to simplify the input-output relationship and achieve high training efficiency (Subasi, 2007; Saravanan and Ramachandran, 2010; Jia et al., 2016; Bakir et al., 2016; Tao et al., 2020). For the neural network-based structural response estimation, the training process can be time-consuming and present convergence issues in the case that the excitation and response histories are directly utilized as the network input and output. For dynamics of civil structures and infrastructure, the structure is mainly excited in a specific frequency range and is insensitive to excitations outside this range. Thus, the excitation-response relationship can be simplified by decomposing them in the frequency domain. Wavelets, carefully designed with compact-support and orthogonal properties, have the advantage of decomposing nonstationary signals in the frequency domain (Daubechies, 1993; Rhif et al., 2019). In this study, the excitation-response relationship is identified in the wavelet domain to reduce the complexity of input and output, and hence accelerate the training process. Specifically, the mapping rules between the wavelet coefficients of the structural response and external forces (e.g., wind excitation) are captured by the stacked LSTM architecture with the KEDL methodology.

In the wavelet-domain projection, the responses $u(t)$, $\dot{u}(t)$, $\ddot{u}(t)$ and the external force $f(t)$ can be approximated as:

$$u(t) \approx \sum_{\tau} W_{u,j,\tau} \varphi_{j,\tau}(t) \tag{12.4}$$

$$\dot{u}(t) \approx \sum_{\tau} W_{u,j,\tau} \dot{\varphi}_{j,\tau}(t) \tag{12.5}$$

$$\ddot{u}(t) \approx \sum_{\tau} W_{u,j,\tau} \ddot{\varphi}_{j,\tau}(t) \tag{12.6}$$

$$f(t) \approx \sum_{\tau} W_{f,j,\tau} \varphi_{j,\tau}(t) \tag{12.7}$$

where W is the wavelet coefficient; j is the wavelet scale; $\varphi_{j,\tau}(t) = \varphi_j(t - \tau)$ is the wavelet scaling function; t is the continuous time; and τ is the wavelet shift parameter. According to Eqs. (12.4)–(12.7), the dynamic equilibrium equation in the time domain [i.e., $m\ddot{u}(i) + c\dot{u}(i) + ku(i) = f(t)$, where m, c, and k are the mass, damping coefficient and stiffness, respectively] can be expressed in the wavelet domain as:

$$m\sum_{\tau} W_{u,j,\tau} \Gamma_{j,\tau,l}^{2,0} + c\sum_{\tau} W_{u,j,\tau} \Gamma_{j,\tau,l}^{1,0} + k\sum_{\tau} W_{u,j,\tau} \delta_{\tau,l} = \sum_{\tau} W_{f,j,\tau} \delta_{\tau,l} \tag{12.8}$$

where $\delta_{\tau,l}$ represents the Kronecker delta; $\Gamma_{j,\tau,l}^{1,0}$ and are $\Gamma_{j,\tau,l}^{2,0}$ connection coefficients expressed as:

$$\Gamma_{j,\tau,l}^{1,0} = \int \dot{\varphi}_{j,\tau}(t) \varphi_{j,l}(t) dt \tag{12.9}$$

$$\Gamma_{j,\tau,l}^{2,0} = \int \ddot{\varphi}_{j,\tau}(t) \varphi_{j,l}(t) dt \tag{12.10}$$

Compared with time-domain response estimation, the wavelet-domain response estimation has the advantage of error-free derivative calculation, which is critical for the KEDL since the knowledge loss L_K is a function of the dynamic velocity and acceleration. In time-domain, the velocity and acceleration errors resulting from numerical differentiation can cause serious accuracy issues and hence affect the training of the network. On the other hand, wavelets can be designed with analytical derivatives of required order. Correspondingly, the error-free derivatives of displacement are available in the wavelet domain and can be used to facilitate the training process of KEDL. The Daubechies wavelet functions with order n is n-times differentiable (Daubechies, 1993; Romine and Peyton, 1997). In this study, the Daubechies wavelet function with order 10 is utilized (Le and Caracoglia, 2015; Zhou et al., 2021).

3. Performance Verification

The performance of KEDL is verified in this section with a linear SDOF system representing an airport tower. ANNs are trained and tested in various scenarios to conduct a comprehensive assessment of the performance of KEDL in terms of the estimation accuracy, data efficiency, and noise robustness. The linear SDOF system is used as a benchmark for the performance verification due to its simplicity. For an appropriately configured neural network with sufficient training (in terms of low loss function values), using linear or nonlinear structural dynamic systems essentially does not affect the performance verification of KEDL.

3.1 Configuration of SDOF System Used for Verification

Figure 12.4 schematically shows the SDOF system used for the performance verification. It is assumed that the SDOF system is excited by the wind load $f(t)$ and has its damping force $f_D(t)$ and the restoring force $f_R(t)$.

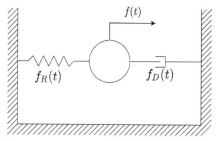

Fig. 12.4: Schematic SDOF system configuration.

The parameters of the SDOF system is set up based on the field measurement of an airport tower (Tamura et al., 1996). Specifically, the mass m, damping coefficient c, and stiffness k are respectively a set as 3.237×10^6 kg, 1.515×10^4 N·m/s, and 1.773×10^5 N/m. Due to the lack of high-quality pairs of wind load and structural response measurements, the synthesized wind loads and corresponding responses calculated with the 4th-order Runge-Kutta (RK4) method are used as the training dataset in this study.

The wind speed is composed of the mean component $\bar{s}$ and the fluctuation component $\tilde{s}(t)$ as:

$$s(t) = \bar{s} + \tilde{s}(t) \tag{12.11}$$

where $\bar{s} = 36.4$ m/s is the mean wind speed at height 46.6 m; $\tilde{s}(t)$ is synthesized using the spectral representation approach with the Karman-type spectrum (Deodatis, 1996; Holmes, 2015; Tamura and Kareem, 2013):

$$\frac{nS_{\tilde{s}}(n)}{\sigma_{\tilde{s}}^2} = \frac{4\left[\dfrac{nL_s}{\bar{s}}\right]}{\left\{1 + 70.8\left[\dfrac{nL_s}{\bar{s}}\right]^2\right\}^{\frac{5}{6}}} \tag{12.12}$$

where n represents frequency; $S_{\tilde{s}}(n)$ is the target power spectral density (PSD); $\sigma_{\tilde{s}}$ is the standard deviation of fluctuation; L_s is the turbulence integral length scale set to be 80 m (Huang et al., 2013). The wind load is obtained from the simulated wind speed $s(t)$ using the quasi-steady theory:

$$f(t) = \frac{1}{2}\rho A C_D s^2(t) \tag{12.13}$$

where ρ is the air density set as 1.25 kg/m³; A is the equivalent area set to be 840 m²; the drag coefficient C_D is set to 1.2 (Tamura and Kareem, 2013). The response samples are generated with RK4 method with zero initial displacement and velocity. The sampling frequency is 50 Hz. Two durations (16 s and 600 s) are selected for different verification purposes. Figure 12.5 depicts the excitation and corresponding response of one sample in the training data. While the excitation is stationary, obvious amplitude modulations are observed in the response. The amplitude modulations essentially result from the fact that the structural response of a specific time instant depends on excitations of previous time instants.

It should be noted that the networks trained with numerically generated training data can only capture the input–output relationship embedded in the simulations, which may deviate from the real airport-tower system. To realistically capture the

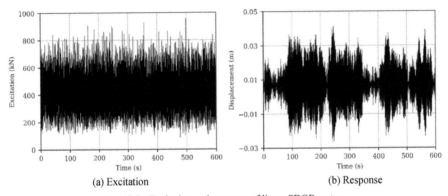

(a) Excitation   (b) Response

Fig. 12.5: Excitation and response of linear SDOF system.

excitation–response relationship of the airport-tower system, the neural network needs to be trained with high-quality measurements. In this study, we aim to explore the application of KEDL to structural response estimation instead of the verification of the realisticity of the network-based estimation. Thus, using numerical simulations is considered as sufficient for the verification purpose here.

3.2 Stacked LSTM Architecture

Various deep RNN architectures have been proposed for different purposes (e.g., Pascanu et al., 2013; Mao et al., 2014; Jain et al., 2016). In this study, the stacked LSTM layers are nested with a dense layer for the estimation of nonlinear dynamic responses (Fig. 12.6). The three LSTM layers are utilized to memorize the structure status and accommodate the nonlinear mapping between input and output datasets. The last dense layer is added to convert the output of LSTM layers to the wavelet coefficients of the structural response. Each LSTM layer has eight LSTM neurons, and the dense layer has one neuron with rectified linear unit (ReLU) activation function (He et al., 2015). The number of neurons of the dense layer is selected to match the output of the SDOF system. Both the excitation and response are normalized following:

$$f_N(i) = \frac{f(i)}{E(|f|_{\max})} \tag{12.14}$$

and

$$u_N(i) = \frac{u(i)}{E(|u|_{\max})} \tag{12.15}$$

where $f_N(i)$ and $u_N(i)$ are respectively the normalized response and excitation; $E(|u|_{\max})$ and $E(|f|_{\max})$ are expected maximum estimated from available samples; and i is the discrete time. The input of the network is set to be the wavelet coefficients of normalized excitation $W_{f,\tau}$ with a shape of B by 1, where B is the number of wavelet coefficients. The output of the network is the wavelet coefficients of normalized response with the same shape of $W_{f,\tau}$.

The neural network model is set up and trained with the widely-used machine learning platform TensorFlow (Abadi et al., 2016). The AdaMax optimization scheme

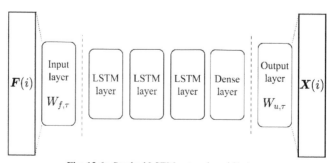

Fig. 12.6: Stacked LSTM network architecture.

is employed with adaptive learning rate (Kingma and Ba, 2014). The data-based and knowledge-based sub-loss functions are respectively calculated as:

$$L_D = \frac{1}{N_D} \sum_{k=1}^{N_D} \left[\sum_{\tau} (W_{\text{out},k,u,j,\tau} - W_{\text{tar},k,u,j,\tau})^2 \right] \qquad (12.16)$$

and

$$\sum_{k=1}^{N_K} \left[\begin{array}{c} m \sum_{\tau} W_{\text{out},k,u,j,\tau} \Gamma_{j,\tau,l}^{2,0} + c \sum_{\tau} W_{\text{out},k,u,j,\tau} \Gamma_{j,\tau,l}^{1,0} + k \sum_{\tau} W_{\text{out},k,u,j,\tau} \delta_{\tau,l} \\ - \sum_{\tau} W_{f,j,\tau} \delta_{\tau,l} \end{array} \right]^2 \qquad (12.17)$$

where N_D and N_K are the samples used for calculating the data-based and knowledge-based sub-loss functions, respectively; W_{out} and W_{tar} are respectively the output and target wavelet coefficients. The a value of Eq. (12.2) is set as 0.5 in this study. It should be noted that the effects of L_K and L_D on the training process can be modified by using another a value in Eq. (12.2). The neural network is treated as sufficiently trained if the loss function stops decreasing.

3.3 Performance of KEDL Methodology

3.3.1 Estimation Accuracy

The accuracy of the trained KEDL is first verified with the SDOF system presented in Fig. 12.4. A hundred 16-s (800-time-step) samples are used for the training process. Another 800 samples are used as the test dataset. The wavelet decomposition level is set to be 4 to cover 99% energies in the structural responses. Figure 12.7a presents the comparison between the RK4-based response and output of the trained neural network. The sample presented in Fig. 12.7 is randomly selected from the test dataset. The absolute error value of the network-based estimation is also shown in Fig. 12.7a using the RK4-based response as a reference. The trained KEDL performs well with a max error is 0.0172 and a mean error of 0.0045. The detailed response between the 10s and 12s is presented in Fig. 12.7b to investigate the relation between the error

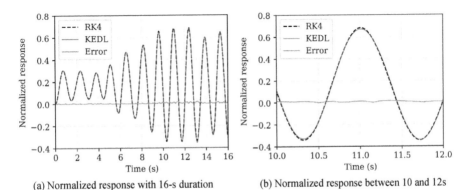

(a) Normalized response with 16-s duration   (b) Normalized response between 10 and 2s

Fig. 12.7: Selected network-based response estimation.

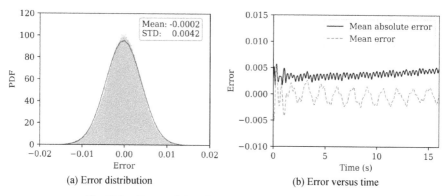

(a) Error distribution   (b) Error versus time

Fig. 12.8:   Error statistics of network-based response estimation.

and response values. It can be found that the error reaches its local maximum when the response reaches its local peak values (e.g., 10.32s and 11.02s), indicating it may be more difficult for the neural network to identify the local peak responses as compared with non-peak responses.

The error statistics of the network-based estimation are obtained from 800 test samples and presented in Fig. 12.8a. The mean and standard deviation of the error are respectively –0.0002 and 0.0042. The corresponding Gaussian distribution is also shown in Fig. 12.8a as black solid line. Compared with the Gaussian distribution, the errors are more concentrated around zero.

For the structural response estimation tasks, the error accumulation is critical. Thus, in addition to the general error probability distribution, it is important to explore the error development with respect to time. The mean error and the mean absolute error values are calculated and presented in Fig. 12.8b. It can be observed that the estimation error increases slightly with time. The time dependency can be attributed to the fact that LSTM neurons recurrently use the last-time-step output for the calculation of the current-time-step output and accumulate estimation error. It should be noted that the error accumulation can be mitigated in the real-world applications by simply training the neural network with longer samples.

3.3.2 Data Efficiency

One advantage of the KEDL methodology is the data efficiency. Specifically, the physics-based and/or empirical knowledge is used to provide guidance for the training process and reduce the demand on training dataset. Compared with a conventional purely data-driven training process, a relatively smaller number of training datasets are needed to achieve the same level of accuracy. To highlight the data-efficiency advantage of KEDL over conventional data-driven training methodology, the network shown in Fig. 12.6 is trained with two a values in Eq. (12.2), namely $a = 0.5$ and $a = 0$. When $a = 0.5$, both data-based and knowledge-based loss functions are used for guiding the training process. When $a = 0$, only the data-based loss function is used for guiding the training process. For each a value, the sample number varies from 1 to 10. Since each sample contains 800 discrete time steps, the size of the training dataset varies from 800 to 8,000.

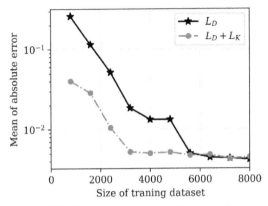

Fig. 12.9: Error versus dataset size for two *a* values.

The trainings for both *a* values are terminated when the loss function stopped decreasing and thus can be treated as sufficient. The mean absolute error values are plotted versus the size of the training dataset in Fig. 12.9 which shows the error of the conventional data-driven training converges when the dataset size exceeds 5,600. By contrast, the KEDL error converges with a size of 3,200. The difference between the two training methodologies shows that the knowledge-based sub-loss function L_K enhanced the regularization mechanism during the training process. It is also observed that, even in the case that the dataset size is insufficient for the convergence (i.e., dataset size smaller than 2400), the error of KEDL is much lower than that of the conventional data-driven training, further proving the data efficiency provided by the knowledge-based loss function.

3.3.3 Noise Robustness

For real-world applications of neural network-based structural response estimation, the robustness to noise is critical. In this section, the noise robustness of the KEDL is assessed by adding various levels of noises to the responses in the training dataset. The white Gaussian noises is used to emulate the noise contamination in measurements (Chen and Li, 2004; Mao and Todd, 2012; Sun and Chang, 2004). The noise level is quantified as the ratio of the standard deviation of noise to that of the response signal. The variation from 0.1 to 2 is adopted (Fig. 12.10a) which shows that noises cannot be visually observed when its level is below 0.2. When the level reaches 2.0, the original response cannot be easily identified.

The neural networks are trained with $a = 0.5$ and $a = 0$, i.e., knowledge-enhanced, and conventional data-driven training methodologies. Figure 12.10b depicts the standard deviations and mean values of errors with respect to the noise levels. For $a = 0$, the error increases fast with the noise level. On the other hand, for $a = 0.5$, the error only increases slightly with the noise level. Even with a noise level of 2.0, the estimation accuracy is controlled within an acceptable level, proving the robustness of KEDL to training dataset noises. The noise robustness is attributed to the fact that knowledge-based sub-loss function L_K is not affected by the contaminated data.

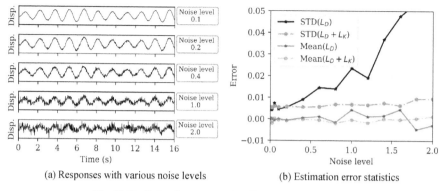

(a) Responses with various noise levels   (b) Estimation error statistics

Fig. 12.10: Estimation errors under different levels of noises.

4. Application of KEDL to Nonlinear Systems

In this section, the neural network-based response estimation is applied to both nonlinear SDOF and MDOF systems to investigate its applicability to realistic nonlinear structural response estimations. The nonlinear SDOF system is used to investigate the applicability of the neural-network-based estimation for different levels of nonlinearity. The nonlinear MDOF system is used to investigate the generalization ability and the efficiency advantage of the network-based estimation.

4.1 Application to Nonlinear SDOF System

To investigate the applicability of the network-based response estimation to systems with various degrees of nonlinearity, the nonlinear SDOF systems with cubic stiffness are utilized:

$$m\ddot{u}(t) + c\dot{u}(t) + ku(t) + k_c u^3(t) = f(t) \tag{12.18}$$

where $m = 3.237 \times 10^6$ kg; $c = 1.515 \times 10^4$ Nm/s; $k = 1.773 \times 10^5$ N/m; and k_c is the cubic stiffness. k_c/k ratios ranging from 500 to 9,000 are selected to represent different degrees of nonlinearity.

The stacked LSTM architecture presented in Fig. 12.6 is trained to estimate the dynamic responses of the nonlinear SDOF systems. The responses are normalized by the corresponding maximum values for the error calculation. Figure 12.11a depicts the mean absolute error versus the k_c/k ratio. It can be observed that the error increases approximately linearly with the k_c/k ratio and reaches 0.066 when the k_c/k reaches 9,000. The positive relation between the error and k_c/k ratio is attributed to the limited approximation ability of the neural network architecture. The number of the LSTM layers are varied from 1 to 5 for the case of $k_c/k = 9000$ to verify this hypothesis. As shown in Fig. 12.11b, the error is negatively related to the number of LSTM layers. For systems with high degree of nonlinearity, more LSTM layers should be added to reach the target accuracy.

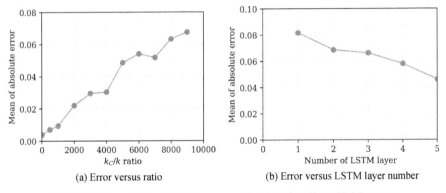

(a) Error versus ratio   (b) Error versus LSTM layer number

Fig. 12.11: Error of network-based response estimation of nonlinear SDOF systems.

4.2 Application to Nonlinear MDOF System

The network-based response estimation is applied to a nonlinear MDOF system to investigate its generalization ability and efficiency advantage. More specifically, the neural network configuration [3 LSTM layers (each with 8 neurons) and 1 dense layer] shown in Fig. 12.6 is employed to estimate the structural response of a lumped-mass model with bilinear stiffness.

4.2.1 System Configuration

The setup of the target building is shown in Fig. 12.12a. The braced frame with a height of 78 m and both width and depth of 25 m. The linear mass distribution is employed (Zhang et al., 2017) as:

$$r_m(z_R) = 1 - 0.5 z_R \tag{12.19}$$

where $z_R = z/H$ is the relative height. The ratio of the top- to-bottom-story masses λ_{MRTB} is 0.5. The density of the building is set to be 220 kg/m³, resulting in a total building mass 1.1×10^7 kg. The stiffness variation can be expressed as:

$$r_s(z_R) = 1 - (1 - 0.55)(z_R)^2 \tag{12.20}$$

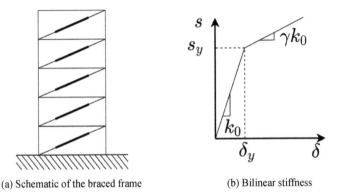

(a) Schematic of the braced frame   (b) Bilinear stiffness

Fig. 12.12: Nonlinear MDOF system configuration.

The ratio of the top- to bottom-level stiffnesses λ_{SRTB} is 0.55 (Lu et al., 2013). The fundamental frequency of the linear superstructure is 1.25 Hz, corresponding to a period of 0.8s (Tamura and Kareem, 2013). The damping ratio of the first mode is assumed to be 0.01 (Feng and Chen, 2018; Siringoringo and Fujino, 2017). The lateral stiffness of the target building is provided by the braces that have a bilinear stiffness (Fig. 12.12b). The building is assumed to remain linear under 300-year mean recurrence interval (MRI) wind loads and have a post-yield stiffness ratio of 0.1. The wind loads are generated according to the procedure described in Section 3, and RK4 method is utilized to obtain the training dataset with a sampling frequency of 50 Hz.

4.2.2 Estimation Accuracy

The accuracy of the network-based response estimation is first verified. One hundred samples with 60-s duration (3,000 discrete time-steps) are used for training the neural network. Figure 12.13a presents the RK4-based and KEDL-based response estimations of different building levels. All the responses are normalized by the standard deviation of the tip response. As shown in Fig. 12.13, the error is small as compared with the RK4-based response. To generally examine the accuracy of the neural network-based estimation, the error of response estimation is estimated from 100 testing samples and presented in Fig. 12.13b. It can be found that the overall error is small. The fitted Gaussian distribution is also presented in Fig. 12.13b. Compared with the Gaussian distribution, the errors are more concentrated near zero.

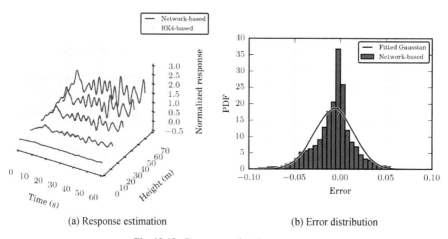

(a) Response estimation   (b) Error distribution

Fig. 12.13: Response estimation accuracy.

4.2.3 Generalization Ability

To verify the generalization ability of the trained neural network, the neural network is tested with samples that have a much longer duration than the training dataset.

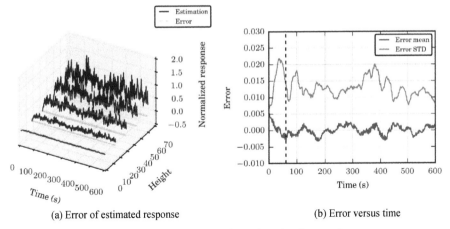

(a) Error of estimated response   (b) Error versus time

Fig. 12.14: Generalization ability on long-duration samples.

Specifically, the neural network is trained with 60-s samples and tested with 600-s samples. The analysis duration of 600s has been widely used in the wind-induced response analyses and is thus selected in this study (e.g., Cui and Caracoglia, 2018; Feng and Chen, 2018). The neural network-based estimations and the corresponding errors are shown in Fig. 12.14a. Although the network is trained with 60-s samples, its application to excitations with a much longer duration of 600s presents low-level errors. To better investigate the effect of a longer duration, the mean and standard deviation (STD) of top displacement estimation error is presented versus time in Fig. 12.14b. It can be observed that the error is generally low (3% of STD). Compared with the 0–60s data (marked with dashed vertical line), the 60–600s data show insignificant error increase. This indicates that the network has successfully learned the structural excitation–response relation.

4.2.4 Estimation Efficiency

The major advantage of the neural network-based nonlinear response estimation is efficiency. To present the efficiency of the network-based response estimation, for each building, 100 response estimations are conducted with 600-s duration. With a single CPU core (Intel Xeon Gold 6130 CPU @ 2.10GHz), the neural network-based estimation is accomplished within 0.47 hours, while the RK4-based estimation takes 2.94 hours. The time consumption distribution for a single sample is presented in Fig. 12.15, where the neural network reduced the mean time consumption from 80 s to 13s. It should be noted that the neural network utilized a batch size of 1 in efficiency comparison. In this case, the data copy operation actually leads to longer time consumption as compared with instruction execution (Jeong et al., 2012). By simply increasing the batch size, the averaged neural network-based estimation time can be further decreased. For example, by increasing the batch size to 200, the average time consumption for each sample can be reduced to 0.62s.

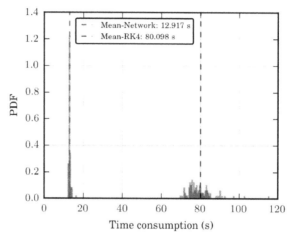

Fig. 12.15: Time consumption for a single sample.

5. Concluding Remarks

The advances in DL shed light on using ANNs for the efficient response estimation of nonlinear structures. In this study, the stacked LSTM architecture was proposed to approximate the excitation-response relationship of nonlinear structures in the wavelet domain. To address the issue of heavy training data demand, the equation-based KEDL methodology was utilized. In KEDL, the physics-based and empirical formulas were jointly employed to regulate the training process and reduce the data demand. The developed network-based method showed great promise in the efficient and accurate estimation of nonlinear structural responses, especially in the case that the training data possessed a smaller duration compared to the testing data. The KEDL methodology also presented high robustness to noise in the training data. It is expected that the KEDL methodology will facilitate the application of DL techniques in the safety analysis of civil structures and infrastructure.

References

Abadi, M., Barham, P., Chen, J., Chen, Z., Davis, A., Dean, J., Devin, M., Ghemawat, S., Irving, G. and Isard, M. 2016. Tensorflow: A system for large-scale machine learning. Presented. *12th USENIX Symposium. Operating Systems Design and Implementation (OSDI 16)*, Savannah, GA, USA, pp. 265–283.

Abiodun, O.I., Jantan, A., Omolara, A.E., Dada, K.V., Mohamed, N.A. and Arshad, H. 2018. State-of-the-art in artificial neural network applications: A survey. *Heliyon*, 4: e00938.

Ajiboye, A., Abdullah-Arshah, R. and Hongwu, Q. 2015. Evaluating the effect of dataset size on predictive model using supervised learning technique. *International Journal of Computer Systems and Software Engineering.* https://doi10.15282/ij.secs.1.2015.6.006.

Bakir, T., Boussaid, B., Odgaard, P.F., Abdelkrim, M.N. and Aubrun, C. 2016. Artificial neural network based on wavelet transform and feature extraction for a wind turbine diagnosis system. Presented. *14th International Conferenceon Control, Automation, Robotics and Vision (ICARCV)*. Phuket, Thailand, pp. 1–6. https://doi.org/10.1109/ICARCV.2016.7838818.

Bengio, Y. 2012. Deep learning of representations for unsupervised and transfer learning In: *Proceedings of ICML Workshop on Unsupervised and Transfer Learning*. JMLR Workshop and Conference Proceedings, pp. 17–36.

Bengio, Y., Simard, P. and Frasconi, P. 1994. Learning long-term dependencies with gradient descent is difficult. *IEEE Transactions on Neural Networks*, 5: 157–166.

Bottou, L. 2012. Stochastic gradient descent tricks. *In: Neural Networks: Tricks of the trade*. Springer, pp. 421–436.

Chaves, V., Sagrilo, L.V.S., da Silva, V.R.M. and Vignoles, M.A. 2015. Artificial neural networks applied to flexible pipes fatigue calculations. Presented. *34th International Conference on Ocean, Offshore and Arctic Engineering*, St. John's, Newfoundland, Canada.

Chen, J. and Li, J. 2004. Simultaneous identification of structural parameters and input time history from output-only measurements. *Computational Mechanics*, 33: 365–374.

Chen, X. 2013. Estimation of stochastic crosswind response of wind-excited tall buildings with nonlinear aerodynamic damping. *Engineering Structures*, 56: 766–778. https://doi.org/10.1016/j.engstruct.2013.05.044.

Chen, X. 2014. Extreme value distribution and peak factor of crosswind response of flexible structures with nonlinear aeroleastic effect. *Journal of Structural Engineering*, 140: 04014091. https://doi.org/10.1061/(ASCE)ST.1943-541X.0001017.

Chuang, W.C. and Spence, S.M.J. 2017. A performance-based design framework for the integrated collapse and non-collapse assessment of wind excited buildings. *Engineering Structures*, 150: 746–758. https://doi.org/10.1016/j.engstruct.2017.07.030.

Cui, W. and Caracoglia, L. 2018. A fully-coupled generalized model for multi-directional wind loads on tall buildings: A development of the quasi-steady theory. *Journal of Fluids and Structures*, 78: 52–68. https://doi.org/10.1016/j.jfluidstructs.2017.12.008.

Daubechies, I. 1993. Orthogonal bases of compactly supported wavelets Ii. Variations on a theme. *SIAM J. Math. Anal.*, 24: 499–519. https://doi.org/10.1137/0524031.

Deodatis, G. 1996. Simulation of ergodic multivariate stochastic processes. *Journal of Engineering Mechanics*, 122: 778–787. https://doi.org/10.1061/(ASCE)0733-9399(1996)122:8(778).

Derkevorkian, A., Hernandez-Garcia, M., Yun, H.-B., Masri, S.F. and Li, P. 2015. Nonlinear data-driven computational models for response prediction and change detection. *Structural Control and Health Monitoring*, 22: 273–288.

Di Matteo, A., Lo Iacono, F., Navarra, G. and Pirrotta, A. 2014. Experimental validation of a direct pre-design formula for TLCD. *Engineering Structures*, 75: 528–538. https://doi.org/10.1016/j.engstruct.2014.05.045.

Dissanayake, M.W.M.G. and Phan-Thien, N. 1994. Neural-network-based approximations for solving partial differential equations. *Communications in Numerical Methods in Engineering*, 10: 195–201. https://doi.org/10.1002/cnm.1640100303

Duda, R.O., Hart, P.E. and Stork, D.G. 2012. *Pattern Classification (2nd Edn.)*. John Wiley and Sons.

Elfadel, I.A.M., Boning, D.S. and Li, X. 2018. *Machine Learning in VLSI Computer-Aided Design*. Springer.

Facchini, L., Betti, M. and Biagini, P. 2014. Neural network based modal identification of structural systems through output-only measurement. *Computers and Structures*, 138: 183–194. https://doi.org/10.1016/j.compstruc.2014.01.013.

Feng, C. and Chen, X. 2018. Inelastic responses of wind-excited tall buildings: Improved estimation and understanding by statistical linearization approaches. *Engineering Structures*, 159: 141–154. https://doi.org/10.1016/j.engstruct.2017.12.041.

Gers, F.A., Schmidhuber, J. and Cummins, F. 1999. Learning to forget: Continual prediction with LSTM. Presented. *9th International Conference on Artificial Neural Networks*, Edinburgh, UK: IET, pp. 1–19.

Gholizadeh, S., Salajegheh, J. and Salajegheh, E. 2009. An intelligent neural system for predicting structural response subject to earthquakes. *Adv. Eng. Softw.*, 40: 630–639. https://doi.org/10.1016/j.advengsoft.2008.11.008.

Gonzalez, J. and Yu, W. 2018. Non-linear system modeling using LSTM neural networks. *IFAC-PapersOnLine*: 51: 485–489.

Goodfellow, I., Bengio, Y. and Courville, A. 2016. *Deep Learning*. Cambridge, Massachusetts: MIT Press.

Graves, A., Mohamed, A. rahman and Hinton, G. 2013. Speech recognition with deep recurrent neural networks. Presented. *2013 IEEE International Conference on Acoustics, Speech and Signal Processing, IEEE*, pp. 6645–6649.

Guarize, R., Matos, N.A.F., Sagrilo, L.V.S. and Lima, E.C.P. 2007. Neural networks in the dynamic response analysis of slender marine structures. *Applied Ocean Research*, 29: 191–198. https://doi.org/10.1016/j.apor.2008.01.002.

He, K., Zhang, X., Ren, S. and Sun, J. 2015. Delving Deep into Rectifiers: Surpassing Human-level Performance on Imagenet Classification. *Preprint arXiv: 1502.01852*.

He, K., Zhang, X., Ren, S. and Sun, J. 2016. Identity mappings in deep residual networks. *In: European Conference on Computer Vision*. Springer, pp. 630–645.

Hochreiter, S. and Schmidhuber, J. 1997. Long short-term memory. *Neural Computation*, 9: 1735–1780. https://doi.org/10.1162/neco.1997.9.8.1735.

Holmes, J.D. 2015. *Wind Loading of Structures (3rd Edn.)*. Boca Raton: CRC Press.

Huang, G., Chen, X., Liao, H. and Li, M. 2013. Predicting of tall building response to non-stationary winds using multiple wind speed samples. *Wind and Structures, an International Journal*, 17: 227–244. https://doi.org/10.12989/was.2013.17.2.227.

Iungo, G.V., Santoni-Ortiz, C., Abkar, M., Porté-Agel, F., Rotea, M.A. and Leonardi, S. 2015. Data-driven reduced order model for prediction of wind turbine wakes. *Journal of Physics: Conference Series*, 625: 012009. https://doi.org/10.1088/1742-6596/625/1/012009.

Jain, A., Zamir, A.R., Savarese, S. and Saxena, A. 2016. Structural-RNN: Deep learning on spatio-temporal graphs. *In: Proceedings of the IEEE Conference on Computer Vision and Pattern Recognition*, pp. 5308–5317.

Jeong, H., Kim, S., Lee, W. and Myung, S.-H. 2012. Performance of Sse and Avx instruction sets. *arXiv preprint arXiv*:1211.0820.

Jia, F., Lei, Y., Lin, J., Zhou, X. and Lu, N. 2016. Deep neural networks: A promising tool for fault characteristic mining and intelligent diagnosis of rotating machinery with massive data. *Mechanical Systems and Signal Processing*, 72-73: 303–315. https://doi.org/10.1016/j.ymssp.2015.10.025.

Jiang, X. and Adeli, H. 2005. Dynamic wavelet neural network for nonlinear identification of high rise buildings. *Computer-Aided Civil and Infrastructure Engineering*, 20: 316–330. https://doi.org/10.1111/j.1467-8667.2005.00399.x.

Joyner, M. and Sasani, M. 2018. Multihazard risk-based resilience analysis of east and west coast buildings designed to current codes. *Journal of Structural Engineering*, 144: 04018156. https://doi.org/10.1061/(ASCE)ST.1943-541X.0002132

Jozefowicz, R., Zaremba, W. and Sutskever, I. 2015. An empirical exploration of recurrent network architectures. *J. Mach. Learn. Res.*, 2342–2350.

Kim, P. 2017. Convolutional neural network. *In: MATLAB Deep Learning*. Springer, pp. 121–147.

Kingma, D.P. and Ba, J. 2014. Adam: A method for stochastic optimization. *arXiv preprint arXiv*:1412.6980.

Krizhevsky, A., Sutskever, I. and Hinton, G.E. 2012. ImageNet classification with deep convolutional neural networks. *Advances in Neural Information Processing Systems*, 1–9. http://dx.doi.org/10.1016/j.protcy.2014.09.007.

Le, T.H. and Caracoglia, L. 2015. Reduced-order wavelet-galerkin solution for the coupled, nonlinear stochastic response of slender buildings in transient winds. *Journal of Sound and Vibration*, 344: 179–208. https://doi.org/10.1016/j.jsv.2015.01.007.

LeCun, Y., Bengio, Y. and Hinton, G. 2015. Deep learning. *Nature*, 521: 436. https://doi.org/10.1038/nature14539.

Li, S., Snaiki, R. and Wu, T. 2021. A knowledge-enhanced deep reinforcement learning-based shape ptimizer for arodynamic mitigation of wind-sensitive structures. *Computer-Aided Civil and Infrastructure Engineering*, 36(6): 733–746.

Li, T., Wu, T. and Liu, Z. 2020. Nonlinear unsteady bridge aerodynamics: Reduced-order modeling based on deep LSTM networks. *Journal of Wind Engineering and Industrial Aerodynamics*, 198: 104116.

Linse, D.J. and Stengel, R.F. 1993. Identification of aerodynamic coefficients using computational neural networks. *Journal of Guidance, Control, and Dynamics*, 16: 1018–1025. https://doi.org/10.2514/3.21122.

Lu, Xiao, Lu, Xinzheng, Guan, H. and Ye, L. 2013. Comparison and selection of ground motion intensity measures for seismic design of super high-rise buildings. *Advances in Structural Engineering*, 16: 1249–1262. https://doi.org/10.1260/1369-4332.16.7.1249.

Luo, J., Wierschem, N.E., Fahnestock, L.A., Spencer, B.F., Quinn, D.D., McFarland, D.M., Vakakis, A.F. and Bergman, L.A. 2014. Design, simulation, and large-scale testing of an innovative vibration mitigation device employing essentially nonlinear elastomeric springs. *Earthquake Engineering and Structural Dynamics*, 43: 1829–1851. https://doi.org/10.1002/eqe.2424.

Mao, J., Xu, W., Yang, Y., Wang, J., Huang, Z. and Yuille, A. 2014. Deep Captioning with multimodal recurrent neural networks (m-Rnn). *arXiv preprint arXiv*:1412.6632.

Mao, Z. and Todd, M. 2012. A model for quantifying uncertainty in the estimation of noise-contaminated measurements of transmissibility. *Mechanical Systems and Signal Processing*, 28: 470–481.

Mead, C. and Ismail, M. 2012. *Analog VLSI Implementation of Neural Systems*. Springer Science and Business Media.

Mittal, S. and Vetter, J.S. 2015. A survey of CPU-GPU heterogeneous computing techniques. *ACM Computing Surveys (CSUR)*, 47: 1–35.

Mogren, O. 2016. C-Rnn-Gan: Continuous recurrent neural networks with adversarial training. *arXiv preprint arXiv*:1611.09904.

Pascanu, R., Gulcehre, C., Cho, K. and Bengio, Y. 2013. How to construct deep recurrent neural networks. *arXiv preprint arXiv*:1312.6026.

Peherstorfer, B. and Willcox, K. 2015. Dynamic data-driven reduced-order models. *Computer Methods in Applied Mechanics and Engineering*, 291: 21–41. https://doi.org/10.1016/j.cma.2015.03.018.

Pei, J.S., Wright, J.P. and Smyth, A.W. 2005. Mapping polynomial fitting into feedforward neural networks for modeling nonlinear dynamic systems and beyond. *Computer Methods in Applied Mechanics and Engineering*, 194: 4481–4505. https://doi.org/10.1016/j.cma.2004.12.010.

Psichogios, D.C. and Ungar, L.H. 1992. A hybrid neural network-first principles approach to process modeling. *AIChE Journal*, 38: 1499–1511. https://doi.org/10.1002/aic.690381003.

Raissi, M., Perdikaris, P. and Karniadakis, G.E. 2017a. Numerical Gaussian processes for time-dependent and non-linear partial differential equations. *arXiv preprint arXiv*:1703.10230.

Raissi, M., Perdikaris, P. and Karniadakis, G.E. 2017b. Physics informed deep learning (Part I): Data-driven solutions of nonlinear partial differential equations. *arXiv preprint arXiv*:1711.10561.

Rhif, M., Ben Abbes, A., Farah, I.R., Martínez, B. and Sang, Y. 2019. Wavelet transform application for/in non-stationary time-series analysis: A review. *Applied Sciences*, 9: 1345.

Romine, C.H. and Peyton, B.W. 1997. *Computing Connection Coefficients of Compactly Supported Wavelets on Bounded Intervals*. United States: Oak Ridge National Laboratory.

Rumelhart, D.E., Hinton, G.E. and Williams, R.J. 1985. *Learning Internal Representations by Error Propagation*. California: California Univ San Diego La Jolla Inst for Cognitive Science.

Saitua, F., Lopez-Garcia, D. and Taflanidis, A.A. 2018. Optimization of height-wise damper distributions considering practical design issues. *Engineering Structures*, 173: 768–786. https://doi.org/10.1016/j.engstruct.2018.04.008.

Saravanan, N. and Ramachandran, K.I. 2010. Incipient gear box fault diagnosis using discrete wavelet transform (DWT) for feature extraction and classification using artificial neural network (ann). *Expert Syst. Appl.*, 37: 4168–4181. https://doi.org/10.1016/j.eswa.2009.11.006.

Sezer, O.B. and Ozbayoglu, A.M. 2018. Algorithmic financial trading with deep convolutional neural networks: Time series to image conversion approach. *Applied Soft Computing*, 70: 525–538. https://doi.org/10.1016/j.asoc.2018.04.024.

Shin, Y., Darbon, J. and Karniadakis, G.E. 2020. On the convergence of physics informed neural networks for linear second-order elliptic and parabolic type pdes. *arXiv preprint arXiv*:2004.01806.

Simonyan, K., Vedaldi, A. and Zisserman, A. 2013. Deep inside convolutional networks: Visualising image classification models and saliency maps. *arXiv preprint arXiv*:1312.6034.

Siringoringo, D.M. and Fujino, Y. 2017. Wind-induced responses and dynamics characteristics of an asymmetrical base-isolated building observed during typhoons. *Journal of Wind Engineering and Industrial Aerodynamics*, 167: 183–197. https://doi.org/10.1016/j.jweia.2017.04.020.

Snaiki, R. and Wu, T. 2019. Knowledge-enhanced deep learning for simulation of tropical cyclone boundary-layer winds. *Journal of Wind Engineering and Industrial Aerodynamics*, 194: 103983.

Spence, S.M.J. and Kareem, A. 2014. Performance-based design and optimization of uncertain wind-excited dynamic building systems. *Engineering Structures*, 78: 133–144. https://doi.org/10.1016/j.engstruct.2014.07.026.

Spiridonakos, M.D. and Chatzi, E.N. 2015. Metamodeling of dynamic nonlinear structural systems through polynomial chaos NARX models. *Computers and Structures*, 157: 99–113. https://doi.org/10.1016/j.compstruc.2015.05.002.

Spottswood, S. and Allemang, R. 2006. Identification of nonlinear parameters for reduced order models. *Journal of Sound and Vibration*, 295: 226–245.

Subasi, A. 2007. EEG signal classification using wavelet feature extraction and a mixture of expert model. *Expert Syst. Appl.*, 32: 1084–1093. https://doi.org/10.1016/j.eswa.2006.02.005.

Suksuwan, A. and Spence, S.M.J. 2018. Performance-based multi-hazard topology optimization of wind and seismically excited structural systems. *Engineering Structures*, 172: 573–588. https://doi.org/10.1016/j.engstruct.2018.06.039.

Sun, Z. and Chang, C.C. 2004. Statistical wavelet-based method for structural health monitoring. *Journal of Structural Engineering*, 130: 1055–1062.

Tabbuso, P., Spence, S.M.J., Palizzolo, L., Pirrotta, A. and Kareem, A. 2016. An efficient framework for the elasto-plastic reliability assessment of uncertain wind excited systems. *Structural Safety*, 58: 69–78. https://doi.org/10.1016/j.strusafe.2015.09.001.

Tamura, Y. and Kareem, A. 2013. *Advanced Structural Wind Engineering*. Japan: Springer.

Tamura, Y., Kohsaka, R., Nakamura, O., Miyashita, K.I. and Modi, V.J. 1996. Wind-induced responses of an airport tower: Efficiency of tuned liquid damper. *Journal of Wind Engineering and Industrial Aerodynamics*, 65: 121–131. https://doi.org/10.1016/S0167-6105(97)00028-7.

Tao, H., Wang, P., Chen, Y., Stojanovic, V. and Yang, H. 2020. An unsupervised fault diagnosis method for rolling bearing using Stft and generative neural networks. *Journal of the Franklin Institute*, 357: 7286–7307.

Vaidyanathan, C.V., Kamatchi, P. and Ravichandran, R. 2005. Artificial neural networks for predicting the response of structural systems with viscoelastic dampers. *Computer-Aided Civil and Infrastructure Engineering*, 20: 294–302. https://doi.org/10.1111/j.1467-8667.2005.00395.

Wang, S. and Mahin, S.A. 2018. High-Performance computer-aided optimization of viscous dampers for improving the seismic performance of a tall building. *Soil Dynamics and Earthquake Engineering*, 113: 454–461. https://doi.org/10.1016/j.soildyn.2018.06.008.

Wu, R.-T. and Jahanshahi, M.R. 2019. Deep convolutional neural network for structural dynamic response estimation and system identification. *Journal of Engineering Mechanics*, 145: 04018125. https://doi.org/10.1061/(ASCE)EM.1943-7889.0001556.

Wu, T. and Kareem, A. 2011. Modeling hysteretic nonlinear behavior of bridge aerodynamics via cellular automata nested neural network. *Journal of Wind Engineering and Industrial Aerodynamics*, 99: 378–388. https://doi.org/10.1016/j.jweia.2010.12.011.

Wu, T. and Snaiki, R. 2022. Applications of machine learning to wind engineering. *Frontiers in Built Environment*, 8: 811460. https://doi: 10.3389/fbuil.2022.811460.

Wu, T., Kareem, A. and Ge, Y. 2013. Linear and nonlinear aeroelastic analysis frameworks for cable-supported bridges. *Nonlinear Dynamics*, 74: 487–516. https://doi.org/10.1007/s11071-013-0984-7.

Xiong, C., Lu, X., Guan, H. and Xu, Z. 2016. A nonlinear computational model for regional seismic simulation of tall buildings. *Bulletin of Earthquake Engineering*, 14: 1047–1069. https://doi.org/10.1007/s10518-016-9880-0.

Zeng, X., Peng, Y. and Chen, J. 2017. Serviceability-based damping optimization of andomly wind-excited high-rise buildings. *Structural Design of Tall and Special Buildings*, 26: 17. https://doi.org/10.1002/tal.1371.

Zhang, L., Lu, Xinzheng, Guan, H., Xie, L. and Lu, Xiao. 2017. Floor acceleration control of super-tall buildings with vibration reduction substructures. *Structural Design of Tall and Special Buildings*, 26: 13. https://doi.org/10.1002/tal.1343.

Zhang, W. and Xu, Y. 2001. Closed form solution for alongwind response of actively controlled tall buildings with LQG controllers. *Journal of Wind Engineering and Industrial Aerodynamics*, 89: 785–807. http://dx.doi.org/10.1016/S0167-6105(01)00068-X.

Zhang, Y., Li, L., Guo, Y. and Zhang, X. 2018. Bidirectional wind response control of 76-story benchmark building using active mass damper with a rotating actuator. *Structural Control and Health Monitoring*, 25: e2216. https://doi.org/10.1002/stc.2216.

Zhao, N., Huang, G., Yang, Q., Zhou, X. and Kareem, A. 2019. Fast convolution Iitegration–based nonstationary response analysis of linear and nonlinear structures with nonproportional damping. *Journal of Engineering Mechanics*, 145: 04019053. https://doi.org/10.1061/(ASCE)EM.1943-7889.0001633.

Zhou, K., Lei, D., He, J., Zhang, P., Bai, P. and Zhu, F. 2021. Real-time localization of micro-damage in concrete beams using DIIC technology and wavelet packet analysis. *Cement and Concrete Composites*, 123: 104198.

Chapter 13

Damage Detection in Reinforced Concrete Girders by Finite Element and Artificial Intelligence Synergy

Hayder A. Rasheed,[1],* Ahmed Al-Rahmani[2] and AlaaEldin Abouelleil[3]

1. Introduction

The structural deterioration of aging infrastructure systems is becoming an increasingly important problem worldwide. To compound this matter, economic strains limit the resources available for repair or replacement of such systems. Over the past several decades, structural health monitoring (SHM) has proven to be a cost-effective tool for the detection and evaluation of damage in structures. Visual inspection and condition rating are one of the most applied SHM techniques. However, the effectiveness of SHM varies depending on the availability and experience of qualified personnel and largely qualitative damage evaluations (Al-Rahmani, 2012).

Most of the researchers agree on relying on reference parameters extracted from healthy beams for comparison. However, they differ in defining these parameters, which could be categorized into dynamic and static parameters. Ndambi et al. (2002) investigated the potential for use of Eigen frequencies and mode shape-derived properties to detect the location and estimate the severity of damage in reinforced concrete beams. The results of their tests suggest that Eigen frequencies of reinforced concrete members can be numerically correlated to changes in stiffness, allowing damage to be quantified with non-destructive test methods. Teughels et al. (2002) presented an approach to damage quantification using Finite Element (FE) model updating. Results of this research demonstrated that modal properties can be used with processing and analysis techniques, such as FE model updating, to reliably quantify

[1] Department of Civil Engineering, Kansas State University, Manhattan, KS, USA-66506.
[2] SK&A Structural Engineers, Potomac, MD, USA-20854. Email: ahmedr@skaengineers.com
[3] AEDA LLC, Manhattan, KS, USA-66503. Email: aladdin10@ksu.edu
* Corresponding author: hayder@k-state.edu

damage in reinforced concrete structures. Ghods and Esfahani (2009) conducted static loading tests and dynamic modal analyses on eight reinforced concrete beams with varying compressive strengths and reinforcement ratios. This experiment highlighted the ability of modal analyses to characterize the specific damage level in reinforced concrete beams. Reynders and De Roeck (2009) applied modal analysis parameters to develop a flexibility-based damage assessment method for beam members. The conclusion was made that the modal data-based flexibility method shows promise as a nondestructive method for quantifying damage in concrete beams. Tan et al. (2020) developed and applied a procedure for detecting damage in a composite slab-on-girder bridge structure comprising of a reinforced concrete slab and three steel I beams, using vibration characteristics and Artificial Neural Network (ANN). The procedure is applied across a range of damage scenarios and the results confirm its feasibility to detect and quantify damage in composite concrete slab on steel girder bridges.

Researchers have also used wave response methods to localize and quantify damage in reinforced concrete structures. One such method, acoustic emission (AE), uses mechanical loading to cause damaged material sections to emit elastic waves that are read by surface sensors. Sagar et al. (2012) utilized the AE technique to assess damage in three experimental reinforced concrete beams subjected to cyclic loading stages. Four AE sensors were attached to each beam to gather wave response data during each loading cycle to evaluate damage severity in the beams. Although the experiment was limited to three experimental beams, the researchers proposed that AE techniques display significant potential for structural damage assessment applications. Shiotani et al. (2012) utilized AE and elastic wave tomography techniques to assess the structural health of an experimental reinforced concrete bridge deck subjected to fatigue damage. Fatigue loading was simulated by load stages of 0, 10,000, and 20,000 passes of a 100 kN moving wheel load, and wave data was recorded for the bridge deck under incremental static load steps after each fatigue loading stage. Elastic wave tomography and AE analyses of two *in situ* bridge decks substantiated their findings, and wave velocities and natural frequencies again showed great promise as damage quantification parameters. Ongpeng et al. (2019) monitored the damage in 18 reinforced concrete beams with varying water cement ratios and reinforcements tested under four-point bending test. Repeated step loads were designed and at each step load AE signals were recorded and processed to obtain the acoustic emission source location (AESL). The computed convex hull volume (CHV) showed good relation to the damage encountered until 60% of the ultimate load at the midspan was reached.

Researchers have also utilized digital imaging techniques such as digital image correlation (DIC), light detection and ranging (LiDAR), and fractal analysis to evaluate damage in reinforced concrete structures. DIC relies on algorithms that process data from high resolution images to measure surface displacements and strains. Li et al. (2008) used DIC to detect cracks and assess damage for several experimental reinforced concrete beams. Loland's model was selected to quantify damage and damage evolution, and data recorded through DIC was applied in FE model updating to define the initial damage and material parameters in the damaged model. The results of this experiment suggested that DIC combined with techniques such as FE model updating can effectively quantify structural damage. Applications of 3D LiDAR technology in damage detection and quantification were investigated in three case studies summarized by Chen et al. (2012). One case study of a bridge in Iowa highlighted the ability of LiDAR to not only detect cracks, but also to describe their precise location and dimensions.

Researchers have also investigated fusion of ANNs and the FE model updating approach. Jeyasehar and Sumangala (2006) investigated the ability of ANNs to quantitatively predict damage in an experiment involving five post-tensioned prestressed concrete beams. Damage in the beams was introduced by snapping a variable percentage of the wires through inducing localized pitting corrosion at a crack former placed at the center of the beams, and the beams were subjected to static loading stages of increasing magnitude. The results showed that the damage level assessment error could be reduced with the introduction of static test data to the ANN instead of dynamic test data. Hasançebi and Dumlupinar (2013) discussed the potential of using ANNs to perform FE model updating operations for reinforced concrete T-beam bridges. The researchers developed analytical FE bridge models with various boundary stiffnesses, elastic moduli, and deck masses. Changes in these properties from an initial state were used to characterize damage and assess structural health. The researchers concluded that ANNs can be successfully used in conjunction with FE model updating to quantitatively evaluate damage, provided they utilize both static and dynamic test results.

This chapter proposes the utilization of a synergy between the Finite Element (FE) simulations and the modeling and prediction powers of Artificial Intelligence (AI) to realize a paradigm shift in the quality of automated and objective structural health monitoring evaluations and intelligent damage detection capabilities. The FE simulations serve as a numerical tool to establish the link and connection between the spatial deterioration of stiffness at discrete nodes in a damaged girder and the actual cracking or damage in such reinforced concrete members. The goal is to model this relationship using AI for the purpose of recovering one set of the parameters in terms of the other one. The beauty of this synergy is that it develops the necessary database relationship for solving two completely independent problems with separate applications, namely:

1. The Forward Problem: In this problem, an objective methodology is established to extract a unique health index parameter from the actual measurement of accessible cracking distribution in the reinforced concrete girders. By using numerical models with physical damage to generate the stiffness degradations along the damaged girder as ratios of their healthy stiffness counterparts, a robust estimate of the stiffness degradation can be recovered. The application of this problem is related to annual inspection of girders for which damage is accessible and can be measured by inspectors. This problem is easier to handle and thus it is left to future developments.

2. The Inverse Problem: In this problem, an objective methodology is established to extract a potential damage distribution based of a certain stiffness deterioration pattern along the girder. By using numerical models with physical damage to

generate the stiffness degradation patterns along the damaged girder as ratios of their healthy stiffness counterparts, a potential estimate of the possible cracking or damage may be proposed if the stiffness degradation pattern is identified. The application of this problem is related to damage detection in non-accessible girders crossing rivers and creeks while the top surface of the girders is within reach. To generate the stiffness deterioration profile, a numerical model is established to mimic the response of the healthy girder while field measurements are applied to extract the response of the actual damaged girder. This can be enabled by using a theodolite, a pointwise scale and a vehicle inducing known weight. This problem does not yield a unique solution and thus it is harder to tackle. Therefore, it is the focus of the attention and treatment of this chapter.

To prove the applicability of this concept, the whole range of numerical models is examined here. First, two dimensional reinforced concrete girders with rectangular cross sections are used to model the damaged members in terms of beam elements and 2D FE meshes. Then, 3D reinforced concrete girders with T cross sections are utilized to model the damaged members in terms of 3D solid FE meshes. The database included varying the geometric and material parameters of the girders as well as the cracking properties in terms of crack location, crack depth and crack width. All beam and finite element numerical simulations are performed in the multi-physics finite element program Abaqus. In the following sections, an introduction to ANN modeling, the methodology of the AI-FE technique and the establishment of the 2D and 3D finite element analysis databases are presented in detail. Sections highlighting the results and discussions, numerical examples and software developments follow. Finally, a chapter recap is provided summarizing the findings and listing recommendations for current and future developments.

2. Artificial Neural Networks

Artificial Neural Network (ANN) is a computational system designed to simulate how the human brain receives, processes, and concludes information. ANN accuracy, like the human brain, highly depends on the number of exposures/datasets experienced. Larger datasets typically produce networks with enhanced prediction accuracies when compared to ANNs produced from limited data. This learning process enables ANN to generalize complex nonlinear relationships for scenarios that were not exercised in the learning phase without extrapolation in the input information.

2.1 Structure and Learning Techniques

ANNs are built with neuron nodes interconnected like a web (Fig. 13.1). These nodes are responsible for processing the information in parallel and they are usually divided in layers or clusters. The distribution of these layers and how they interact with each other, as well as the interconnection inside each layer, results in a countless number

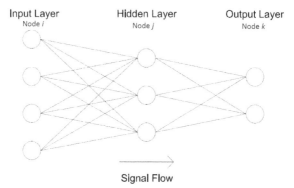

Fig. 13.1: General feedforward ANN architecture.

of ANN structures and learning techniques. Most network structures consist of at least three layers of nodes: an input layer, one or more hidden layers, and an output layer, Fig. 13.1. The input layer provides defining or introductory information to the ANN. On the other hand, the hidden layer(s) are responsible of the learning process through several mathematical operations performed within these hidden layer(s) and their connection weights to the input and output layers. Finally, the output layer extracts the results from the ANN. The two primary types of ANN structures are feedforward (static) and recurrent (dynamic) architectures. In feedforward networks, signal flow is unidirectional from the input layer to the output layer, and nodes within the different layers are not interconnected, see Fig. 13.1. Nodes within layers can be interconnected for recurrent networks, and output signals are transmitted back into the ANN in a variety of loop configurations. It is decided in this chapter to refer to any node in the input layer by node (i), any node in the hidden layer by node (j), and any node in the output layer by node (k) for the purpose of clarity of presentation, see Fig. 13.1. It is also important to note that the words input (I) and output (O) refer in this treatment to the start and end of a process, respectively.

There are three common learning techniques typically followed by ANN processes. These are classified based on the evaluation algorithm of the accuracy of the outputs. These three learning techniques are supervised, reinforced, and unsupervised learning. The outputs/predictions in the supervised learning are judged against the actual data in the training phase. Errors associated with these predictions are then used to adjust the connection weights for a higher accuracy. Reinforced learning is generally akin to the supervised learning, however, only evaluations of the quality of the predicted outputs are used to adjust connection weights and actual outputs are not provided to the ANN. In the unsupervised method, learning occurs as the network explores patterns in the input data and build correlations. The inverse method treated in this chapter adapts a feedforward supervised learning-based ANN.

2.2 *Data Preparation and Training Procedure*

The datasets which are used to build the ANN are divided into three phases: training, testing, and validating sets. The training dataset serves as the first step of the learning process. In this step, the ANN goes through a series of iterations of connection weight adjustments. It is vital to assign the datasets entries containing the minimum and maximum values of input parameters in the training step to ensure a full exposure for the ANN and a meaningful interpolation, rather than extrapolation, of new input events. After assigning the appropriate connection weights, the ANN undergoes evaluation to predict the accuracy of the network over the testing set. These two steps are applied for many ANNs with different architectures (e.g., number of hidden layers) and configurations (e.g., learning rates). To choose the optimum ANN, the top-performing network architectures from the training and testing stages are reevaluated over the validating datasets.

The ANN training process could be described as a trial-and-error process. Each node in the input layer is connected to every node in the hidden layer, and connection weights between the nodes are randomly assigned during the first iteration. The jth hidden node receives an output signal (O_j) equal to the summation of the values from the input layer (I_i) multiplied by their corresponding connection weights (w_{ji}) plus the bias (b_j), or threshold, associated with the hidden node (j) as represented by the following equation:

$$O_j = \sum_{i=1}^{n} w_{ji} I_i + b_j \qquad (13.1)$$

where:

O_j = output value for node j

W_{ji} = connection weight between nodes i in the input layer and j in the hidden layer

I_i = Input value at node I of the input layer

b_j = bias for node j of the hidden layer

An activation function then processes the new signal of the hidden nodes (j) to eliminate large or negative values and expose the ANN to nonlinearity. The activation function utilized by the ANN in this chapter was the sigmoidal function. This function normalizes new values in the hidden nodes below (–5) to approximately –0 and above (5) to approximately 1, see Fig. 13.2. Within the range of values of (–5, 5), the function yields a nonlinear variation in the range of (0, 1). The sigmoidal function is applied using following equation while setting each O_j from Eq. (13.1) to be named I_j now:

$$O_j = f(I_j) = \frac{1}{1 + e^{-I_j}} \qquad (13.2)$$

where:

O_j = activation function output at node j of the hidden layer

$f(I_j)$ = sigmoidal activation function

I_j = computed value for node j of the hidden layer, originally O_j.

Outputs from the activation function in hidden layer (O_j) are first named (I_j) then multiplied by their corresponding connection weights, summed, and added to the

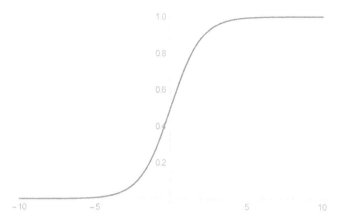

Fig. 13.2: Sigmoidal function plot.

biases associated with the nodes of the output layer to obtain the new values for the output layer (O_k), see Eq. (13.3):

$$O_k = \sum_{j=1}^{n} w_{kj} I_j + b_k \qquad (13.3)$$

These resulting values are then processed by the activation function and de-normalized to provide the predicted output values for the ANN, after setting (O_k) to be (I_k).

$$O_k = f^{-1}(I_k) = \ln(\frac{I_k}{1 - I_k}) \qquad (13.4)$$

Error factors are calculated and used to compute the necessary incremental connection weight adjustments according to the following equation:

$$\Delta w_{kj} = \eta \delta_k O_k + \mu \Delta w_{kj}^{previous} \; (no \; sum \; on \; k) \qquad (13.5)$$

where:

$$\delta_k = (y_k - O_k)(1 - O_k)O_k \qquad (13.6)$$

Δw_{kj} = incremental connection weight adjustment between nodes j of the hidden layer and k of the output layer

η = learning rate

δ_k = error factor of node k of the output layer

μ = momentum rate

$\Delta w_{kj}^{previous}$ = incremental adjustment in connection weight from the previous iteration

y_k = actual value at output node k of the output layer

Incremental changes in the biases of the output layer are also calculated by the following equation:

$$\Delta b_k = \eta \delta_k + \mu \Delta b_k^{previous} \qquad (13.7)$$

where:

Δb_k = incremental bias adjustment for node k of the output layer

$\Delta b_k^{previous}$ = incremental adjustment in bias for node k from the previous iteration

Similarly, error factors are calculated for the hidden nodes using the following equation:

$$\delta_j = (\Sigma_{k=1}^n \delta_k w_{kj})(1 - O_j)O_j \qquad (13.8)$$

where:

δ_k = error factor of output node k calculated by Eq. (13.6)

w_{ki} = connection weight between hidden node and output nodes

O_j = predicted value at node j of the hidden layer from Eq. (13.1)

δ_j = error factor of hidden node j calculated by Eq. (13.8)

Accordingly, the incremental connection weight adjustments between the input nodes and the hidden nodes are:

$$\Delta w_{ji} = \eta \delta_j O_j + \mu \Delta w_{ji}^{previous} \ (no \ sum \ on \ j) \qquad (13.9)$$

Similarly,

$$\Delta b_j = \eta \delta_j + \mu \Delta b_j^{previous} \qquad (13.10)$$

Connection weights and biases are updated for all nodes in the network, and the entire process is repeated for every training dataset until a predetermined number of training iterations is reached or the output errors are reduced to an acceptable level. The ANN completes these iterations over a range of hidden nodes specified by the user. Utilizing too few hidden nodes can result in a network architecture that is unable to solve the problem depending on its complexity, while using too many hidden nodes may result in overtraining, or memorization, of the data. The absolute maximum number of hidden nodes is calculated by the following equation:

$$MHN = \frac{TR - 0}{I + 0 + 1} \qquad (13.11)$$

where:

MHN = maximum number of hidden nodes

TR = number of training datasets

O = number of output parameters

I = number of input parameters

2.3 Optimum Model Selection Criteria

The ANN models are, typically, evaluated according to some statistical measures include the coefficient of determination (R^2), the mean absolute relative error

(MARE), and the average of squared error (ASE). These measures are calculated by the following equations:

$$R^2 = \Sigma^o \left(\frac{n\Sigma^n xy - \Sigma^n x \Sigma^n y}{\sqrt{n\Sigma^n x^2 - (\Sigma^n x)^2} \sqrt{n\Sigma^n y^2 - (\Sigma^n y)^2}} \right)^2 / o \qquad (13.12)$$

$$MARE(\%) = \frac{\Sigma^o \Sigma^n (\frac{|y - x|}{x} \times 100)}{o * n} \qquad (13.13)$$

$$ASE = \frac{\Sigma^o \Sigma^n (y - x)^2}{o * n} \qquad (13.14)$$

where:

x = actual value of a parameter

y = ANN predicted value of a parameter

o = number of outputs

n = number of datasets

More effective ANNs are typically characterized by low MARE and ASE values and high R^2 values. The prediction capabilities of ANNs that display optimum statistical performance and utilize the fewest number of hidden nodes during the training and testing phase are reassessed with the validation datasets. Provided the optimum models perform well in the validation phase, the ANNs are then retrained on all datasets so that predictions can be made using the optimal network architecture trained with the entirety of the available data.

3. Methodology of AI-FE Synergy

To apply the AI treatment to the inverse problem of damage detection described earlier in the introduction section of this chapter, FE analysis is conducted to model the stiffness deterioration profile of those girders. For this work, Abaqus FEA is selected to be the most appropriate structural analysis software package due to its extendibility and scripting capabilities (Abaqus 6.10 Online Documentation, 2010). As it will be covered later in this chapter, this will be critical due to the immense number of models that need to be generated to obtain the datasets to be used in the artificial neural network training process. Cracks in girders are modeled by a change in the cross-section of the simply supported beam as shown in Fig. 13.3. Cracking

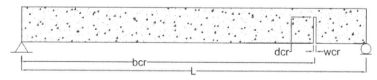

Fig. 13.3: Concrete girder elevation showing cracking parameters considered in the analysis.

parameters used in modeling include the depth (d_{cr}), width (w_{cr}), and location (b_{cr}) of the cracks. This procedure can be expanded to other beam configurations, such as multi-span continuous beams (Al-Rahmani et al., 2013).

Both 2D and 3D FE models are generated in this work. The relative simplicity of the 2D models offer an excellent starting point to establish the damage databases for this method in less time. On the other hand, 3D models allow to further expand this approach and more accurately develop models that represent real-world bridge girders. Starting with 2D modeling, beams are modeled in three phases. Beams with a single and two cracks are modeled in phases I and II, respectively. For these two phases, the 2-node cubic beam in a plane element (B23), which utilizes the classical Euler-Bernoulli assumption and uses cubic interpolation functions, is chosen to model the beam segments in Abaqus FEA. It is understood that this assumption does not hold for non-slender elements; however, due to the constraints of time and computational power, the loss of accuracy expected due to the usage of this element is considered within the tolerance for the purpose of the 2D analysis. Additionally, it is expected that the introduced error would not dramatically affect the results since the obtained values will be normalized as it will be mentioned later in this chapter. For this reason, steel reinforcement is not included in these two phases. The FE mesh developed for these models includes two types of elements: healthy and cracked elements. Healthy elements, which represent the healthy parts of the beam, have the same depth as the beam, while cracked elements have a reduced depth to represent the crack. A visual representation of the mesh is shown in Fig. 13.4.

Fig. 13.4: Finite element mesh for phases I, II.

For phase III, 8-node biquadratic plane stress quadrilateral elements (CPS8) and 6-node quadratic plane stress triangle elements (CPS6) are used in modeling beams with a single crack. Additionally, steel reinforcement is included in the analysis this time. A fixed steel ratio (ρ) of 1% is assumed for the analyzed beams. To determine the optimal mesh size, a sensitivity analysis is performed. Evaluated mesh sizes are 100 mm, 50 mm, 25 mm, and 10 mm. A healthy beam with a width of 250 mm, depth of 500 mm, span length of 3.5 m, and compressive strength of 21 MPa is analyzed as a reference. Next, two cracked beams with the same geometric and material parameters and different cracking parameters are modeled. The first beam (C1) has a 200 mm deep, and a 2.5 mm-wide crack located 1.4 m away from the left support of the beam, while the second beam (C2) has a 375 mm deep, and 5 mm wide crack located 0.7 m away from the left support of the beam. The stiffness ratios are calculated at 5 stiffness nodes for each beam while varying the mesh size. The percentage change in the stiffness ratios (defined later in this chapter) moving from mesh size 1 (MS1) to mesh size 2 (MS2) is shown in Table 13.1.

Overall, the obtained variations are less than 1%. It should be noted that running the analysis for a model meshed with 100 mm elements took quarter the time it took

Table 13.1: 2D-modeling sensitivity analysis results.

	MS1	MS2	Δk1 (%)	Δk2 (%)	Δk3 (%)	Δk4 (%)	Δk5 (%)
C1	100	50	0.32722%	1.03696%	0.85103%	0.43108%	0.05413%
	50	25	0.33583%	0.83787%	0.72907%	0.36756%	0.01824%
	25	10	0.26779%	0.71102%	0.66568%	0.36886%	0.28775%
C2	100	50	0.58492%	0.70983%	0.33187%	0.17025%	0.04872%
	50	25	0.85723%	0.44899%	0.22606%	0.06457%	−0.04472%
	25	10	0.69113%	0.86437%	0.43349%	0.23939%	0.13328%

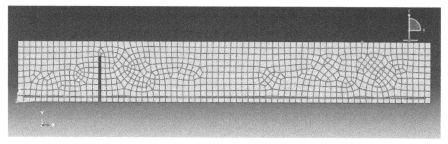

Fig. 13.5: Finite element mesh for a beam in phase III.

the model with 10 mm elements to complete. The mesh size of 50 mm is chosen as a middle ground as it provided acceptable accuracy with reasonable running time. Figure 13.5 shows the mesh generated automatically by Abaqus for a beam with a mesh size of 50 mm.

Elastic material models are used to model steel and concrete in this work. Steel is modeled as a linear elastic material with a Poisson's ratio (v) of 0.3 and a modulus of elasticity (E) of 200 GPa. Concrete is also modeled using a linear elastic material model with a Poisson's ratio (v) of 0.2 and a modulus of elasticity (E) calculated by the following equation:

$$E = 4723\sqrt{f'_c}$$
(13.15)

where: E = *modulus of elesticity (MPa)*

f'_c = 28-*day compressive strength of concrete (MPa)*

In all phases, a specified number of stiffness nodes are added to the mesh as shown in Fig. 13.6 (in phase III, the stiffness nodes are added to the top surface of the beam). A defined load is applied at each stiffness node and the resulting displacement is obtained from the analysis as shown in Fig. 13.6 through Fig. 13.8. Finally, the stiffness at the node is calculated by dividing the applied load by the obtained displacement at that node according to the following equation:

$$k_n = P_n / \Delta n$$
(13.16)

where: k_n = *stiffness value at node n*

P_n = *load applied at node n*

Δn = *deflection obtained at node n*

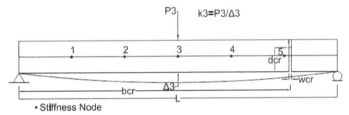

Fig. 13.6: Beam elevation with 5 stiffness nodes in phase 1.

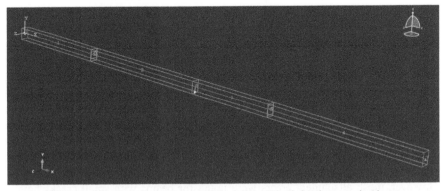

Fig. 13.7: A 3D view of a concrete beam from phase II loaded at the center in Abaqus.

Fig. 13.8: Elevation view of a beam from phase III in Abaqus.

This is done for the healthy and the cracked beams to determine the stiffness ratios, which are obtained by dividing the cracked stiffness by the healthy stiffness at each node for beams with the same geometric and material parameters, as shown in the following formula:

$$k_n\% = k_{n_{cr}} / k_{n_h} \tag{13.17}$$

where: $k_n\%$ = *stiffness ratio at node n*

$k_{n_{cr}}$ = *stiffness value at node n in the cracked beam*

k_{n_h} = *stiffness value at node n in the healthy beam*

Stiffness ratios ($k_n\%$) serve as a localized indicator of the severity of the damage in the beam and can reveal where the crack could be in the beam. Lower stiffness ratios are expected in beams with deeper and wider cracks. Stiffness ratios at nodes closer to the location of the crack are expected to be lower compared to the ratios at nodes further away. After performing a parametric study, a total of nine nodes was needed to offer the optimum computational performance.

Moving on to the 3D analysis, conventionally reinforced concrete T-beams were modeled. These beam models were composed of 3D solid parts and were meshed with 10-node quadratic tetrahedral (C3D10) elements utilizing an automatic meshing technique. Reinforcement was modeled using 3D wire parts meshed with 3-node truss elements (T3D3), and three equally spaced reinforcing bars were embedded in all concrete beam models with a perfect non-slip bond between the concrete and the reinforcing bars. Figures 13.9 and 13.10 show 3D ABAQUS views of a sample of reinforced concrete T-beam. Reinforcement was present in the sample beam but was not depicted in these figures.

The same linear elastic model assumptions made for the 2D analysis was used for the 3D analysis. To determine the optimal mesh size for the 3D models, a mesh sensitivity analysis was performed. The midspan deflection results for an uncracked sample 3D T-beam under a constant load with mesh sizes ranging from 50 mm–25 mm are plotted in Fig. 13.11. Sample ABAQUS analysis times for the beam with mesh sizes of 48 mm, 35 mm, and 26 mm were 0:40, 1:00, and 2:13 (hours: minutes), respectively. From Fig. 13.11, the variation in the deflection results is insignificant with respect to the variation in ABAQUS analysis time. However, mesh attachment errors were noted for beams with small cracks having

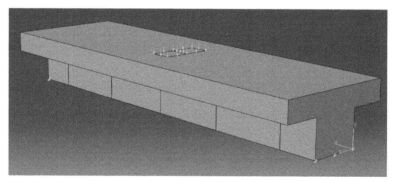

Fig. 13.9: ABAQUS view of 3D reinforced concrete T-beam showing five cracks along span, support conditions and applied load.

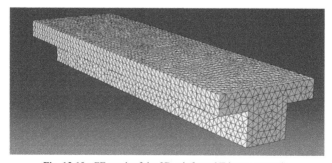

Fig. 13.10: FE mesh of the 3D reinforced T-beam example.

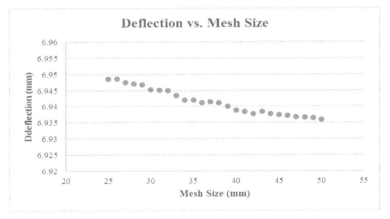

Fig. 13.11: Variation in beam deflection with reduction of mesh size.

mesh sizes of 50 mm. Testing revealed that the frequency of mesh attachment errors reduced with decreasing mesh size. Therefore, the mesh size is set at 35 mm for both the concrete and steel elements in the 3D beam models.

For the 3D models, nine stiffness nodes were inserted along the top face of the beam at equidistant spacing. This number of stiffness nodes was suggested by the 2D analysis Phase III to yield very accurate beam health models. Each stiffness node was subjected to a static load in turn, and deflection results at the stiffness nodes were extracted from the ABAQUS analysis. The load was applied as an area load over a small section of the top face of the beam, centered on the stiffness node, for the 3D analyses. This scheme of loading configurations was necessary because the application of a point load for 3D beam models resulted in very high relative deflections at the stiffness node, suggesting that localized deformation was significant. For all models, the stiffness at a given node was found by dividing the total applied force by the deflection at that node according to Eq. (13.16).

As earlier described, nodal stiffness ratios were used to quantitatively describe the localized damage level, or residual structural health, of the reinforced concrete beams. In general, beams with low nodal stiffness ratios have deep, wide, or an extensive number of cracks. Stiffness ratios of nodes near existing cracks are typically lower than stiffness ratios of nodes far from damage locations. It was observed that cracks placed very close to the beam supports resulted in unrealistically low nodal stiffness ratios in 3D models. This can be, in part, attributed to be a failure to meet Saint-Venant's principle, which suggests that the stress distribution in a material may be assumed to be independent of the manner of load application, except in the immediate vicinity of the applied loads (Beer et al., 2012). The presence of both an applied load and a discontinuity (crack) in the beam close to the supports likely introduced stress and strain concentrations that contributed to the unexpected deflection of the beam. Therefore, crack placement was limited to 0.06–0.94 of the beam span length in 3D models; a range in which the nodal stiffness ratios were generally observed to stabilize to reasonable values.

With the FE modeling procedure established, the next section will cover the generation of the databases and the training process.

4. Database Generation and AI Implementation

4.1 2D-Analysis Databases

Concrete beams with different parameters are modeled to generate the 2D-analysis damage databases. These parameters include geometric parameters such as the width of the cross-section (b), the depth of the cross-section (d) and the span length of the beam (L), a material parameter represented by the concrete compressive strength (f'_c), and cracking parameters including the depth (d_{cr}), width (w_{cr}) and location (b_{cr}) of the cracks. Most parameters are normalized so that the database could be generalized to beams not included in this study but are within the range of the normalized modeled data.

The 2D modeling study is conducted in three phases. In phases I and III, damage databases are generated for beams with a single crack in the first half of the beam. In phase II, damage databases are generated for beams with two cracks. To regulate the possibilities of this case, beams' spans in phase II are divided into two regions. Only a single crack could exist in each region, thus a beam can have two, one or no cracks at all. Figure 13.12 shows an example of a beam with two cracking regions. Lists of the parameters and the associated values used to generate the mentioned damage databases are given in Tables 13.2 and 13.3.

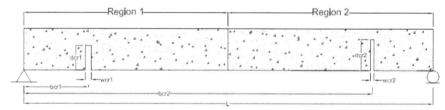

Fig. 13.12: Elevation view of a sample concrete beam with two cracks.

Table 13.2: Phase I, III modeling parameters (beams with a single crack).

b/h	L/h	f'$_c$ (MPa)	d$_{cr}$/h	w$_{cr}$ (mm)	b$_{cr}$/L
0.5	7	21	0.25	0.5	0.1
0.7	10	31	0.4	1	0.2
0.9	13	41	0.5	2.5	0.3
–	–	–	0.75	5	0.4
–	–	–	–	–	0.5

Table 13.3: Phase II modeling parameters (beams with two cracks).

b/h	L/h	f'$_c$ (MPa)	Crack 1			Crack 2		
			d$_{cr}$/h	w$_{cr}$ (mm)	b$_{cr}$/L	d$_{cr}$/h	w$_{cr}$ (mm)	b$_{cr}$/L
0.5	7	21	0.25	0.5	0.167	0.25	0.5	0.667
0.7	10	31	0.5	2.5	0.333	0.5	2.5	0.833
0.9	13	41	0.75	5	0.5	0.75	5	–

Table 13.2 shows that for phases I and III, cracks were modeled in the first half of the beam's span only, utilizing the symmetry of the cracks' distribution, i.e., a beam with a crack located at 0.1 L will have the same stiffness values as a beam with the same crack located at 0.9 L, except that they will be in a reversed order. This was done to increase the accuracy of the ANN's predictions as well as to reduce the computation time. This does not apply in phase II, as the cracks are independently distributed in each half of the beam. For phases I and III, a healthy beam was identified by having a crack location to beam span ratio (b_{cr}/L) of 0. For phase II, beams without a crack in the first region had a crack location to beam span ratio (b_{cr}/L) of 0, while beams with no cracks in the second region had a crack location to beam span ratio (b_{cr}/L) of 0.5. As previously mentioned, a specific number of stiffness nodes was included in the finite element mesh. For phases I and III, databases for 3, 5, 7, and 9 stiffness nodes were generated. Utilizing the results from phase I, databases were generated for the optimum only at 9 stiffness nodes. An MS Excel VBA (Visual Basic for Applications) macro was written to generate Abaqus input files (*.inp files) by varying the parameters mentioned earlier for phases I and II. For phase III, a Python script was written to generate the input files directly by interfacing with Abaqus. For all three phases, a Python script was used to batch run the created input files in Abaqus, and another Python script was written to extract the output deflections from Abaqus binary output databases (*.odb files), determine the stiffness values corresponding to the stiffness nodes, normalize them with the healthy beam stiffness values and store them in the stiffness database. The normalized stiffness values, or stiffness ratios (kn%), were calculated as the ratio of the stiffness at a node in the cracked beam to the stiffness at the same node in the healthy beam. These ratios serve as an indicator of the damage in the beam and can reveal in which half of the beam the crack is located.

Moving to ANN modeling and training, the first step was to label the datasets and parameters. ANNs with the stiffness ratios (k_n%) and the beam parameters (b/h, L/h, and f'$_c$) as inputs and the crack parameters (d_{cr}/h, w_{cr}, and b_{cr}/L) as outputs were created. This is an inverse problem for which no unique solution exists. The accuracy of the ANNs is expected to improve with the increase in the number of stiffness nodes. Also, it is expected that better accuracy will be obtained by decreasing the number of required output parameters. This problem was solved in all three phases. A sample header file is shown in Table 13.4. Inputs and outputs are identified by 'I', and 'O', respectively. Flag values of 1, 2, or 3 indicate whether the dataset is used in training, testing or validation, respectively.

Following the ANN modelling methodology described earlier in this chapter, several ANN models were evaluated. For all phases, the maximum number of hidden nodes was taken to be 20. Also, the initial number of hidden nodes was taken from 2 to 10. In phase I, each damage database contained 2,187 datasets that corresponded to the generated Abaqus beam models. The datasets included 27 healthy beams, in

Table 13.4: Sample database header for the inverse problem with 3 stiffness nodes in phase I.

b/h	L/h	f'$_c$ (MPa)	k1%	k2%	k3%	d_{cr}/h	w_{cr} (mm)	b_{cr}/L	Flag
I	I	I	I	I	I	O	O	O	1, 2 or 3

addition to 2,160 damaged beams obtained by varying the previously mentioned modeling parameters. The ANNs were trained and tested on 1,093 and 550 datasets, respectively, to obtain the optimal number of hidden nodes and iterations. The training sets included the maximum and minimum values for each parameter to capture the full range of the datasets. The ranges used to normalize the parameters were intentionally expanded to facilitate better mapping. The input parameters were mapped to fit 10%–90% of the expanded range, while the output parameters were mapped to fit 20%–80% of the expanded range. This was done so that the parameters are within the sensitive region of the sigmoid function. The expanded ranges used in each phase are shown in Table 13.5. The top performing ANN models were then chosen based on statistical measures such as the Averaged-Squared-Error (ASE), coefficient of determination (R^2), and Mean Absolute Relative Error (MARE), which have been defined earlier in this chapter.

Table 13.5: Expanded normalization ranges.

Phase	Phase I		Phase II		Phase III	
Parameter	Max	Min	Max	Min	Max	Min
b/h	0.95	0.45	0.95	0.45	0.95	0.45
L/h	13.750	6.250	13.750	6.250	13.750	6.250
f'_c	43.500	18.500	43.500	18.500	43.500	18.500
$k_1\%$	1.041	0.634	1.041	0.634	1.014	0.878
$k_2\%$	1.041	0.633	1.041	0.633	1.016	0.857
$k_3\%$	1.041	0.633	1.044	0.600	1.017	0.847
$k_4\%$	1.041	0.633	1.041	0.633	1.018	0.841
$k_5\%$	1.039	0.652	1.052	0.536	1.039	0.652
$k_6\%$	1.032	0.716	1.050	0.547	1.015	0.868
$k_7\%$	1.025	0.776	1.051	0.541	1.011	0.903
$k_8\%$	1.020	0.821	1.043	0.612	1.008	0.931
$k_9\%$	1.016	0.853	1.045	0.593	1.007	0.958
d_{cr}/h_1	1.000	−0.250	1.000	−0.250	1.000	−0.250
w_{cr1}	6.667	−1.667	6.667	−1.667	6.667	−1.667
b_{cr}/L_1	0.667	−0.167	0.667	−0.167	0.667	−0.167
d_{cr}/h_2	–	–	1.000	−0.250	–	–
w_{cr2}	–	–	6.667	−1.667	–	–
b_{cr}/L_2	–	–	1.111	−0.278	–	–

4.2 3D-Analysis Databases

The 3D reinforced concrete T-beam database was established by variation of the geometric and material parameters for all beams as well as cracking parameters for cracked beams. Geometric parameters included the width of the beam web cross-section (b_w), the depth of the entire beam cross-section (h), the height of the beam flange, or slab (h_s), and the beam span length (L). The beam flange was

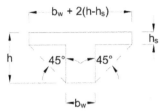

Fig. 13.13: T-beam dimension.

modeled according to the provisions of American Concrete Institute (ACI, 318-14) Section 8.4.1.8 for two-way slabs, which indicates that the total width of the T-beam flange should be equal to the sum of the web width and twice the depth of the beam extending below the flange as shown in Fig. 13.13 (ACI Committee 318, 2014).

The size of the reinforcing bars in each beam depended upon the steel ratio (ρ) and the beam's cross-sectional area according to the relationship: $A_s = \rho b_w d$, where A_s is the total cross-sectional area of steel reinforcement and d is the effective depth of the beam section, taken as the beam depth minus the cover. The 28-day concrete compressive strength (f'_c) was the only variable material parameter. Cracking parameters included the location (b_{cr}), depth (d_{cr}), and width (w_{cr}) of each crack, and all crack locations were measured from a constant (left) support of the beam. Except for f'_c, ρ, and w_{cr}, all parameters were directly normalized with respect to a beam web width of b_w = 250 mm (9.8 in), resulting in the normalized parameters b_w/h, h_s/h, L/h, b_{cr}/L, and d_{cr}/h. This normalization was performed so that the results of this study could be applied to beams of various sizes. The slab height was also set to remain constant at h_s = 100 mm (3.9 in) to limit the number of beams created through parametric variation, although the normalized parameter h_s/h fluctuated due to variability in the beam depth. Figure 13.14 displays an elevation view of a beam with two cracks and highlights several geometric property dimensions and all stiffness node locations (Fletcher et al., 2017).

All beams were divided into five equal segments that could each contain up to one crack. Seventy-two healthy beams were first created to provide reference results for computation of the nodal stiffness ratios for cracked beams. Most of the beam database was composed of beams with a single crack; a total of 6,624 singly cracked beams were modeled. These beams were allowed to have a crack in any segment, and crack locations were varied between the beginning, middle, and end of each segment. Significantly fewer beams (3,600) with five cracks were generated. Two

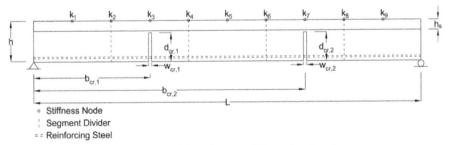

Fig. 13.14: 3D reinforced concrete T-beam elevation view.

crack location configurations were allowed for these beams, and crack locations were set at the center of each segment. Crack location configurations are defined as a unique combination of cracked segments and b_{cr}/L values. Only 384 beams with two cracks, 448 beams with three cracks, and 320 beams with four cracks were created and a crack existed in or was missing from each of the five segments. By exposing the ANNs to a large amount of data for singly-cracked beams with sample data for beams with two and five cracks, the neural networks were anticipated to provide reasonably accurate damage predictions for beams with 0–5 cracks in any configuration.

One ANN was trained to develop a structural damage detection model for 3D reinforced concrete T-beams. The network was trained with the geometric, material, and stiffness parameters (b_w/h, h_s/h, L/h, ρ, f'_c, $k\%_n$, where n is the nth stiffness node at the nine equidistant stiffness nodes) as inputs, see Table 13.6, and the crack parameters ($b_{cr,m}/L$, $d_{cr,m}/h$, and $w_{cr,m}$, where m denotes the mth crack) as outputs.

Table 13.6: Parameter variability for inputs.

Parameters	Values								
b_w/h	0.5		0.7		0.9				
L/h	7		10		13				
ρ	0.005		0.01						
f'_c MPa (ksi)	20 (2.9)		30 (4.35)			40 (5.8)		50 (7.25)	
$K\%_n$	$K\%_1$	$K\%_2$	$K\%_3$	$K\%_4$	$K\%_5$	$K\%_6$	$K\%_7$	$K\%_8$	$K\%_9$

Microsoft Excel spreadsheets were used to maintain the beam database and format the datasets for ANN analyses. In addition to the 14 geometric, material, and nodal stiffness input parameters and the fifteen crack parameters, the spreadsheets also contained an identification number (*ID*) and a set identifier (*Set*). The identification number served as a term by which beams were referenced between the spreadsheets and ANNs. The set identifiers had values of 1, 2, or 3, and indicated whether the beam data was used for training, testing, or validation of the ANN, respectively. These identifiers were assigned manually in a semi-random fashion, but some beams with parameters having extreme values were assigned to the training dataset. Also, according to this procedure, 50%, 25%, and 25% of the original unique beam datasets were assigned to the training, testing, and validation sets, respectively. Table 13.7 displays a sample header for the Microsoft Excel spreadsheets where 'I' indicates an input parameter, 'O' indicates an output parameter, 'm' indicates the mth crack, and 'n' indicates the nth stiffness node.

The final reinforced concrete beam database contained data for 42,804 beams. A total of 37,116 beams (5,688 original and 31,428 duplicated by mirroring the location, depth, and width of cracks) were used for training. On the other hand, one half of the

Table 13.7: Sample Microsoft Excel database header for damage detection problem.

ID	b_w/h	h_s/h	L/h	ρ	f'_c (MPa)	$k\%_n$	$b_{cr,m}/L$	$d_{cr,m}/h$	$w_{cr,m}$ (mm)	Set
#	I	I	I	I	I	I	O	O	O	1, 2, or 3

original number of beams (2,844 beams) was employed in each of the testing and validation of the ANNs. Minimum and maximum parameter value ranges supplied to the ANNs were expanded to keep the actual parameter values in the sensitive region of the sigmoidal activation function of the network. The input parameter range was expanded to allow the input parameters to fit within 10%–90% of the expanded range. Nodal stiffness ratios were limited to those of a healthy beam (<=1.0), so the output parameter range was expanded so that only the minimum values were modified, and the output parameters fit within 20%–100% of the expanded range. The expanded parameter ranges are displayed in Table 13.8.

Table 13.8: ANN expanded parameter ranges for the damage detection problem of 3D-Analysis.

Inputs			Outputs		
Parameter	**Minimum**	**Maximum**	**Parameter**	**Minimum**	**Maximum**
b_w/h	0.45	0.95	b_{cr1}/L	0.013	0.247
h_s/h	0.18	0.38	d_{cr1}/h	0.000	0.852
L/h	6.25	13.75	w_{cr1}	0.000	6.667
ρ	0.00	0.01	b_{cr2}/L	0.133	0.467
f'_c MPa (ksi)	16.25 (2.35)	53.75 (7.79)	d_{cr2}/h	0.000	0.852
$k\%_1$	0.635	1.00	w_{cr2}	0.000	6.667
$k\%_2$	0.509	1.00	b_{cr3}/L	0.333	0.667
$k\%_3$	0.477	1.00	d_{cr3}/h	0.000	0.852
$k\%_4$	0.434	1.00	w_{cr3}	0.000	6.667
$k\%_5$	0.412	1.00	b_{cr4}/L	0.533	0.867
$k\%_6$	0.432	1.00	d_{cr4}/h	0.000	0.852
$k\%_7$	0.483	1.00	w_{cr4}	0.000	6.667
$k\%_8$	0.512	1.00	b_{cr5}/L	0.753	0.987
$k\%_9$	0.639	1.00	d_{cr5}/h	0.000	0.852
			w_{cr5}	0	6.667

5. Results and Discussion

5.1 AI Modeling of 2D-Analysis Database

Several networks were created for each database following the previously described methodology. Each network has the beams' geometric and material parameters (b/h, L/h, and f'$_c$) in addition to the stiffness ratios ($k_n\%$) as inputs, while the cracking parameters (d_{cr}/h, w_{cr}, and b_{cr}/L) are the required outputs. From training, testing and validation, the optimal ANN structure for each damage database is obtained. Starting with phase I, for the 3 stiffness nodes database, the minimum MARE and ASE values and the maximum R^2 value are obtained with 19 hidden nodes at 18200 iterations. Similarly, the optimal structure is obtained with 18 hidden nodes at 20,000 iterations, 8 hidden nodes at 18,100 iterations, and 11 hidden nodes at 20,000 iterations for the 5, 7, and 9 stiffness nodes' databases, respectively. It is observed that, generally, the number of hidden nodes decreases as the number of stiffness nodes increases. It appears that providing the ANN with more inputs reduces the need for more hidden nodes for this application. This effect is desirable as decreasing the number of hidden nodes decreases the complexity of the ANN and lowers the possibility of abnormalities occurring for different predictions. Table 13.9 shows the best models obtained for each database in phase I and their statistics after training on all datasets. Additionally, Figs. 13.15(a) and (b) shows how ASE and R^2, respectively, vary with the increase in the number of stiffness nodes. As expected, the noted trend indicates an increase in the ANNs predictions' accuracy as the number of stiffness nodes increases, with the rate of change in error stabilizing at 9 stiffness nodes. The statistics for training, testing, validation, and training on all datasets for 9 stiffness-node ANNs are summarized in Table 13.10. From training and testing, the optimal network structures obtained are 9, 11 and 9 hidden nodes again, all at 20,000 iterations. Models 1 and 2 provide better results than model 3 in validation. Finally, model 2 is chosen as it has provided similar results to model 1 in validation and better results after training on all datasets.

Moving to phase II, ANNs are created to solve the inverse problem of damage detection for beams with two cracks modeled using beam elements following the previously described methodology. These ANNs are trained, tested, and validated on the generated 9 stiffness-node database. For this phase, the inputs were the beams' geometric and material parameters (b/h, L/h, and f'$_c$) in addition to the stiffness ratios (k_n%), while the outputs were the two cracks' parameters (d_{cr}/h_1, w_{cr1}, b_{cr}/L_1, d_{cr}/h_2, wcr_2, and bcr/L_2). Table 13.11 shows the best two models obtained in phase II and their detailed statistics for training, testing, validation, and training on all datasets.

Table 13.9: Inverse problem ANNs' results in phase I.

# of Stiffness Nodes	Model (INP-HN-OUT)	Iterations	MARE (%)	R^2	ASE
3	6-19-3	18200	69.919	0.499	0.024
5	8-18-3	20000	64.681	0.592	0.018
7	10-8-3	18100	52.858	0.674	0.012
9	12-11-3	20000	52.338	0.678	0.012

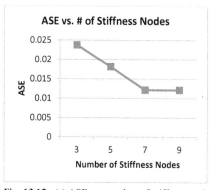
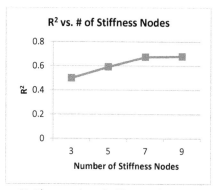

Fig. 13.15: (a) ASE vs. number of stiffness nodes curve (b) R^2 vs. number of stiffness nodes curve for phase I.

Table 13.10: Inverse problem ANNs detailed results in phase I (9 stiffness nodes).

		Model 1	Model 2	Model 3
		12_(4-9_20000)_3	12_(5-11_20000)_3	12_(6-9_20000)_3
Training	MARE %	52.065	**51.835**	53.166
	R^2	0.677	**0.682**	0.666
	ASE	0.012	**0.012**	0.012
Testing	MARE %	53.879	**53.432**	54.414
	R^2	0.679	**0.671**	0.672
	ASE	0.012	**0.012**	0.012
Validation	MARE %	48.081	**48.084**	49.089
	R^2	0.680	**0.678**	0.672
	ASE	0.012	**0.012**	0.012
All Data	MARE %	52.583	**52.338**	54.273
	R^2	0.661	**0.678**	0.665
	ASE	0.012	**0.012**	0.013

Table 13.11: Inverse problem ANNs detailed results in phase II.

		Model 1	Model 2
		12_(4-18_20000)_6	12_(5-16_20000)_6
Training	MARE %	**88.174**	89.314
	R^2	**0.609**	0.580
	ASE	**0.022**	0.023
Testing	MARE %	**87.913**	88.936
	R^2	**0.607**	0.579
	ASE	**0.022**	0.024
Validation	MARE %	**85.971**	86.963
	R^2	**0.603**	0.575
	ASE	**0.023**	0.024
All Data	MARE %	**84.568**	88.522
	R^2	**0.652**	0.610
	ASE	**0.018**	0.024

From training and testing, the optimal structures obtained were 18 hidden nodes in model 1 and 16 hidden nodes in model 2, both obtained at 20,000 iterations. Next, validation is performed and model 1 provided the least MARE and ASE values and the highest R^2 value, thus it is chosen as the best model. Model 1 also provides better statistics when trained on all datasets.

A slight decrease in the accuracy of the inverse problem in ANN is experienced in phase II. Statistical measures for ANN II are MARE = 84.568%, R^2 = 0.65207, ASE = 0.01812, compared to ANNIi-1, where MARE = 52.338%, R^2 = 0.67834, ASE = 0.012113 in phase I. The percentage differences from ANNIi-1 to ANN II are

61.58%, –3.87%, and 49.59%, for MARE, R^2, and ASE, respectively. This decrease is expected as the ANN is trying to predict three additional outputs in phase II compared to phase I. Even though the obtained errors were relatively high, the predictions of this ANN could still be considered reasonable for practical applications. This is because the cracking parameters are very small in magnitude, so even a large error value can only cause a variation of fractions of millimeters, especially in the crack width.

Finally, in phase III, ANNs are created to solve the inverse problem of damage detection for beams with a single crack modeled using plane stress FEs following the previously described methodology. For this phase, the inputs are the beams' geometric and material parameters (b/h, L/h, and f'$_c$) in addition to the stiffness ratios (kn%), while the outputs were the cracking parameters (d_{cr}/h, w_{cr}, and b_{cr}/L). These ANNs were trained, tested, and validated on the generated 3, 5, 7, and 9 stiffness-nodes' databases. From training and testing, the optimal structure obtained is 12 hidden nodes at 20,000 iterations, 9 hidden nodes at 4,100 iterations, 10 hidden nodes at 11,200 iterations, and 9 hidden nodes at 19,600 iterations for the 3, 5, 7, and 9 stiffness-node databases, respectively. Compared to networks obtained in phase I, the ANNs obtained in phase III generally has less hidden nodes, especially for networks with a lower count of stiffness nodes. This further reinforces the implication that the quality of datasets within the damage databases in phase III is higher. Table 13.12 shows the best models obtained for each database in phase III and their statistics after training on all datasets. Additionally, Figs. 13.16(a) and (b) shows how ASE and R^2, respectively, vary with the increase in the number of stiffness nodes. As expected, the

Table 13.12: Inverse problem ANNs' results in phase III.

# of Stiffness Nodes	Model (INP-HN-OUT)	Iterations	MARE (%)	R^2	ASE
3	6-12-3	20000	63.645	0.576	0.017
5	8-9-3	4100	67.240	0.625	0.016
7	10-10-3	11200	62.514	0.645	0.015
9	12-9-3	19600	60.693	0.651	0.015

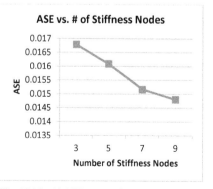
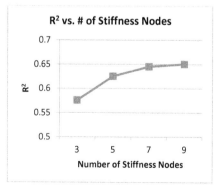

Fig. 13.16: (a) ASE vs. number of stiffness nodes curve (b) R^2 vs. number of stiffness nodes curve for phase III.

Table 13.13: Inverse problem ANNs' detailed results in phase III (9 stiffness nodes).

		Model 1	Model 2	Model 3
		12_(5-9_19600)_3	12_(3-7_7100)_3	12_(4-7_19100)_3
Training	MARE %	**60.801**	62.040	62.711
	R²	**0.655**	0.645	0.640
	ASE	**0.015**	0.015	0.015
Testing	MARE %	**57.994**	58.468	58.788
	R²	**0.633**	0.627	0.631
	ASE	**0.016**	0.016	0.016
Validation	MARE %	**58.779**	59.159	59.756
	R2	**0.634**	0.625	0.626
	ASE	**0.015**	0.015	0.015
All Data	MARE %	**60.693**	63.941	60.966
	R²	**0.651**	0.640	0.642
	ASE	**0.015**	0.015	0.015

noted trend indicates an increase in the ANNs predictions' accuracy as the number of stiffness nodes increases. The statistics for training, testing validation, and training on all datasets for 9 stiffness-node ANNs are summarized in Table 13.13. From training and testing, the optimal network structures obtained are 9 hidden nodes at 19,600 iterations, 7 hidden nodes at 7,100 iterations, and 7 hidden nodes again at 19,100 iterations. Model 1 is chosen as it provided better results than models 2 and 3 in all modeling stages, including training on all datasets.

5.2 AI Modeling of 3D-Analysis Database

An ANN program with a single hidden layer was utilized in this study, and the initial number of hidden nodes was set to vary between 2 and 10. Although the maximum number of hidden nodes calculated exceeded 270, the use of 20 hidden nodes as a maximum was found to sufficiently facilitate learning within the neural network. In addition to reducing program run time, using fewer than the calculated maximum number of hidden nodes helped the ANNs avoid memorization. The training and testing datasets were used to train nine network architectures, each corresponding to a different number of initial hidden nodes. The ANNs were first evaluated by their ASE (average-squared errors) and R^2 (coefficient of determination). The optimal initial and final numbers of hidden nodes and the number of iterations at the final numbers of hidden nodes were recorded for the most effective three networks in terms of their statistical measure performance. These optimal networks were then tested using the validation datasets and trained on the entire reinforced concrete T-beam database to expose them to all available data and establish a robust structural damage detection model. Table 13.14 presents the statistical results for these optimum models used during the selection of the final network structures for damage detection prediction applications.

Table 13.14: Statistical results for optimal ANN crack prediction parameters of 3D-Analysis

		Model 1 (4-15-2000)	Model 2 (8-10-2000)	Model 3 (9-16-18900)
Training	R^2	0.405	0.365	0.422
	ASE	0.046	0.050	0.041
Testing	R^2	0.377	0.289	0.422
	ASE	0.043	0.044	0.032
Validation	R^2	0.365	0.298	0.404
	ASE	0.046	0.049	0.035
All Data	R^2	0.391	0.351	0.426
	ASE	0.045	0.049	0.040

Similarly, the ANN training process and optimal model selection procedure was applied to the FE-reinforced concrete beam model database to establish the damage detection prediction model. Model identification was based on the following nomenclature: initial number of hidden nodes-final number of hidden nodes-iterations at the final number of hidden nodes. Statistical results from the initial training and testing phase revealed the best-performing network architectures to be models 4-15-2000, 8-10-2000, and 9-16-18900. These ANNs were then tested on the validation sets and trained on the entire database (all data). Detailed statistical results for these optimal networks are displayed in Table 13.14.

Several general observations and trends are evident in Table 13.14. First and foremost, the networks displayed weak prediction accuracies with high squared errors (ASE) in all phases and coefficients of determination (R^2) exceeding 0.42 for all data analysis. As expected, the damage detection network, which predicts 15 unknowns, performed modestly according to the statistical measures due to the non-uniqueness nature of the problem. In addition, the voids in the dataset established in terms of the beams with missing two, three, and four cracks are expected to contribute to slightly lowering the statistical measures of the ANN model. Enriching the database with additional information, which is very time consuming, is expected to boost the resulting statistical measures marginally. Nevertheless, these measures are expected to remain relatively modest due to the lack of uniqueness of the damage prediction process.

5.3 Examples

5.3.1 2D-Analysis Example

Two examples for each phase are given next to illustrate the application of the AI-FE synergy approach developed in this chapter. One of the sample cases is part of the datasets used to train the ANN, while the second case is for a beam generated using finite element analysis which the network was not exposed to earlier. See Tables 13.15–13.21 for the input parameters and the outputs obtains from the ANN models for phase I, II, and III.

Table 13.15: Input parameters for example beams for phase I.

ID	b/h	L/h	f'$_c$ (MPa)	k_1 (%)	k_2 (%)	k_3 (%)	k_4 (%)	k_5 (%)	k_6 (%)	k_7 (%)	k_8 (%)	k_9 (%)
B1	0.7	13	21	0.961	0.990	0.996	0.998	0.998	0.999	0.999	0.999	1.000
B2	0.6	12	35	0.997	0.996	0.997	0.998	0.999	0.999	0.999	1.000	1.000

Table 13.16: Example beams' obtained predictions for phase I.

ID	Actual			Predicted with ANN1			Predicted with ANN8, 9, 10		
	d_{cr}/h	w_{cr} (mm)	b_{cr}/L	d_{cr}/h	w_{cr} (mm)	b_{cr}/L	d_{cr}/h	w_{cr} (mm)	b_{cr}/L
B1	0.75	1	0.1	0.714	1.239	0.108	0.701	0.860	0.099
B2	0.45	2	0.25	0.426	2.48	0.252	0.400	2.458	0.215

Table 13.17: Geometric, material, and cracking parameters for example beams for phase II.

ID	b/h	L/h	f'$_c$ (MPa)	d_{cr}/h^1	w_{cr} (mm)^1	b_{cr}/L^1	d_{cr}/h^2	w_{cr} (mm)^2	b_{cr}/L^2
B1	0.9	7	21	0.25	2.5	0.333	0.75	5	0.667
B2	0.7143	10	28	0.55	4	0.22	0.22	2	0.68

Table 13.18: Stiffness ratios for example beams for phase II.

ID	k_1 (%)	k_2 (%)	k_3 (%)	k_4 (%)	k_5 (%)	k_6 (%)	k_7 (%)	k_8 (%)	k_9 (%)
B1	0.935	0.919	0.897	0.867	0.821	0.747	0.694	0.747	0.789
B2	0.968	0.960	0.959	0.952	0.936	0.906	0.868	0.896	0.916

Table 13.19: Example beams' obtained predictions for the inverse problem for phase II.

ID	Crack	Actual			Predicted		
		d_{cr}/h	w_{cr} (mm)	b_{cr}/L	d_{cr}/h	w_{cr} (mm)	b_{cr}/L
B1	1	0.25	2.5	0.333	0.232	2.357	0.328
	2	0.75	5	0.667	0.671	5.160	0.681
B2	1	0.55	4	0.22	0.243	3.87	0.154
	2	0.22	2	0.68	0.723	2.835	0.668

Table 13.20: Input parameters for example beams for phase III.

ID	b/h	L/h	f'$_c$ (MPa)	k_1 (%)	k_2 (%)	k_3 (%)	k_4 (%)	k_5 (%)	k_6 (%)	k_7 (%)	k_8 (%)	k_9 (%)
B1	0.9	10	21	0.946	0.926	0.961	0.977	0.985	0.990	0.993	0.995	0.997
B2	0.65	11	28	0.970	0.957	0.943	0.947	0.965	0.976	0.982	0.987	0.991

Table 13.21: Example beams' obtained predictions for phase III.

ID	Actual			Predicted		
	d_{cr}/h	w_{cr} (mm)	b_{cr}/L	d_{cr}/h	w_{cr} (mm)	b_{cr}/L
B1	0.5	2.5	0.2	0.499	2.118	0.208
B2	0.75	3	0.35	0.684	2.162	0.340

5.3.2 3D-Analysis Example

Two examples are given below to illustrate the applications of the ANNs developed in this analysis. The first example is a beam which was included in the datasets used in developing the ANNs. The second example is a beam that the network was not exposed to earlier. The first beam parameters are: $b = 250$ mm, $h = 500$ mm, $L = 3500$ mm, and $f'_c = 20$ MPa. The beam normalized geometric parameters and obtained stiffness ratios are shown in Table 13.22. The developed ANN model was used to predict the crack parameters and the results are shown in Table 13.23. The actual beam contains two cracks at the first third of the beam, the first crack parameters are $d_{cr}/h = 0.639$, $w_{cr} = 5$ mm, and $b_{cr}/L = 0.1$, while the second crack parameters are $d_{cr}/h = 0.1$, $w_{cr} = 0.05$ mm, and $b_{cr}/L = 0.3$. The predicted parameters for the first crack are $d_{cr}/h = 0.705$, $w_{cr} = 2.88$ mm, and $b_{cr}/L = 0.114$ and for the second crack are $d_{cr}/h = 0.149$, $w_{cr} = 1.58$ mm, and $b_{cr}/L = 0.338$. The ANN predicted an additional small crack at $b_{cr}/L = 0.87$ with a small $d_{cr}/h = 0.01$, and $w_{cr} = 0.056$ mm. The predicted third crack is a direct result of the non-uniqueness nature of the inverse problem. The second beam parameters are: $b = 250$ mm, $h = 332$ mm, $L = 3863.6$ mm, and $f'_c = 43.16$ MPa. The beam normalized geometric parameters and obtained stiffness ratios are shown in Table 13.22. The developed ANN model was used to predict the crack parameters and the results are shown in Table 13.23. The actual beam contains three cracks at the last half of the beam, the first crack parameters are $d_{cr}/h = 0.5$, $w_{cr} = 1.3$ mm, and $b_{cr}/L = 0.6$, while the second crack parameters are $d_{cr}/h = 0.56$, $w_{cr} = 3.2$ mm, and $b_{cr}/L = 0.64$, and the third crack parameters are $d_{cr}/h = 0.029$, $w_{cr} = 0.54$ mm, and $b_{cr}/L = 0.8$. The ANN predicted four cracks: a minor crack with $d_{cr}/h = 0.018$ at $b_{cr}/L = 0.076$ and three major cracks at the last half of the beam. The

Table 13.22: Example beams' parameters.

ID	b/h	L/h	f'$_c$ (MPa)	k$_1$ (%)	k$_2$ (%)	k$_3$ (%)	k$_4$ (%)	k$_5$ (%)	k$_6$ (%)	k$_7$ (%)	k$_8$ (%)	k$_9$ (%)
B2C1	0.5	7	20	0.905	0.918	0.945	0.964	0.976	0.984	0.989	0.992	0.996
B3CT	0.753	11.64	43.16	0.957	0.939	0.919	0.892	0.852	0.794	0.809	0.851	0.892

Table 13.23: Example beams' results.

ID	Actual			Predicted		
	d$_{cr}$/h	w$_{cr}$ (mm)	b$_{cr}$/L	d$_{cr}$/h	w$_{cr}$ (mm)	b$_{cr}$/L
First Beam						
First Crack	0.639	5	0.1	0.705	2.88	0.114
Second Crack	0.1	0.05	0.3	0.149	1.58	0.338
Third Crack	0	0	0	0.01	0.056	0.87
Second Beam						
First Crack	0	0	0	0.018	0.44	0.076
Second Crack	0.5	1.3	0.6	0.32	1.73	0.48
Third Crack	0.565	3.2	0.64	0.54	2.53	0.65
Fourth Crack	0.029	0.543	0.8	0.5	2.4	0.9

ANN managed to exactly predict the location and the depth of the second crack of the major cracks, while approximately predicting the locations of the other two major cracks within a 10% error. Even though the overall ANNs statistical measures showed a relatively high error, the results obtained here were reasonable regarding the locations and the depth of the cracks and show good agreement with the actual data.

5.4 Software Interface

5.4.1 2D-Analysis Software Interface

After completing training, testing, validation, and finally training on all datasets, the optimal ANN models were obtained. These ANN models consist of the nodes, their thresholds, and the connection weights between them. To facilitate damage evaluation and detection using the developed ANN models, Excel interfaces have been created. These interfaces were created using Microsoft Excel 2010 and Visual Basic for Application (VBA) code was utilized for their backends.

The first Excel interface integrates phases I and III ANNs. The user can choose to use either ANNIi-1 model or ANNIi-8, 9, 10 models to predict a single crack in a simply supported beam. ANN1 predicts all output parameters in a single model, while ANN8–10 predict each output parameter within an independent ANN. Alternatively, the user can choose the more accurate ANNIII model developed based on plane stress finite elements. The model descriptions were also provided within the interface for the user to review. Next, the user can choose to input the beam's parameters in absolute or normalized value formats. The inputs for the exact format are the compressive strength (f'$_c$), width (b), depth (h), and span length (L) of the beam, while the inputs for the normalized format are the width to depth ratio (b/h), span length to depth ratio (L/h), and the compressive strength (f'$_c$). Similarly, the user can also choose to input the stiffness values in either exact or normalized value formats. The inputs for the exact format are the stiffness values for the beam in question (kna) and the stiffness values for the same beam in its initial or healthy state (knh), while the inputs for the normalized format are the stiffness ratios (kn%) as defined earlier. Next, the interface immediately provides the ANNs predicted cracking parameters for these inputs in normalized and/or exact formats, depending on the format of the provided input beam parameters. Also, the interface can determine which half of the span the crack is in and informs the user if the stiffness values should be reversed. Additionally, the interface provides a graphical representation of the beam's profile and shows where the crack is located within the beam's span and how deep it is relative to the depth of the beam. The plot also represents the locations of the stiffness nodes with red 'x' marks. Figure 13.17 shows the interface developed for phases I and III.

For phase II, the interface is modified to facilitate prediction for two cracks. First, the user chooses the desired operation mode. The user can choose either "Damage Prediction" mode to predict cracks' depth, width, and location by providing beam parameters and stiffness values, or choose "Damage Evaluation" mode to determine the health index by providing beam and crack parameters. Next, the user can choose to input the beam parameters in absolute or normalized value formats. The inputs

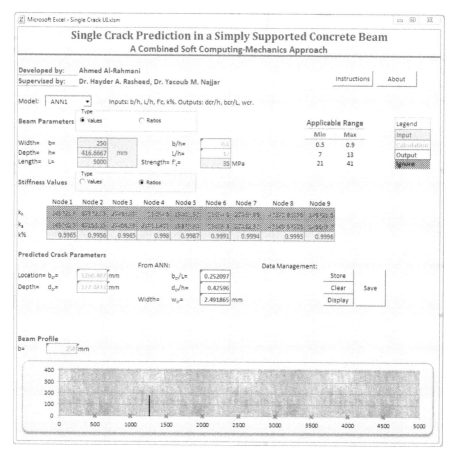

Fig. 13.17: Excel user interface for phases I and III operating with ANNIi-1 model.

for the exact format are the compressive strength (f'$_c$), width (b), depth (h), and span length (L) of the beam, while the inputs for the normalized format are the width to depth ratio (b/h), span length to depth ratio (L/h), and the compressive strength (f'$_c$). The next step depends on the operation mode chosen by the user. If the interface is operating in "Damage Prediction" mode, which is the focus of this chapter, the user then either chooses to input the stiffness values in exact or normalized value format. The inputs for the exact format are the stiffness values for the beam in question (k$_{na}$) and the stiffness values for the same beam in its initial or healthy state (k$_{nh}$), while the inputs for the normalized format are the stiffness ratios (k$_n$%) as defined earlier. The interface then immediately provides the ANN's predicted cracking parameters for these inputs in normalized and/or exact formats, depending on the format of the provided input beam parameters. Additionally, the interface provides a graphical representation of the beam profile and shows where the two cracks are located within the beam's span and how deep they are relative to the depth of the beam. The plot also represents the locations of the stiffness nodes with red 'x' marks. Figure 13.18 shows the interface developed for phase II in "Damage Prediction" mode.

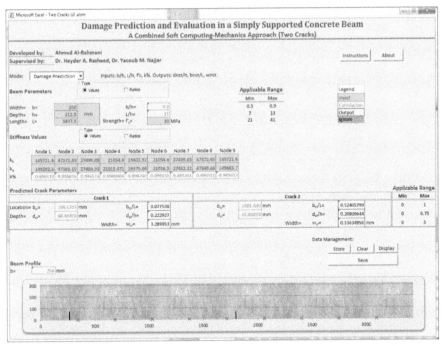

Fig. 13.18: Excel user interface developed for ANNIIi in "Damage Prediction" mode.

5.4.2 3D-Analysis Software Interface

After applying the ANN training process and optimum model selection procedure and evaluating the top-performing models' prediction accuracies on additional testing datasets, the optimal ANN architectures for cracking parameters prediction application were established. Biases and connection weights for the input layer-hidden layer and hidden layer-output layer were retrieved after these network architectures were trained on all the data in the main database. Application of these biases and connection weights to the summation and activation functions described allowed for calculation of the ANN-predicted cracking parameters for any simply supported reinforced concrete T-beam having parameters within the range of those used to create the finite element (FE) model database. To automate these calculations and facilitate on-site damage evaluation, in the field, using the optimal ANN models, a touch user-interface driven computer application was developed.

The application was developed with Visual Studio 2015 using .net framework version 4.5 on a desktop running PC Windows operating system. Figure 13.19 displays the final version of the graphical user interface that may be loaded on a tablet PC and used for field prediction.

As shown in Fig. 13.19, the user interface accepts geometric and material parameters (bw, h, hs, L, ρ, and f'c) in the "Beam Data" section and the nodal stiffness ratios (Kn %) in the "Nodal Stiffness" section as inputs. Upon activation of the 'Solve' button, the input parameters are normalized to function as input nodes for the ANN calculations. The biases and connection weights from the optimum cracks

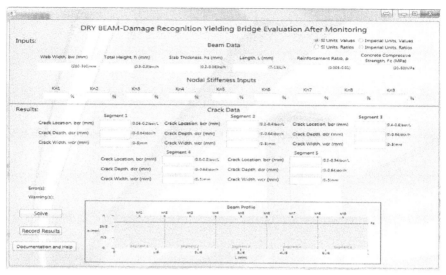

Fig. 13.19: Cracks damage prediction graphical user interface (DRY BEAM).

parameters ANN models are applied to pass the input parameters to the hidden nodes and ultimately to the output nodes. The output parameters are then de-normalized, and the predicted cracking parameters are displayed in the 'Results' section at the lower half of the user interface. Also, after pressing the 'Solve' button, the "Beam Profile" is updated to display the location of the flange of the T-beam and the location and depth of each crack. The height and length labels are also changed to reflect the user-input height and length of the beam.

6. Conclusions

A series of comprehensive studies was conducted on 2D simply supported rectangular beams and 3D simply supported T beams utilizing the AI-FE synergy approach described in detail in this chapter. The results generated at all levels of AI modeling and prediction support the following conclusions:

1. The proposed methodology of AI-FE synergy offers a robust framework to detect damage in reinforced concrete bridge and building beams.

2. Due to the use of the ANN trained, tested, and validated models, it is possible to produce a very fast prediction tool applicable for field inspections provided that the FE database of damage parameters is predefined or generated *a priori.*

3. The accuracy measures of the ANN models built were all within reasonable prediction range regardless of the complexity of the analysis framework used and the number of damage parameters involved.

4. Although the accuracy measures were modest in values due to the non-uniqueness of possible solutions, but the detected damage was practically very reasonable compared with the actual damage due to the small values of the

cracking parameters compared to the size of the beams. This was especially true for the crack width.

5. The results showed that the ANN model can predict the most important cracking parameter (crack location) very accurately while it can predict the crack depth with less but reasonable accuracy.

6. The developed software using the touch-enabled devices could prove to be very powerful in offsetting or reducing reliance on subjective damage detection techniques by paving the way towards adopting more objective tools.

References

Abaqus 6.10 Online Documentation. 2010. Dassault Systèmes.

Al-Rahmani, A. 2012. *A Combined Soft Computing-mechanics Approach to Damage Evaluation and Detection in Reinforced Concrete Beams.* MS thesis. Kansas State University, Manhattan, KS.

Al-Rahmani, A., Rasheed, H. and Najjar, Y. 2013. Intelligent damage detection in bridge girders: A hybrid approach. *J. Eng. Mech.*, 139(3). doi: 10.1061/(ASCE)EM.1943-7889.0000536.

American Concrete Institute Committee 318. 2014. *Building Code Requirements for Structural Concrete and Commentary.* (ACI No. 318R-14). Farmington Hills, MI: American Concrete Institute.

Beer, F.P., Johnston, E.R., Dewolf, J.T. and Mazurek, D.F. 2012. *Mechanics of Materials* (6th Edn.). New York, NY: McGraw-Hill Companies, Inc.

Chen, S., Liu, W., Bian, H. and Smith, B. 2012. *3D LiDAR Scans for Bridge Damage Evaluation.* Paper presented at the 6th Congress on Forensic Engineering: Gateway to a Better Tomorrow. San Francisco, CA, United States. doi: 10.1061/9780784412640.052.

Fletcher, E., Abouelleil, A. and Rasheed, H. 2017. *FE-ANN Based Modeling of 3D Simple Reinforced Concrete Girders for Objective Structural Health Evaluation.* Institute for Transportation, Iowa State University, Final Research Report, June 2017, Sponsored by Kansas Department of Transportation, Midwest Transportation Center, and U.S. Department of Transportation Office of the Assistant Secretary for Research and Technology.

Ghods, A.S. and Esfahani, M.R. 2009. *Damage Assessment of Reinforced Concrete Beams by Modal Test.* Paper presented at the 4th International Conference on Structural Health Monitoring of Intelligent Infrastructure. (22–24 July). Zurich, Switzerland. Retrieved from Compendex database.

Hasançebi, O. and Dumlupinar, T. 2013. Linear and nonlinear model updating of reinforced concrete T-beam bridges using artificial neural networks. *Computers & Structures*, 119: 1–11. doi: 10.1016/j.compstruc.2012.12.017.

Jeyasehar, C.A. and Sumangala, K. 2006. Damage assessment of prestressed concrete beams using artificial neural network (ANN) approach. *Computers & Structures*, 84(26): 1709–1718. doi: 10.1016/j.compstruc.2006.03.005.

Li, L., Ghrib, S. and Lee, S. 2008. *Damage Identification of Reinforced Concrete Beams by Digital Image Correlation and FE-updating.* (10–13 June). Paper presented at the Canadian Society for Civil Engineering Annual Conference: Partnership for Innovation. Quebec City, QC, Canada. Retrieved from Compendex database.

Ndambi, J., Vantomme, J. and Harri, K. 2002. Damage assessment in reinforced concrete beams using eigenfrequencies and mode shape derivatives. *Engineering Structures*, 24(4): 501–515. doi: 10.1016/S0141-0296(01)00117-1.

Ongpeng, J.M.C., Oreta, A.W.C. and Hirose, S. 2018. Monitoring damage using acoustic emission source location and computational geometry in reinforced concrete beams. *Applied Sciences*, 8(2): 189.

Reynders, E. and De Roeck, G. 2009. A local flexibility method for vibration-based damage localization and quantification. *Journal of Sound and Vibration*, 329(12): 2367–2383. doi: 10.1016/j.jsv.2009.04.026.

Sagar, R.V., Prasad, B.K.R. and Sharma, R. 2012. Evaluation of damage in reinforced concrete bridge beams using acoustic emission technique. *Nondestructive Testing and Evaluation*, 27(2): 95–108. doi: 10.1080/10589759.2011.610452.

Shiotani, T., Ohtsu, H., Momoki, S., Chai, H.K., Onishi, H. and Kamada, T. 2012. Damage evaluation for concrete bridge deck by means of stress wave techniques. *Journal of Bridge Engineering*, 17(6): 847–856. doi: 10.1061/(ASCE)BE.1943-5592.0000373.

Tan, Z.X., Thambiratnam, D.P., Chan, T.H., Gordan, M. and Abdul Razak, H. 2020. Damage detection in steel-concrete composite bridge using vibration characteristics and artificial neural network. *Structure and Infrastructure Engineering*, 16(9): 1247–1261.

Teughels, A., Maeck, J. and De Roeck, G. 2002. Damage assessment by FE model updating using damage functions. *Computers & Structures*, 80(25), 1869–1879. doi: 10.1016/S0045-7949(02)00217-1.

Chapter 14

Deep Learning in Transportation Cyber-Physical Systems

Zadid Khan,[1] Sakib Mahmud Khan,[1,] Mizanur Rahman,[2]*
Mhafuzul Islam[3] and Mashrur Chowdhury[4]

1. Introduction

Transportation Cyber-Physical Systems (TCPS) is defined as a system that ensures "higher efficiency and reliability by enabling increased feedback-based interactions between the cyber system and the physical system in transportation" (Deka et al., 2018; Zeng et al., 2020). However, as discussed in recent studies, the social domain, which consists of the human stakeholders that the TCPS serves, must be one of the focal points of the overall system design. Thus, TCPS integrates the interaction between the connected transportation infrastructure, users, and computing and communication services, either at the edge or cloud infrastructure through feedback loops to improve social acceptance of overall systems. Figure 14.1 shows an example scenario of bridge condition monitoring TCPS where data are collected from bridge monitoring sensors, connected trucks, and smartphone users. The connected truck will share the real-time status of the truck, and other smartphone users will be connected through smartphone applications or apps (including social media apps or navigation apps like Waze, where users can provide input). This overall system is supported by an edge-centric architecture, where roadside units are the fixed-edge devices communicating with the other fixed-edge devices (bridge monitoring sensors) and mobile-edge devices (connected trucks). The bridge monitoring center (BMC) is the system edge that collects data from the bridge health monitoring sensors, such as cameras, whereas data from the connected trucks and smartphone apps are forwarded to the traffic management center (TMC). Both BMC and TMC

---
[1] 311 N Walton Blvd, Walmart, Inc., Bentonville, AR 72712, USA-29634. Email: mdzadik@clemson.edu
[2] 3015 Cyber Hall, University of Alabama, Tuscaloosa, AL, USA-35487. Email: mizan.rahman@ua.edu
[3] General Motors R&D, Warren Technical Center, Warren, MI 48092, USA. Email: mdmhafi@clemson.edu
[4] 216 Lowry Hall, Clemson University, Clemson, SC, USA-29634. Email: mac@clemson.edu
* Corresponding author: sakibk@clemson.edu

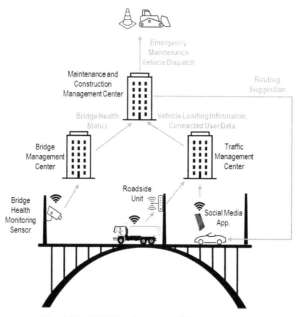

Fig. 14.1: TCPS for Bridge condition monitoring.

analyze the collected data and further notify the emergency management center if any imminent bridge repairment is needed. The emergency management center could suggest alternate routes or route warnings, which will benefit the approaching traffic and other road users. Such TCPS along with many others help the society by:

- Improving safety and mobility for vulnerable road users such as pedestrians
- Adapting the transportation systems operations based on the real-time and non-real-time input from the users and other stakeholders
- Enabling intelligent human-machine interaction
- Maximizing user acceptance.

2. Artificial Intelligence, Machine Learning, and Deep Learning

Machine learning (ML) and deep learning (DL) concepts evolved from Artificial Intelligence (AI). If a machine can produce intelligent behavior, it can be referred to as an AI-enabled machine. As a computer program, ML is used to learn the relationship between inputs and outputs through data observations. In ML, the program is not explicitly designed by the programmer. The program can learn from: (1) input data, (2) error calculation matrices that compute the difference between the predicted data and real data, and (3) a feedback loop based on the error calculation metrics. Neural Network is a type of ML. Deep learning is a specific type of ML, particularly neural networks, where additional layers are added for the program to learn more details through training. Each layer in DL contains many neurons. The original machine learning domain had shallow neural network, logistic regression and support vector machines. To keep up with the growth of available data and the recent improvement in hardware capabilities (such as the introduction of graphical processing units (GPUs) (Haghighat et al., 2020), DL has become a powerful method of tackling complex real-world problems as these models can automatically extract features from complex data (Pouyanfar et al., 2018). Traditional ML models require external feature extractors to extract relevant features from the data. An example use case is image classification, where the accuracy has improved to about 96–97%, which is even better than the accuracy achieved by humans (Pouyanfar et al., 2018). DL can be used with large-scale real-world sensor data, such as images, videos, speech, audio, and text. Most importantly, DL is the closest known technology that closely mimics the functionality of human brains.

3. Brief History of Deep Learning

The history of the deep neural network goes back to an electronic cognition device, perceptron. After that, the invention of the backpropagation technique paved the way for modern neural networks (Dong et al., 2021). In the 1980s, the basic concept behind convolutional and recurrent neural networks was introduced. LeNet was the first neural network developed in the 1990s, but it was not practical for large datasets due to hardware limitations (Pouyanfar et al., 2018). Restricted Boltzmann machines (RBM) were also introduced around that time, which was an unsupervised two-layer model. A breakthrough came in 2006 when the deep belief network (DBN) was proposed. DBNs allowed the capability to add more layers and train one layer at a time, which gave rise to the class of 'deep' models (Pouyanfar et al., 2018). As a result, deep neural networks were born. After that, convolution and recurrent networks were applied to create deep neural networks of different kinds. The initial training method was modified to train all the layers at once rather than training layer by layer (Dong et al., 2021). In recent times, the two biggest revolutions have been big data and computing hardware. The availability of vast amounts of heterogeneous data and the introduction of large-scale computing hardware have enabled the DL models to outperform previous ML models. An example of the progress made by DL is Google's AlphaGo, which won 60 straight games of "Go" against human players, including the champion "Go" player during 2016 "and 2017.

4. Deep Learning Models for TCPS Applications

4.1 Early Deep Learning Models

The major difference between the early/traditional and advanced neural networks is that the traditional networks are trained in a greedy layer-wise method. There are many different variations of such networks, which are described below.

4.1.1 Restricted Boltzmann Machine (RBM)

In RBM, all neurons are binary variables, and the entire probability distribution follows a Boltzmann distribution. The RBM network is made up of an input layer and a hidden layer. The hidden layer activations are independent of the input layer. Using ample hidden layers, all discrete distributions can be fitted to a Boltzmann machine, which is a network of neuron-like units, which are symmetrically connected.

4.1.2 Deep Belief Network (DBN)

DBN contains multiple unsupervised networks, which can include autoencoder or RBMs. There are many hidden layers in a DBN. The hidden layers can be trained in a greedy layer-wise manner. Another aspect of DBN is the absence of connection between the neurons of each layer, unlike modern neural networks. The layers contain only one connection, which passes the output of the last layer as input to the next layer. The top and bottom layers are connected to provide feedback. The target is input reconstruction, so it is also an unsupervised model. After the learning phase, DBNs can also be used for classification (Wang et al., 2019).

4.1.3 Deep Boltzmann Machine (DBM)

DBM is like DBN because it also consists of multiple RBMs. The differences are unidirectional connections and the absence of feedback from the top to the bottom layer. DBMs contain many hidden layers. The estimated expectations include data-dependent expectations using variational approximation and data-independent expectations using Markov chains. These two expectations are entered into the gradient of the log-likelihood function. The learning is implemented using an unsupervised method, where the model is trained in one forward pass layer by layer. A DBM is more powerful and robust than a DBN because it can extract the features from more vague data and complex data, such as voice or speech signal (Dong et al., 2021).

4.2 Multilayer Perceptron (MLP)

MLP is the simplest form of deep neural network. It has several hidden layers between the input and output layer (Wang et al., 2019). If all the neurons are interconnected, then this network is known as a fully connected MLP. The connection between each layer consists of a set of weight parameters. Let's consider m as the input layer, n as the output layer, and h_1 and h_2 as hidden layers. The weights between the layers are W_1, W_2, and W_3, respectively. Each layer contains an activation function σ. Some popular activation functions are *tanh*, *sigmoid*, and rectified linear unit or ReLU (Veres and Moussa, 2020). This function adds non-linearity to the model, which is required to model the nonlinearity of complex real-world problems. MLP combines matrix multiplications, additions, and nonlinear activations, as shown using the following Eqs. (14.1–14.3).

$$h_1 = \sigma(W_1 m + b_1) \tag{14.1}$$

$$h_2 = \sigma(W_2 h_1 + b_2) \tag{14.2}$$

$$\hat{n} = \sigma(W_3 h_2 + b_3) \tag{14.3}$$

The weights are updated during training based on the input layer data. The backpropagation technique can be used to calculate the gradient of the loss function for updating the weights. As a result, the weights are modified to minimize the training loss (Wang et al., 2019).

4.3 Autoencoder

An automatic encoder or autoencoder is a combination of an encoder and a decoder network. The encoder network extracts the latent features using the training data, and the decoder network uses these features to reconstruct the input (Haghighat et al., 2020). The reconstruction loss is minimized in the training process to get the best representation of encoded features from the input. As the learning process is automatic, it is called an automatic encoder. The first usage of autoencoders was as denoisers. Recently, autoencoders have been very popular for dimensionality reduction and feature extraction. Autoencoders are also useful when labeled data is not available since the training process is automatic and does not require output labels (Haghighat et al., 2020). Another type of autoencoders are variational autoencoders. In this model, a variational inference method is used for minimizing the input reconstruction loss.

4.4 Convolutional Neural Network (CNN)

CNN is specially designed to fully exploit the properties of 2D data structure (i.e., images) (Khan et al., 2020). The major characteristic of CNN is the convolution operation (Pouyanfar et al., 2018). Small filters are used to apply the convolution operation to small sections of images rather than the whole image. These small filters are known as kernels. This enables the extraction of spatially local correlation in the data. This process results in a reduction in weights, making the network faster. A typical CNN model contains several convolution and pooling layers (i.e., max pooling or average pooling). The pooling layer are used for reducing the dimensionality of the input using down sampling. A flattened layer and several fully connected layers follow the sequence of multiple convolution and pooling layers. These layers are required to get the desired number of labels or predictions in the output layer. Let's assume I is the single-channel input image of size (P, Q), K is a kernel of size (X, Y), and h is an output image. Then, the convolution operation for each pixel in h can be expressed using the following Eq. (14.4) (Veres and Moussa, 2020)

$$h_{i,j} = \sum_{x=1}^{X} \sum_{y=1}^{Y} I(i+x-1, j+y-1) K(x,y) \tag{14.4}$$

where $h_{i,j}$ is obtained after the nonlinear activation function is used on top of the convolution. The dimension of $h_{i,j}$ will be (P–M+1, Q–N+1). If n such kernels are used to generate samples, (P–M+1, Q–N+1, n) dimensional data will be generated. If the pooling filter size of the pooling layer is R, then the pooling layer's output will have the dimension [(P–M+1)/R, (Q–N+1)/R, n]. Basically, the pooling layer divides the input into blocks of size (R, R) and then performs an aggregation operation, such as average or max. CNN has the last few layers as fully connected layers. If the model is a classification model, then the final layer contains an activation function that outputs the probability score of each class. Examples of such activation functions are sigmoid, SoftMax, etc. The images and kernels will have an additional dimension for RGB (red, green, blue) images since the input images will have three channels for three colors.

The CNN architecture was first proposed by LeCun et al. (1989) for a vision-based cat recognition system. In their study, they used only one kernel for convolution operation on the input image and extracting local features within the image (LeCun et al., 1989). CNN models have become very popular recently with the growth of labeled data. The most popular datasets related to image classification and object detection within images are ImageNet, Common Objects In Context (COCO) and Pascal Visual Object Classes (VOC) datasets (Lin et al., 2014). For classification, the examples of the most popular architectures are AlexNet, VGGNet, Inception, and Resnet. For object detection and bounding box (an imaginary rectangle used for object detection in an image) within images, there are other types of models, such as Region-CNN (R-CNN) (Girshick et al., 2014), Fast R-CNN (Girshick, 2015), Faster R-CNN (Ren et al., 2015), and You Only Look Once–Version 3 (YOLOv3) (Redmon and Farhadi, 2018). For other tasks (such as semantic segmentation, action recognition, and text recognition), CNN models can also be used.

A VGG-16 model is shown in Fig. 14.2 (Simonyan and Zisserman, 2015). The input is a 224x224x3 image. The architecture starts with two convolutional layers and a max-pooling with filter size 2. After these three layers, the new dimension is 112x112x128. The same process is repeated to get a new dimension of 56x56x256. After that, there are three convolution layers and maxpooling with filter size 2. The final dimension before the flattening layer is 7x7x512. After flattening, the

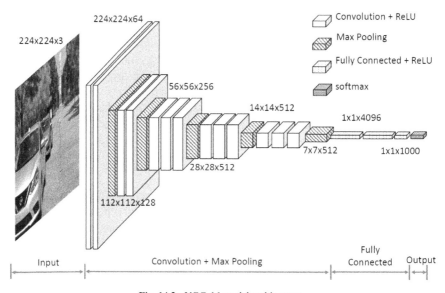

Fig. 14.2: VGG-16 model architecture.

flattened layer contains 4,096 neurons. Then, the flattened layer is connected to a fully connected layer with 1,000 neurons (Simonyan and Zisserman, 2015). Finally, a SoftMax layer is used to convert the output to probability values.

4.5 Recurrent Neural Network (RNN)

Just like CNNs, RNNs are specially designed for sequential data, such as time series data. It is particularly useful for time-series forecasting, natural language processing (NLP), and speech processing, etc. RNN stores the representation of short-term and long-term inputs and uses a feedback connection to pass information for the prediction of future steps (Pouyanfa et al., 2018). The sequential processing of input data allows for variable-length input, which is another advantage of RNN models. The basic Eq. (14.5) to describe the RNN models is given below (Veres and Moussa, 2020).

$$h_t = \sigma(Wm_t + Uh_{t-1})\tag{14.5}$$

where W and U are the weight matrices for connection to current and previous time-steps, respectively. The traditional RNN suffers from two major problems, which are vanishing gradients and exploding gradients. Vanishing gradients occur when the multiplication of lots of small gradients may make the resultant gradient close to zero. As a result, the effect of long-term variations might be lost for current timesteps. The opposite problem is the exploding gradient problem, where the multiplication of lots of large gradients may result in a very high gradient value. This makes the network unstable, and the model is unable to learn anything from the data (Haghighat et al., 2020). The result for both issues is that the network gradually forgets initial inputs when new data comes in. To solve this, Long Short-Term Memory (LSTM) neural networks have been developed (Khan et al., 2019). The additional operations used in an LSTM neuron to overcome the vanishing gradient problem is shown in Fig. 14.3. The cell state (c) acts as the LSTM model's memory. The cell state solves the vanishing and exploding gradient issues by using pointwise additive and multiplicative operations. These operations are used to update memory information. An LSTM neuron contains input gate, forget gate, and output with "*tanh*" activation function. The mathematical operations are shown using the following Eqs. (14.6–14.10) (Khan et al., 2019). m_t, c_{t-1}, and n_{t-1} are the input to the model. n_t and c_t are the model output. σ refers to the sigmoid activation function. From forget, input, and output gates, f_t, i_t, and o_t are generated, respectively. The weights associated with each gate are represented using W, U, and b. Element-wise multiplication is represented by dots (.) in the equations.

$$f_t = \sigma(W_f m_t + U_f n_{t-1} + b_f)\tag{14.6}$$

$$i_t = \sigma(W_i m_t + U_i n_{t-1} + b_i)\tag{14.7}$$

$$o_t = \sigma(W_o m_t + U_o n_{t-1} + b_o)\tag{14.8}$$

$$c_t = f_t.c_{t-1} + i_t.tanh(W_c m_t + U_c n_{t-1} + b_c)\tag{14.9}$$

$$n_t = o_t.tanh(c_t)\tag{14.10}$$

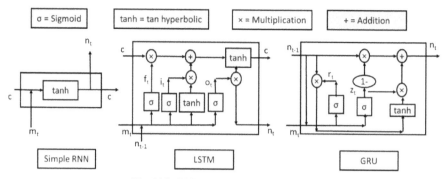

Fig. 14.3: RNN, LSTM, and GRU models.

A modified version of the LSTM neuron was created to reduce its complexity, which is known as a gated recurrent unit (GRU). A new "update gate" is created in GRU, which combines the input gate and forget gate. The cell state and hidden states are combined for faster operation. Equations 14.11–14.13 describe the mathematical operations behind GRU, where the intermediate states are z_t and r_t.

$$z_t = \sigma(W_z m_t + U_z n_{t-1} + b_z) \qquad (14.11)$$

$$r_t = \sigma(W_r m_t + U_r n_{t-1} + b_r) \qquad (14.12)$$

$$n_t = z_t . n_{t-1} + (1 - z_t).tanh(W_n m_t + U_n(n_{t-1}.r_t) + b_n) \qquad (14.13)$$

Simple RNN, LSTM, and GRU all have bidirectional variants, and they are trained similarly compared to unidirectional ones.

4.6 Deep Reinforcement Learning

Reinforcement learning or RL is another special branch of ML where an agent is trained to take the appropriate actions based on observations from the environment. The agent starts with an arbitrary action and then observes the effect. The observed reward and current state are fed to the model to identify the next step. Maximization of the long-term reward is considered as the goal. RL algorithms are usually based on Markov Decisions Processes (MDP) (Haghighat et al., 2020). MDPs are systems containing a set of agents, states, actions, and some randomness, where each action takes the agent from one state to another with some uncertainty. RL algorithms use the state value function and quality function (Q-function, also referred to as an action-value function). In the Q-function approach, the function gives the expected return for a given state and action. The Q-learning approach can be used to approximate the Q-value, and deep neural networks can be used to predict the optimal Q value (Haghighat et al., 2020). Deep Q Network (DQN) is a popular DL method for RL. In this method, the CNN model and fully connected layers are used to predict the Q value, but it has shown some instability for RL solutions. RL network stability can be improved using experience relay and target networks. Improvements on the DQN model have also been proposed, including double DQN and dueling DQN models (Haghighat et al., 2020). In double DQN, two separate networks are used for approximating the Q value, one is the online network, and the other is the target network. The online network is updated in real-time, but the target network is updated intermittently, which solves the issues of overestimation of Q value and instability in the target values. The dueling DQN consists of a non-sequential model architecture, where the core DL network branches out into two paths. One of the branches outputs the value function corresponding to a state and the other function outputs the advantage of a particular action compared to no action. Essentially, the combination of these two values can give the Q value.

4.7 Other Models

4.7.1 Attention Models

Attention models are focused on context rather than the actual prediction. It is one type of DNN that gathers context and then makes a prediction. The computation at a particular time is focused only on the layer-wise features instead of the whole model. It has already been applied to NLP problems (Veres and Moussa, 2020).

4.7.2 Generative Adversarial Network

Generative adversarial network (GAN) is a completely different type of DL method compared to all other models. This model extracts the statistical distribution from training data and creates new synthetic data (Dong et al., 2021). The property of these synthetic data is that they are very similar to the real data, so they can be used for data augmentation, or denoising. The model tries to maximize the likelihood of training data (Pouyanfar et al., 2018). This model has two parts: generative and discriminative. The generative model creates synthetic data by passing random noise or perturbations through a multilayer network. The noise or perturbation is usually minuscule since the target is to create data like real-world data. The discriminative model identifies whether the sample is real or fake (Haghighat et al., 2020). The basic GAN model uses fully connected layers in the generative and discriminative models. In deep convolutional GAN (DCGAN), batch normalization is used in all layers, and no pooling layers are used. Another variation is conditional GAN, where the sample generation and discrimination are conditional to specific classes (Haghighat et al., 2020).

5. Deep Learning Model Training and Development Considerations for TCPS Applications

5.1 Activation Function

An activation function is an important factor in the development of DL models. Sigmoid, Hyperbolic tangent, ReLU are some popular activation functions. For example, the sigmoid function converts any input to a value between 0 and 1, which is usually very helpful for binary classification tasks (Dong et al., 2021).

5.2 DL Overfitting and Solutions

DL overfitting is a common problem during training. Sometimes the model is trained so that it is fine-tuned to the training data only, rather than to unknown data, which results in very low-test accuracy, using unknown data. The solution to this problem is regularization. Regularization refers to the slight modification of the learning algorithm to reduce overfitting and improve model performance. The common pattern of regularization is to add some randomness to the training process and then average out the randomness in the testing process (Pouyanfar et al., 2018).

5.3 Transfer Learning

Access to powerful computing hardware is not always possible, so researchers have found a way to use existing models and their pre-trained weights and apply them to domains specific to the researchers. This technique is called transfer learning (Pouyanfar et al., 2018). The deep neural network is pre-trained on a large dataset, and the model and its weights can be saved. After that, for a different application, this model and its pre-trained weights can be used as starting points. Then, anyone can re-train for the new type of data without a significant computational burden. Transfer learning is popular because it allows the usage of existing computational techniques to develop DL models efficiently. It is also useful when there is a shortage of training data (Pouyanfar et al., 2018).

5.4 Hyperparameter Optimization

The most popular optimization technique to reduce the loss function is stochastic gradient descent (SGD) which can show oscillating and unstable behavior with variation in learning rate (Dong et al., 2021). Momentum was introduced to improve SGD performance. Weight decay and learning decay are two techniques for extracting the learning rate for the SGD model. Adaptive learning rates were introduced using the Adagrad model (Duchi et al., 2011). After that, improvements were made in the Adadelta, Adamax, and Adam models (Dogo et al., 2018). Adam is the most popular optimization technique for minimizing the loss function in DL models (Dong et al., 2021).

5.5 Other Considerations

Processing large amounts of data with different characteristics (such as in Big Data) in real-time is one of the primary challenges for the real-world implementation of DL models. It usually requires high-end computing hardware to process a large amount of data in real-time within the corresponding computational latency constraint of different TCPS applications (Dong et al., 2021). As the DL model computational performances are primarily associated with the related hardware, providing high-end hardware to meet a real-time performance requirement will increase the cost of the TCPS deployment. Thus, a combination of public cloud infrastructure and edge infrastructure could be a way to address the issue of handling and distribution of the Big Data processing functions among cloud and edge computing infrastructure.

Although cloud computing resources can be used for DL model training and validation of model performance, the communication latency between the cloud (where the model is trained and validated) and edges (where the functions are performed) could limit complete reliance on the cloud technology. Transferring a large amount of data over the communication network from the cloud to edges will increase the communication latency, which may not satisfy the real-time requirements of DL-based TCPS applications. The federated learning technique can be used to distribute a portion of the DL model training in the vehicles, a portion in the roadside edges and another portion in a cloud, in a decentralized manner. However, it is a challenge to deal with heterogeneous and massive distributed networks because of the complexity of large-scale DL model implementation, distributed model optimization, and privacy-preserving data analysis.

In order to achieve the best performance of a DL model, it is necessary to tune DL model parameters, which include two tasks: neural architecture search and hyperparameter adjustment. The neural architecture search could be an automated or manual process in any DL model architecture development to improve the performance by finding an optimal model structure. However, there are two primary challenges in finding the neural architecture. First, the super-structure and sub-structures of a model are usually designed heuristically using prior human knowledge, which limits the search space (Dong et al., 2021). Second, the large search space also leads to expensive computation in the performance evaluation process due to an increasing number of models that need to be tested. In addition, it is a challenge to efficiently explore optimal hyperparameter sets by selecting the best optimization method, and appropriate batch size and learning rate for different DL models.

6. Hardware and Software Frameworks for Deep Learning

6.1 Software

The popular software frameworks for DL models are Caffe, Torch, Neon, Theano, MXNet, TensorFlow, and Microsoft Cognitive Toolkit (CNTK) (Pouyanfar et al., 2018). Python and C++ are the two most popular programming languages for DL models. All the frameworks support GPU through CuDNN (Chetlur et al., 2014). Among the frameworks, Torch and TensorFlow are very popular among Python users (Pouyanfar et al., 2018). Tensorflow provides a high-level model development layer named Keras, which is very useful for quick model development and implementation.

6.2 Hardware

Cloud computing has made significant progress, and DL models can be deployed in the cloud. However, for edge computing (where computing nodes are placed close to the edge data sources (Khan et al., 2019), powerful computing devices are required (Haghighat et al., 2020). Currently, Nvidia Jetson Xavier is a popular edge computing device for DL models (Mittal, 2019). It consists of the Volta GPU, which contains integration with Tensor Cores and dual DL Accelerators (DLAs) (Haghighat et al., 2020). Xavier can deliver 32 TeraFLOPs of computing performance. Nvidia Jetson

TX2 is another popular device with Pascal GPU. The TX2 can deliver more than 1 TeraFlop of performance (Haghighat et al., 2020). There are many other embedded systems for DL model deployment, such as Intel Neural Computing Stick 2 (NCS 2) and Raspberry Pi 3. The Intel NCS 2 uses the Myriad X vision processing unit (VPU) for computation. The advantage of NCS 2 devices is low power consumption and inherent support for DL frameworks like Tensorflow and Caffe.

7. Vulnerabilities of Deep Learning in TCPS

As discussed in previous sections, using DL, it is possible to solve problems and improve the efficiency of TCPS. However, the DL-based system itself can be vulnerable to different cyberattacks, which, in turn, can reduce the utility of DL in TCPS. Several studies show that it is now possible to fool a DL-based system (Papernot et al., 2017; Qayyum et al., 2020; Yao et al., 2019; Li et al., 2019). For example, as shown in Fig. 14.4, an attacker can attack a physical traffic sign and introduce small perturbations. The impurity injected into the traffic sign can compromise the AI based traffic sign recognition module of a vehicle. If the traffic sign module is compromised this way, it will produce the incorrect classification of the traffic sign (Sitawarin et al., 2018; Eykholt et al., 2017). In this example, a stop sign is identified as a 45-mph traffic speed sign. This attack is known as an adversarial attack (Akhtar and Mian, 2018; Yuan et al., 2019). An attacker can also create a cyberattack including noise on the data captured by a camera sensor. Once the data in compromised, a wrong decision made by the traffic sign module will change the vehicle planning and can lead to crashes. This section focuses on the types of cyberattacks that can happen on DL, and the areas of protecting the DLs against such cyberattacks are explored.

Generally, DL models are typically used for performing detection, prediction, and classification. Therefore, the goal of an attacker is to produce incorrect output by compromising the DL model to make it produce an incorrect targeted or untargeted output (Akhtar and Mian, 2018). In a broader sense, the cyberattacks on DL models can be categorized into three categories: (1) Whitebox threat model, (2) Blackbox threat model, and (3) Gray-box threat model. Figure 14.5 shows the three categories

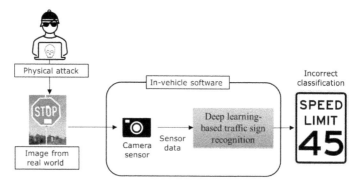

Fig. 14.4: Example of a cyberattack on DL-based traffic sign recognition module of a connected automated vehicle.

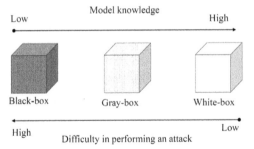

Fig. 14.5: Cyberattack types and difficulty in performing attacks.

and the relative sequence based on an attacker's knowledge about the DL model used in the target system and the challenges to perform an attack.

7.1 Whitebox Threat Model

In a Whitebox attack, an attacker has access to the knowledge of the DL models that the attacker wants to compromise. The adversaries can be crafted easily as the attacker has insider knowledge of the system. This knowledge may include the model structure, hyperparameters, training data of the DL models.

7.2 Blackbox Threat Model

In a blackbox model setup for an attacker, the attacker has no knowledge of the DL model running in a system. Let us consider a scenario where an automated vehicle uses a DL model for classifying traffic signs. In this setup, an outside attacker may not know what the model architecture is and what hyperparameters are running inside the vehicle. However, an attacker can make an educated guess that it will recognize certain types of traffic signs, such as speed limit sign and stop sign, and perform an attack on the blackbox model. However, these types of attacks are difficult to perform, as the attacker has no knowledge of the system.

7.3 Gray-box Threat Model

A gray box attack happens when an attacker knows less about the model than a white box but more than a black box. The exact definition of a gray-box threat varies. A gray-box attack is commonly discussed as an attack scenario where an attacker knows the architecture of the target DL model; but does not know the weights of the DL network. The attacker uses a surrogate DL model with the same network architecture to perform an attack.

8. Applications of Deep Learning in TCPS

8.1 Traffic Characteristics Prediction

The evolution of TCPS has made it possible to inform road users about existing and future traffic conditions utilizing real-time vehicle location and motion information (e.g., speed and acceleration) (Hua et al., 2018; Khan et al., 2017). The current and future traffic conditions rely on the accurate prediction of traffic flow parameters. These predicted flow parameters can also be used for real-time route diversion, route planning, and scheduling to improve operational conditions and traffic safety and for energy optimization to reduce vehicle fuel consumption. Real-time traffic flow parameter prediction is challenging as traffic flow is dynamic in nature. The TCPS provides an opportunity to solve this challenge by allowing vehicles to broadcast Basic Safety Messages (BSM)s. In a TCPS, BSMs include trajectory data (i.e., time, location, speed, acceleration). The Society of Automotive Engineers standard, SAE J2735, has defined the format and elements of the BSMs. Estimating and predicting real-time traffic flow parameters from BSMs require the inclusion of temporal and spatial variations of the data, which could be addressed effectively through the DL models. These models capture the nonlinearity of traffic patterns. Deep RNNs capture temporal and spatial variation and predicts time series data with a high level of accuracy. Deep RNNs have been used to predict traffic volume (Li et al., 2019).

8.2 Driver Behavior Prediction

Deep ML-based driver behavior modeling is popular in autonomous vehicles. These models can be used to capture the human driving pattern using trajectory data. DL-based car-following models have shown high accuracy in capturing the longitudinal movement patterns of human drivers. There are two aspects of these models which are important (Chetlur et al., 2014): (1) parameter calibration depending on the driving scenarios; and (2) model training with adequate data for different possible driving scenarios, including unexpected situations. Deep RL or supervised learning has shown the ability to reduce the training data requirements and improve learning ability in unexpected scenarios. An ML-based car-following model can leverage the longitudinal movement data related to unexpected driving scenarios to make better decisions, which future studies need to address.

8.3 Visual Recognition in TCPS

DL methods are commonly used for visual recognition tasks, such as recognizing vehicle, pedestrian, and traffic signs. For example, autonomous vehicles use visual sensors, such as cameras, to sense the surrounding environment and utilize DL models for detecting objects around the vehicles. CNN models are commonly used for visual recognition. For real-time applications, R-CNN, Faster R-CNN, or YOLOv3 can be used. Based on the visual detection's accuracy, the YOLOv3 shows superior performance with approximately 20 frames per second, exceeding other state-of-the-art DL models. Real-time performance of DL models is necessary for the in-vehicles system and transportation infrastructure-focused applications. With the availability of higher processing GPU-enabled devices, DL models' accuracy and computational time are improving for TCPS applications.

8.4 Applications for Transportation Network Company and Public Transportation

DL methods have been used for predicting the optimal routes for transportation network companies (TNC) and public transportation services by considering traffic demand. DL models have shown their capability of capturing spatial, temporal, and exogenous features. These features are important for predicting passenger demand. For example, a deep convolutional LSTM network (FCL-Net) has been developed considering spatiotemporal variables (i.e., historical demand intensity, travel time), and nonspatial time-series variables (i.e., day of the week, time of the day, weather) (Ke et al., 2017). FCL-Net is used to predict optimal routes using New York City taxi record dataset. In addition, TNCs require to minimize passenger waiting time and cost by modifying the schedule of the driver routes to pick up passengers. Recent research shows a deep reinforcement learning model able to suggest TNC drivers the best route considering the current vehicle location, time, and competition between drivers. This DRL model helps reduce the TNC vehicles' idle time and ensures long-term revenue for drivers (Shi et al., 2018).

8.5 TCPS Cybersecurity Enhancement

Transportation cyberinfrastructure, such as connected traffic signals, roadside data infrastructure, and cloud infrastructure, are integral parts of a smart city. However, intricate and highly connected transportation cyber infrastructure systems are subject to widespread failure from cyberattacks, leading to devastating impacts, such as corporate and personal data breaches, disruptions to system and vehicle performance, and ultimately compromised public safety. A cyberattack on connected and automated transportation systems may well lead to disastrous consequences. For example, attackers could access the control system of connected vehicles and traffic signals and cause catastrophic multi-vehicular crashes. As a vast amount of data are available from different sensors in a connected transportation system, DL models in a TCPS can learn threats and guard against cybersecurity attacks. However, as cybersecurity attacks are dynamic, the detection and mitigation of cyber threats in real-time for connected transportation systems is a challenging task (Ren et al., 2015). There are extensive applications of DL models, such as deep recurrent neural networks and deep reinforcement learning, in communication network security, in-vehicle security, and infrastructure security. Deep recurrent neural networks can learn temporal and spatial variations of the data and predict unusual behavior for threat detection. Also, because of its adaptability and usefulness in real-time and adversarial environments, DRL demonstrates excellent suitability for cybersecurity applications where cyberattacks are increasingly sophisticated.

9. Conclusions

Transportation cyber-physical systems (TCPS) will become mainstream in every facet of the multi-modal transportation system with the advancement of digital infrastructure, which consists of sensing, computing, and communication technologies. DL will become an integral part of TCPS to support the analytics associated with different safety, mobility, and environmental applications. TCPS must encompass the human and social systems in its conception, planning, design, operation, and management. Thus, DL should become transparent to serve every human and society and provide equal mobility, safety, and environmental benefits to all. DL algorithms must be explainable, so the inherent characteristics and goals are not in a blackbox. As DL will play a major role in TCPS, it must be trustworthy to all stakeholders, such as the users, transportation system providers, and vendors. One important aspect of building trustworthiness is to protect the DL in TCPS. Adversarial attacks on TCPS are a threat that can compromise the safety and privacy of all stakeholders. Thus, it is critical to protect the DL program residing in different components of TCPS, such as in the infrastructure, vehicles, and personal information devices. The success of the DL integration in TCPS will lie in building trustworthy, equitable, and effective machine learning.

References

Akhtar, N. and Mian, A. 2018. Threat of adversarial attacks on deep learning in computer vision: A survey. *IEEE Access*, 6: 14410–14430. https://doi.org/10.1109/ACCESS.2018.2807385.

Chetlur, S., Woolley, C., Vandermersch, P., Cohen, J., Tran, J., Catanzaro, B. and Shelhamer, E. 2014. CuDNN: Efficient primitives for deep learning. *arXiv*:1410.0759 cs].

Deka, L., Khan, S.M., Chowdhury, M. and Ayres, N. 2018. Transportation cyber-physical system and its importance for future mobility. pp. 1–20. *In*: Deka, L. and Chowdhury, M. (eds.). *Transportation Cyber-Physical Systems*, Elsevier.

Dogo, E.M., Afolabi, O.J., Nwulu, N.I., Twala, B. and Aigbavboa, C.O. 2018. A comparative analysis of gradient descent-based optimization algorithms on Convolutional Neural Networks. Presented at the *2018 International Conference on Computational Techniques, Electronics and Mechanical Systems (CTEMS)*.

Dong, S., Wang, P. and Abbas, K. 2021. A survey on deep learning and its applications. *Computer Science Review*, 40: 100379. https://doi.org/10.1016/j.cosrev.2021.100379.

Duchi, J., Hazan, E. and Singer, Y. 2011. Adaptive subgradient methods for online learning and stochastic optimization. *The Journal of Machine Learning Research*, 12(null): 2121–2159.

Eykholt, K., Evtimov, I., Fernandes, E., Li, B., Rahmati, A., Xiao, C., Prakash, A., Kohno, T. and Song, D. 2017. Robust physical-world attacks on deep learning models. https://doi.org/10.1109/CVPR.2018.00175.

Girshick, R. 2015. Fast R-CNN. Presented at the *Proceedings of the IEEE International Conference on Computer Vision*.

Girshick, R., Donahue, J., Darrell, T. and Malik, J. 2014. Rich feature hierarchies for accurate object detection and semantic segmentation. *arXiv*:1311.2524 cs].

Haghighat, A.K., Ravichandra-Mouli, V., Chakraborty, P., Esfandiari, Y., Arabi, S. and Sharma, A. 2020. Applications of deep learning in intelligent transportation systems. *Journal of Big Data Analytics in Transportation*, 2(2): 115–145. https://doi.org/10.1007/s42421-020-00020-1.

Hua, Y., Zhao, Z., Li, R., Chen, X., Liu, Z. and Zhang, H. 2018. Traffic prediction based on random connectivity in deep learning with long short-term memory. *arXiv*:1711.02833 cs].

Ke, J., Zheng, H., Yang, H. and (Michael) Chen, X. 2017. Short-term forecasting of passenger demand under on-demand ride services: A spatio-temporal deep learning approach. *Transportation Research Part C: Emerging Technologies*, 85: 591–608. https://doi.org/10.1016/j.trc.2017.10.016.

Khan, S.M., Chowdhury, M., Morris, E.A. and Deka, L. 2019. Synergizing roadway infrastructure investment with digital infrastructure for infrastructure-based connected vehicle applications: *Review of current status and future directions. Journal of Infrastructure Systems*, 25(4): 03119001. https://doi.org/10.1061/(ASCE)IS.1943-555X.0000507.

Khan, S.M., Dey, K.C. and Chowdhury, M. 2017. Real-time traffic state estimation with connected vehicles. *IEEE Transactions on Intelligent Transportation Systems*, 18(7): 1687–1699. https://doi.org/10.1109/TITS.2017.2658664.

Khan, S.M., Fournier, N., Mauch, M., Patire, A.D. and Skabardonis, A. 2020. *Hybrid Data Implementation: Final Report for Task Number 3643.*

Khan, Z., Khan, S.M., Dey, K. and Chowdhury, M. 2019. Development and evaluation of recurrent neural network-based models for hourly traffic volume and annual average daily traffic prediction. *Transportation Research Record*, 2673(7): 489–503. https://doi.org/10.1177/0361198119849059.

LeCun, Y., Boser, B., Denker, J.S., Henderson, D., Howard, R.E., Hubbard, W. and Jackel, L.D. 1989. Backpropagation applied to handwritten zip code recognition. *Neural Computation*, 1(4): 541–551. https://doi.org/10.1162/neco.1989.1.4.541.

Li, J., Schmidt, F.R. and Kolter, J.Z. 2019. Adversarial camera stickers: A physical camera-based attack on deep learning systems. *arXiv*:1904.00759 cs, stat].

Lin, T.-Y., Maire, M., Belongie, S., Hays, J., Perona, P., Ramanan, D., Dollár, P. and Zitnick, C.L. 2014. *Microsoft COCO: Common Objects in Context.* Cham.

Mittal, S. 2019. A survey on optimized implementation of deep learning models on the NVIDIA Jetson platform. *Journal of Systems Architecture*, 97: 428–442. https://doi.org/10.1016/j.sysarc.2019.01.011.

Papernot, N., McDaniel, P., Goodfellow, I., Jha, S., Celik, Z.B. and Swami A. 2017. Practical black-box attacks against machine learning.

Pouyanfar, S., Sadiq, S., Yan, Y., Tian, H., Tao, Y., Reyes, M.P., Shyu, M.-L., Chen, S.-C. and Iyengar, S.S. 2018. A survey on deep learning: Algorithms, techniques, and applications. *ACM Computing Surveys*, 51: 92:1–92:36. https://doi.org/10.1145/3234150.

Qayyum, A., Usama, M., Qadir, J. and Al-Fuqaha, A. 2020. Securing connected & autonomous vehicles: Challenges posed by adversarial machine learning and the way forward. *IEEE Communications Surveys & Tutorials*, 22(2): 998–1026. https://doi.org/10.1109/COMST.2020.2975048.

Redmon, J. and Farhadi, A. 2018. YOLOv3: An incremental improvement. *arXiv*:1804.02767 cs].

Ren, S., He, K., Girshick, R. and Sun, J. 2015. Faster R-CNN: Towards real-time object detection with region proposal networks. *Advances in Neural Information Processing Systems*, 28.

Shi, D., Ding, J., Errapotu, S.M., Yue, H., Xu, W., Zhou, X. and Pan, M. 2018. Deep Q-network based route scheduling for transportation network company vehicles. *Presented at the 2018 IEEE Global Communications Conference (GLOBECOM).*

Simonyan, K. and Zisserman, A. 2015. Very deep convolutional networks for large-scale image recognition. *arXiv*:1409.1556 cs].

Sitawarin, C., Bhagoji, A.N., Mosenia, A., Chiang, M. and Mittal, P. 2018. DARTS: Deceiving autonomous cars with toxic signs. *arXiv*:1802.06430.

Veres, M. and Moussa, M. 2020. Deep learning for intelligent transportation systems: A survey of emerging trends. *IEEE Transactions on Intelligent Transportation Systems*, 21(8): 3152–3168. https://doi.org/10.1109/TITS.2019.2929020.

Wang, Y., Zhang, D., Liu, Y., Dai, B. and Lee, L.H. 2019. Enhancing transportation systems via deep learning: A survey. *Transportation Research Part C: Emerging Technologies*, 99: 144–163. https://doi.org/10.1016/j.trc.2018.12.004.

Yao, Y., Li, H., Zheng, H. and Zhao, B.Y. 2019. Latent backdoor attacks on deep neural networks. *Proceedings at 2019 ACM Conference on Computer and Communications Security.* New York, NY, USA, pp. 2041–2055.

Yuan, X., He, P., Zhu, Q. and Li, X. 2019. Adversarial examples: Attacks and defenses for deep learning. *IEEE Transactions on Neural Networks and Learning Systems*, 30(9): 2805–2824. https://doi.org/10.1109/tnnls.2018.2886017.31.

Zeng, J., Yang, L.T., Lin, M., Ning, H. and Ma, J. 2020. A Survey: Cyber-physical-social systems and their system-level design methodology. *Future Generation Computer Systems*, 105: 1028–1042. https://doi.org/10.1016/j.future.2016.06.034.

Chapter 15

Artificial Intelligence in the Construction Industry

Theory and Emerging Applications for the Future of Work

Amir H. Behzadan,[1,*] *Nipun D. Nath*[2] and *Reza Akhavian*[3]

1. Introduction

Construction is one of the largest global industries and has a major impact not only on economic growth and urban development, but also on the environment and well-being of the people who live, work, and commute in and between various parts of the built environment. With 7.4 million employees, representing almost 5% of the overall workforce, construction is one of the largest sectors of the U.S. economy (BLS, 2021). In 2019, the total construction expenditure was USD $1.3 trillion, or approximately 6.3% of the nation's GDP (BLS, 2019). It is predicted that the global construction market will grow at the rate of 4.2% and reach $10.5 trillion by 2023 (Lucintel, 2021). Particularly, residential construction is estimated to develop at a faster rate due to rapid urbanization and population growth. There is also an increasing demand for green construction, affordable housing, digital design and prefabrication, and new materials such as fibre-reinforced polymer composites (Doumbouya et al., 2016; Hao et al., 2020; Businesswire, 2021). Although projections show signs of a prosperous future for the industry, major roadblocks are expected to slow down the modernization of traditional construction practices in the 21st century. For example, compared to other industries such as manufacturing and healthcare, the construction industry has been continuously suffering from slow growth in labor productivity (Economist, 2017). Also, construction is one of the least digitalized industries, which

[1] Associate Professor, Department of Construction Science, Texas A&M University, College Station, Texas, USA-77843-3137.
[2] Department of Civil and Environmental Engineering, Texas A&M University, College Station, Texas, USA-77843-3137. Email: nipundebnath@tamu.edu
[3] Assistant Professor, Department of Civil, Construction, and Environmental Engineering, San Diego State University, San Diego, California, USA-92182-1326. Email: rakhavian@sdsu.edu
* Corresponding author: abehzadan@tamu.edu

contributes to high volumes of generated waste and lost resources due to deviations from plan and rework (Gandhi et al., 2016). In fact, numbers show that only about 30% of all construction projects finish within the estimated budget and time (Mieritz, 2012). A study of 258 transportation projects, from 20 different countries, revealed that 9 out of 10 projects had cost overruns (Aljohani et al., 2017). Furthermore, frequent work-related illnesses, injuries, and fatalities have made this industry home to some of the most hazardous occupations in the U.S. and globally (BLS, 2014).

The future landscape of construction work is nevertheless changing. Rapid advancements in computational techniques, engineering and project knowledge, and sensor technologies are revolutionizing many of the current practices and putting the industry back on track for major disruption. According to the McKinsey Global Institute (2017), construction labor productivity could potentially double through the implementation of new technologies such as building information modeling (BIM), 3D printing, virtual and augmented reality (VR/AR), internet of things (IoT), and robotics. Also, those engaged in frontier-level scientific research anticipate that the next-generation construction processes will bring new opportunities for collaboration between humans and intelligent machines (Autodesk, 2020; NSF, 2020). To make this a reality, humans and machines will have to delegate tasks and share responsibility. In most cases, while high-risk or routine tasks will be carried out by autonomous robots, for cognitively demanding tasks, robots will work hand-in-hand with humans to maximize the productivity and quality of work, while minimizing cost, completion time, and operational risks. To achieve this goal, emerging areas of research and practice are increasingly utilizing computer vision and artificial intelligence (AI) due to their proven success in numerous real-world applications across various domains, e.g., detecting everyday objects (Redmon and Farhadi, 2017), developing self-driving cars (Rao and Frtunikj, 2018), and analyzing medical images (Shen et al., 2017).

2. Construction Safety

Construction is a hazardous occupation, characterized by a high number of workplace accidents and injuries. In 2016–17, compared to all other industries in the U.S., the total number of fatal occupational injuries was the highest in construction. In 2018, the industry accounted for 21% of all worker fatalities in the U.S. private sector (OSHA, 2019). Moreover, in 2019, fatalities in the private construction industry increased by 5% to a total of 1,061 which was the largest since 2007 (BLS, 2020). The Occupational Safety and Health Administration (OSHA) has identified four leading causes of construction fatalities that are responsible for nearly 60% of all worker deaths (OSHA, 2019). These causes are collectively termed the "fatal four" and include fall, struck by an object, electrocution, and caught-in/between objects.

The industry also exhibits frequent non-fatal illnesses and injuries among its workforce. As construction projects are becoming more complex and challenging, workers in various trades often find themselves having to go beyond their natural physical limits when performing repetitive tasks for long periods of time. Such sustained physical demand on workers' bodies is shown to cause health issues and bodily injuries, referred to as musculoskeletal disorders (MSDs). Besides their adverse physical and psychological implications, MSDs can also lead to significant financial losses. For example, the total financial loss resulting from MSDs is estimated to be around $50 billion per year in the U.S. (Middlesworth, 2020).

2.1 Work-related MSDs (WMSDs) and Prevention through Design (PtD)

Work-related musculoskeletal disorders (WMSDs) refer to the MSDs caused by activities performed in the workplace. Manual construction work involves activities such as handling and lifting heavy objects, body twisting, and awkward postures, all major contributors to WMSDs. The most common types of WMSDs in construction are sprain, strain, carpal tunnel syndrome (CTS), tendonitis, and back pain. In 2019 alone, 231 construction laborers in every 10,000 were injured from WMSDs in the U.S. In the same year, the total number of reported WMSDs cases was 19,790 with 11 median days away from work (BLS, 2020). These and similar statistics provide a glimpse into the loss of productivity in construction due to WMSDs.

Prevention through Design (PtD), an initiative introduced by the National Institute for Occupational Safety and Health (NIOSH), refers to anticipating and designing out ergonomic-related hazards in facilities, work methods, operations, processes, equipment, tools, products, new technologies, and the organization of work (NIOSH, 2014). A primary goal of PtD is to minimize WMSDs by preventing and confining occupational injuries, illnesses, and fatalities. According to NIOSH, this can be achieved by: (1) eliminating or reducing potential risks to workers to an acceptable level at the source or as early as possible in a project life cycle; (2) including design, redesign, and retrofit of new and existing work premises, structures, tools, facilities, equipment, machinery, products, substances, work processes, and the organization of work; and (3) enhancing the work environment through the integration of prevention methods in all designs that affect workers and others (NIOSH, 2014).

OSHA identifies eight risk factors that can lead to WMSDs, including force exertion, repetition, awkward postures, static postures, quick motion, compression or contact stress, vibration, and extreme temperatures (OSHA, 2000). Following the PtD principles, these risk factors can be eliminated by rearranging the workplace and/or selecting appropriate tools for the workers. However, different jobs impose different types and magnitudes of risk, adding an extra level of difficulty to proper identification of ergonomic risks associated with a particular job. Although a thorough job hazard analysis can identify workplace risks, it may be challenging to conduct such analysis because of the complexity of the job and the manual effort required to monitor the work processes on the jobsite (Alwasel et al., 2011).

2.2 Fatal Four and Personal Protective Equipment (PPE) Requirements

In 2017, "fatal four" were responsible for 59.8% of all 971 construction worker deaths (OSHA, 2019). Particularly, falls were the most frequent cause accounting for 39.2% (391 cases) of all fatalities, while struck-by-object, electrocution, and caught-in/between were the root causes of 8.2% (80 cases), 7.3% (71 cases), and 5.1% (50 cases) of all 971 fatalities, respectively (OSHA, 2019). Regardless of the cause, the majority of these incidents could have been prevented if workers had worn appropriate PPE, e.g., hard hat, safety vest, gloves, safety goggles, and steel toe shoes (OSHA, 2019). In fact, OSHA, and similar agencies in other parts of the world (e.g., Centre for Occupational Health and Safety in Canada, EU-OSHA in Europe, State Administration of Work Safety in China) require that all personnel, working in proximity of site hazards, wear proper PPE to minimize the risk of being exposed to or getting injured by hazards. Particularly, researchers and practitioners emphasize on properly wearing hard hats to minimize injuries and fatalities. For example, between 2003 and 2010, a total of 2,210 construction fatalities (25% of all construction fatalities) occurred because of traumatic brain injury (TBI) (CDC, 2019). The most common causes for these fatalities include fall from height, fall of objects on the head, and electrocution by overhanging wires (CDC, 2019; OSHA, 2019). In light of these, OSHA mandates workers to wear hard hats when there are overhead objects, or if there is a possibility that workers bump their heads against objects, or have accidental contact with overhanging electrical hazards (OSHA, 2019). Safety vest is another important PPE component that can minimize the risk of injuries. For example, struck by objects can occur due to site traffic or heavy equipment (e.g., trucks, bulldozers, graders) operating close to workers (OSHA, 2019). Therefore, to ensure high visibility in all lighting and weather conditions, OSHA requires that workers wear safety vests (generally, red or orange in color, and reflective if employee works at night) (OSHA, 2019).

Despite these provisions, there is a sizeable percentage of construction employees who fail to comply with the industry safety measures. Research has identified root causes of this noncompliance to be the lack of awareness, discomfort of wearing PPE, and the feeling that PPE interferes with one's work or duties on the jobsite (Akbar-Khanzadeh, 1998; Huang and Hinze, 2003; Fang et al., 2018). For example, not using proper PPE (i.e., eye and face protection) was one of the most violated regulations in 2017–18 (OSHA, 2019). Governing laws and safety regulations (e.g., OSHA-1926.28a) almost always hold employers responsible for enforcing, monitoring, and maintaining appropriate PPE on the job (OSHA, 2019). To enforce PPE-related compliance on the jobsite, OSHA fines the employer by imposing a penalty for each employee who has failed to comply with PPE requirements (OSHA, 2019). In a 12-month period ending September 2018, OSHA issued 895 citations for not wearing hard hat (total penalty of $1,916,511) and 78 citations for violating other PPE requirements (total penalty of $309,474) in construction sites (OSHA, 2019). The burden of such compensations often falls onto the employers and/or employees since insurance companies do not cover any damages caused by improper PPE practice.

3. Technical Overview

3.1 Data Collection Techniques

A safe workplace requires not only the timely identification of the risk factors but also taking appropriate actions to eliminate or confine them. To achieve this goal, relevant and reliable data must be collected from the field for continuous monitoring of workers' safety. This data also helps the safety management team to make informed decisions to proactively prevent undesired incidents which may occur due to unsafe practices. In general, there are three different approaches to data collection, namely self-assessment, observation-based, and direct measurements (David, 2005; Nath et al., 2017).

3.1.1 Self-assessment Approach

In the self-assessment approach, data are collected through workers' diaries, self-reports, interviews, and questionnaires. For example, Li et al. (2017) surveyed 445 construction workers in China to evaluate their perception of construction safety. Gerdle et al. (2008) interviewed 9,952 participants in Sweden through mail questionnaires to study the effect of work status on anatomical pain. Campo et al. (2008) surveyed 882 U.S. participants on WMSDs, and Östergren et al. (2005) studied 4,919 people in Sweden to correlate work-related physical and psychological factors with shoulder and neck pains. More recently, researchers have also utilized video- and web-based questionnaires to gather self-assessment data. For example, Sunindijo and Zou (2012) collected the opinions of 353 construction personnel on the importance of soft skills for construction safety management through a web-based online survey. Although the self-assessment approach is straightforward, it often adds to the cost of data collection, particularly in cases where a large sample size is desired, as well as calls for special skills for analyzing and interpreting the results (David, 2005). Moreover, previous studies have cited that workers' self-assessments may be imprecise, unreliable, and biased. For example, Balogh et al. (2004) found that for a similar level of exposure (e.g., physical exertion) to WMSDs, workers who had previous complaints about body pain (during the last 12 months) reported the exposure to be significantly high, compared to those who did not have prior complaints.

3.1.2 Observation-based Approach

In the observation-based approach, data are collected by an expert inspector who systematically evaluates activities on the jobsite. For example, Teschke et al. (2009) employed trained observers to assess postures of construction workers for an epidemiological study of back disorders. However, this approach is disruptive in nature, particularly, when data must be collected in real-time with physical presence on the jobsite (David, 2005). One way to remedy this problem is to analyze the recorded videos of site activities. For example, in a study by Dartt et al. (2009), two trained observers estimated the upper limb postures of manufacturing workers from two synchronized videos (capturing the views of the sagittal and frontal plane of the workers, respectively). A key advantage of the video-based observation method over the direct observation method is that the videos can be analyzed multiple times to minimize intra-observer variability (i.e., disagreement among the measurements taken by the same individual) (Mathiassen et al., 2013). However, this method may be prone to inter-observer variability (i.e., disagreement among the measurements taken by different individuals) and could not be practical due to substantial cost, time, and technical training required (David, 2005).

3.1.3 Direct Measurement Approach

In the direct measurement approach, various peripherals are used to directly collect data from the workers or the workspace for subsequent analyses. Examples include, but are not limited to, using radio frequency identification (RFID), inertial measurement unit (IMU), smartphone sensors, RGB camera, RGB-D camera (i.e., visual plus depth information), laser detection and ranging (LADAR), and light detection and ranging (LiDAR). For example, Lee et al. (2012) proposed using RFID tags on workers or equipment for real-time localization and tracking in a complex construction site. Li et al. (2012) developed an RFID-based system to monitor stationary and mobile occupants in a building. Kelm et al. (2013) proposed using RFID tags on PPE (e.g., hard hat, safety vest, safety goggles) worn by workers that could be tracked by an RFID reader gate when workers entered the construction site. Aryal and Becerik-Gerber (2019) used wrist-worn sensors to measure occupants' thermal comfort levels. Yang et al. (2016) developed a system for detecting near-miss falls of iron workers using wearable IMUs (e.g., accelerometer, gyroscope). IMUs have been also used to continuously collect kinematic data on workers' motion (Yan et al., 2017) and gait (Jebelli et al., 2014; Seel et al., 2014) to prevent work-related illnesses and injuries. Moreover, Rashid and Louis (2019) proposed using IMU data to recognize construction equipment activities. Recently, consumer-grade smartphones have turned into promising data collection tools due to their ubiquity, ease of use, and inclusion of built-in IMUs alongside other sensors (e.g., camera, GPS, magnetometer, barometer). For example, to monitor WMSD-related risks, Nath et al. (2017) used smartphone sensors to analyze postures of construction workers, and Nath et al. (2018) developed a smartphone-based method to monitor activities with the risk of overexertion. Also, Rashid and Behzadan (2018) used a body-mounted smartphone to predict the movement trajectories of a worker to prevent proximity-related collisions with moving equipment. Akhavian and Behzadan (2016), and Akhavian and Behzadan (2015) utilized smartphone sensors to monitor the activities of construction workers and equipment, respectively.

The direct measurement approach can also utilize image or video (i.e., visual) data captured by an RGB camera. In contrast to video-based observation, this approach leverages visual data directly using an automated system with minimum human involvement. For example, Nath et al. (2020) and Fang et al. (2018) utilized visual data from RGB cameras to recognize the PPE of construction workers. Also, to support the safety inspection process, Kolar et al. (2018) proposed a method for recognizing the presence of safety guardrails in construction sites using RGB images. Ding et al. (2018) developed an automated video analysis method for identifying the unsafe behavior of workers while climbing a ladder. Tang et al. (2020) developed a path prediction model to track the movements of workers and equipment from videos to proactively prevent collision accidents. Han and Lee (2013) used a marker-based motion capture system to construct 3D skeleton models of construction workers from video frames for monitoring their unsafe behavior. Researchers have also used depth sensors integrated with RGB cameras (a.k.a., RGB-D camera) to collect more sophisticated data from the jobsite. For example, Han et al. (2012) used Microsoft Kinect (a depth-based motion capture system) to analyze the unsafe climbing behavior of workers. Similarly, Ray and Teizer (2012) used Kinect to evaluate the postures of construction workers. Moreover, LADAR and LiDAR technology (which can also sense depth by emitting light) have been deployed for collecting data. To improve workplace awareness, Teizer et al. (2007) proposed to use LADAR to detect and track construction resources (e.g., workers, equipment, materials) in real-time. However, the application of depth cameras is usually limited to indoor spaces due to their limited accuracy especially in the presence of bright sunlight, and having a relatively short working range (Seo et al., 2015). Nonetheless, in general, the direct measurement approach can be used to collect highly accurate data compared to the other approaches (David, 2005; Teschke et al., 2009). Finally, it must be noted that the upfront investment in purchasing and maintaining the direct measurement systems, and the technical knowledge needed to subsequently analyze the data could be very high (David, 2005).

3.2 Artificial Intelligence (AI) for Data Analysis

To analyze the large amounts of complex multi-modal data collected, using one or more of the above methods, various statistical and computational methods have been proposed. These methods include data mining (e.g., regression, pattern recognition, clustering), computer vision (image classification, object detection, object tracking, semantic and instance segmentation), text mining, spatio-temporal analysis, and natural language processing (NLP). Several of these methods are built on the ability of an algorithm to effectively identify and process informative pieces of information (a.k.a., features) in the raw data using a set of rules and/or instructions that are either known to the system upfront (a.k.a., hand-coded) or can be discovered through training on the data of interest. The former category of algorithms is commonly called heuristics (Romanycia and Pelletier, 1985) while the latter constitutes the foundation of artificial intelligence (AI). Given the scope of this chapter, in the following paragraphs, an overview of some of the prevailing AI techniques is presented. This discussion is followed by a description of relevant domain applications in the next section.

3.2.1 Machine Learning (ML)

When a machine, without being explicitly programmed, learns from its experience (E) to perform a particular task (T), the term machine learning (ML) refers to the ability of the machine to consistently improve its performance (P) at task T following the experience E (Mitchell, 2006). In ML, generally, the task could be a regression (i.e., given the input data, outputting a continuous quantitative value), a classification (i.e., assigning the input data into discrete categories), or a clustering (i.e., partitioning the input data into several different groups). ML approaches can be broadly divided into two categories—supervised learning (i.e., learning from given examples), and unsupervised learning (i.e., dividing the input dataset in a meaningful manner). While regression and classification tasks fall into the former category, clustering belongs to the latter. Examples of the most prevailing supervised ML algorithms include linear regression, logistic regression, naïve Bayes, decision tree, random forest, support vector machine (SVM), and *k*-nearest neighbour (kNN) (Dunham, 2006). On the other hand, *k*-means clustering, principal component analysis (PCA), and singular value decomposition (SVD) are examples of unsupervised ML techniques (Bishop, 2006).

Among all ML models, neural network (NN), also known as artificial neural network (ANN), have been adopted in several application domains due to their capability to learn complex nonlinear functions. The idea of NN is derived from neurobiology where the network mimics a human brain containing neurons and synaptic connections (Gardner, 1993). As shown in Fig. 15.1, generally, a NN has three types of layers: input, hidden, and output layers—each containing multiple neurons (a.k.a., nodes). Each node in the hidden and output layers independently uses the inputs on the node and weights on edges, coming from the nodes in the previous layer, and after processing the result with an activation function, passes the output to the nodes in the next layer. At the conclusion of these sequential mathematical operations from layer to layer (a.k.a., forward pass), the node(s) in the output layer finally yields the desired result (e.g., a real value in a regression problem, or a class designation for a classification problem). During the training phase (when the machine learns), the weights of the NN are updated using a backpropagation algorithm to minimize a cost function (a.k.a., loss) (Dunham, 2006).

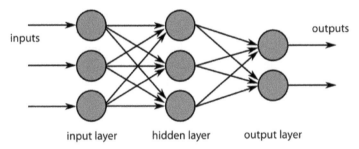

Fig. 15.1: Schematic diagram of a neural network.

Apart from the two traditional learning schemes (i.e., supervised and unsupervised), there are also other ML techniques including deep learning (Simonyan and Zisserman, 2014), semi-supervised learning (Zhu and Goldberg, 2009), active learning (Settles, 2009), reinforcement learning (Sutton and Barto, 2018), adversarial learning (Goodfellow et al., 2014), and imitation learning (Billard and Grollman, 2012). Some of these methods are described in the following sections due to their recent successful use cases in the construction domain.

3.2.2 Deep Learning (DL)

Deep learning (DL) is a branch of ML where a sophisticated NN is trained to perform intricate tasks, e.g., computer vision, audio processing, and NLP. Examples of DL architectures include deep neural network (DNN) (i.e., a NN with multiple layers) and recurrent neural network (RNN) (i.e., a NN with memory blocks to remember sequential data). However, one of the most popular DL architectures is the convolutional neural network (CNN), which is widely used in computer vision problems to analyze image or video data. In general, a CNN architecture includes convolutional layers, batch-normalization layers, pooling layers, and fully connected layers. This form of architecture has gained much popularity, particularly, due to its ability to self-learn features from a given dataset without demanding exorbitant computational power (Kolar et al., 2018). A precursor to the modern CNN algorithms, proposed by LeCun et al. (1998), can recognize handwritten digits in an image. This architecture was developed to automatedly recognize U.S. zip codes handwritten on mail envelopes. More recent CNN models are capable of identifying 1,000 different categories of objects (e.g., dog, cat, flowers, printer, umbrella, bicycle, and many more) in a variety of images (Krizhevsky et al., 2012, Simonyan and Zisserman, 2014). Examples of some popular CNN architectures include VGG-16 (Simonyan and Zisserman, 2014), ResNet-50 (He et al., 2016), and Xception (Chollet, 2017).

In computer vision, CNN is commonly used to perform image classification, object detection, and semantic segmentation (Fig. 15.2). Image classification refers to classifying the entire image into one or multiple classes. In contrast, object detection refers to localizing objects of interest in an image and identifying the class of each object. Due to the complexity of the object detection task, more sophisticated CNN-based algorithms have been developed over the last years. One of the most prevailing examples is region-based CNN (a.k.a., R-CNN) (Girshick et al., 2014) which uses selective search to identify regions of interest (RoI), followed by a CNN to extract features from each region, and finally applying SVM to classify the object in that region. However, as this algorithm performs multiple time-consuming operations on each candidate region of the image, the required computing time and space are excessively high. Therefore, several faster variants of R-CNN, namely Fast R-CNN (Girshick, 2015) and Faster R-CNN (Ren et al., 2017), have been also proposed. Particularly, Faster R-CNN comprises region proposal network (RPN) (a fully convolutional network for proposing RoIs) followed by the Fast R-CNN algorithm for performing the final object detection (Ren et al., 2017). While R-CNN and Fast/Faster R-CNN can output only rectangular bounding boxes for each detected object, another variant of R-CNN, namely Mask R-CNN (He et al., 2017), can output segmentation masks of irregular shapes (a.k.a., semantic segmentation). This architecture has an extra branch to output the segmentation masks in addition to the existing branches of Faster R-CNN that output classification labels and rectangular bounding boxes (He et al., 2017). Another variant, namely region-based fully convolutional network (R-FCN), eliminates the computationally extravagant fully connected layers in R-CNN and uses only the convolutional layers for faster yet accurate object detection (Dai et al., 2016).

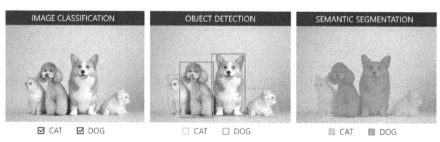

Fig. 15.2: Image classification, object detection, and semantic segmentation.

The region proposal-based methods detect objects using multiple networks, each designed to perform a specific subtask. In contrast, you-only-look-once (YOLO) (Redmon and Farhadi, 2018) and single shot multibox detector (SSD) (Liu et al., 2016) algorithms combine the classification and localization tasks into one single NN which significantly reduces the computational burden. A comparison of the performance of different algorithms performed by Liu et al. (2016) shows that only the YOLO algorithm can perform object detection in real-time, i.e., more than 30 frame per second (FPS).

3.2.3 Generative Adversarial Network (GAN)

Generative adversarial network (GAN) is a ML framework that generates new data from a given input (Goodfellow et al., 2014). A GAN model internally consists of two independent networks, a generator G, and a discriminator D, as shown in Fig. 15.3. The job of G is to generate some data (Y_G) which is indistinguishable from real data (Y_R). On the other hand, given a Y_G or Y_R, the task of D is to correctly distinguish Y_G from Y_R. During training, G and D compete to improve their individual performance. In particular, G tries to generate realistic Y_G so that D fails to differentiate it from real Y_R. Meanwhile, D tries to improve itself to learn more subtle differences between Y_G and Y_R so that G cannot deceive it. This dynamics can be technically viewed as a minimax game, from the perspective of the game theory (Freund and Schapire, 1996), where G and D are two agents and the game settles when each agent achieves the minimum level of competency that is perceived as maximum by the other agent (Goodfellow et al., 2014). After the training step, which is commonly termed adversarial training, discriminator D learns to differentiate Y_G and Y_R even if they are closely similar, while generator G learns to generate realistic Y_G that is difficult to distinguish from Y_R.

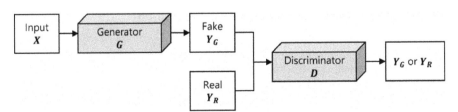

Fig. 15.3: Schematic diagram of a GAN model.

3.2.4 Reinforcement Learning (RL)

Reinforcement learning (RL), another branch of ML, is a trial-and-error method of learning behavior (i.e., what action to take in which situation) through continuous interaction with the environment (Sutton and Barto, 2018). Examples include but are not limited to Google's AlphaGo learning to play a game named Go (Silver et al., 2016; Silver et al., 2017), self-learning robots learning how to walk (Haarnoja et al., 2018; Yang et al., 2020), and self-driving cars learning how to drive on roads (Pan et al., 2017; Liang et al., 2018).

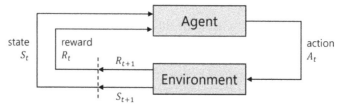

Fig. 15.4: Schematic diagram of an RL process.

In an RL problem, the learner is often referred to as an agent. Everything surrounding the agent, that can be interacted with, is called the environment. At any given time, the part of the environment that is accessible by the agent can be described in a mathematical form which is called the state of the environment. As shown in Fig. 15.4, an RL agent can perform an action to alter the state of the environment. By performing such action, the agent receives feedback from the environment which indicates the merit of the action in meeting the ultimate learning goal of the problem. This feedback is described in a numerical form and is termed the reward.

It must be noted that RL is fundamentally different from commonly used supervised ML techniques. In supervised ML, for example, the machine learns from a given set of training data labeled by a supervisor (e.g., human annotator). However, in RL, the agent actively searches for the training data (i.e., state and reward) in the environment that is not labeled. In practice, RL can be considered as online learning since not all training data are available at once. The RL agent explores the environment and collects the training data. In each iteration of training, the agent considers only the training data available up to that point.

4. AI Applications in Construction

4.1 Successful Use Cases with Practical Implications

4.1.1 Monitoring Ergonomic Risks and Crew Productivity

As previously stated, construction workers are increasingly exposed to ergonomic risks which subsequently lead to WMSDs. One of the major sources of WMSDs is awkward posture while conducting labor-intensive activities. To monitor workers' ergonomic risks related to awkward posture, as shown in Fig. 15.5(a), Nath et al. (2017) proposed to mount smartphones on the worker's body and utilize the built-in IMU sensors to unobtrusively measure the posture and autonomously identify potential work-related ergonomic risks. This low-cost and ubiquitous approach was validated through experiments and results indicated that measurements of trunk and shoulder flexions of a worker are very close to ground truth measurements obtained by video-based observation. The method is applicable not only to construction workers (e.g., laborers, carpenters, welders, material handlers) but also field workers and employees in various other occupations (e.g., farmers, health assistants, teachers, and office workers) who are regularly exposed to WMSDs due to awkward postures.

Fig. 15.5: Awkward posture (a) and overexerted activity (b) recognition using body-mounted smartphones.

Akin to awkward posture, overexertion (i.e., putting excessive force to perform an activity) can lead to severe WMSDs in the long run. Therefore, in another study, Nath et al. (2018) used body-mounted smartphones to recognize overexerted activities (e.g., push, pull, lift, carry, load, unload) in a construction site. Particularly, as shown in Fig. 15.5(b), the time series data collected from bodily-worn smartphone sensors (accelerometer, gyroscope) were analyzed using traditional ML algorithms (e.g., SVM) to recognize workers' activities in 3-second window intervals. Next, by combining the predictions for a longer time interval, frequency and duration information of each activity were measured, and this information was used in conjunction with OSHA provisions (OSHA, 2012) to measure the ergonomic risk associated with overexertion. Moreover, the duration of the recognized activities provides insight into crew productivity and can be used to create discrete event simulation models to further optimize the workflow, allocate resources, and design site layout.

4.1.2 Content Retrieval from Visual Data

Visual data is one of the most common data modalities for recording and documenting construction fieldwork. Photos and videos are frequently used in progress reports, requests for information (RFIs), safety training, productivity monitoring, and claims and litigation. Visual data can be also utilized for generating point clouds and reconstruct building elements in 3D (Brilakis et al., 2011; Feng et al., 2016; Yang et al., 2018). In some applications, the emergence of AI has allowed the automatic analysis of visual data for volumetric measurement of materials (Kamari and Ham, 2020), tracking of workers and resources (Brilakis et al., 2011; Chen and Tang, 2019), monitoring workplace safety (Wang et al., 2019; Tang et al., 2020), and recognizing construction materials (Dimitrov and Golparvar-Fard, 2014).

Recently, the ubiquity of digital cameras, mobile devices (e.g., smartphones) with internet connectivity, and unmanned aerial vehicles (UAVs) (a.k.a., drones) equipped with onboard cameras has exponentially increased the size and quality of visual data collected on a daily basis (Ham and Kamari, 2019). For example, an estimated 1.43 trillion images were taken in 2020 (Canning, 2020) which is 43% more than what was captured in 2018. In construction sites, a large volume of visual data is collected on a daily basis. Han and Golparvar-Fard (2017) collected more than 400,000 images during the construction of a 750,000-sf commercial facility. Fang et al. (2018) collected more than 100,000 images from 25 construction sites in a one-year period. When properly processed and annotated, this big visual data can help increase the accuracy and timeliness of decision-making in construction (Bilal et al., 2016). However, captured images often do not contain rich metadata other than date, time, and (in some cases) geolocation information. Therefore, retrieving specific visual contents from a large volume of images (a.k.a., content retrieval) may turn into a non-trivial and resource-intensive task. A potential solution to this challenge is to create a semantic structure for the collection of images by using metadata tags that, among others, describe the content (e.g., objects, scenes) and appearance (e.g., color, context). In pattern recognition and ML, automated generation of metadata tags for visual data using AI models is termed image content retrieval (Datta et al., 2005). To train an AI model for content retrieval, Nath et al. (2019) developed an in-house dataset (Pictor-v1) which contains ~2,000 images of general construction-related objects—e.g., "building under construction", "construction equipment", 'truck', 'dozer', 'excavator', 'crane', and "construction worker" (Fig. 15.6). By training VGG-16-based CNN models, they were able to successfully assign single or multiple labels (buildings, equipment, worker) to each image. Particularly, transfer learning (Shin et al., 2016) was used to train the CNN model, i.e., the model was pre-trained on a large ImageNet dataset (containing millions of images) and then re-trained (i.e., fine-tuned) on the Pictor-v1 dataset. The trained model was able to successfully assign single class labels with ~90% accuracy and multiple class labels with ~85% accuracy.

Fig. 15.6: Examples of construction image classification and object detection (image datasets can be accessed at https://github.com/ciber-lab).

4.1.3 Construction Object Detection

As shown in Fig. 15.6, compared to image classification, object detection provides more semantically enriched information, such as the number of objects of interest, and the spatial relationships among them, which can be utilized for more refined content retrieval. Furthermore, high-fidelity object detection outlines the basis for scene understanding by helping extract informative descriptors (e.g., type, color, position) of individual objects in the scene. Nath and Behzadan (2020a) developed Pictor-v2, a visual dataset of ~2,500 images with instance-level annotations of construction-related objects (e.g., building, equipment, and worker). Particularly, the dataset contains images obtained from web mining (retrieved through the Internet) and crowdsourcing (collected from construction sites). In general, web-mined images are well structured and have good brightness and contrast. However, they may also contain manipulated backgrounds or other post-processed visual effects. On the other hand, crowdsourced images, despite having a lower visual appeal, may contain more real-world construction-specific information. Therefore, to investigate the influence of these factors, YOLO models were trained and tested on different combinations of the Pictor-v2 dataset (i.e., web-mined, crowdsourced, or combined) with results indicating that models tend to perform better on test images that are visually like the training images. For example, as shown in Fig. 15.7, when a model is trained on only web-mined images, it performs considerably well when tested on web-mined images, and the performance deteriorates when the same model is tested on crowdsourced images. This also reveals a major pitfall in training AI models; since web-mined images are relatively easier and faster to collect, it may be more convenient to train and test the models only on web-mined images and expect similar performance when these models are tested in a real-world scenario. However, such models perform disappointingly poorly in those scenarios, as indicated in Fig. 15.7. To remedy this issue, Nath and Behzadan (2020a) proposed to use a combination of web-mined and crowdsourced images. Their investigation concluded that the model, when trained on the combined dataset, performs well on both web-mined and crowdsourced datasets and exhibits higher accuracy on any testing subset compared to the case where the model was trained on either web-mined or crowdsourced dataset only (Fig. 15.7).

| | | YOLO-v2 | | | YOLO-v3 | | |
| | | Testing subset | | | Testing subset | | |
		Crowdsourced	Webmined	Combined	Crowdsourced	Webmined	Combined
Training subset	Crowdsourced	52.1	46.7	48.5	72.5	53.6	61.1
	Webmined	36.9	62.5	52.3	50.8	70.0	62.3
	Combined	59.6	65.0	62.6	78.2	76.6	77.3

Fig. 15.7: Mean average precision (%) of YOLO-v2 and YOLO-v3 models in detecting construction-related objects, when trained and tested on crowdsourced and web-mined images.

4.1.4 Verification of PPE Compliance

In another construction application, researchers have used AI to automatically monitor workers' compliance with PPE guidelines in real-time. Previous studies, however, either used computationally expensive models (e.g., R-CNN) (Fang et al., 2018; Mneymneh et al., 2018) or trained models to detect only one type of PPE (primarily hard hat) (Fang et al., 2018; Mneymneh et al., 2018; Xie et al., 2018; Wu et al., 2019). In contrast, the study conducted by Nath et al. (2020) designed a comprehensive framework to automatically monitor workers' compliance with PPE guidelines by detecting two PPE components at once (e.g., hard hat, safety vest) along with their colors. To train and test the AI models, another dataset, Pictor-v3, containing ~1,500 annotated images and ~4,700 instances of workers wearing various combinations of PPE components, was developed, and multiple approaches to verify PPE compliance were tested. For example, as shown in Fig. 15.8, in Approach-1, a YOLO-v3-based model detected workers and different PPE types, e.g., hard hat and safety vest, individually. However, such individual detection lacked the contextual relationship between the detected objects (Rabinovich et al., 2007). Therefore, a subsequent ML algorithm (e.g., decision tree) was applied to verify if each of the detected workers is 'properly' wearing any of the detected PPE (i.e., hard hat, safety vest). In Approach-2 (Fig. 15.8), another YOLO-v3-based model was designed to localize workers and directly classify them based on their PPE attire – no hard hat or safety vest (W), only hard hat (WH), only safety vest (WV), and both hard hat and safety vest (WHV). Although this approach was more straightforward, the model exhibited higher interclass confusion since all four classes (i.e., W, WH, WV, WHV) were visually very similar (i.e., all contain persons). In Approach-3 (Fig. 15.8), only workers (regardless of the PPE attire) were detected and cropped. Then, cropped parts of the image that contained workers were fed to a subsequent algorithm to verify the PPE attire. For this, various CNN classifiers were trained to classify the worker's PPE attire into one of the four classes: W, WH, WV, and WHV. It was found that applying the Bayesian ensemble model to combine predictions from multiple CNNs, e.g., VGG-16, ResNet-50, and Xception, yields better accuracy.

Nath and Behzadan (2020b) also described another innovative way through which not only the type of PPE components but also their color was detected with high accuracy. The particular motivation to recognize the color of the PPE components is that, in construction sites, color-coding rules are used to differentiate roles, trades, or access rights (England Highways, 2016). Therefore, the ability to identify the PPE color can provide additional insight into the type of activities that are taking place in a particular location within the site, as well as monitoring site security. In particular, Nath and Behzadan (2020b) adopted a similarity-based technique where a query image of a worker (with unknown PPE attire) was compared with the gallery images of other workers (with known PPE attires). Next, as shown in Fig. 15.9, from the best-matching gallery images, the PPE attire, and their color, for the query worker, was predicted. This technique could detect PPE attire with 90% accuracy while recognizing the color of the PPE component with 77% accuracy.

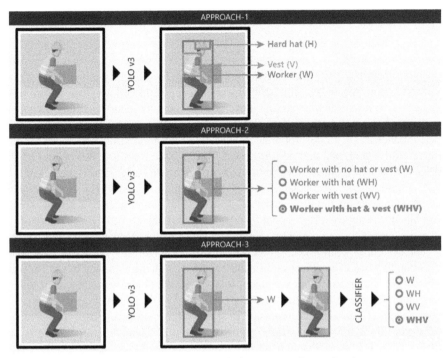

Fig. 15.8: Three approaches for verifying PPE compliance of workers.

Fig. 15.9: Identification of PPE components (hard hat and safety vest) and their colors based on similarity measures.

4.2 Key Challenges in Developing DL Models and Potential Remedies

The first challenge in developing robust DL models for construction applications is the dynamic nature and complexity of construction sites, which means that visual data captured from these sites are intermittent and visually diverse. For example, when viewed in multiple images, the same object may appear in various sizes, at different distances from the camera, and/or be occluded by other objects. Therefore, to better understand the influence of data quality on the quality of visual recognition, the performance of DL models for images taken under different visual conditions should be carefully examined and benchmarked. To this end, Nath and Behzadan (2019) divided object instances in the Pictor-v2 dataset into two categories—'smaller' if the size is smaller than the median size of all objects, and 'larger' otherwise. Moreover, construction sites generally consist of crowded spaces and occluded scenes which can hinder the ability of the model to accurately find objects of interest. To quantify visual congestion in an image, Nath and Behzadan (2019) further split object instances in the Pictor-v2 dataset into 'more crowded' and 'less crowded' groups. If the bounding box of an object significantly overlaps with the bounding box of another object (i.e., the intersection over union, IoU > 0%), the object was considered 'more crowded'. Furthermore, images captured in a poorly lit construction site may not contain content-rich information (due to less brightness and contrast). The amount of useful information in an image can be measured by Shannon entropy (Wu et al. 2013). Objects having Shannon entropy larger than the median value for all objects were tagged as 'well-lit', and all other objects were considered 'poorly-lit'. Given this grouping scheme, it was found that the YOLO-v3 model performs substantially better in detecting large objects. Particularly, for detecting workers, the precision and recall of the model were 98%, indicating near human-level accuracy. Also, the model was able to successfully detect less crowded and well-lit objects more accurately.

The second challenge in developing robust DL models for construction applications is the quality of visual data taken by image capturing devices. Generally, all vision-based algorithms (including object detection) perform better when the quality of the input image is high. For example, researchers have achieved better results in recognizing faces in images and retrieving biological information from medical images by using higher-quality images (Trinh et al., 2014; Li et al., 2018). However, the fact that construction projects often take place in harsh environments might hinder the ability to collect good quality and well-lit imagery, which in turn, can significantly lower the performance of DL models, as was reported by Nath and Behzadan (2019) and Hendrycks and Dietterich (2019). Therefore, in another study, Nath and Behzadan (2020c) proposed an AI-based image enhancement technique that particularly uses a GAN to increase image resolution for fast and reliable object detection. As shown in Fig. 15.10, a GAN model was trained to increase the resolution of an input image (low-resolution) by a factor of 4 in each dimension (width and height). In doing so, the missing information in intermediate pixels was replaced with synthetic high-quality information.

Fig. 15.10: Example of low-resolution images, GAN-generated images, and corresponding high-resolution images.

Next, two previously developed YOLO-v3 models were applied to the GAN-generated images to measure the quality of object detection when performed on enhanced images. Particularly, one YOLO-v3 model was trained on the Pictor-v2 dataset to detect common construction objects, namely buildings, equipment, and workers, as described earlier. Another YOLO-v3 model was trained on the Pictor-v3 dataset to detect workers and PPE, e.g., hard hat (H) and safety vest (V), also described in the previous sections. Results illustrated that for varying input resolutions, both models performed 4%–21% better in detecting objects when the input image was improved by the GAN model.

The third challenge in developing robust DL models for construction applications is the reliance of the previously mentioned methods for object detection and safety monitoring on the visibility of the objects of interest in the scene. The problem of visual occlusion is one of the most prevailing barriers to successful visual recognition (Hoiem et al., 2011). For example, in a previous study, a significant drop (2%–7%) in the performance of facial recognition was observed when the face was partially occluded (Ekenel and Stiefelhagen, 2009). When DL models are used for safety monitoring, a worker's PPE compliance may not be verifiable with high certainty from a particular camera vantage point due to parts of the worker's body remaining occluded by other objects. A trivial solution to this problem is to revisit the construction site and capture images from a different viewing angle in which the worker appears adequately visible. However, data collected from various parts of a large construction site is ephemeral especially when multiple crews and equipment occupy and work in the same area simultaneously. Additionally, when objects move in front of one another and partially block the camera view, it may not be feasible to always collect occlusion-free images. Short of installing multiple cameras, which requires extensive maintenance and calibration effort, a potential solution could be to develop an intelligent system where a camera with active vision capability (i.e., ability to navigate in the scene and operate autonomously) searches for the optimal viewpoint from which sufficiently good and occlusion-free visual data can be collected. Nath (2021) demonstrated the feasibility of this approach by training an AI-enabled flying camera (e.g., mounted on a drone) using RL, and testing it in a simulated environment to validate the proposed method. They found that starting from the randomly selected initial positions, in almost 90% of the times, the camera was able to find a position from which the object of interest (in this case, a construction worker) was better visible. This framework has the potential to support and promote human-machine collaboration in construction sites by delegating the routine-task of continuous monitoring of safety compliance to the autonomous robots.

5. Practical Considerations for the Use of AI and Automation in Construction

AI is revolutionizing the construction domain by leading the industry toward automation and intelligent workflows. Particularly, the application of AI in human-machine collaboration in the construction site is expected to lead to safer workspaces, more productive crews, and higher quality of work. To bring the full potential of AI into practice and outlining the theoretical knowledge for successful human-machine collaboration, researchers and major technology developers are leaning toward construction robotics. Beyond technical challenges (some of which were described in this chapter), there are some practical considerations that must be adequately addressed for these technologies to be fully accepted and adopted by industry users. This section provides a discussion of some of these challenges and potential remedies in the future of construction.

5.1 Technology Acceptance and Adoption

Successful industry-wide implementation of any new technological advancement depends on addressing a major question: will workers and company leadership trust the new system? Construction practitioners are generally distrustful of new technologies, and obtaining management buy-ins, adjusting the established engineering workflows, and creating trust among workers have proven extremely difficult (Son et al., 2012). While the introduction of new technologies is a continuous endeavor, there is an urgent need to identify the proper psychological infrastructure required to adopt them in construction. AI, in particular, is perceived by many end-users as a 'blackbox' (Castelvecchi, 2016) that to a large extent, overlooks trust factors such as technology transparency, reliability, privacy and security, ethics, and cultural elements, all of which need to be properly addressed before AI solutions can be adopted. Robots, as another example, may mitigate the risk of physical safety while worker-robot teams' psychological safety must be also considered. It is important that workers see robots as teammates who enable transitioning their role to higher-level tasks rather than outsiders waiting to ambush them for their jobs. Currently, the latter is a more widespread theory among trade workers if not companies' leadership (Dahlin, 2019; Ivanov et al., 2020).

5.2 Industry Trends and Emergence of the 'Contech' Entrepreneurial Ecosystem

Currently, in the construction industry, technologies are rarely developed in-house, and the common trend is to purchase off-the-shelf products, or license them from or partner up with outside developers. This has resulted in the emergence of small but fast-growing start-up companies founded by individual enthusiasts with AI-based value propositions who have chosen to focus on construction as a niche market. This relatively new ecosystem is known as 'Contech' (Braesemann and Baum, 2020) in the entrepreneurial community and has been a welcome news for venture capitalists who can invest in companies that enter a profitable market given the severe need for technology to accelerate project durations, enhance safety and productivity issues, and provide predictive capacity for project bidding or performance measurement. In the meantime, construction companies have also embraced such partnerships as they see the competitive edge that is brought about. The business models of Contech companies are similar and simple (i.e., subscription- or project-based revenues steams) and generally work well with the unique cashflow system in construction. This is in part due to the heavy use of subcontractors in construction which is a known model for general contractors that they adopt to Contech start-ups.

5.3 Construction Collaborative Robotics through Imitation Learning

Industrial robots that perform necessary but tedious tasks alongside humans are a staple of science fiction. Rosie the Robot from the Jetsons is a stereotypical model of a robot maid capable of autonomously manipulating objects and accomplishing complex tasks (Anderson and Oates, 2007). There is a consensus that industrial robots with the abilities of Rosie could save significant time and effort from human workers.

One strategy for making robots autonomous and capable of manipulation is to program them to follow a specific trajectory and hardcode their joints. This approach is feasible for position-controlled industrial robots in well-structured assembly lines. However, such robotic systems cannot adapt to changing environments such as construction jobsites that are highly dynamic and unstructured. In such workspaces, a robot needs to understand its environment and workpieces to adjust its manipulation task(s) accordingly. This requires the robot to learn the tasks as opposed to being programmed to perform them. A large portion of construction processes involves repetitive tasks that occur in less structured environments than those of other industries such as manufacturing. Construction robots must, therefore, understand their environment and be capable of manipulation without being preprogrammed to follow a specific trajectory. What makes this even more complex is that in each given task, there might be an infinite number of conceivable scenarios making it practically impossible to pre-program or make explicit and real-time commands to a non-adaptive robot. Therefore, adaptation to the work, workplace, and workers' way of doing things is key in construction robotics. A trending robot training approach that enables this adaptation is learning from demonstration (LfD) or imitation learning (Argall et al., 2009; Billard and Grollman, 2012). The key idea behind LfD is for the robot to learn the job by observing demonstrations of a task. LfD is equivalent to behavior cloning loss in the space of the primary latent encoding (Rahmatizadeh et al., 2018). This technique enhances the transfer of knowledge from the user (e.g., construction worker, supervisor) to the robot through the natural means of demonstration in place of other methods such as reward specification (e.g., reinforcement learning). For example, Finn et al. (2017) demonstrated a one-shot meta-imitation learning approach where a robot learns to place an object into a container (both the object and the container are not previously seen) from only one single demonstration. In another work, Bonardi et al. (2020) showed that a robot can infer hand actions by analyzing simulated demonstrations and learns a new action (e.g., placing or pushing objects) from a single real-world demonstration. Sieb (2019) showed that the robot can learn such actions just by observing a human in a video. Also, Codevilla et al. (2018) demonstrated an end-to-end imitation learning approach where an autonomous vehicle learns to follow a road and avoid obstacles by imitating an expert human driver.

5.4 Trust Building and Calibration

Trust is a significant concern in deploying AI and robotics technologies in any context (Siau and Wang, 2018). Recent research effort on topics such as explainable AI (XAI) with the goal of substantial improvement in transparency stems from this concern and proves the need for further investigation (Holzinger, 2018; Samek et al., 2019). Construction is a uniquely sensitive industry in this regard, and confronts a real socio-technical issue with respect to adopting any new technology but especially those that lack transparency (Singh and Ashuri, 2019; Pan and Zhang, 2021). The reasons can be summarized as follows:

1. The construction industry is a conglomerate of many small businesses: 82.3% of the construction employment belongs to small businesses (compared to manufacturing at 44.4% and retail at 35.1%) (Kobe, 2002), and small companies tend to be slower in adopting technologies (Peltier et al., 2012).

2. There is a heavy use of subcontractors: lack of consistency in technology adoption requires workers to switch back-and-forth between different methods of work when employed as a subcontractor of different general contractors who have varied practices in technology adoption. Phased and inconsistent adoption of workplace technologies is known as a stressor due to the increased confusion and mistake-making as a result (Ragu-Nathan et al., 2008).

3. Construction is a competitive-bid-based industry: the typically very small profit margin (i.e., 3%–5%) makes construction practitioners more conservative in

using new technologies with unproven reliability, privacy/security guarantees, or cost-saving performance.

4. The industry lacks technology and innovation leaders: a leading industry advocate often paves the way and shows why, when, and how to benefit from technology. In the retail, Internet, or automotive industries there are Amazon, Google, Tesla, respectively, but construction lacks such a role model. This is a major barrier to creating trust toward innovation (Audretsch et al., 2018).

In a recent survey of the literature, researchers identified transparency and explainability, privacy and security, safety and reliability, and ethics and fairness among the main drivers of trust in AI and robotics technologies. They also concluded that these elements play an important role in engendering trust in AI and robotics in construction industry regardless of the project phase, project type, application objective (e.g., safety, productivity), and specific technology being leveraged (e.g., BIM, blockchain) (Emaminejad et al., 2021). There is a dearth of research on the topic of construction technology acceptance and adoption and particularly about AI and robotics. A survey of industry professionals suggests that the lack of personnel training and the abundance of technology-averse practitioners are among the most important barriers to technology adoption while safety, productivity, and progress monitoring are the top three application areas that can benefit from data-driven technology integration (Mansouri et al., 2020). A vast majority of the previous work in the area of technology acceptance investigated the adoption barriers and enablers of only one technology or related advancements; BIM (Ademci and Gundes, 2018; Nguyen and Akhavian, 2019). Research on BIM implementation indicates that usability metrics such as a perceived usefulness (PU) and perceived ease of use (PE) are among the strong antecedents of intentions to adopt technology in construction (Wang and Song, 2017). PU and PE are the main determinants of users' attitudes and behavioral intention and the core constructs of the most widely applied behavioral model in the information systems (Davis, 1989; Hess et al., 2014; Wicaksono and Maharani, 2020) called the technology acceptance model (TAM) (Rauniar et al., 2014). Other popular models include the social network threshold (SNT) (Valente, 1996) and diffusion of innovation (DoI) theory (Mustonen-Ollila and Lyytinen, 2003; Rogers, 2003). While these are meant to be used as generic and consistent tools (Legris et al., 2003), research has shown that extending their constructs with domain-specific (e.g., construction-specific) factors to derive curated models yields improved outcome (Ishak, 2013). Finally, it is imperative to not overlook the psychological safety and the negative connotation that words such as robotics and AI carry in the context of future workers' job security. The somehow stressful presence of AI-enabled robots alongside construction workers could have serious negative effects on their health (Gihleb et al., 2020). Furthermore, eventhough building trust is essential to increasing adoption levels, placing unjustified trust in AI will result in disuse/misuse due to under-trust/over-trust (Parasuraman and Riley, 1997; Hoffman et al., 2013). Real-world applications and laboratory-scale experiments indicate that both cases are extremely common (Hoffman et al., 2013). The process of appropriately adjusting human level of trust with the actual reliability of the AI system is called trust calibration (Okamura and Yamada, 2020; Zhang et al., 2020). Research has shown that creating transparency about AI's abilities significantly increases the likelihood of building calibrated trust and increasing adoption levels (Wang et al., 2016). In construction, many studies have shown that perceived added value of new technologies in terms of safety, efficiency, and gains in cost or time saving have a substantial impact on adoption levels of those technologies (Nikas et al., 2007; Ahuja et al., 2009; Sepasgozar and Bernold, 2013; Delgado et al., 2019).

5.5 Future Workforce

The demand for jobs in the construction, infrastructure, and building sectors is growing due to labor shortages and to meet affordable housing challenges. This new demand is projected to create 80–200 million jobs by 2030 (Manyika et al., 2017). While such growth in other sectors may mean a substantial increase in the number of freelancers or remote workers, construction workers will continue to work collaboratively and on sites. The future workers are Millennials and Gen Z who are more 'tech-savvy' compared to the current workforce in construction who are characterized by being 'tech-averse' (Singh and Dangmei, 2016). Additionally, the existing construction workforce is aging. The number of construction workers aged 55 and up has increased from just under 17% in 2011 to almost 22% in 2018 (BLS, 2018). Meanwhile, the number of construction workers aged 25–54 has declined from about 75% to 69% (BLS, 2018). The National Academies of Sciences, Engineering, and Medicine has indicated a major gap in adequately developing and sustaining a technical workforce with the skills needed to compete in the 21st century (National Academies, 2017). This skill gap can lead to major consequences in practice particularly when workers are tasked with complex jobs. For example, while in 1992, workers aged 55 and over accounted for only 20% of fatal injuries, in 2019, workers aged 55 and over accounted for 38% of all workplace fatalities (BLS, 2019). Therefore, the problem is two-fold – the current workforce is aging which means that they need to be reskilled, and the future workforce is tech-savvy which means that they demand more technology in their workspace for the job to be appealing to them.

Workers are the cornerstone of production and thus central to the growth of the construction industry. Therefore, workforce development and upskilling through technology should be at the forefront of efforts to address the grand challenges this faced by the construction industry. This means that AI and robotics should be socially adaptive to the workflows and technology culture established in the industry. For example, the social adeptness of collaborative robots is needed so that team members can coordinate their actions to follow the shared plan to assess the efficacy of worker-robot collaboration. Towards this goal, theoretical foundations built on social and behavioral science notions such as shared cooperative activity (Bratman, 1992), joint intention (Jennings, 1993), and common ground (Park et al., 2012) can help achieve the required level of co-adaptation in adopting AI by the future construction workforce.

6. Conclusion

Despite being one of the largest global industries and having a major impact on the world economy, the construction industry is suffering from slow productivity growth. Additionally, the large number and frequency of illnesses, injuries, and fatalities have characterized construction as one of the most hazardous occupations. As one of the least digitalized industries, a substantial portion of the construction industry lags in adopting new technologies. Furthermore, the increasing problem of an aging workforce necessitates significant reskilling efforts to train a competitive workforce for the 21st century. Fortunately, there are indications that the construction industry will continue to boom despite its major drawbacks. Particularly, new technologies (e.g., BIM, 3D printing, VR, AR, IoT, robotics) hold the key to revolutionizing the industry and improving productivity and safety by a significant margin. More specifically, human-machine collaboration has the potential to disrupt the industry by reforming current practices and increasing the quality of work, ultimately leading to savings in project cost and time, while confining risks.

To achieve this objective, researchers and practitioners are increasingly leaning on AI, computer vision, and robotics due to their remarkable success in other domains. This chapter provided a narrative about the authors' successful past work that has demonstrated highly efficient AI-based methods to improve construction safety and productivity. For example, ML techniques were utilized to analyze body-mounted smartphone sensor data to prevent WMSD-related risks on the jobsite. Also, DL-based fast algorithms were designed and validated to identify common construction objects in real-time which laid the foundation for scene understanding by autonomous robots. Moreover, realizing the importance of wearing proper PPE (particularly, hard hat and safety vest) to prevent jobsite injuries, DL methods were designed to verify PPE compliance of workers in real-time. While these AI applications have shown a remarkable potential to autonomously monitor the safety and productivity of workers, key challenges were also identified that might impede the large-scale adoption and implementation of these methods in practice. For example, the dynamic and complex nature of a construction site could limit the ability to collect good quality and occlusion-free images for visual recognition. Firstly, it was found that AI models perform object detection better if the image contains larger (closer to the camera), less crowded, and well-lit objects. Therefore, it is recommended that to obtain best results, similar visual conditions must be met in practice. Even then, it might not be feasible to continuously capture high-quality images due to, for example, weather conditions and hardware limitations. Therefore, GAN-based techniques were developed to enhance the quality of low-resolution images that led to better object detections. Also, to address the occlusion problem, autonomous cameras (with multiple degrees of freedom) were trained through RL to hover in a construction site and find locations from which better visibility of the objects of interest can be achieved.

While researchers continue to lay the theoretical foundations for successful human-machine collaborations in construction, there are several important responsibilities for the practitioners to effectively promote these forthcoming technologies in the jobsite. This chapter discussed some of the most important barriers to the full acceptance and adoption of these technologies in the field. Firstly, trust in AI needs to be built among the business owners and employees. Particularly, this can be achieved by promoting the usefulness and ease of use of the AI technologies and demonstrating its benefits in tangible terms such as cost and time savings and reduced work-related injuries, as well as through fostering transparency and explainability, privacy and security, safety and reliability, and ethics and fairness. Moreover, AI developers (particularly Contech start-ups with strong ties to the industry and labor unions) should come forward to establish themselves as role models for others. At the same time, the construction workforce needs to be reskilled and upskilled to work with new technologies and be prepared to adopt AI and robotics for future construction. Ultimately, by promoting and enabling successful human-machine partnerships in the construction industry, a safer and more productive workplace can be established that can bring about socio-economic benefits to the employers, employees, and stakeholders.

References

Academies, N. 2017. United States' Skilled Technical Workforce is Inadequate to Compete in Coming Decades: Actions Needed to Improve Education, Training, and Lifelong Learning of Workers. https://www.nationalacademies.org/news/2017/05/united-states-skilled-technical-workforce-is-inadequate-to-compete-in-coming-decades-actions-needed-to-improve-education-training-and-lifelong-learning-of-workers.

Ademci, E. and Gundes, S. 2018. Review of studies on BIM adoption in AEC industry. *In*: Ademci, E. and Gundes, S. (eds.). *Review of Studies on BIM Adoption in AEC Industry, 5th International Project and Construction Management Conference (IPCMC) Proceedings*.

Ahuja, V., Yang, J. and Shankar, R. 2009. Study of ICT adoption for building project management in the Indian construction industry. *Automation in Construction*, 18(4): 415–423.

Akbar-Khanzadeh, F. 1998. Factors contributing to discomfort or dissatisfaction as a result of wearing personal protective equipment. *Journal of Human Ergology*, 27(1-2): 70–75.

Akhavian, R. and Behzadan, A.H. 2015. Construction equipment activity recognition for simulation input modeling using mobile sensors and machine learning classifiers. *Advanced Engineering Informatics*, 29(4): 867–877.

Akhavian, R. and Behzadan, A.H. 2016. Smartphone-based construction workers' activity recognition and classification. *Automation in Construction*, 71: 198–209.

Aljohani, A., Ahiaga-Dagbui, D. and Moore, D. 2017. Construction projects cost overrun: What does the literature tell us? *International Journal of Innovation, Management and Technology*, 8(2): 137.

Alwasel, A., Elrayes, K., Abdel-Rahman, E.M. and Haas, C. 2011. Sensing construction work-related musculoskeletal disorders (WMSDs). *ISARC*.

Anderson, M.L. and Oates, T. 2007. A review of recent research in metareasoning and metalearning. *AI Magazine*, 28(1): 12–12.

Argall, B.D., Chernova, S. Veloso, M. and Browning, B. 2009. A survey of robot learning from demonstration. *Robotics and Autonomous Systems*, 57(5): 469–483.

Aryal, A. and Becerik-Gerber, B. 2019. A comparative study of predicting individual thermal sensation and satisfaction using wrist-worn temperature sensor, thermal camera and ambient temperature sensor. *Building and Environment*, 160: 106223.

Audretsch, D.B., Seitz, N. and Rouch, K.M. 2018. Tolerance and innovation: The role of institutional and social trust. *Eurasian Business Review*, 8(1): 71–92.

Autodesk. 2020. Future of work. Retrieved 19 January 2021. https://www.autodesk.com/future-of-work.

Balogh, I., Ørbæk, P., Ohlsson, K., Nordander, C., Unge, J., Winkel, J. and Hansson, G.Å. 2004. Self-assessed and directly measured occupational physical activities: Influence of musculoskeletal complaints, age and gender. *Applied Ergonomics*, 35(1): 49–56.

Bilal, M., Oyedele, L.O. Qadir, J., Munir, K., Ajayi, S.O., Akinade, O.O., Owolabi, H.A., Alaka, H.A. and Pasha, M. 2016. Big Data in the construction industry: A review of present status, opportunities, and future trends. *Advanced Engineering Informatics*, 30(3): 500–521.

Billard A. and Grollman D. 2012. Imitation learning in robots. *In*: Seel, N.M. (ed.). *Encyclopedia of the Sciences of Learning*. Springer, Boston, MA. https://doi.org/10.1007/978-1-4419-1428-6_758.

Bishop, C.M. 2006. *Pattern Recognition and Machine Learning.* Berlin-Heidelberg, Germany: Springer. http://www.academia.edu/download/30428242/bg0137.pdf.

BLS. 2014. *Nonfatal Occupational Injuries and Illnesses Requiring Days Away from Work.* https://www.bls.gov/news.release/osh2.nr0.htm.

BLS. 2018. *Labor Force Statistics from the Current Population Survey.* https://www.bls.gov/cps/cpsaat18b.htm.

BLS. 2019. *Industries at a Glance: Construction.* Retrieved 20 January 2019. https://www.bls.gov/iag/tgs/iag23.htm.

BLS. 2019. *Injuries, Illnesses, and Fatalities.* https://www.bls.gov/iif/oshcfoi1.htm.

BLS. 2020. *Employer-reported Workplace Injuries and Illnesses: 2019.* Retrieved 20 May 2021. https://www.bls.gov/news.release/pdf/osh.pdf.

BLS. 2020. *National Census of Fatal Occupational Injuries in 2019.* Retrieved 20 May 2021. https://www.bls.gov/news.release/pdf/cfoi.pdf.

BLS. 2021. *Industries at a Glance: Construction.* Retrieved 17 June 2021. https://www.bls.gov/iag/tgs/iag23.htm.

Bonardi, A., James, S. and Davison, A.J. 2020. Learning one-shot imitation from humans without humans. *IEEE Robotics and Automation Letters*, 5(2): 3533–3539.

Braesemann, F. and Baum, A. 2020. *PropTech: Turning real estate into a data-driven market?* Available at SSRN (Social Serives Research Networks) 3607238.

Bratman, M.E. 1992. Shared cooperative activity. *The Philosophical Review*, 101(2): 327–341.

Brilakis, I., Fathi, H. and Rashidi, A. 2011. Progressive 3D reconstruction of infrastructure with videogrammetry. *Automation in Construction*, 20(7): 884–895.

Brilakis, I., Park, M.-W. and Jog, G. 2011. Automated vision tracking of project related entities. *Advanced Engineering Informatics*, 25(4): 713–724.

Businesswire. 2021. *Global Construction Industry Report 2021: $10.5 Trillion Growth Opportunities by 2023.* ResearchAndMarkets.com. https://www.businesswire.com/news/home/20210111005587/en/Global-Construction-Industry-Report-2021-10.5-Trillion-Growth-Opportunities-by-2023---ResearchAndMarkets.com.

Campo, M., Weiser, S., Koenig, K.L. and Nordin, M. 2008. Work-related musculoskeletal disorders in physical therapists: A prospective cohort study with 1-year follow-up. *Physical Therapy*, 88(5): 608–619.

Canning, J. 2020. 1.43 trillion photos were taken in 2020 but how many of them were captured on our mobile phones? Retrieved 5 April 2021. https://www.buymobiles.net/blog/1-43-trillion-photos-were-taken-in-2020-but-how-many-of-them-were-captured-on-our-mobile-phones.

Castelvecchi, D. 2016. Can we open the black box of AI? *Nature News*, 538(7623): 20.

CDC. 2019. Traumatic brain injuries in construction. Retrieved 20 January 2019. https://blogs.cdc.gov/niosh-science-blog/2016/03/21/constructiontbi/.

Chen, J. and Tang, P. 2019. A computational framework for characterizing multiple object tracking methods in construction field applications. *In*: *Computing in Civil Engineering 2019: Data, Sensing, and Analytics, American Society of Civil Engineers.* Reston, VA, 290–297.

Chollet, F. 2017. Xception: Deep learning with depthwise separable convolutions. *Proceedings of the IEEE Conference on Computer Vision and Pattern Recognition.*

Codevilla, F.M., Müller, P., López, A., Koltun, V. and Dosovitskiy, A. 2018. End-to-end driving via conditional imitation learning. *2018 IEEE International Conference on Robotics and Automation (ICRA), IEEE.*

Dahlin, E. 2019. Are robots stealing our jobs? *Socius*, 5: 2378023119846249.

Dai, J., Li, Y., He, K. and Sun, J. 2016. R-fcn: Object detection via region-based fully convolutional networks. *Advances in Neural Information Processing Systems.*

Dartt, A., Rosecrance, J., Gerr, F., Chen, P., Anton, D. and Merlino L. 2009. Reliability of assessing upper limb postures among workers performing manufacturing tasks. *Applied Ergonomics*, 40(3): 371–378.

Datta, R., Li, J. and Wang, J.Z. 2005. Content-based image retrieval: Approaches and trends of the new age. *Proceedings of the 7th ACM SIGMM International Workshop on Multimedia Information Retrieval.*

David, G.C. 2005. Ergonomic methods for assessing exposure to risk factors for work-related musculoskeletal disorders. *Occupational Medicine*, 55(3): 190–199.

Davis, F.D. 1989. Perceived usefulness, perceived ease of use, and user acceptance of information technology. *MIS Quarterly*, 13(3): 319–339.

Delgado, J.M.D., Oyedele, L., Ajayi, A., Akanbi, L., Akinade, O., Bilal, M. and Owolabi H. 2019. Robotics and automated systems in construction: Understanding industry-specific challenges for adoption. *Journal of Building Engineering*, 26: 100868.

Dimitrov, A. and Golparvar-Fard, M. 2014. Vision-based material recognition for automated monitoring of construction progress and generating building information modeling from unordered site image collections. *Advanced Engineering Informatics*, 28(1): 37–49.

Ding, L., Fang, W., Luo, H., Love, P.E., Zhong, B. and Ouyang, X. 2018. A deep hybrid learning model to detect unsafe behavior: Integrating convolution neural networks and long short-term memory. *Automation in Construction*, 86: 118–124.

Doumbouya, L., Gao, G. and Guan, C. 2016. Adoption of the Building Information Modeling (BIM) for construction project effectiveness: The review of BIM benefits. *American Journal of Civil Engineering and Architecture*, 4(3): 74–79.

Dunham, M.H. 2006. *Data Mining: Introductory and advanced concepts.* Upper Saddle River, NJ. Pearson Education.

Economist. 2017. The construction industry's productivity problem. https://www.economist.com/leaders/2017/08/17/the-construction-industrys-productivity-problem.

Ekenel, H.K. and Stiefelhagen, R. 2009. Why is facial occlusion a challenging problem? *International Conference on Biometrics.* Springer.

Emaminejad, N., North, A.M. and Akhavian, R. 2021. Trust in AI and Implications for the AEC Research: A Literature Analysis. *International Conference on Computing in Civil Engineering (i3CE).* Reston, VA, ASCE.

England, H. 2016. Health and safety for major road schemes: Safety helmet colours. Retrieved 24 July 2019. https://www.gov.uk/government/publications/health-and-safety-for-major-road-schemes-safety-helmet-colours.

Fang, Q., Li, H., Luo, X., Ding, L., Luo, H., Rose, T.M. and An, W. 2018. Detecting non-hardhat-use by a deep learning method from far-field surveillance videos. *Automation in Construction*, 85: 1–9.

Feng, C., Taguchi, Y. and Kamat, V. 2016. *Method for Extracting Planes from 3D Point Cloud Sensor Data.* Google Patents.

Finn, C., Yu, T., Zhang, T., Abbeel, P. and Levine, S. 2017. One-shot visual imitation learning via meta-learning. *Conference on Robot Learning, PMLR.*

Freund, Y. and Schapire, R.E. 1996. Game theory, on-line prediction and boosting. *Proceedings of the Ninth Annual Conference on Computational Learning Theory.*

Gandhi, P., Khanna, S. and Ramaswamy, S. 2016. Which industries are the most digital (and why). *Harvard Business Review*, 1: 45–48.

Gardner, D. 1993. *The Neurobiology of Neural Networks.* MIT Press.

Gerdle, B., Björk, J., Cöster, L., Henriksson, K.-G., Henriksson, C. and Bengtsson A. 2008. Prevalence of widespread pain and associations with work status: A population study. *BMC Musculoskeletal Disorders*, 9(1): 1–10.

Gihleb, R., Giuntella, O., Stella, L. and Wang, T. 2020. *Industrial Robots, Workers' Safety, and Health.* Institute of Labour Economics (IZA), Bonn.

Girshick, R. 2015. Fast R-CNN. *Proceedings of the IEEE International Conference on Computer Vision.* arXiv:1504.08083.

Girshick, R., Donahue, J., Darrell, T. and Malik, J. 2014. Rich feature hierarchies for accurate object detection and semantic segmentation. *Proceedings of the IEEE Conference on Computer Vision and Pattern Recognition.*

Goodfellow, I., Pouget-Abadie, J., Mirza, M., Xu, B., Warde-Farley, D., Ozair, S., Courville, A. and Bengio, Y. 2014. Generative adversarial nets. *Advances in Neural Information Processing Systems.*

Haarnoja, T., Ha, S., Zhou, A., Tan, J., Tucker, G. and Levine, S. 2018. Learning to walk via deep reinforcement learning. *arXiv preprint arXiv*:1812.11103.

Ham, Y. and Kamari, M. 2019. Automated content-based filtering for enhanced vision-based documentation in construction toward exploiting big visual data from drones. *Automation in Construction,* 105: 102831.

Han, K.K. and Golparvar-Fard, M. 2017. Potential of big visual data and building information modeling for construction performance analytics: An exploratory study. *Automation in Construction,* 73: 184–198.

Han, S. and Lee, S. 2013. A vision-based motion capture and recognition framework for behavior-based safety management. *Automation in Construction,* 35: 131–141.

Han, S., Lee, S. and Peña-Mora, F. 2012. Vision-based detection of unsafe actions of a construction worker: Case study of ladder climbing. *Journal of Computing in Civil Engineering,* 27(6): 635–644.

Hao, J.L., Cheng, B., Lu, W., Xu, J., Wang, J., Bu, W. and Guo, Z. 2020. Carbon emission reduction in prefabrication construction during materialization stage: A BIM-based life-cycle assessment approach. *Science of the Total Environment,* 723: 137870.

He, K., Gkioxari, G., Dollár, P. and Girshick, R. 2017. Mask R-CNN. *2017 IEEE International Conference. on Computer Vision (ICCV).*

He, K., Zhang, X., Ren, S. and Sun, J. 2016. Deep residual learning for image recognition. *Proceedings of the IEEE Conference on Computer Vision and Pattern Recognition.*

Hendrycks, D. and Dietterich, T. 2019. Benchmarking neural network robustness to common corruptions and perturbations. *arXiv preprint arXiv*:1903.12261.

Hess, T.J., McNab, A.L. and Basoglu, K.A. 2014. Reliability generalization of perceived ease of use, perceived usefulness, and behavioral intentions. *MIS Quarterly,* 38(1): 1–28.

Hoffman, R.R., Johnson, M., Bradshaw, J.M. and Underbrink, A. 2013. Trust in automation. *IEEE Intelligent Systems,* 28(1): 84–88.

Hoiem, D., Efros, A.A. and Hebert, M. 2011. Recovering occlusion boundaries from an image. *International Journal of Computer Vision,* 91(3): 328–346.

Holzinger, A. 2018. From machine learning to explainable AI. *2018 World Symposium on Digital Intelligence for Systems and Machines (DISA), IEEE.*

Huang, X. and Hinze, J. 2003. Analysis of construction worker fall accidents. *Journal of Construction Engineering and Management,* 129(3): 262–271.

Ishak, S.S.M. 2013. *Resistance Factors to Technology Innovation in Construction Organisations.* University of New South Wales Sydney, Australia.

Ivanov, S., Kuyumdzhiev, M. and Webster, C. 2020. Automation fears: Drivers and solutions. *Technology in Society,* 63: 101431.

Jebelli, H., Ahn, C.R. and Stentz, T.L. 2014. The validation of gait-stability metrics to assess construction workers' fall risk. http://ascelibrary.org/doi/abs/10.1061/9780784413616.124.

Jennings, N.R. 1993. Specification and implementation of a belief-desire-joint-intention architecture for collaborative problem solving. *International Journal of Intelligent and Cooperative Information Systems,* 2(03): 289–318.

Kamari, M. and Ham, Y. 2020. Vision-based volumetric measurements via deep learning-based point cloud segmentation for material management in jobsites. *Automation in Construction,* 121: 103430.

Kelm, A., Laußat, L., Meins-Becker, A., Platz, D., Khazaee, M.J., Costin, A.M., Helmus, M. and Teizer, J. 2013. Mobile passive Radio Frequency Identification (RFID) portal for automated and rapid control of Personal Protective Equipment (PPE) on construction sites. *Automation in Construction,* 36: 38–52.

Kobe, K. 2002. Small business GDP. *Update,* 2010: 46.

Kolar, Z., Chen, H. and Luo, X. 2018. Transfer learning and deep convolutional neural networks for safety guardrail detection in 2D images. *Automation in Construction,* 89: 58–70.

Krizhevsky, A., Sutskever, I. and Hinton, G.E. 2012. Imagenet classification with deep convolutional neural networks. *Advances in Neural Information Processing Systems.*

LeCun, Y., Bottou, L., Bengio, Y. and Haffner, P. 1998. Gradient-based learning applied to document recognition. *Proceedings of the IEEE*, 86(11): 2278–2324.

Lee, H.-S., Lee, K.-P., Park, M., Baek, Y. and Lee, S. 2012. RFID-based real-time locating system for construction safety management. *Journal of Computing in Civil Engineering*, 26(3): 366–377.

Legris, P., Ingham, J. and Collerette, P. 2003. Why do people use information technology? A critical review of the technology acceptance model. *Information & Management*, 40(3): 191–204.

Li, N., Calis, G. and Becerik-Gerber, B. 2012. Measuring and monitoring occupancy with an RFID based system for demand-driven HVAC operations. *Automation in Construction*, 24: 89–99.

Li, P., Prieto, L., Mery, D. and Flynn, P. 2018. Face recognition in low quality images: A survey. *arXiv preprint arXiv*:1805.11519.

Li, Q., Ji, C., Yuan, J. and Han, R. 2017. Developing dimensions and key indicators for the safety climate within China's construction teams: A questionnaire survey on construction sites in Nanjing. *Safety Science*, 93: 266–276.

Liang, X., Wang, T., Yang, L. and Xing, E. 2018. CIRL: Controllable imitative reinforcement learning for vision-based self-driving. *Proceedings of the European Conference on Computer Vision* (*ECCV*).

Liu, W., Anguelov, D., Erhan, D., Szegedy, C., Reed, S., Fu, C.-Y. and Berg, A.C. 2016. SSD: Single shot multibox detector. *European Conference on Computer Vision*. Springer.

Lucintel. 2021. *Growth Opportunities in the Global Construction Industry.* https://www.lucintel.com>.

Mansouri, S., Castronovo, F. and Akhavian, R. 2020. Analysis of the synergistic effect of data analytics and technology trends in the AEC/FM industry. *Journal of Construction Engineering and Management*, 146(3): 04019113.

Manyika, J., Lund, S., Chui, M., Bughin, J., Woetzel, J., Batra, Ko, R. and Sanghvi, S. 2017. *What the future of work will mean for jobs, skills, and wages.* McKinsey Global Institute.

Mathiassen, S.E., Liv, P. and Wahlström, J. 2013. Cost-efficient measurement strategies for posture observations based on video recordings. *Applied Ergonomics*, 44(4): 609–617.

Middlesworth, M. 2020. The Cost of Musculoskeletal Disorders (MSDs). Retrieved 23 May 2021. https://ergo-plus.com/cost-of-musculoskeletal-disorders-infographic.

Mieritz, L. 2012. Gartner Survey Shows Why Projects Fail. https://thisiswhatgoodlookslike.com/2012/06/10/gartner-survey-shows-why-projects-fail/.

Mitchell, T.M. 2006. *The Discipline of Machine Learning.* School of Computer Science, Carnegie Mellon University, Pittsburg, USA.

Mneymneh, B.E., Abbas, M. and Khoury, H. 2018. Vision-based framework for intelligent monitoring of hardhat wearing on construction sites. *Journal of Computing in Civil Engineering*, 33(2): 04018066.

Mustonen-Ollila, E. and Lyytinen, K. 2003. Why organizations adopt information system process innovations: a longitudinal study using Diffusion of Innovation theory. *Information Systems Journal*, 13(3): 275–297.

Nath, N.D. 2021. Human-Centered Computing and Visual Analytics for Future of Work in Construction. *Doctoral dissertation, Texas A&M University.* https://hdl.handle.net/1969.1/195090.

Nath, N.D. and Behzadan, A. 2019. *Deep Learning Models for Content-Based Retrieval of Construction Visual Data.* ASCE International Conference on Computing in Civil Engineering, Atlanta, Georgia.

Nath, N.D., Chaspari, T. and Behzadan, A. 2019. Single-and multi-label classification of construction objects using deep transfer learning methods. *Journal of Information Technology in Construction*, 24(28): 511–526.

Nath, N.D., Akhavian, R. and Behzadan, A.H. 2017. Ergonomic analysis of construction worker's body postures using wearable mobile sensors. *Applied Ergonomics*, 62: 107–117.

Nath, N.D. and Behzadan, A.H. 2020a. Deep convolutional networks for construction object detection under different visual conditions. *Frontiers in Built Environment*, 6: 1–22.

Nath, N.D. and Behzadan, A.H. 2020b. Deep learning detection of personal protective equipment to maintain safety compliance on construction sites. *ASCE Construction Research Congress 2020: Computer Applications, Reston, VA.*

Nath, N. and Behzadan, A. 2020c. Deep generative adversarial network to enhance image quality for fast object detection in construction sites. *IEEE Winter Simulation Conference*, Orlando, Florida.

Nath, N.D., Behzadan, A.H. and Paal, S.G. 2020. Deep learning for site safety: Real-time detection of personal protective equipment. *Automation in Construction*, 112: 103085.

Nath, N.D., Chaspari, T. and Behzadan, A.H. 2018. Automated ergonomic risk monitoring using body-mounted sensors and machine learning. *Advanced Engineering Informatics*, 38: 514–526.

Nguyen, P. and Akhavian, R. 2019. Synergistic effect of integrated project delivery, lean construction, and building information modeling on project performance measures: A quantitative and qualitative analysis. *Advances in Civil Engineering*, 2019.

Nikas, A., Poulymenakou, A. and Kriaris, P. 2007. Investigating antecedents and drivers affecting the adoption of collaboration technologies in the construction industry. *Automation in Construction*, 16(5): 632–641.

NIOSH. 2014. *The State of the National Initiative on Prevention through Design*. Washington DC.

NSF. 2020. Future of Work at the Human-Technology Frontier. Retrieved 19 January 2021. https://www.nsf.gov/news/special_reports/big_ideas/human_tech.jsp.

Okamura, K. and Yamada, S. 2020. Adaptive trust calibration for human-AI collaboration. *PlOS ONE*, 15(2): e0229132.

OSHA. 2000. Ergonomics: The Study of Work.

OSHA. 2012. Ergonomics for Trainers, OSHA.

OSHA. 2019. Cited Standard in Construction. Retrieved 20 January 2019. https://www.osha.gov/pls/imis/citedstandard.naics?p_naics=23&p_esize=&p_state=FEFederal.

OSHA. 2019. Commonly used statistics. Retrieved 20 January 2019. https://www.osha.gov/oshstats/commonstats.html.

OSHA. 2019. Employer payment for Personal Protective Equipment; Final rule. Retrieved 20 January 2019. https://www.osha.gov/laws-regs/federalregister/2007-11-15-0.

OSHA. 2019. Safety and health regulations for construction. Retrieved 20 January 2019. https://www.osha.gov/laws-regs/regulations/standardnumber/1926/1926.28.

OSHA. 2019. Safety vest requirements to protect flaggers from traffic hazards during construction work. Retrieved 20 January 2019. https://www.osha.gov/laws-regs/standardinterpretations/2002-03-11.

OSHA. 2019. Top 10 most frequently cited standards. Retrieved 20 January 2019. https://www.osha.gov/Top_Ten_Standards.html.

OSHA. 2019. Worker safety series: Construction. Retrieved 20 January 2019. https://www.osha.gov/Publications/OSHA3252/3252.html.

Östergren, P.-O., Hanson, B.S., Balogh, I., Ektor-Andersen, J., Isacsson, A., Örbaek, P., Winkel, J. and Isacsson, S.-O. 2005. Incidence of shoulder and neck pain in a working population: Effect modification between mechanical and psychosocial exposures at work? Results from a one year follow up of the Malmö shoulder and neck study cohort. *Journal of Epidemiology & Community Health*, 59(9): 721–728.

Pan, X., You, Y., Wang, Z. and Lu, C. 2017. Virtual to real reinforcement learning for autonomous driving. *arXiv preprint arXiv:*1704.03952.

Pan, Y. and Zhang, L. 2021. Roles of artificial intelligence in construction engineering and management: A critical review and future trends. *Automation in Construction*, 122: 103517.

Parasuraman, R. and Riley, V. 1997. Humans and automation: Use, misuse, disuse, abuse. *Human Factors*, 39(2): 230–253.

Park, S.I., Quek, F. and Cao, Y. 2012. Modeling social groups in crowds using common ground theory. *Proceedings of the 2012 Winter Simulation Conference (WSC), IEEE*.

Peltier, J.W., Zhao, Y. and Schibrowsky, J.A. 2012. Technology adoption by small businesses: An exploratory study of the interrelationships of owner and environmental factors. *International Small Business Journal*, 30(4): 406–431.

Rabinovich, A., Vedaldi, A., Galleguillos, C., Wiewiora, E. and Belongie, S. 2007. *Objects in Context. 2007 IEEE 11th International Conference on Computer Vision, IEEE*.

Ragu-Nathan, T., Tarafdar, M., Ragu-Nathan, B.S. and Tu, Q. 2008. The consequences of technostress for end users in organizations: Conceptual development and empirical validation. *Information Systems Research*, 19(4): 417–433.

Rahmatizadeh, R., Abolghasemi, P., Bölöni, L. and Levine, S. 2018. Vision-based multi-task manipulation for inexpensive robots using end-to-end learning from demonstration. *2018 IEEE International Conference on Robotics and Automation (ICRA), IEEE*.

Rao, Q. and Frtunikj, J. 2018. Deep learning for self-driving cars: Chances and challenges. *Proceedings of the 1st International Workshop on Software Engineering for AI in Autonomous Systems.*

Rashid, K.M. and Behzadan, A.H. 2018. Risk behavior-based trajectory prediction for construction site safety monitoring. *Journal of Construction Engineering and Management*, 144(2): 04017106.

Rashid, K.M. and Louis, J. 2019. Times-series data augmentation and deep learning for construction equipment activity recognition. *Advanced Engineering Informatics*, 42: 100944.

Rauniar, R., Rawski, G., Yang, J. and Johnson, B. 2014. Technology acceptance model (TAM) and social media usage: An empirical study on Facebook. *Journal of Enterprise Information Management.*

Ray, S.J. and Teizer, J. 2012. Real-time construction worker posture analysis for ergonomics training. *Advanced Engineering Informatics*, 26(2): 439–455.

Redmon, J. and Farhadi, A. 2017. YOLO9000: Better, faster, stronger. *arXiv preprint arXiv:1612.08242.*

Redmon, J. and Farhadi, A. 2018. Yolov3: An incremental improvement. *arXiv preprint arXiv:1804.02767.*

Ren, S., He, K., Girshick, R. and Sun, J. 2015. Faster R-CNN: Towards real-time object detection with region proposal networks. *Advances in Neural Information Processing Systems.*

Rogers, E.M. 2003. *Diffusion of Innovations*. New York: Free Press.

Romanycia, M.H. and Pelletier, F.J. 1985. What is a heuristic? *Computational Intelligence*, 1(1): 47–58.

Samek, W., Montavon, G., Vedaldi, A., Hansen, L.K. and Müller, K.-R. 2019. *Explainable AI: Interpreting, Explaining and Visualizing Deep Learning.* Springer Nature.

Seel, T., Raisch, J. and Schauer, T. 2014. IMU-based joint angle measurement for gait analysis. *Sensors*, 14(4): 6891–6909.

Seo, J., Han, S., Lee, S. and Kim, H. 2015. Computer vision techniques for construction safety and health monitoring. *Advanced Engineering Informatics*, 29(2): 239–251.

Sepasgozar, S.M. and Bernold, L.E. 2013. Factors influencing the decision of technology adoption in construction. *In: ICSDEC 2012: Developing the Frontier of Sustainable Design, Engineering, and Construction*, pp. 654–661.

Settles, B. 2009. *Active Learning Literature Survey.* Computer Sciences Technical Report. University of Wisconsin-Madison, USA.

Shen, D., Wu, G. and Suk, H.-I. 2017. Deep learning in medical image analysis. *Annual Review of Biomedical Engineering*, 19: 221–248.

Shin, H.-C., Roth, H.R., Gao, M., Lu, L., Xu, Z., Nogues, I., Yao, J., Mollura, D. and Summers, R.M. 2016. Deep convolutional neural networks for computer-aided detection: CNN architectures, dataset characteristics and transfer learning. *IEEE Transactions on Medical Imaging*, 35(5): 1285–1298.

Siau, K. and Wang, W. 2018. Building trust in artificial intelligence, machine learning, and robotics. *Cutter Business Technology Journal*, 31(2): 47–53.

Sieb, M. 2019. *Visual Imitation Learning for Robot Manipulation.* Robotics Institute, Carnegie Mellon University.

Silver, D., Huang, A., Maddison, C.J., Guez, A., Sifre, L., Van Den Driessche, G., Schrittwieser, J., Antonoglou, I., Panneershelvam, V. and Lanctot, M. 2016. Mastering the game of Go with deep neural networks and tree search. *Nature*, 529(7587): 484–489.

Silver, D., Schrittwieser, J., Simonyan, K., Antonoglou, I., Huang, A., Guez, A., Hubert, T., Baker, L., Lai, M. and Bolton, A. 2017. Mastering the game of go without human knowledge. *Nature*, 550(7676): 354–359.

Simonyan, K. and Zisserman, A. 2014. Very deep convolutional networks for large-scale image recognition. *arXiv preprint arXiv:1409.1556.*

Singh, A.-P. and Dangmei, J. 2016. Understanding the generation Z: The future workforce. *South-Asian Journal of Multidisciplinary Studies*, 3(3): 1–5.

Singh, S. and Ashuri, B. 2019. *Leveraging Blockchain Technology in AEC Industry during Design Development Phase. Computing in Civil Engineering 2019: Visualization, Information Modeling, and Simulation*, American Society of Civil Engineers Reston, VA, pp. 393–401.

Son, H., Park, Y., Kim, C. and Chou, J.-S. 2012. Toward an understanding of construction professionals' acceptance of mobile computing devices in South Korea: An extension of the technology acceptance model. *Automation in Construction*, 28: 82–90.

Sunindijo, R.Y. and Zou, P.X. 2012. Political skill for developing construction safety climate. *Journal of Construction Engineering and Management*, 138(5): 605–612.

Sutton, R.S. and Barto, A.G. 2018. *Reinforcement Learning: An introduction.* MIT Press.

Tang, S., Golparvar-Fard, M., Naphade, M. and Gopalakrishna, M.M. 2020. Video-based motion trajectory forecasting method for proactive construction safety monitoring systems. *Journal of Computing in Civil Engineering*, 34(6): 04020041.

Teizer, J., Caldas, C.H. and Haas, C.T. 2007. Real-time three-dimensional occupancy grid modeling for the detection and tracking of construction resources. *Journal of Construction Engineering and Management*, 133(11): 880–888.

Teschke, K., Trask, C., Johnson, P., Chow, Y., Village, J. and Koehoorn, M. 2009. Measuring posture for epidemiology: Comparing inclinometry, observations and self-reports. *Ergonomics*, 52(9): 1067–1078.

Trinh, D.-H., Luong, M., Dibos, F., Rocchisani, J.-M., Pham, C.-D. and Nguyen, T.Q. 2014. Novel example-based method for super-resolution and denoising of medical images. *IEEE Transactions on Image Processing*, 23(4): 1882–1895.

Valente, T.W. 1996. Social network thresholds in the diffusion of innovations. *Social Networks*, 18(1): 69–89.

Wang, G. and Song, J. 2017. The relation of perceived benefits and organizational supports to user satisfaction with building information model (BIM). *Computers in Human Behavior*, 68: 493–500.

Wang, N., Pynadath, D.V. and Hill, S.G. 2016. Trust calibration within a human-robot team: Comparing automatically generated explanations. *2016 11th ACM/IEEE International Conference on Human-Robot Interaction (HRI), IEEE*.

Wang, Y., Liao, P.-C., Zhang, C., Ren, Y., Sun, X. and Tang, P. 2019. Crowdsourced reliable labeling of safety-rule violations on images of complex construction scenes for advanced vision-based workplace safety. *Advanced Engineering Informatics*, 42: 101001.

Wicaksono, A. and Maharani, A. 2020. The effect of perceived usefulness and perceived ease of use on the technology acceptance model to use online travel agency. *Journal of Business Management Review*, 1(5): 313–328.

Wu, J., Cai, N., Chen, W., Wang, H. and Wang, G. 2019. Automatic detection of hardhats worn by construction personnel: A deep learning approach and benchmark dataset. *Automation in Construction*, 106: 102894.

Wu, Y., Zhou, Y., Saveriades, G., Agaian, S., Noonan, J.P. and Natarajan, P. 2013. Local Shannon entropy measure with statistical tests for image randomness. *Information Sciences*, 222: 323–342.

Xie, Z., Liu, H., Li, Z. and He, Y. 2018. A convolutional neural network based approach towards real-time hard hat detection. *2018 IEEE International Conference on Progress in Informatics and Computing (PIC), IEEE*.

Yan, X., Li, H., Li, A.R. and Zhang, H. 2017. Wearable IMU-based real-time motion warning system for construction workers' musculoskeletal disorders prevention. *Automation in Construction*, 74: 2–11.

Yang, K., Ahn, C.R., Vuran, M.C. and Aria, S.S. 2016. Semi-superised near-miss fall detection for ironworkers with a wearable inertial measurement unit. *Automation in Construction*, 68: 194–202.

Yang, Y., Caluwaerts, K., Iscen, A., Zhang, T., Tan, J. and Sindhwani, V. 2020. Data efficient reinforcement learning for legged robots. *Conference on Robot Learning, PMLR*.

Yang, Y., Feng, C., Shen, Y. and Tian, D. 2018. Foldingnet: Point cloud auto-encoder via deep grid deformation. *Proceedings of the IEEE Conference on Computer Vision and Pattern Recognition*.

Zhang, Y., Liao, Q.V. and Bellamy, R.K. 2020. Effect of confidence and explanation on accuracy and trust calibration in AI-assisted decision making. *Proceedings of the 2020 Conference on Fairness, Accountability, and Transparency*.

Zhu, X. and Goldberg, A.B. 2009. Introduction to semi-supervised learning. *Synthesis Lectures on Artificial Intelligence and Machine Learning*, 3(1): 1–130.

Chapter 16

The Use of Machine Learning in Heat Transfer Analysis for Structural Fire Engineering Applications

Yavor Panev, *Tom Parker* and *Panagiotis Kotsovinos**

1. Introduction

The discipline of structural fire engineering (SFE) deals with the safe and efficient design of structures against thermal exposure due to a fire in an internal or external environment. Most structural fire engineering assessments usually comprise of 3 components: (1) Definition of the fire scenario and the thermal conditions caused by the fire on the structure, (2) Assessment of the thermal response of the structure because of the fire exposure, and (3) Assessment of the mechanical response of the structural system because of the thermal propagation. The complexity of the analysis of each of these three components will depend on the assessment goals.

With regards to the second component of the SFE process, accurate understanding of how heat propagates through a construction detail when subject to fire exposure allows the engineer to determine among others:

- which components are likely to experience thermal degradation of material properties,
- the thermal gradient within a structural detail and therefore the potential for thermal bowing to be induced,
- risk of ignition of combustible elements,
- need for specification of additional fire protection, and
- classification rating of construction details against standardised fire tests.

Arup, UK.
* Corresponding author: panos.kotsovinos@arup.com

To assess the thermal response of a construction detail, it is usually required to solve a form of the heat transfer partial differential equation for the proposed construction detail geometry and to use appropriate boundary conditions. When complexities arise like non-symmetrical geometry, temperature dependent material properties, multiple modes of heat transfer, etc., that make the solution of the equation impossible via simple numerical scheme, the use of more powerful tools like Finite Element Analysis (FEA) is required using a dedicated software package. Over the past 20 years it has been shown that the FEA method can correctly predict the thermal response of various construction details when subject to fire exposure when the tool has been appropriately verified and validated for the conditions considered (Gernay and Kotsovinos, 2021). As a result, FEA has been widely adopted by the structural fire engineering community.

The main drawback of FEA-based heat-transfer analysis is that it is a time-intensive exercise requiring specialised experience in numerical computation when compared to using analytical expressions. It can take anywhere between a couple of minutes to a couple of hours for a typical construction detail to be analysed depending on the size of the geometry, mesh coarseness, and simulation time. In addition, all inputs must be defined prior to the start of the simulation. Therefore, given that in a building many different types of construction details are possible, and they can vary from design project to design project, performing an FEA analysis for each design case will be impractical. It would also carry a risk in the design process if changes to the design are necessary before the FEA simulations have been completed.

To address the challenges, it is proposed to use a machine learning based methodology for prediction of thermal response using data from previously conducted FEA analyses. The proposed system performs the function of a surrogate model which is more computationally efficient than a complete FEA simulation. This allows designers to have a preliminary understanding of whether a construction detail is likely to achieve the thermal response performance requirements in advance of progressing with a more detailed modelling approach.

2. Machine Learning in SFE

Although common in other engineering fields, the use of machine learning and data science in the field of structural fire engineering is quite recent. Naser (Naser, 2018) adopted Artificial Intelligence (AI) and machine learning (ML) tools to propose new temperature-dependent thermal and mechanical material models for structural steel and validated them against several case studies. Naser (Naser, 2019) extended his previous work to also consider timber structures and adopted AI to propose temperature-dependent material models for wood to derive the thermal and mechanical performance of structural timber members. Cai et al. (Cai et al., 2019) presents an artificial neural network to determine the residual shear resistance of reinforced concrete (RC) beams after fire using finite element analyses in ABAQUS. The researchers derived 384 data entries for training and testing of their approach. Lazarevska et al. (Lazarevska et al., 2018) also used fuzzy neural networks to predict the fire resistance of eccentrically loaded reinforced concrete columns using analyses of reinforced concrete columns and the ANFIS model available in MATLAB®. In another study Ryu et al. (Ryu et al., 2020) propose a machine learning technique based on a convolutional neural network (CNN) approach, which is a deep learning technique, to detect crack information of concrete beams due to a fire. The approach analyses data from surface images of the fire damaged beam taken with digital cameras. Liu et al. (Liu and Zhang, 2020) use neural network models to predict the explosive spalling of reinforced concrete with PP fibres in fire. The authors present two models, a concrete mix-based and a concrete strength-based that were trained and validated, with a prediction accuracy of 100% and 90%, respectively. Al-Jabri and Al l Alawai (Al-Jabri and Al-Alawi, 2007) used ANN to predict the moment rotation response of semi-rigid joints under fire condition. The network was trained and tested on time series data obtained from five fire experiments. A similar technique was successfully implemented by Daryan and Yahyai (Daryan and Yahyai, 2018) for predicting the moment rotation response of welded connections. Feng (Feng, 2020) proposed a data acquisition framework based on the Monte Carlo method and Random Sampling to generate dataset for failure patterns within steel framed buildings. The dataset was then subsequently used to train an artificial neural network (ANN) classifier to rapidly investigate the potential of failure for different design arrangements.

As seen from the above examples the focus of current research is primarily based on developing machine-building models using experimental data acquired in a laboratory setting. The framework proposed in this chapter is based on numerical models and it is tailored for use by the structural fire engineering practitioner.

3. Framework for Development of Surrogate Models

3.1 Inspiration

The use of surrogate models based on pre-computed data is not a new approach in engineering. Before the development of the (FE) method and the advancement in computing from the late 1960s, engineers used various sizing charts, which by acting as a surrogate represented approximate solutions to the heat transfer partial differential equation for various basic geometries like slab, spheres, cylinders (Incropera, 2011). Engineering judgement was used to approximate the shape and thermal boundary conditions of the studied detail to a combination of representative solutions from the sizing chart.

With the development of the FE method, however, this approach fell out of popularity, as FEA allowed for more accurate prediction of the thermal response of the analysed geometry without the need of approximation to simpler forms. As discussed above, FEA is a computationally intensive approach, and it is often impractical to analyse all potential geometric combinations. ML-based tools built on top of experimental and numerical data can greatly enhance computational efficiency and early decision making as discussed by Salehi and Burgueno (Salehi and Burgueno, 2018) in their review of emerging ML trends in structural engineering. This chapter presents an "equivalent modern version" of the traditional sizing chart by training an ML model on pre-computed heat transfer data. This is a novel application of ML-based approaches to structural fire engineering problems.

3.2 Appropriateness

The proposed methodology is appropriate for cases where the parametric space under investigation has been initially thoroughly explored with adequate verified simulations or there are other sources of relevant data (e.g., empirical evidence from experiments). There should also be a consistent and considerable design queries to benefit the overtime of the faster speed of assessment. Guidance to judge the amount of data needed is discussed within each of the case studies of this chapter. It is important, however, that the parameters of the provided dataset capture sufficiently large range of values, so that future queries fall within the data ranges that the ML algorithm has been trained for. As expected, most ML algorithms would perform poorly when required to extrapolate outside the bounds of the trained data.

It is noted that significant effort is dedicated to acquiring the initial dataset where often a separate simulation will be needed for each data point. However, experience from practical application of the system for design has shown that once the ML system has been constructed the savings in time offered from the rapid assessment outweigh the initial efforts to develop the system in first place.

3.3 System Components

The proposed ML surrogate model is comprised of three components:

(1) Data validator which assesses whether the input data is representative of the data the model has been trained on.
(2) ML algorithm which makes a prediction with sufficient accuracy.
(3) Estimator of the prediction confidence.

While the second component is responsible for delivering the result, the simultaneous function of the first and third components ensure that the system is used appropriately, and the prediction is valid. This is essential for any engineering calculation where it needs to be demonstrated that the output achieves a certain degree of reliability.

It is also noted that maximum benefit of such a system comes when it is well integrated within a larger workflow. It is often the case that efficient integration brings larger benefits than the choice of the ML algorithm provided that sufficient accuracy is achieved since processes like input data management, processing, and results reporting become time-consuming when done at a large scale. When applied to construction, detailing appropriate integration to a common data management platform like the BIM (Building Information Model) is a potential solution.

3.4 Development Process

Building an ML based surrogate model is a multistage process that can be roughly separated in the following stages: data acquisition, model development, development of components for validation, and verification. Each of these stages might include all or some of the processes as shown in Fig. 16.1.

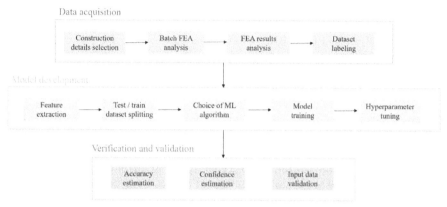

Fig. 16.1: Development process for ML surrogate model.

How each stage is practically achieved is discussed via two case studies in Section 4 of this chapter. Both case studies address typical heat transfer problems in structural fire engineering. They vary in complexity and demonstrate various substages.

The algorithms for both the case studies described within this chapter were implemented with Python 3.6 and sci-kit—learn library.

4. Case Studies

Two case studies that illustrate the ML framework covered in Section 2 are presented.

The first case study features the use of ML for the prediction of the fire classification rating for a composite beam construction detail.

The second case study presents the ML methodology for the prediction of the fire classification rating for a composite shallow floor construction detail.

4.1 Composite Beam Protection

4.1.1 Problem Introduction

A composite beam is defined as a hot rolled steel I or H section connected via shear connectors to a precast concrete or composite floor slab where the steel section and floor slab are designed to act together/compositely.

In the event of fire, the steel I or H section could be exposed to fire from below. Fire protection is usually applied to the beam to limit the rise in steel temperature hence allowing it to maintain its structural stability. A common method of providing this fire protection is to encase the exposed parts of the beam with non-combustible materials such as boarding or mineral wool.

Verification of stability can be demonstrated by limiting the temperature rise of the steel beam below a *critical temperature*. This critical temperature is dependent on the type of structural element (e.g., beam or column) and the ratio of the applied load to the structural capacity of the element. According to *Association of Fire Protection*

Specialists Yellow Book, the limiting temperature for composite beam construction is usually taken as 550°C in the absence of specific loading information and section details.

The fire resistance rating of a steel beam that forms part of composite construction is defined as the time at which the steel beams exceed the 550°C limiting temperature when heated to a standard fire time temperature curve ISO 834 rounded up to the nearest 30 min.

In industry, BS EN 1993-1-2 provides a commonly used lumped-capacity heat transfer model to calculate the temperature rise of a protected steel beam and subsequently evaluate its fire rating. The model assumes a one-dimensional heat transfer through the fire protection encapsulating the beam using an iterative timestep calculation.

Figure 16.2 presents an example of a temperature rise for a protected UKB 305x105x25 beam. The limiting temperature of 620°C is exceeded at 60 min, hence the fire resistance rating is 60 min.

Traditionally, the BS EN 1993-1-2 calculation needs to be repeated for every change of the section size or the thickness of the insulating material. To circumvent this and allow for more rapid assessment, it is proposed to use an ML model trained on a dataset comprised of pre-computed results.

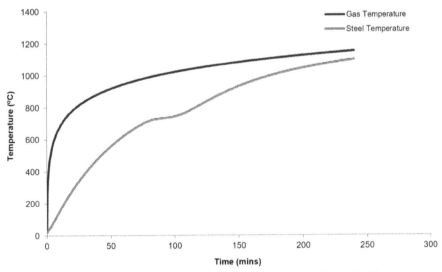

Fig. 16.2: Temperature development in protected beam exposed to standard fire.

4.1.2 Data Acquisition and Feature Selection

The fire resistance rating of a steel beam depends on two main variables when using the Eurocode 3 calculation model.

Firstly, the ratio of the heated surface area (A) of the beam to the volume (V) of the beam. This ratio is known as the "Section Factor". Structural elements with a low section factor will take longer to heat than those with a high section factor. These factors are dependent on the geometry of the section. The values for standard section sizes such as those chosen for this case study are available in publicly available technical datasheets.[1]

The other main variable is the type and thickness of the applied fire protection. In this example, mineral wool insulation was used as the fire protection method. Based on review of suppliers' technical data sheets, it was found that this would typically be in the range of 25–55 mm.

BS EN 1993-1-2 calculation was undertaken to determine the temperature rise of the 26 steel sections with 25–55 mm of applied mineral wool protection. This resulted in a total of 156 unique details. Sections were chosen to be representative of the full range of sizes of the UK standard steel section catalogue (Steel Construction Institute, 2021). Key size ranges are summarised in Table 16.1.

Table 16.1: Composite beam dataset summary.

Parameter	Range
Height	127 mm–1026 mm
Width	76 mm–424 mm
Flange thickness	7.6 mm–77 mm
Web thickness	4 mm–47.6 mm
Section factor	31 m^{-1}–325 m^{-1}
Insulation thickness	25–55 m

Each of the analyzed details was then classified according to the insulation criteria discussed in Section 4.1.1. The classification distribution is summarised in Table 16.2.

Table 16.2: Classification distribution.

Class	Count	Percentage
30 mins	8	5%
60 mins	25	16%
90 mins	33	21%
120 mins	44	28%
180 mins	22	14%
240 mins	24	15%

The scatter plot between the section factor and the thickness of the provided protection is presented in Fig. 16.3. The color of the datapoints present the fire rating classification in accordance with the BS EN 1993-1-2 methodology.

As seen from Fig. 16.3, all data points can be easily spatially separated according to their class. This is depicted by the red dashed lines signifying potential decision boundaries. The ability to visually separate data classes indicates a likely good performance of a fitted ML classifier.

---

[1] https://www.steelforlifebluebook.co.uk/ub/ec3-ukna/section-properties-detailing-fire-parameters/; Accessed: 20/10/2021.

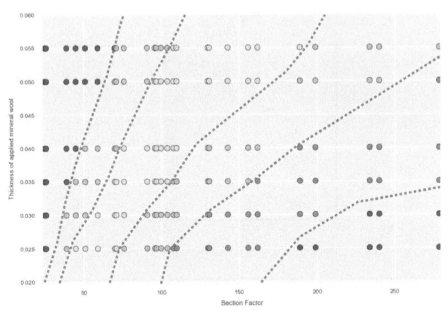

Fig. 16.3: Potential decision boundaries indicated with red dashed lines.

4.1.3 Model Selection

The ML model fitted to the acquired data was based on Support Vector Machine (SVM) algorithm which is a supervised learning classification algorithm aimed at finding a hyperplane that separates two or more classes of data points. The algorithm iterates to find the most optimal hyperplane, i.e., the plane which has the maximum margin width between the data points of the different classes, known as support vectors (Fig. 16.4).

The following considerations were given for the choice of SVM over other supervised learning classification algorithms:

- SVM is a geometric algorithm and does not require large amounts of data points to achieve acceptable accuracy since the decision boundary is only affected by the data points which are closest to the decision boundary, i.e., the support vectors. It will later be shown that this property can be used to significantly reduce the data required to train a model (Ng, 2017).

- SVM hyperplane transforms to a line for 2D datasets which offers a great degree of results' visualisation and interpretability.

- SVM, as a maximum margin classifier, performs well with linearly separable data. The investigated data set is expected to be linearly separable (under some combination of features) as it is produced by a deterministic process (FEA simulation) over controlled array of exactly defined inputs. Therefore, the noise within the output dataset is expected to be negligible.

- SVMs can be enhanced with kernel functions to find nonlinear decision boundaries. A kernel function is a transformation function that is applied to the existing features of the data so that the transformed dataset can be separated by a

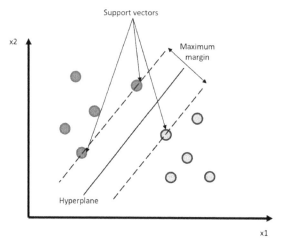

Fig. 16.4: SVM demonstration of principle of operation.

hyperplane. The projection of the fitted hyperplane within the original dataspace forms the nonlinear decision boundary (Ng, 2017).

- SVMs have been demonstrated to work well with data related structural engineering problems as discussed by Cevic et al. (Cevik et al., 2014).

4.1.4 Model Setup

As seen from the exploratory data analysis discussed in 4.1.2, the data appears to be linearly separable when represented as a combination of *section factor* and *applied mineral wool thickness* features. This indicates that training an SVM algorithm on these two features will potentially achieve good performance.

As the two features span different orders of magnitude (insulation: 10^{-2}, section factor: 10^2) normalization was implemented to improve the performance. This is because geometric classifiers are sensitive to the distance between the different data points within the parametric space (Ng, 2017). Therefore, when features are measured across different scales one can become improperly dominant over the other.

Radial basis kernel function (RBF) was also considered to accommodate the nonlinear decision boundaries seen from Fig. 16.2. RBF is a transformation function that transforms the existing features of the data to a higher dimensional space where an optimal hyperplane can be fitted. Specifically, the RBF kernel is a transformation function that generates new features by measuring the closeness of each data point to a fixed center within the data space. Usually, centers are chosen to be the data points themselves. More in-depth discussion to the functioning of RBF can be found in (Ng, 2017).

The performance of the SVN model also depends on the specific choice of hyperparameters. They do not depend on the provided data and are defined before the learning stage. Optimal hyperparameters are usually chosen by an extensive parametric search.

The hyperparameters considered in this case study are presented in Table 16.3.

Table 16.3: Description of considered hyperparameters.

Hyperparameter	Description	Tested Range
C-value	C-value is a measure of the degree of tolerance the algorithm takes for misclassifying points that lie close to the decision boundary. Low C- value means large tolerance (soft margin), i.e., the algorithm is lenient toward misclassification. On the contrary, a high C-value (hard margin) imposes a large penalty for misclassification. Too stringent a penalty for misclassification will cause the model to overfit, i.e., it will fit very well to the training set but will fail to produce accurate results on the test set.	10^{-4}–10^{15}
gamma	Gamma controls the strength of the RBF kernel function in influencing the location of the decision boundary as discussed above.	10^{-3}–10^{6}

4.1.5 Performance Estimation

For each set of hyperparameters the average accuracy of the model was estimated following the 5-fold cross-validation procedure. The accuracy is measured as the fraction of correct classification on the cross-validation set. The results from the parametric search are presented on the heatmap in Fig. 16.4.

The best SVM RBF kernel model from the investigated cohort achieves a maximum accuracy score of 0.96 (C-value = 10000000, gamma = 0.001). The model is visualized in detail in Fig. 16.5. As seen from Fig. 16.6, the achieved fit is acceptably good for the stated purpose. Estimated accuracy of 96% on unseen data is reasonably close to the human level performance of 100%. This can be further improved by inclusion of additional data points around the boundaries. Reliance is

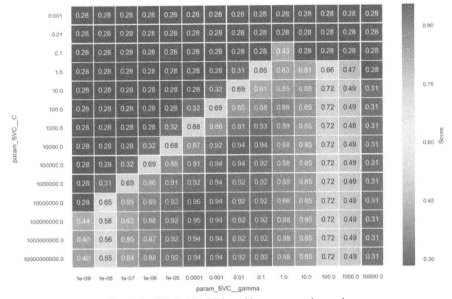

Fig. 16.5: SVM with RBF kernel hyperparametric search.

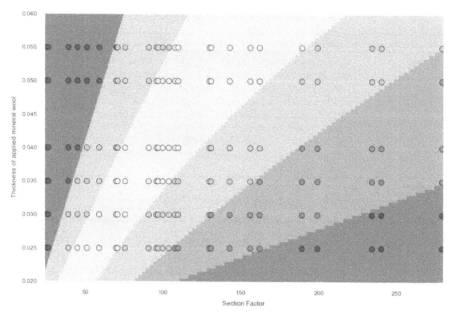

Fig. 16.6: Visualisation of optimal SVM model.

also enhanced by the fact that results can be easily visualised in an interpretable way due to the 2dimensional nature of the dataset.

4.2 Slim Floor System Protection

4.2.1 Problem Introduction

Slim floor systems are a flooring composite system where the steel section is partially embedded in the concrete slab (Fig. 16.7) to reduce the total depth of the floor construction allowing for more effective use of space. Embedding the steel however facilitates an efficient way of heat transfer through the slab which can impact the fire resting performance of the construction. A potential solution to this problem is to completely encapsulate all exposed parts of the steel section with a suitable fire protection system. However, it is often difficult to achieve this for shallow floor systems which could have exposed steelwork on their top face because of space constraints in the floor void that sometimes make it difficult for the top flange to be protected.

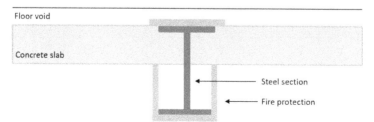

Fig. 16.7: Diagram of slim-floor system.

Heat transfer finite element analysis is often implemented to determine the fire resistance rating of the detail if the top flange is left unprotected. The analysis monitors the heat propagation through the detail when subject to the following boundary conditions:

- Exposure of the top face pf the detail to ISO 834 standard fire temperature time curve via convective and radiative boundary conditions in accordance with BS 1993-1-2.
- Bottom face of the detail is exposed to a constant ambient temperature.
- Concrete and steel temperature-dependent material properties are assumed as per the recommendations of BS 1993-1-2 and BS EN 1992-1-2.

Example thermal distribution of a slim-floor detail after 120 min standard fire exposure is shown in Fig. 16.8.

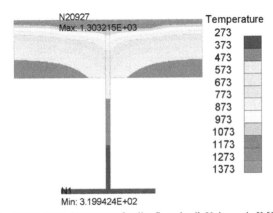

Fig. 16.8: Thermal response of a slim floor detail. Units are in K [16].

The fire resistance rating is determined as the exposure time at which either of the following limiting criteria are breached in accordance with Clause 11.3 BS EN 1363-1:2012:

- Average temperature on the unexposed side exceeds 140° above ambient temperature (20°C).
- Maximum temperature on the unexposed side exceeds 180°C above ambient temperature (20°C).

For qualification purposes of the fire resistance, the time is then rounded down off to one of the standard fire resistance rating categories: 60 min, 90 min, 120 min, and 180 min.

Traditionally such analysis should be ideally performed for each different geometry. Project time-constraints often allow only for a small number of geometries to be checked. Engineering judgement must be exercised in selecting which details should undergo further analysis. This approach is subjective and might lead to errors.

To mitigate this risk, it is proposed to use the ML algorithm trained on results of previous simulation results to predict the expected fire performance of the details without the need of explicit simulation.

4.2.2 Data Acquisition

The dataset is comprised of computed FEA results from models approximating slim-floor construction. LS-DYNA v6.1.1, multipurpose FEA software package was used for each computation.

Twenty-six section sizes were chosen to span along the whole range of sizes of the UKB/UKC catalogue. For modelling, the geometry of the doubly symmetric 'I' section has been simplified to four parameters: section height, section width, flange thickness, and web thickness.

Each of the sections is embedded in a concrete slab with thickness ranging between 100 mm–250 mm. The total number of analyzed cases are 182. The parameters of the dataset including the range of the values (minimum and maximum values) are summarised in Table 16.4. An indicative range of the generated model geometries are visualised in Fig. 16.9.

Table 16.4: Summary of dataset features.

Feature	Designation	Value Range
Concrete thickness	concT	100–250 @ 25 mm increments
Section height	sech	127.5–1026 mm
Section breadth	secW	76–424 mm
Flange thickness	secTw	7.4–77.5 mm
Web thickness	secTf	4–47.8 mm

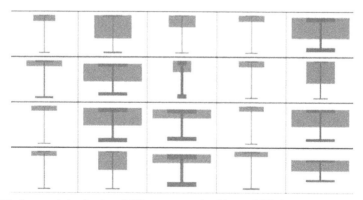

Fig. 16.9: Representative fraction of different geometries (20 out of 182 shown) (Panev et al., 2021).

It is noted that some of the investigated details may not be typically used in construction practice. However, they were included as part of this study to aid the completeness of the dataset and ensure that details used in construction lie within the boundaries of the investigated dataset. Asymmetric sections are not included in the current study, but they could be considered in a similar manner using the presented ML-based approach.

Each of the analyzed details is classified according to the insulation criteria discussed in Section 5.3. The classification distribution is summarised in Table 16.5.

Table 16.5: Classification distribution.

Class	Count	Percentage
60 min	75	41
90 min	48	26
120 min	32	18
180 min	27	15

4.2.3 Feature Selection

The acquired dataset was initially examined visually by plotting 10 scatter plots between each of the key 5 input parameters presented in Table 16.4. The classification of each data point is displayed with different color. The scatter plots are presented in Fig. 16.10.

The purpose of this exercise is by visually inspecting the position of the different classes on various scatter plots to infer which parameters are most likely to govern the classification problem. Scatter plots between variables where the classes appear to be well separated indicate that the classification is likely to be significantly dependent on these parameters. Vice versa, plots where the spatial separation between the categories is not apparent suggest that the combination of variables is not a good indicator classification.

Closer inspection indicates that *concrete slab thickness* and section *web thickness* appear to be good determinants for the fire resting insulation class. Figure 16.11, shows that all data points can be easily spatially separated according to their class. This is depicted by the red dashed lines signifying potential decision boundaries.

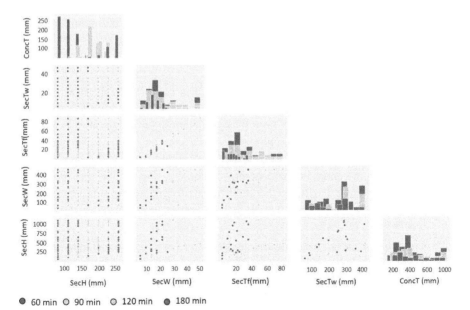

Fig. 16.10: Scatter plots between dataset features (Panev et al., 2021).

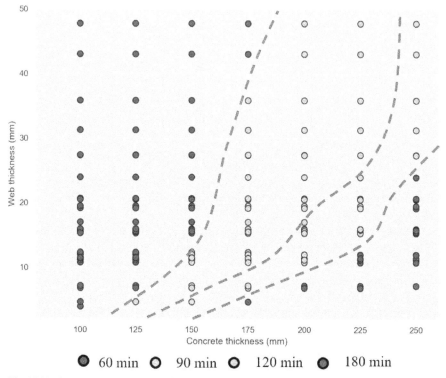

Fig. 16.11: Scatter plot between web thickness and concrete slab thickness demonstrating good degree of separability (Panev et al., 2021).

The inference that web thickness and concrete thickness are the main parameters that govern this classification problem is also supported with the following two principles from the heat conduction theory which suggest the following casual relationships:

- The thickness of the concrete slab determines the distance the heat needs to conduct through to reach the back side of the detail. As seen in Fig. 16.9, the thicker slab requires greater time for conduction to occur resulting in a higher rating of insulation classification.

- For the considered 2D analysis, the area of the highly conductive steel material (relative to the surrounding concrete) is equal to the section web thickness. More efficient conduction is facilitated by higher conductive surface area. This is evident in Fig. 16.9 as the classification rating decreases with increase of web thickness.

Considering these two fundamental principles of thermal behavior and the results of the exploratory data analysis, the next step is to conduct further training of a supervised ML model adopting the concrete thickness and web thickness as input features. This is also supported by the following additional considerations:

- Including features of lesser significance to the investigated problem might introduce unnecessary noise which will decrease the accuracy of the model (Ng, 2017).

- The need for inclusion of additional features will become apparent in the process of model testing.
- It is a good practice to use as few input features as possible to enhance model interpretability. Furthermore, for the special case of two features, all results can be visualised with contour plots which significantly enhances readability and understanding of the model by designers aiming to use it as a tool.

4.2.4 Model Training and Performance

As seen from the exploratory data analysis discussed in 4.2.3, the data appears to be linearly separable when represented as a combination of *concrete slab thickness* and section *web thickness* features. Following similar consideration as in composite beam case study, the SVM algorithm was adopted for supervised classification. Inclusion of RBF was also considered to accommodate the potentially non-linear decision boundaries.

The hyperparameters of the SVM model are tuned using grid-search over the range of values indicated in Table 16.6.

For each set of hyperparameters the average accuracy of the model was estimated following a 5- fold cross-validation procedure. The results from the parametric search are presented on the heatmap in Fig. 16.12. Visualisation of the fit for six different models with various combination of hyperparameters are shown on Fig. 16.13. The

Table 16.6: Hyperparameter ranges for slim-floor case study.

Hyperparameter	Tested Range
C-value	10^{-4}–10^{5}
gamma	10^{-3}–10^{6}

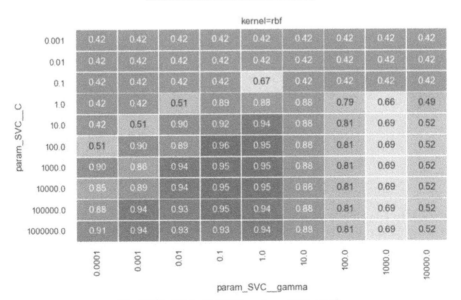

Fig. 16.12: SVM with RBF kernel parametric search.

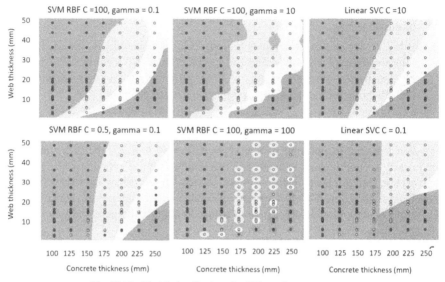

Fig. 16.13: Model visualisations for different hyperparameters.

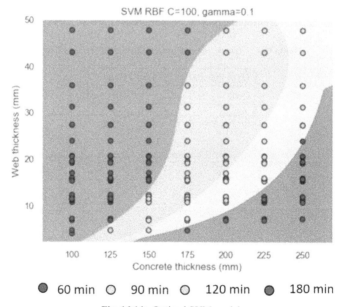

Fig. 16.14: Optimal SVM model.

graphs represent the data used for training as dots and the predictions of the model as a background color. The interplay between the effect of the various hyperparameters can be easily observed.

The best SVM RBF kernel model from the investigated cohort achieves a maximum accuracy score of 0.96 (C-value = 100, gamma = 0.1). The model is visualized in detail in Fig. 16.14, which shows that the achieved fit is acceptably good for the stated purpose. Estimated accuracy of 96% on unseen data is reasonably close to the human level performance of 100%. This can be further improved by inclusion of additional data points around the boundaries. Reliance is also enhanced by the fact that results can be easily visualised in an interpretable way due to the 2D nature of the dataset.

4.2.5 Confidence Estimation

To boost the reliability of the ML model, it was decided to include an estimation of reliability alongside the prediction of the classification rating.

As seen from Fig. 16.4, SVMs are developed to assess the Euclidian ("straight line") distance between the prediction and the fitted decision boundary (hyperplane). However, this distance is model-specific which can make it difficult for the user to gauge its magnitude without having profound knowledge of the contents of the used dataset. To address this model dependency, the Platt scaling transformation technique (Platt John, 2000) was used to map the distances between the data points and the decision boundaries between values ranging from 0.5 (achieved at the boundary) and 1 (achieved further away from the boundary). In basic principle, Platt scaling works by fitting a logistic regression to the scores produced by the SVM. It should be noted that the technique was adopted only to provide a "consistent measure" of relative strength of prediction based on distance from a decision boundary within the same dataset. This implementation was not intended to compare prediction results between different ML models or to infer an empirical probability distribution.

Figure 16.15 visualizes the best performing SVM RBF model (left) enhanced with Platt scaling technique (right). The intensity of the background color of the right figure represents the scaling rating, i.e., how close a prediction is close to a decision boundary. The contour lines represent the numerical value of the rating which is a value between 0.5–1.

It is noted that for 2D dataset the use of Platt scaling technique to inform prediction reliability can be obsolete for the work presented in this paper as the user can visually inspect how close the prediction is to the decision boundary. However,

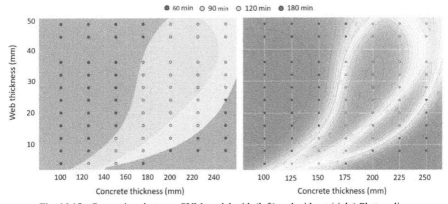

Fig. 16.15: Comparison between SVM model with (left) and without (right) Platt scaling.

accurate visualisation is difficult or impossible in more complex problems where more than three features are involved (Ng, 2017). Platt scaling, on the other hand can be easily applied for higher dimensional datasets.

5. Limitations of the Proposed Machine Learning Approach

The limitations discussed in this section were observed in both case studies and conclusions drawn for mitigating them are considered. Relevant examples from each study are provided where appropriate.

5.1 Input Data Validation

Since geometric-based ML models draw information from the relative geometric position between the data points, they perform poorly in regions away from the dataset. Figure 16.16 shows poor extrapolation capabilities for an SVM model trained on the slim-floor dataset.

This illustrates that any input data into the predictive system must be initially verified to be representative of the dataset which the predictive model has been trained on. Any outliers shall be flagged as anomalies and notification shall be issued to the user.

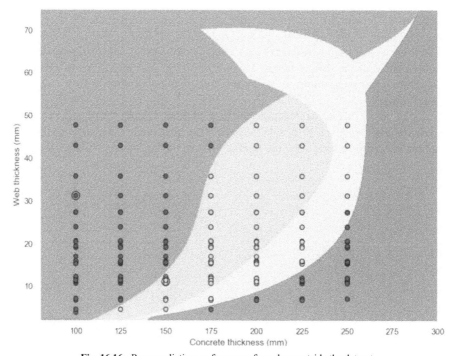

Fig. 16.16: Poor prediction performance for values outside the dataset.

For the slim-floor case study, a basic anomaly detection system was set up to check the validity of input data based on the following selection criteria:

- Input data falls within the range of each input parameter.

- Ratios between different section parameters fall within specified threshold which can be inferred from analyzing the UK standard section catalogue. For example, flange thickness is well correlated with web thickness for both column and beam sections. The ratio falls between 0.5–0.85 as observed on the histogram on Fig. 16.17.

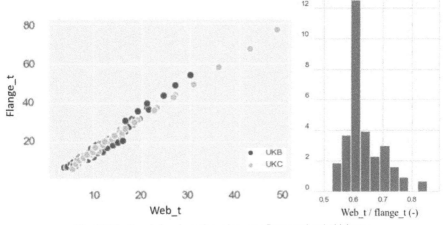

Fig. 16.17: Correlation dependence between flange and web thickness.

5.2 Effects of Dataset Size

The accuracy of the best performing model of the slim-floor case study was analyzed with respect to the size of the training data. The model was retrained with 10 randomly chosen subsets of the available training data with a varying size of 10%–100%. Each model was then tested on cross-validation test forming 20% of the original sample size. This was repeated five times for each training subset.

In addition, model training was conducted with 60 data points (42% of the available training data) which were specifically selected to be close to the potential decision boundary following inspection of Fig. 16.10.

Average model accuracy on both the cross-validation and training set is presented in Fig. 16.18. The standard deviation is presented with a lighter shading.

As seen from the graph increasing the number of training examples has a strong beneficial effect for improving accuracy up to approximately 100 examples. Past this point the gains of adding more training data diminish. Model accuracy can be further improved by carefully selecting the training examples which are expected to be situated closely to the decision boundaries. 92% accuracy was achieved with only 60 pre-selected examples. Similar accuracy is achieved with 100 examples if chosen at random. This demonstrates an important property of the SVM algorithm which can be applied to other similar datasets—performance depends on the data points

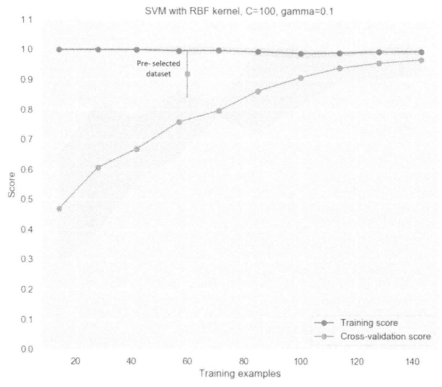

Fig. 16.18: Effect on size of training dataset on model performance.

positioned closest to the decision boundary which act the function of the supporting vectors.

5.3 *Effects of Noise*

The datasets used in both case studies have been generated by a deterministic FEA solver using input parameters which are either exact geometric measurements or conservative characteristics values for thermal properties taken from the Eurocode technical standards. The dataset was also labelled to different insulation classes using an exact threshold value for unexposed side temperature rise based on the BS EN 1363-1:2012. Therefore, the noise in the data, arising from the distribution of the input parameters has been eliminated at the data acquisition stage. Testing the proposed ML model at different 'noisier' dataset has been outside the scope of this study.

According to Ng (Ng, 2017), as the SVM is a geometrically based classifier noise within the data it reduces the model performance only if uncertainty arises in data points which are close to the decision boundaries. Uncertainty in datapoints away from the decision boundaries does not affect the results. Model accuracy can be improved with a good choice of a C-value hyperparameter or kernel function.

6. Conclusions

This chapter describes a novel approach of using an SVM machine learning algorithm as a surrogate model for undertaking heat transfer analysis for structural fire engineering assessments. The processes of data acquisition, performing exploratory data analysis, model training, and validation were explored by two case studies. Final versions of the proposed solution achieved acceptable degrees of accuracy of 95% and 96% for the composite beam and the slim-floor case study, respectively.

The proposed learning approach achieves satisfactory accuracy only for inputs which are representative of the dataset used for training the model. It is not suitable for extrapolation to sections outside the range to those studied. Therefore, an additional module for validating input values must be considered for sections outside the range that the tool has been developed for.

Based on the two case studies, it is shown that the proposed ML approach will be suitable for the structural fire engineering applications as it:

(1) allows for rapid prediction of thermal response and corresponding fire resistance rating with sufficient accuracy needed for a preliminary/concept study

(2) can return a high and measurable degree of reliability of predictions

(3) does not require prior advanced FEA simplification of input geometry

(4) is not computationally expensive and does not require specialised skills in numerical modelling after the creation of the initial dataset and training the ML model

(5) can be readily implemented in preliminary optimisation studies and to inform inputs for more detailed models.

References

Al-Jabri, S.K. and Al-Alawi, M.S. 2007. Predicting the behaviour of semi-rigid joints in fire using an artificial neural network. *Steel Structures*, 7: 209–217.

Cai, B., Xu, L.F. and Fu, F. 2019. Shear resistance prediction of post-fire reinforced concrete beams using artificial neural network. *International Journal of Concrete Structures and Materials*, 13(1): 46.

Cevik, A., Kurtoglu, A., Bilgehan, M., Gulsan, E.M. and Albegmprli, M.H. 2014. Support vector machines in structural engineering: A review. *Journal of Civil Engineering and Management*, 21(3).

Cevik, A., Kurtoglu, A., Bilgehan, M., Gulsan, E.M. and Albegmprli, M.H. 2014. Support vector machines in structural engineering: A review. *Journal of Civil Engineering and Management*, 21 (3).

Daryan, S.A. and Yahyai, M. 2018. Predicting the behaviour of welded angle connections in fire using artificial neural network. *Journal of Structural Fire Engineering*, 9(1).

Feng, F. 2020. Fire induced progressive collapse potential assessment of steel framed buildings using machine learning. *Journal of Construction Steel Research*, 105918.

Gernay, T. and Kotsovinos, P. 2021. Advanced analysis. In: *International Handbook of Structural Fire Engineering*. Springer, pp. 413–467.

Incropera, F.P. 2011. *Fundamentals of Heat and Mass Transfer* (7th Edn.). John Wiley.

Lazarevska, M., Gavriloska, A.T., Laban, M., Knezevic, M. and Cvetkovska, M. 2018. Determination of fire resistance of eccentrically loaded reinforced concrete columns using fuzzy neural networks. *Complexity*, 2018.

Liu, J.C. and Zhang, Z. 2020. Neural network models to predict explosive spalling of PP fiber reinforced concrete under heating. *Journal of Building Engineering*, 101472.

Naser, M.Z. 2018. Deriving temperature-dependent material models for structural steel through artificial intelligence. *Construction and Building Materials*, 191: 56– 68.

Naser, M.Z. 2019. Fire resistance evaluation through artificial intelligence: A case for timber structures. *Fire Safety Journal*, 105: 1–18.

Ng, A. 2017. Coursera Online Course: Machine Learning. University of Stanford. Available at: https://www.coursera.org/learn/machine-learning/home/welcome.

Panev, Y., Kotsovinos, P., Deeny, S. and Flint, G. 2021. The use of machine learning for the prediction of fire resistance of composite shallow floor systems. *Fire Technology*, 57: 3079–3100.

Pedregosa et al. 2011. Scikit-learn: Machine Learning in Python. *JMLR*, 12: 2825–2830.

Platt, John. 2000 Probabilistic outputs for support vector machines and comparisons to regularized likelihood methods. *Advances in Large Margin Classifiers*, 10(3): 61–74.

Ryu, E., Kang, J., Lee, J., Shin, Y. and Kim, H. 2020. Automated detection of surface cracks and numerical correlation with thermal-structural behaviors of fire damaged concrete beams. *International Journal of Concrete Structures and Materials*, 14: 1– 12.

Salehi, H. and Burgueno, R. 2018. Emerging artificial intelligence methods in structural engineering. *Engineering Structures*, 171.

Steel Construction Institute. 2021. *Interactive Blue Book*. Available at: https://www.steelforlifebluebook.co.uk/ub/ec3-ukna/section-properties-detailing-fire-parameters/.

Chapter 17

Using Artificial Intelligence to Derive Temperature Dependent Mechanical Properties of Ultra-High Performance Concrete

Srishti Banerji

1. Introduction

Ultra-high performance concrete (UHPC) is an advanced type of concrete with improved compressive and tensile strength, durability, ductility (post-cracking), and energy absorption properties as compared to conventional concretes. For the realization of these enhanced properties, UHPC batch mixes are designed to have a low water-to-binder ratio, contain steel fibers and high fineness admixtures, and require specialized mixing procedures and high-temperature curing (Banerji et al., 2020; Dong, 2018; Yoo et al., 2016). Due to its unique mix design and fabrication procedure, the microstructure of UHPC is different from that of conventional concretes, which essentially results in its improved mechanical performance. As a result of its superior properties, UHPC is gaining attention for building leaner, aesthetic, and long-lasting structures.

Structural members, when used in buildings, must satisfy fire resistance requirements as fire safety is one of the key considerations in building design (Kodur and Dwaikat, 2007). Concrete structures made of traditional normal strength concrete (NSC) exhibit excellent fire resistance without the need for any additional fire protection measures. This is attributed to the low thermal conductivity and high thermal capacity of NSC, as well as to slower degradation of its strength and modulus properties with temperature. However, recent studies have indicated that structural members fabricated using UHPC do not exhibit the same level of fire resistance as that of members made using traditional concretes (Banerji et al., 2019; Qin

Applied Engineering Sciences, Michigan State University, East Lansing, MI, USA-48824.
Email: banerjis@msu.edu

et al., 2021). This is mainly due to the faster degradation of mechanical properties of UHPC with temperature, as well as its high susceptibility to fire-induced spalling. Spalling leads to loss of the concrete section from the structural member, which in turn results in higher heat progression and further increases the rate of mechanical property deterioration.

Typically, the sectional temperatures in a concrete structural member under building fires are in the range of 20–800°C. Therefore, for analytically assessing the fire resistance of a member, the knowledge of material property variation is required in that temperature range. The mechanical properties of primary interest in fire resistance evaluation of UHPC members are compressive strength, elastic modulus, and tensile strength. Typically, only compressive strength and elastic modulus are included in fire resistance calculations of traditional NSC members, whereas tensile strength is discounted. This is due to the fact that the tensile strength of conventional NSC is substantially lower than its compressive strength (Khaliq and Kodur, 2011). On the other hand, the tensile strength of UHPC cannot be ignored from strength calculations because it has a notably higher tensile strength than NSC, and contributes to the attainment of higher member capacity. In addition, tensile strength is critical in controlling crack propagation in concrete. Further, tensile strength is an important property under fire conditions as it helps in resisting fire-induced spalling (LaMalva, 2018) (https://ascelibrary.org/doi/book/10.1061/9780784415047).

Mechanical properties of all concrete types deteriorate at elevated temperatures due to temperature-induced microstructural changes, which are mainly governed by the moisture content, mix proportions, and volume of admixtures in concrete. High-temperature mechanical properties have been extensively studied for NSC and high strength concrete (HSC). The most widely adopted temperature-dependent concrete codal property relations are specified by ASCE Manual (1992) and Eurocode 2. However, these relations were developed in the 1990–2000s by testing conventional concretes, and hence, no models for the temperature dependence of mechanical properties for UHPC are specified. Moreover, there are no standardized test methods and procedures for measuring the mechanical properties of any concrete type, including UHPC, at elevated temperatures (Kodur et al., 2020). Due to this, high-temperature mechanical properties of UHPC in the published research have been evaluated using inconsistent test setups and test procedures including heating rate, specimen size, concrete mix proportions, etc. This has led to a broad range of variance in the available test data on high-temperature mechanical properties of UHPC in the literature. Owing to these reasons, a unified model for describing the temperature-induced strength and modulus degradation for UHPC is currently lacking.

Since the last decade, artificial intelligence (AI) and machine learning (ML)-based techniques are being increasingly applied to solve intricate problems in structural engineering fields. Some of its prominent applications have been in the domain of material property modeling and prediction and structural health monitoring (Cheung et al., 2008; Dantas et al., 2013). AI has been shown to solve diverse and complex problems by mimicking the cognitive skills of humans to perform tasks in a smart manner. The implementation of these innovative AI tools is efficient in terms of cost and time since the AI models do not involve the setting up of elaborate experiments or the execution of complicated numerical models with several parameters. The majority of UHPC studies applying AI in the literature have centered on optimization of mix design, prediction of compressive strength, and evaluation of member capacity at room temperature (Fan et al., 2021; Zhang et al., 2020). There have been a few AI-based studies targeted at developing high-temperature material property models for construction materials, but with a limited concentration on UHPC (Iqbal et al., 2021; Naser, 2018; Naser and Uppala, 2020).

On these accounts, this paper hypothesizes that AI could serve as a potential solution for deriving accurate unified high-temperature mechanical property models for UHPC. To develop these property relations, published experimental studies on high-temperature mechanical property tests on UHPC in the literature are reviewed and all the test data from these studies are compiled. The collated UHPC material property database is then analyzed using an integration of artificial neural network (ANN) and genetic programming (GP). This AI-based analysis is used to trace the loss of mechanical properties at high temperatures and develop simplified mathematical expressions for capturing this temperature-induced degradation in UHPC. The following sections present the steps involved in the development of AI-based high-temperature mechanical property relations for UHPC in the 20–800°C temperature range.

2. Temperature-dependent Mechanical Properties of UHPC

The critical mechanical properties for fire resistance evaluation of concrete are compressive strength (f'_c), tensile strength (f_t), and modulus of elasticity (E). A databasen for UHPC is developed by compiling results from a number of high-temperature mechanical property tests obtained from open literature to create the AI models (Abid et al., 2019; Banerji and Kodur, 2021; Behloul et al., 2002; Chadli et al., 2021; Li and Liu, 2016; Liang et al., 2019; Mindeguia et al., 2007; Sanchayan and Foster, 2016; So et al., 2015; Tai et al., 2011; Yang and Park, 2019; Zhang et al., 2021; Zheng et al., 2013, 2012a, 2012b). Typically, UHPC is made with steel fibers, which contributes to its high post-cracking ductility and tensile strength properties. Some UHPC mixes also contain polypropylene (PP) fibers in addition to steel fibers, for preventing fire-induced spalling by increasing concrete permeability at elevated temperatures. A review of the literature shows that the sole presence of polypropylene fibers in UHPC does not significantly influence the variation in degradation of relative strength and modulus at high temperatures. The database was generated by compiling property tests on UHPC mixes with only steel fibers as well as UHPC mixes containing both steel and polypropylene fibers, due to the inconsequential effect on property variance. It should be noted that UHPC specimens without steel fibers were not considered since they do not adhere to the typical ductile and tensile characteristics of UHPC.

For measuring the mechanical properties of concrete at room temperature, specific test procedures are well-established in test standards. On the contrary, the current test standards do not specify any guidance at elevated temperatures on test procedures, equipment, or parameters for evaluating the mechanical properties of concrete. Adding to this complexity, three different types of testing regimes can be followed to assess high-temperature mechanical properties of concrete: unstressed, stressed, and residual. As per the unstressed testing regime, the specimen is to be heated to a predetermined target temperature without applying any preload. Once the specimen has reached a uniform temperature throughout, the loading is to be applied on the specimen until it fails. In the stressed testing regime, the specimen is preloaded before heating, and that preload is maintained throughout the heating phase. The specimen is then loaded to failure once it has reached thermal equilibrium. As per the residual test regime, the specimen is heated to a target temperature (with or without any preload) until it reaches a steady state. After the temperature in the specimen has stabilized, it is allowed to cool down to ambient temperature and then the specimen is loaded till failure. While the stressed and unstressed test regimes simulate the behavior of heated concrete during fire, the residual test method represents the behavior of concrete following cool down after fire exposure. Despite the fact that property measurements are dependent on the test setup and conditions, the test standards do not provide any guidance on the particulars of the testing procedure for any of these aforementioned assessing regimes. To this end, researchers have followed various non-standardized test procedures and specimen conditions (such as varying sample size and shape, heating rate, load level, and loading rate) for carrying out high-temperature studies in the literature. Since researchers used self-judgment to select and apply unstandardized evaluating procedures, there is significant variation in the measured test results for high-temperature mechanical property characterization of concrete.

The reported test data on the variation of compressive strength with temperature is compiled for UHPC and plotted in Fig. 17.1 along with codal and widely used relations (Eurocode 2; ASCE manual; Kodur et al., 2008) for NSC and HSC. Overall, the compressive strength of UHPC diminishes with an increase in temperature. It can be noted from the Fig. 17.1 that in some studies, an initial increase in compressive strength takes place up to a temperature of 200°C, followed by a declining trend

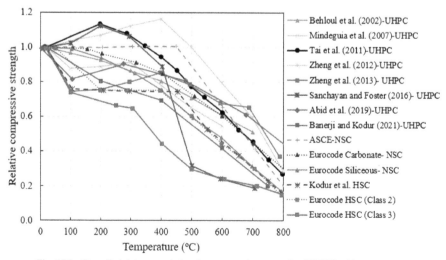

Fig. 17.1: Compiled data on variation in compressive strength of UHPC with temperature.

as temperatures rise further. This increase in compressive strength of UHPC upon initial heating till 200°C is due to the completion of pozzolanic reactions and hydration of the unhydrated cementitious products in the microstructure. Further, there is a notable range of variation in the measured compressive strength of UHPC at elevated temperatures. This is mainly owing to the adoption of different test setups and procedures including specimen size, heating rate, concrete mix proportions. For instance, Tai et al. (2011) tested 50x100 mm cylinders by heating at a rate of 2°C/min following the residual testing regime, whereas Abid et al. (2019) tested 70x70x70 mm cubes by heating at a rate of 5°C/min following the unstressed evaluating regime. In addition to significant variation in test data collected from studies by different researchers, material models adopted by Eurocode 2 and ASCE manual also exhibit substantial disparity among their individual models as well as with the UHPC test data.

The published test data on the temperature-dependent elastic modulus for UHPC is plotted in Fig. 17.2 along with Eurocode 2 and ASCE manual relations for NSC. The compiled data show that the elastic modulus of UHPC degrades rapidly with temperature rise. Fewer studies have been carried out on the variation of elastic modulus of UHPC with temperature as compared to the number of studies focused on capturing compressive strength variation. Similar to compressive strength data, disparity can be observed in measured loss of elastic modulus with temperature. Because there are fewer data points, the fluctuation in elastic modulus seems to be smaller. Discrepancies persist in the testing procedure of elastic modulus for UHPC at high temperatures, such as Zheng et al. (2012) measured elastic modulus using 70x70x228 mm prisms subjected to a heating rate of 4°C/min following residual test regime, whereas Banerji and Kodur (2021) utilized 75x150 mm cylinders with a heating rate of 2°C/min following the unstressed assessing regime.

The compiled test data on the high-temperature tensile strength of UHPC, as well as Eurocode 2 and ASCE manual relations for NSC, are presented in Fig. 17.3.

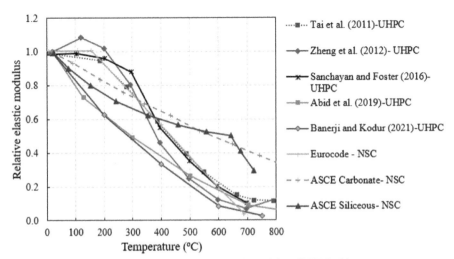

Fig. 17.2: Compiled data on variation in elastic modulus of UHPC with temperature.

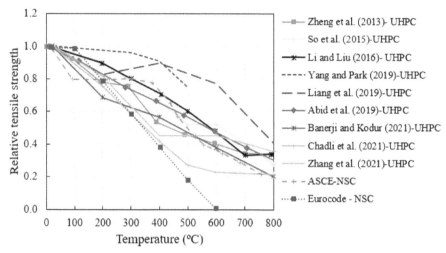

Fig. 17.3: Compiled data on variation in tensile strength of UHPC with temperature.

The tensile strength of concrete can be evaluated in three forms: flexural, direct, and splitting. Flexural tensile strength involves subjecting a small concrete beam (prism) to third-point flexural loading. The direct tensile strength can be measured through testing dogbone specimens by applying axial tensile load in a suitable test machine until the sample breaks in direct tension. Splitting tensile strength is evaluated by applying a diametrical compressive load on a cylindrical concrete specimen along its length till failure occurs through the splitting of the specimen along the vertical diameter. Due to the application of different test procedures in the absence of standardized guidelines, there is significant variation in measured test data from the reported studies as can be seen in Fig. 17.3. For example, Li and Liu (2016) tested 150x75 mm dogbone specimens by heating at 4°C/min under the residual test regime, and Chadli et al. (2021) tested 40x40x160 mm prisms by heating at 3°C/min following an unstressed test regime.

Aside from the aforementioned variances in gathered data associated with respect to specimen size, heating rate, and testing regime, inconsistencies in the collected test results can be attributed to additional factors as well. UHPC mixes are typically designed to contain steel fibers for attaining their characteristic high tensile strength and improved ductility properties. However, there are no established recommendations on the amount of steel fibers in the UHPC mix. As a result, variable dosages of steel fibers have been added to UHPC mixes across different experiments in the literature, resulting in variations in measured temperature-induced degradation of mechanical properties. Likewise, some studies also incorporated polypropylene fibers in addition to steel fibers to prevent fire-induced spalling. Due to a lack of guidance, the dosage of polypropylene fibers was varied across the reported studies, which lead to variation in their corresponding rate of property deterioration. Further, steam curing is crucial during the fabrication of UHPC for facilitating the complete hydration of cement products and the development of superior mechanical properties in UHPC, many of the reported experimental studies neglected to include it. Lastly, several of the high-temperature mechanical property tests on UHPC dried the test specimens in an oven before heating them for the property test in the furnace. Such pre-oven-drying procedure was incorporated to remove moisture from the specimen for minimizing the risk of fire-induced spalling. However, pre-oven-drying is not a standardized test procedure and could have contributed to exacerbating the anomalies in the measured data.

The large variation in the collected test data can lead to overpredicting or underpredicting the fire resistance of a UHPC member. Overprediction can lead to unsafe engineering design, whereas underprediction can result in an uneconomical structural design. To remove ambiguity from the assessment of fire resistance, unified property relations for predicting high-temperature mechanical properties of UHPC are needed. To realize this, the current study aims to develop a simplified temperature-dependent relation for each mechanical property of UHPC by employing the AI model on the collated test data. The material models in the design codes are not considered for this AI analysis since they were developed using property tests carried out on conventional concretes. The details on the development of the AI model and the analysis are described in the following sections.

3. Development of Artificial Intelligent Model

AI utilizes algorithms to uncover patterns in complex datasets by replicating the cognitive processes that human beings apply for problem-solving. In this study, a combination of ANN and GP was applied to develop the AI model for formulating temperature-dependent mechanical property relations for UHPC. Similar ANN-GP hybrid models have been used in the literature for deriving material property models for other civil engineering materials (Naser, 2018; Naser and Uppala, 2020). The ANN was developed using MATLAB® and is structured as feedforward, wherein neurons are organized into three layers: input, hidden, and output as illustrated in Fig. 17.4. Neurons in all the successive layers are interconnected to each other. For

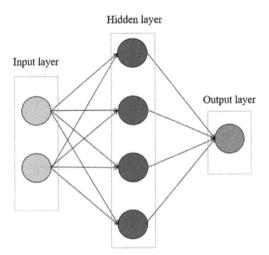

Fig. 17.4: Layout of an artificial neural network (ANN).

this research, the input layer comprised of the reduction factors assembled from published test data for the strength and modulus properties of UHPC at various high temperatures. The input layer is then connected to 10 hidden layers and weights are internally assigned to each connection. The weights are multiplied by the input data for activating transfer functions (such as sigmoid function) and the transformed outputs are summed together to generate the final answer from the output layer. The performance of the developed neural network is determined by comparing its predictions with the measured values from the reported high-temperature tests. The comparison is measured and adpated by learning through multiple iterations for each mechanical property until the performance of the ANN is satisfactory.

Once an ANN model for a property is developed, that model is exported as a MATLAB function and formulated into a mathematical expression through tree-based GP using an open-source software, Heuristic Lab (Wagner et al., 2014). The GP analysis involves creating a random population of potential solutions symbolically in the form of an expression tree, which is made up of tree nodes and terminal (leaf) nodes. The tree nodes contain operators (addition, division, etc.) or functions (log, power, etc.) whereas the terminal nodes contain variables (a, b, c, etc.) or constants. The expression for capturing any property data is represented in the tree structure and is analyzed recursively to yield a multivariate expression with better fitness in each iteration. This recursive procedure of GP analysis employs an algorithm, based on the fundamentals of biological evolution, that combines random mutation, crossover, and evolution resulting in the survival of the fittest expression. Finally, the performance of the GP-derived property relation is assessed by comparing it with the input values as predicted from the developed ANN model.

4. Derivation of the Mechanical Property Models

The measured data points corresponding to compressive strength, elastic modulus, and tensile strength reduction factors (e.g., 1, 0.96, 0.87, etc.) for UHPC at various temperature points in the 20–800°C range (e.g., 20°C, 100°C, 200°C, etc.) were the input to the ANN model. The input data consisted of 406 data points collated from the experimental studies in the open literature as discussed in Section 2. The material models recommended by the design codes (ASCE and Eurocode 2) were not inputted in the ANN since these models were derived decades ago by carrying out high-temperature strength tests on conventional types of concrete (such as normal strength concrete) and do not include any measurements on UHPC. The compiled input data set was randomly divided into three parts: 70% for training, 15% for validating, and 15% for testing the developed ANN. The successfully trained ANN was applied to formulate generalized temperature-dependent mechanical material property models. The predictions from ANN are plotted together with the measured data for each mechanical property of UHPC in Fig. 17.5 which shows that the ANN is successful in capturing the integrated trend of the reduction factors with temperature and the ANN-predicted values lie within the range of the compiled test data.

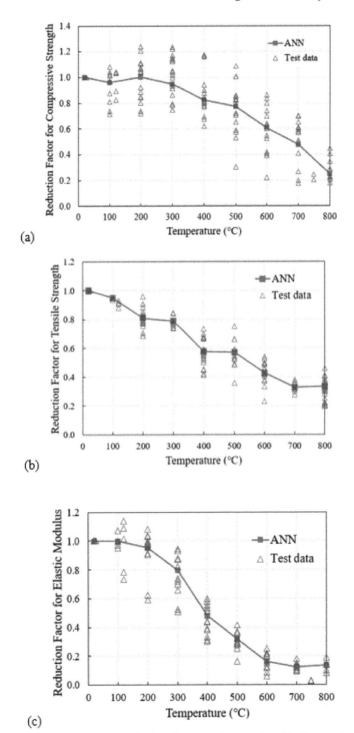

Fig. 17.5: Comparison between AI-derived and measured properties: (a) Compressive strength; (b) Tensile strength; (c) Elastic modulus.

The performance of the ANN model is also evaluated by computing two statistical fitness metrics. The first metric is mean squared error (MSE) which is defined as:

$$MSE = \frac{1}{z}\sum_{i=1}^{z}(m_i - p_i)^2 \tag{17.1}$$

where z = the total number of data points, m_i = the *i-th* measured output, and p_i = the *i-th* predicted output. The second performance evaluation metric is coefficient of determination (R^2) and it can be calculated as:

$$R^2 = \frac{\Sigma_i(p - \frac{1}{z}\Sigma_{i=1}^{z}m)^2}{\Sigma_i(m - \frac{1}{z}\Sigma_{i=1}^{z}m)^2} \tag{17.2}$$

where z = the total number of data points, i = input number, m = the measured output, and p = the predicted output. Lower values of MSE and higher values of R² indicate favorable performance. The computed MSE and R² for the developed ANN are presented in Table 17.1, in which the performance evaluation metrics indicate that the developed ANN agrees with the measured data and is suitable for tracing degradation in temperature-dependent mechanical properties of UHPC. The performance of the ANN model can be enhanced even further by using a larger dataset, which necessitates additional testing and research efforts. Based on this testing and validation, the developed ANN is deemed fit to develop expressions for temperature-dependent mechanical property models with confidence.

Table 17.1: Performance metrics obtained through ANN.

Property	R²	MSE
Compressive strength	91.82	0.0043
Tensile strength	93.41	0.0044
Elastic modulus	93.45	0.0091

The predictions from the ANN model are fitted into nonlinear symbolic expressions using a tree-based GP algorithm. The initial expression trees from the GP analysis were complex and lengthy. Iterations and simplifying operations were carried out to compress the expression trees for deriving easy-to-use compact mechanical property relations. To illustrate, the simplified outcome from the GP analysis in the symbolic representation for high-temperature compressive strength is presented in Fig. 17.6. The nodes of the expression tree are assembled together to obtain the property relations. The unified temperature-dependent mechanical property expressions derived from GP analysis are listed in Table 17.2, along with their coefficient of determination (R²). The generated expressions for compressive strength, tensile strength, and elastic modulus property have high R² values (as shown in Table 17.2) which imply the accuracy of these proposed equations.

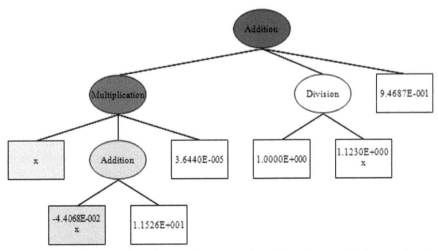

Fig. 17.6: Symbolic regression using genetic programming (GP) to derive unified expression for compressive strength of UHPC (x is Temperature in °C).

Table 17.2: GP-derived expressions of the reduction factors for mechanical properties of UHPC.

Property	Derived Expressions	R² (%)
Compressive strength	$-1.60584*10^{-6}T^2 + 4.20007*10^{-4}T + \dfrac{1}{1.123T} + 0.94687$	99.26
Tensile strength	$-0.00089267T + \dfrac{1}{-0.37299T + 135.13} + 1.0057$	98.43
Elastic modulus	$7.5598*10^{-9}T^3 - 8.9033*10^{-6}T^2 + 1.2608*10^{-3}T + 0.97514$	99.33

5. Current Challenges and Future Research

For realistic fire performance calculations of UHPC members, reliable high-temperature mechanical properties of UHPC are to be used. Currently, there is a lack of simplified expressions in the fire design codes for high-temperature strength and elastic modulus properties of UHPC. Moreover, there is a considerable scatter in the material models proposed in the available literature due to variances in the adopted fabrication process and high-temperature testing methods (including test regime, test setup, specimen shapes and sizes, heating and loading rates, etc.). As a consequence, it is challenging for a fire researcher or practitioner to adopt a specific material model for undertaking a proper fire resistance evaluation. To bridge this knowledge gap and to facilitate fire design of UHPC structures, the current research demonstrates the applicability of utilizing AI models for developing unified high-temperature mechanical property relations. However, there are some impediments with the current scenario of AI implementation for deriving material properties of UHPC.

The proper training and performance of any AI model are strongly dependent on the size of the selected data set. A challenge faced in the current research was since UHPC is a relatively new type of concrete, a lesser amount of test data is available for UHPC as compared to the amount of data available for more thoroughly researched conventional concrete types, such as NSC and HSC. Although the developed AI model in this study is deemed capable of providing accurate mechanical property models, the derived results could possibly be refined by adding more test data points into the current database through further testing efforts in the future.

Undertaking property investigations at elevated temperatures is complex and requires significant effort and resources. The complexity is further increased by an absence of standardized test methods for conducting high-temperature concrete mechanical property assessments. This led to another challenge of substantial non-uniformity in the collected data due to different test conditions and procedures followed in the reported mechanical property tests as discussed earlier. To achieve more uniformity in examined results, test standards need to be devised for measuring the mechanical properties of UHPC at high temperatures.

6. Conclusions

The present study applies an AI model involving ANNs and GP to derive temperature-dependent mechanical property models UHPC in the 20–800°C temperature range. The studied mechanical properties included compressive strength, tensile strength, and elastic modulus. The following conclusions can be drawn from this paper:

- Standardization of test procedures for assessing temperature-dependent mechanical properties of UHPC is urged, as are further future experiments, in order to compile a more consistent and richer database.

- The developed ANN model is capable of comprehending and expressing the rate of mechanical property deterioration of UHPC at elevated temperatures.

- The proposed GP model can fittingly derive generalized simple expressions for high-temperature mechanical properties of UHPC.

- The developed mechanical property relations can be used for practical design purposes for evaluating the fire resistance of UHPC structures.

References

Abid, M., Hou, X., Zheng, W. and Hussain, R.R. 2019. Effect of fibers on high-temperature mechanical behavior and microstructure of reactive powder concrete. *Materials*, 12: 329.

Banerji, S. and Kodur, V. 2021. Effect of temperature on mechanical properties of ultra-high performance concrete. *Fire and Materials*. https://doi.org/10.1002/fam.2979.

Banerji, S., Kodur, V. and Solhmirzaei, R. 2020. Experimental behavior of ultra high performance fiber reinforced concrete beams under fire conditions. *Engineering Structures*, 208: 110316. https://doi.org/10.1016/j.engstruct.2020.110316.

Banerji, S., Solhmirzaei, R. and Kodur, V.K.R. 2019. Fire response of ultra high performance concrete beams. *In: International Interactive Symposium on Ultra-High Performance Concrete*. Albany, NY: Iowa State University Digital Press. https://doi.org/10.21838/uhpc.9658.

Behloul, M., Chanvillard, G., Casanova, P., Orange, G. and France, F.F.F. 2002. Fire resistance of Ductal ultra high performance concrete, *In: 1st Fib Congress*, Osaka Japan, pp. 421–430.

Chadli, M., Tebbal, N. and Mellas, M. 2021. Impact of elevated temperatures on the behavior and microstructure of reactive powder concrete. *Construction and Building Materials*, 300: 124031. https://doi.org/10.1016/j.conbuildmat.2021.124031.

Cheung, A., Cabrera, C., Sarabandi, P., Nair, K.K., Kiremidjian, A. and Wenzel, H. 2008. The application of statistical pattern recognition methods for damage detection to field data. *Smart Materials and Structures*, 17: 065023.

Dantas, A.T.A., Leite, M.B. and de Jesus Nagahama, K. 2013. Prediction of compressive strength of concrete containing construction and demolition waste using artificial neural networks. *Construction and Building Materials*, 38: 717–722.

Dong, Y. 2018. Performance assessment and design of ultra-high performance concrete (UHPC) structures incorporating life-cycle cost and environmental impacts. *Construction and Building Materials*, 167: 414–425.

EN 1992-1-2, 2004. Eurocode 2: *Design of Concrete Structures*, Part 1–2: General rules-structural fire design. CEN, Brussels.

Fan, D., Yu, R., Fu, S., Yue, L., Wu, C., Shui, Z., Liu, K., Song, Q., Sun, M. and Jiang, C. 2021. Precise design and characteristics prediction of Ultra-High Performance Concrete (UHPC) based on artificial intelligence techniques. *Cement and Concrete Composites*, 122: 104171. https://doi.org/10.1016/j.cemconcomp.2021.104171.

Iqbal, M., Zhang, D., Jalal, F.E. and Javed, M.F. 2021. Computational AI prediction models for residual tensile strength of GFRP bars aged in the alkaline concrete environment. *Ocean Engineering*, 232: 109134.

Khaliq, W. and Kodur, V.K.R. 2011. Effect of high temperature on tensile strength of different types of high-strength concrete. *ACI Materials Journal*, 108: 394–402.

Kodur, V. and McGrath, R. 2003. Fire endurance of high strength concrete columns. *Fire Technology*, 39: 73–87.

Kodur, V.K.R. and Dwaikat, M. 2007. Performance-based fire safety design of reinforced concrete beams. *Journal of Fire Protection Engineering*, 17: 293–320. https://doi.org/10.1177/1042391507077198.

Kodur, V.K.R., Dwaikat, M.M.S. and Dwaikat, M.B. 2008. High-temperature properties of concrete for fire resistance modeling of structures. *ACI Materials Journal*, 105: 5: 517–527.

Kodur, V.K.R., Banerji, S. and Solhmirzaei, R. 2020. Test methods for characterizing concrete properties at elevated temperature. *Fire and Materials*, 44: 381–395.

LaMalva, K.J. (ed.). 2018. Manual of Practice 138: Structural Fire Engineering. *American Society of Civil Engineers*, Reston, Virginia. https://ascelibrary.org/doi/book/10.1061/9780784415047. https://doi.org/10.1061/9780784415047.

Li, H. and Liu, G. 2016. Tensile properties of hybrid fiber-reinforced reactive powder concrete after exposure to elevated temperatures. *International Journal of Concrete Structures and Materials*, 10: 29–37.

Liang, X., Wu, Chengqing, Yang, Y., Wu, Cheng and Li, Z. 2019. Coupled effect of temperature and impact loading on tensile strength of ultra-high performance fibre reinforced concrete. *Composite Structures*, 229: 111432.

Lie, T.T. 1992. *Structural Fire Protection*. NY, New York: American Society of Civil Engineers. https://doi.org/10.1061/9780872628885.

Mindeguia, J.-C., Pimienta, P., Simon, A. and Atif, N. 2007. Experimental and numerical study of an UHPFRC at very high temperature. Consec07 (4–6 June), Paris. *Concrete under Severe Conditions: Environment and Loading*.

Naser, M.Z. 2018. Deriving temperature-dependent material models for structural steel through artificial intelligence. *Construction and Building Materials*, 191: 56–68. https://doi.org/10.1016/j.conbuildmat.2018.09.186.

Naser, M.Z. and Uppala, V.A. 2020. Properties and material models for construction materials post exposure to elevated temperatures. *Mechanics of Materials*, 142: 103293. https://doi.org/10.1016/j.mechmat.2019.103293.

Qin, H., Yang, J., Yan, K., Doh, J.-H., Wang, K. and Zhang, X. 2021. Experimental research on the spalling behaviour of ultra-high performance concrete under fire conditions. *Construction and Building Materials*, 303: 124464. https://doi.org/10.1016/j.conbuildmat.2021.124464.

Sanchayan, S. and Foster, S.J. 2016. High temperature behaviour of hybrid steel–PVA fibre reinforced reactive powder concrete. *Material Structures*, 49: 769–782. https://doi.org/10.1617/s11527-015-0537-2.

So, H., Jang, H., Khulgadai, J. and So, S. 2015. Mechanical properties and microstructure of reactive powder concrete using ternary pozzolanic materials at elevated temperature. *KSCE Journal of Civil Engineering*, 19: 1050–1057.

Tai, Y.-S., Pan, H.-H. and Kung, Y.-N. 2011. Mechanical properties of steel fiber reinforced reactive powder concrete following exposure to high temperature reaching 800°C. *Nuclear Engineering and Design*, 241: 2416–2424. https://doi.org/10.1016/j.nucengdes.2011.04.008.

Wagner, S., Kronberger, G., Beham, A., Kommenda, M., Scheibenpflug, A., Pitzer, E., Vonolfen, S., Kofler, M., Winkler, S. and Dorfer, V. 2014. Architecture and design of the HeuristicLab optimization environment. *In: Advanced Methods and Applications in Computational Intelligence*. Springer, pp. 197–261.

Yang, I.-H. and Park, J. 2019. Mechanical and thermal properties of UHPC exposed to high-temperature thermal cycling. *Advances in Materials Science and Engineering*, 2019.

Yoo, D.-Y., Banthia, N. and Yoon, Y.-S. 2016. Flexural behavior of ultra-high-performance fiber-reinforced concrete beams reinforced with GFRP and steel rebars. *Engineering Structures*, 111: 246–262. https://doi.org/10.1016/j.engstruct.2015.12.003.

Zhang, D., Liu, Y. and Tan, K.H. 2021. Spalling resistance and mechanical properties of strain-hardening ultra-high performance concrete at elevated temperature. *Construction and Building Materials*, 266: 120961.

Zhang, G., Ali, Z.H., Aldlemy, M.S., Mussa, M.H., Salih, S.Q., Hameed, M.M., Al-Khafaji, Z.S. and Yaseen, Z.M. 2020. Reinforced concrete deep beam shear strength capacity modelling using an integrative bio-inspired algorithm with an artificial intelligence model. *Engineering with Computers*, 1–14.

Zheng, W., Li, H. and Wang, Y. 2012a. Compressive behaviour of hybrid fiber-reinforced reactive powder concrete after high temperature. *Materials and Design*, 41: 403–409.

Zheng, W., Li, H. and Wang, Y. 2012b. Compressive stress–strain relationship of steel fiber-reinforced reactive powder concrete after exposure to elevated temperatures. *Construction and Building Materials*, Complete: 931–940. https://doi.org/10.1016/j.conbuildmat.2012.05.031.

Zheng, W., Luo, B. and Wang, Y. 2013. Compressive and tensile properties of reactive powder concrete with steel fibres at elevated temperatures. *Construction and Building Materials*, 41: 844–851. https://doi.org/10.1016/j.conbuildmat.2012.12.066.

Appendix

This appendix provides an illustration of the followed procedure of an AI-derived model for one of the studied mechanical properties of UHPC, which is elastic modulus at high temperatures. The collected data points from published experiments are listed in Table 17A.1 as reduction factors for elastic modulus at different temperatures in the 20–800°C range.

Table 17A.1: Collated input data for elastic modulus.

Temperature (°C)	Reduction Factor for Elastic Modulus
25	1.00
200	0.98
300	0.74
400	0.58
500	0.30
600	0.25
700	0.18
800	0.08
25	1.00
200	0.96
300	0.72
400	0.44
500	0.29
600	0.22
700	0.14
800	0.10
25	1.00
200	0.98
300	0.66
400	0.39
500	0.32
600	0.21
700	0.15
800	0.14
20	1.00
120	1.14
200	1.08

Table 17A.1 contd. ...

...Table 17A.1 contd.

Temperature (°C)	Reduction Factor for Elastic Modulus
300	0.87
400	0.44
500	0.29
600	0.13
700	0.10
800	0.19
900	0.26
20	1.00
120	1.09
200	1.03
300	0.88
400	0.38
500	0.28
600	0.12
700	0.09
800	0.14
20	1.00
120	1.01
200	0.97
300	0.70
400	0.31
20	1.00
200	1.00
300	0.93
20	1.00
100	0.95
200	0.91
300	0.87
400	0.60
500	0.37
600	0.22
700	0.11
20	1.00

Table 17A.1 contd. ...

...Table 17A.1 contd.

Temperature (°C)	Reduction Factor for Elastic Modulus
100	1.00
200	0.99
300	0.87
400	0.55
500	0.17
600	0.15
700	0.13
20	1.00
100	1.07
200	1.04
300	0.94
400	0.53
500	0.42
600	0.20
700	0.12
20	1.00
100	0.97
200	0.91
300	0.82
400	0.51
500	0.36
20	1.00
120	0.78
300	0.51
500	0.25
700	0.10
800	0.07
20	1.00
120	0.73
300	0.53
500	0.28
700	0.12
800	0.08

Table 17A.1 contd. ...

...Table 17A.1 contd.

Temperature (°C)	Reduction Factor for Elastic Modulus
20	1.00
200	0.63
400	0.33
600	0.08
750	0.03
20	1.00
200	0.59
400	0.30
600	0.06
750	0.03

This collated data serves as the input in the ANN environment in MATLAB. Out of the entire data 70% is randomly selected for testing and the remaining 30% is used for testing and validating. The ANN model is developed following the training and validation process iteratively in MATLAB and outputted as a function as shown below:

```
function [Y,Xf,Af] = myNeuralNetworkFunction (X,~,~)
% [Y] = myNeuralNetworkFunction(X,~,~) takes these arguments:
% X = 1xTS cell, 1 inputs over TS timesteps
% Each X{1,ts} = Qx1 matrix, input #1 at timestep ts.
% and returns:
% Y = 1xTS cell of 1 outputs over TS timesteps.
% Each Y{1,ts} = Qx1 matrix, output #1 at timestep ts.
% where Q is number of samples (or series) and TS is the number of timesteps.
%#ok<*RPMT0>
% ===== NEURAL NETWORK CONSTANTS =====
% Input 1
x1_step1.xoffset = 20;
x1_step1.gain = 0.00227272727272727;
x1_step1.ymin = -1;

% Layer 1
b1 = [-14.039113863056696729;-
10.866364582994942012;7.8597186096997759819;-4.5198750977821102737;-
1.8123812854266617028;-1.6467926129040495464;-5.2774474739817813784;8.20
54917404224223532;10.497002943071436221;13.349337512103298664];
IW1_1 = [13.960934314838093329;14.015708010746221746;-
13.954411513638259379;14.046310367718517043;13.975317362403755439;-
13.987436928605090358;-13.776699378932498519;13.742958513539955234;14.
369445144717680307;14.612350815182994168];
```

```
% Layer 2
b2 = -0.076623505967661015292;
LW2_1 = [-0.17454168317482388062 0.24147024523723298328
0.16937979639947120525 0.061844625277168288613 -0.22984946972986358693
0.13793768224471977812 0.64034474458553547471 0.58860082292853133309
-0.39449015233464557184 0.030805115776667928856];
% Output 1
y1_step1.ymin = -1;
y1_step1.gain = 1.8001800180018;
y1_step1.xoffset = 0.027;
% ===== SIMULATION ========
% Format Input Arguments
isCellX = iscell(X);
if ~isCellX
    X = {X};
end
% Dimensions
TS = size(X,2); % timesteps
if ~isempty(X)
    Q = size(X{1},1); % samples/series
else
    Q = 0;
end
% Allocate Outputs
Y = cell(1,TS);

% Time loop
for ts=1:TS
    % Input 1
    X{1,ts} = X{1,ts}';
    Xp1 = mapminmax_apply(X{1,ts},x1_step1);
    % Layer 1
    a1 = tansig_apply(repmat(b1,1,Q) + IW1_1*Xp1);
    % Layer 2
    a2 = repmat(b2,1,Q) + LW2_1*a1;
    % Output 1
    Y{1,ts} = mapminmax_reverse(a2,y1_step1);
    Y{1,ts} = Y{1,ts}';
end
% Final Delay States
Xf = cell(1,0);
Af = cell(2,0);
% Format Output Arguments
if ~isCellX
    Y = cell2mat(Y);
end
end
```

```
% ===== MODULE FUNCTIONS ========
% Map Minimum and Maximum Input Processing Function
function y = mapminmax_apply(x,settings)
y = bsxfun(@minus,x,settings.xoffset);
y = bsxfun(@times,y,settings.gain);
y = bsxfun(@plus,y,settings.ymin);
end

% Sigmoid Symmetric Transfer Function
function a = tansig_apply(n,~)
a = 2 ./ (1 + exp(-2*n)) - 1;
end
% Map Minimum and Maximum Output Reverse-Processing Function
function x = mapminmax_reverse(y,settings)
x = bsxfun(@minus,y,settings.ymin);
x = bsxfun(@rdivide,x,settings.gain);
x = bsxfun(@plus,x,settings.xoffset);
end
```

The temperature-dependent elastic modulus values predicted by the ANN model are entered in the GP program using Heuristic Lab and the generated symbolic expression is presented in Fig. 17A.1.

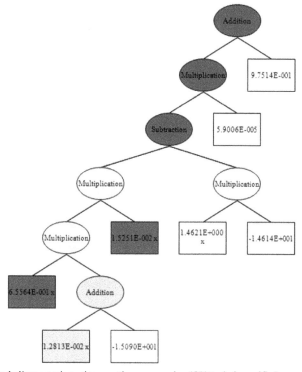

Fig. 17A.1: Symbolic regression using genetic programming (GP) to derive unified expression for elastic modulus of UHPC.

Chapter 18

Smart Tunnel Fire Safety Management by Sensor Network and Artificial Intelligence

Xinyan Huang,* Xiqiang Wu, Xiaoning Zhang and *Asif Usmani*

1. Introduction to Tunnel Fire Safety

Tunnels have played an essential role in modern transportation systems since the mid-20th century, owing to their high utility and flexibility in mountainous areas and their effectiveness in tackling the tight land supply of crowded metropolitan areas. The construction of underground space and tunnels has been a favorable option, especially in the fast-developing countries in Asia. These road, train, and underground tunnels help meet the increasing demand for a more efficient urban transportation system for megacities of a high population density (Dindarloo and Siami-Irdemoosa, 2015).

Today, although the probability of building fire incidents is smaller (Ingason et al., 2015; Li and Ingason, 2018; Carvel, 2019), the absolute annual number of tunnel fire accidents is still very high, considering thousands of tunnels around the world, the high traffic densities and accidents. Because the tunnel is a long and confidence space that is difficult to evacuate, once a tunnel fire occurs, it could be fatal and cause a catastrophic economic loss (Beard, 2009; Beard and Carvel, 2012; Li and Liu, 2020; Casey, 2020). In 1995, a catastrophic tunnel fire incident in Azerbaijan was induced by an electrical fault and finally caused 289 deaths and many injuries (Fig. 18.1a) (Haack, 2002). In 2020, a fire occurred in an ascending train in the tunnel of the Gletscherbahn Kaprun 2 funicular in Kaprun, Austria and killed 155 people (Fig. 18.1b). In 2014, a coal truck collided with a methanol-tanker truck inside the Yanhou Tunnel, China. The liquid methanol flame triggered a rapid fire spread and a series of explosions, causing more than 30 deaths (Fig. 18.1c). In

Research Centre for Fire Safety Engineering, Department of Building Environment and Energy Engineering, The Hong Kong Polytechnic University, Hong Kong.
Emails: xiqiang.wu@polyu.edu.hk; xiaon.zhang@connect.polyu.hk; asif.usmani@polyu.edu.hk
* Corresponding author: xy.huang@polyu.edu.hk

Fig. 18.1: Well-known tragical tunnel fire accidents, (a) Baku Metro Fire, Azerbaijan, 1995, (b) Kaprun Fire Disaster, Austria, 2000, (c) Yanhou Tunnel fire, China, 2014, and (d) Samae 2 Tunnel fire, Korea, 2020.

Table 18.1: Selected major tunnel fire accidents with severe casualties over the last 50 years (Ingason et al., 2015; Casey, 2020; Vianello et al., 2012; Ren et al., 2019).

Year	Tunnel Location	Accident Type	Casualties
1972	Hokuriku Tunnel, Japan	Short Circuit	744
1983	Pecrile, Italy	Collision	31
1984	San Benedetto Tunnel, Italy	Bomb Attack	137
1995	Baku Underground Railway, Azerbaijan	Electrical Malfunction	289
1999	Mont Blanc, France-Italy	Oil leakage Motor	39
1999	Tauren Tunnel, Austria	Multi-car Collision	61
2000	Gletscherbahn Kaprun, Austria	Electric Fan Heater	155
2003	Vicenza, Italy	Bus Turnover	56
2003	Daegu Subway, South Korea	Subway Fire	340
2010	Huishan Tunnel, China	Man-made Arson	43
2014	Yanhou, China	Collision	31
2019	Maoliling Tunnel, China	Self-ignition of Tire	36
2020	Samae 2 Tunnel, Korea	Collision	47

2020, a fire accident occurred in the Samae 2 Tunnel, Korea, after the collision of dozens of tanks and trucks, which killed four people and injured more than 40 others (Fig. 18.1d). Table 18.1 lists some recent severe tunnel fire accidents that caused more than 30 deaths, and a more detailed database of tunnel fire accidents are available online (Zhang et al., 2021).

The frequent occurrence of tunnel fires around the world re-emphasizes the importance of tunnel fire safety design, early detection, and initial fire suppression. The severe consequences of tunnel fire can attribute to challenging evacuation from poorly ventilated spaces with high-temperature and high-density smoke and toxic gases. Furthermore, once a tunnel fire occurs, to prompt a safe evacuation, initial suppression is the most important thing due to the fast fire development and limited regress time (Beard, 2009; Zhang et al., 2021). In addition to the optimization of tunnel fire safety design and the early detection and alarming of fire, it is also essential to identify the real-time fire scenarios, and predict the fire evolution based on toxic substances and flame propagation in various scenarios by using fire detectors, sensors, and other more advanced methods, such as the Internet-of-things (IoT) sensor networks and artificial intelligence (AI) methods (Zhang et al., 2021; Wu et al., 2021; Wu et al., 2021; Wu et al., 2022; Naser, 2021).

Like most of the building fire accidents, statistics have also confirmed that most casualties of tunnel fire were induced by smoke inhalation (Beard, 2009). In tunnel fires, heat, hot smoke, and toxic gases from the burning will spread along the tunnel and impose life-threatening impact to the occupants in the tunnels (Barbato et al., 2014; Cong et al., 2017; Wang and Wang, 2016). The fires in tunnels burn fiercely in the enclosed space with the influence of heat feedback from the tunnel linings, and the chemical goods carried by heavy-duty trucks in tunnels add to the possibility to trigger explosions (Seike et al., 2016). Therefore, designing a reliable ventilation system is one of the most important safety features in tunnels. This is fundamentally different from the fire protection in conventional buildings, where the compartment of fire and smoke are the priority, and the active fire protection systems, such as sprinklers and water mist, are either required or recommended. This is mainly because in the tunnel environment, applying the principle of the compartment will inevitably reduce the capacity of evacuation.

Then, controlling the smoke spread in the tunnel is the first principle in tunnel fire protection, and most of the tunnel fire protection regulations require the smoke ventilation system to ensure a sufficient available safe egress time (ASET) (Ingason et al., 2015). The natural and mechanical ventilation systems can be installed in the tunnel. Depending on the methods and configuration of smoke extraction systems, it can be defined as semi-transverse, fully transverse, side extraction, vertical extraction, etc. Additional jet fans in the ceiling have also been applied to control the longitudinal flows. Each ventilation system has its pros and cons in performance, cost, and reliability. In practice, the tunnel fire protection design is often performance-based rather than prescribed, which requires a thorough examination and optimization assisted by numerical simulations of tunnel fire and evacuation processes.

The evacuation systems of a tunnel usually include multiple escape routes distributed at equal intervals or rescue stations. This is usually arranged as a bypass between two parallel tunnel tubes or safe havens built specifically for evacuees. The specific designs of the separation distance and capacity of egress vary considerably, depending on the local laws, code, standards, and guidelines. Occupants have to evacuate to the outside or find shelters at the early stage of the fire before it gets worsen (Santos-Reyes and Beard, 2017). During the fire evacuation, people suffer difficulty in finding correct exits and paths in a dark environment because they are unfamiliar with tunnel layouts and seldom have the chance to walk in tunnels in daily life (Caliendo et al., 2012). Therefore, performance-based engineering analysis is often required to simulate the fire scenarios and smoke motions to assist the evacuation modelling. Overall, the evacuation in tunnel fires is a relatively new research area due to the insufficient understanding of human behavior in the tunnel fire environment (Yamamoto et al., 2018; Boonmee and Quintiere, 2005).

2. Tunnel Fire Dynamics and Data

Figure 18.2 shows (a) a diagram of typical tunnel fire and the key parameters of fire and smoke layers, and (b) the typical ventilation strategies, such as the longitudinal and transverse smoke extraction. Compared to building compartment fires that have been studied for almost a century, tunnel fire research is still a relatively new area that started in the late 1950s (Anderson, 1936; Ole and Cruthers, 1947; Sevcik, 1928), a relatively limited number of full-scale tunnel fire tests have been conducted. It is because these large-scale tests are dangerous, and it is difficult to guarantee researchers' life safety in the large-scale fire test. For tunnel fire safety, the main research and design questions are: (1) to understand the fire dynamics and critical events related to fire and smoke evolution in the confided tunnel space, and (2) to design an optimal Fixed Fire Fighting System (FFFS), particularly the ventilation system that ensures the safety of evacuees.

To study the tunnel fire dynamics, there are three typical methods: (1) full-scale fire test in a real tunnel or a real-scale tunnel model, (2) experiments in a reduced-scale tunnel in the laboratory, and (3) the numerical simulation based on computational fluid dynamics (CFD). The results of full-scale tunnel fire tests are considered as most reliable and valuable, which are used to verify the results of reduced scale fire tests or guide the tunnel fire-safety design. Some well-known full-scale tests include the EUREKA EU499 tests (Norway, 1990–92) (Haack, 1998), Memorial tunnel tests (USA, 1993–95) (Giblin, 1997), and METRO tests (Sweden, 2009–12) (Ingason et al., 2012). However, real-scale tests are costly and dangerous, so to date, full-scale tunnel fire test data are still quite limited (Ingason et al., 2015). (See recent review in (Zhang et al., 2021).)

Comparatively, the model-scale or reduced-scale laboratory tests, based on the scaling laws, provide a greater number of experimental data (Ingason and Zhen, 2010). By analyzing the correlation between fire heat release rate (HRR) or fire size, smoke temperature and velocity, and the scale of the tunnel, the empirical correlations and scaling laws can be derived and corrected from those of the compartment fires.

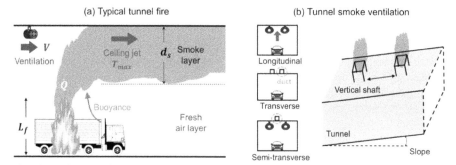

Fig. 18.2: Diagrams of (a) typical tunnel fire and smoke motion, and (b) smoke ventilation systems in tunnel.

Because the tunnel fire and smoke dynamics are also mainly controlled by buoyancy, the classical analysis based on the Zukoski number and Froude number (Quintiere, 2006) is still valid. Note that the reduced-scale tunnel fire tests and the scale analysis are most valid to understand the steady-state conditions. It is difficult to evaluate the transient process in a reduced-scale tunnel, e.g., the fire suppression and evacuation model, because the time scale is neglected in the scale modelling.

More recently, the improving computational capacity drives the CFD modelling techniques to be more widely applied in tunnel fire research and tunnel fire safety design. The numerical simulations can potentially provide much more information that is difficult to measure in experiments, and the most widely used CFD codes include Fire Dynamics Simulator (FDS) (McGrattan et al., 2019) and ANSYS Fluent. Today, combining the experiments of various scales and numerical simulations have become a common approach in recent tunnel fire researches (Li et al., 2012; Weng et al., 2015), as well as the performance-based design for tunnel fire safety (Ingason et al., 2015; Meacham et al., 2005).

Although there are several international standards and handbooks (Kuesel et al., 2012; Blennemann and Girnau, 2005; Cote, 2008) related to tunnel fire safety, such as NFPA 130 and NFPA 502, thanks to the decades of research, many tunnel-fire problems remain that need further research, like early detection, emergency evacuation, and the prediction of tunnel fire behaviors. Once a fire occurs in the tunnel, the real-time information on site like fire location and size, as well as the location and number of people, are essential for the firefighting and emergency decision making. Various fire detection technologies have been adopted for tunnel fire engineering, and generally, their installation and maintenance costs are high (Jevtić and Blagojević, 2014). Although existing techniques such as line-type heat detectors and cameras can locate the fire, these techniques are designed for detecting the fire rather than continuously monitoring the fire event and smoke motion. Inevitably, these systems are fragile that become invalid in a short time due to the rapid-fire development and smoke transport. Thus, developing the smart fire detection, monitoring, and real-time forecast capability in tunnel firefighting and rescue is an emerging research area.

2.1 Critical Fire Events in Tunnel

In this section, the key parameters in tunnel fire and the associated critical events, including the fire HRR, flame length, maximum ceiling temperature, smoke layer thickness, critical ventilation velocity for smoke, the smoke back-layering length, and plug-holing will be briefly reviewed. Compared to the conventional building compartment fire, the tunnel fire does not have the phenomenon of flashover because it is almost impossible to ignite every fuel (usual vehicles) inside a long tunnel. The tunnel fire also behaves differently from the conventional travelling fire in a large open-space floor (Rackauskaite et al., 2021; Gupta et al., 2021), because the oxygen supply is much more limited. Therefore, the key fire events in tunnel are usually related to the smoke motion. For the same reason, although the local structure failure (e.g., spalling in concrete and the deforming of steel) could occur in case of a tunnel fire, the complete collapse of a tunnel due to fire is extremely rare, so the tunnel structure fire resilience is not discussed in this chapter.

Fire HRR is one of the most important parameters in fire to describe the size and severity of the fire, which is also applied to the tunnel fire. The value of HRR is also closely related to other parameters, such as flame length and critical ventilation velocity. Because the primary fuel load in the tunnel is the vehicle and goods, for these real-scale tests, the fire source often uses real burning vehicles or simulated by liquid pool fires or gas burners, based on the measurements of the vehicle fire. The peak heat release rate (pHRR) of a vehicle depends on the type of vehicle and ventilation conditions, which requires the burning of full-scale vehicles. In general, the pHRR is 1–5 MW for small passage cars, 1–10 MW for large passage cars, 10–50 MW for buses and heavy goods vehicles (HGVs), and 300–430 MW for oil tankers (Ingason et al., 2015; Zhang et al., 2021). Burning real vehicles is not often tested in the tunnel; instead, most often, researchers use the well-controlled gas fuel burner and liquid fuel fire to mimic the burning of vehicles.

The flame length refers to the characteristic size of the flame (Drysdale, 2011). For the tunnel fire, the flame height (L_f) is most used (see Fig. 18.2a), but sometimes, the flame could impinge the ceiling to generate an extended impinging flame. The flame length not only indicates the fire HRR and turbulent behavior but also determines the fire spread rate and fire impact on the structure. The fire flame length could be correlated with burner size and proposed a widely accepted formula of $L_f/D \sim 1.7$ (Blinov and Khudiakov, 1961).

The maximum ceiling gas temperature is a significant parameter in tunnel fire, because it can be used to design the threshold values of the heat and smoke detectors (Alpert and Science, 1975) and affects the structural resistance to fire (Kurioka et al., 2003). Many factors, such as bifurcation of structure, portal sealing and blockage, will affect the smoke motion and the maximum temperature under different ventilation conditions. Similar to the ceiling jet temperature in a room, empirical formulas are often used to calculate the maximum gas temperature $\Delta T_{max} = \alpha \dot{Q}^{2/3}/H_{ef}^{5/3}$, where coefficient α usually around 16.9–17.9 (Yao et al., 2018).

The back-layering is generally defined as a phenomenon in which smoke flows in the direction of ventilation despite the operation of fans, which has a negative effect on refugees. As shown in Fig. 18.3(a), the back-layering length (L_b) can be defined as the length between the smoke leading edge and the fire source opposite to the ventilation wind. The *critical velocity* of smoke ventilation (V_c) is defined as the minimum longitudinal ventilation velocity that prevents the upstream movement of the combustion product from fire. Thomas (Thomas, 1958) first proposed the concept of back-layering and modified the Froude number to consider the effect of friction resistance on the smoke back-layering distance. Based on many recent experiments, various equations were proposed to calculate the dimensionless back-layering distance and the critical velocity (Guo et al., 2019; Lee and Ryou, 2005).

Plug-holing phenomenon occurs in the tunnel Natural Ventilation System (NVS). When the lower fresh air layer mixes with the upper smoke layer, the plug-holing phenomenon occurs near the shaft region, as shown in Fig. 18.3(b). Very limited studies have investigated this phenomenon, and most of the existing studies used the theoretical analysis or reduced-scale model (Ji et al., 2012; Baek et al., 2017; Takeuchi et al., 2017).

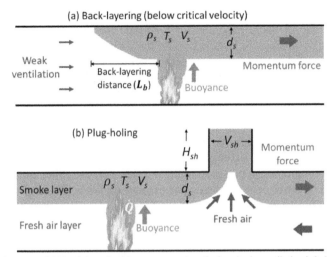

Fig. 18.3: Diagrams of (a) back-layering phenomenon when the longitude ventilation is below the critical velocity, and (b) plug-holing phenomenon in the vertical smoke ventilation systems in tunnel.

2.2 Establishment of Tunnel Fire Database

A large and reliable training database is crucial to train the AI model and enable it to learn the hidden features of fire phenomena, so that the quantity and diversity of the database should be sufficiently large (Fig. 18.4) (Zhang et al., 2021). The database of tunnel fire can be constructed based on the sensor data from previous fire tests/events and numerical simulations. In practice, neither the sensor data nor numerical data are perfect, and they all have pros and cons.

Sensor data: For a typical fire test or a real fire incidence, a several of parameters are commonly measured by distributed point sensors, such as the temperature by thermocouples or other heat detectors, the presence of smoke-by-smoke detectors,

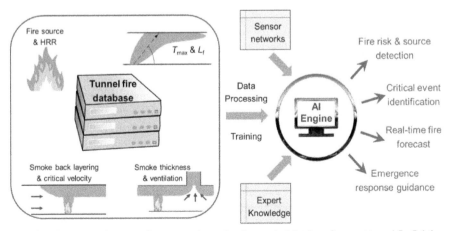

Fig. 18.4: Processes of constructing an experimental and numerical database for smart tunnel firefighting (Zhang et al., 2021).

CO and CO_2 by gas detectors, and heat flux by a radiometer. Point sensors are often installed at one or many locations for long-term measurement. Imaging sensors are also widely commonly installed inside the tunnel, including the closed-circuit television (CCTV) cameras, flame detectors, infrared cameras.

Unfortunately, the database of sensor data for tunnel fire is very limited. There are two major reasons. First, although many studies have published their tunnel fire tests to the journal publications and technical reports, most of the raw data are either not open access or difficult to extract from the plots. Essentially, most of the existing fire test data initially are not produced and presented under the principle of sharing; and most studies are only interested in producing empirical correlations for steady-state conditions, so that most of the transient data are either ignored or poorly documented. In terms of video data, most of them are not available online, and even if the researchers want to share, the video files are often too large to share easily. More importantly, the whole fire research community has not recognized the value and importance of data.

Secondly, for the existing tunnels, the sensor data are also not available to form a database. On the one hand, the fire incidence is considered a rare event, so that most of the daily operational data are not related to fire. On the other hand, most of the sensors, such as the heat detector and smoke detectors, do not record, store, or share the data, because their systems are not designed to do so. Once a fire occurs, most of the key sensors and wiring near the fire could be damaged by fire. Therefore, new data collection systems are needed to support smart firefighting in the tunnel (see Section 3.1).

Numerical fire database: Compared with conducting tunnel fire experiments, the computational simulations (or numerical experiments) for tunnel fire are more cost-effective. Moreover, the CFD modelling results can provide much more detailed temperature information than any experiment. Today, CFD simulations have been widely used in fire research, engineering analysis, or performance-based design of complex infrastructure, like the atrium, underground space, and tunnel. A database formed by massive CFD-based numerical simulation could enable AI-based fire forecasting.

For numerical simulations, the visual data refer to the contours and videos generated from computational results. These time-sequence data (e.g., images and videos) can directly show the real-time scene and scale of the fire, evacuation process, distributions of smoke and temperature, and firefighting activities. Compared to sensor data in fire tests, the 2D or 3D numerical data are several orders of magnitude larger, which can show more details about the fire process. In contrast, the video data in fire tests and real tunnels are quite rare and mostly unpublic. Even if there are fire-test videos, they are less uniform because of various angles of the shot, and their large size and complexity prohibit further data processing and scientific analysis. Although there have been thousands of tunnel fire simulations by researchers and engineers, very few of them are available to access and combine into a large database. The same issue as the experimental database, the community has not recognized the importance of numerical data, and there is no guideline to store, organize, and share the data.

It would be ideal if adequate sensor data could be obtained from full-scale fire experiments or real fire incidents, and all past numerical fire simulation data were stored in a large open cloud database. However, the existing tunnel-fire data from the literature are not large enough or sufficiently well organized to support the training of AI, particularly the training of multidimensional data (e.g., images and videos) of the deep learning algorithm. Despite all kinds of challenges, a global fire database should be established step by step. Also, it is urgent to establish a global standard and guideline for organizing, presenting, and sharing fire research data as a community effort.

Challenges and perspectives: In short, there are four changes to form a database for the tunnel fire or other fire types.

(1) Insufficient experimental and numerical data are recorded, reported, and analyzed in the literature, especially the transient data (i.e., the data changes with time).

(2) The data size is enormous, not only including the transient data from dozens or hundreds of sensors, but also the videos and CFD results that can easily reach the size of 100 GB or 1 TB.

(3) Researchers lack the analytical skills to handle such a large amount of data. Most physics-based empirical correlations and conventional analytical methods can only interpret the steady-state fire conditions. New data analytical approaches are needed (e.g., data-driven methods).

(4) The community is also lacking the habit and standard of recording, organizing, and sharing the data, because it takes additional time without any immediate rewards, and most researchers have not realized the importance of data.

Currently, the literature data of tunnel fire can only be categorized by key parameters in tunnel fire, such as the fire size, smoke layer thickness, critical velocity, back-layer length, and plug-holing, through a thorough literature review (Zhang et al., 2021). Although more experimental and numerical data can be extracted from the published documents or requested from some authors, it is difficult to reorganize everything by a small group of researchers.

Comparatively, conducting new tunnel fire simulations is the easiest way to form a database that can be used to demonstrate the training and application of AI models. If the numerical model is validated by some full-scale tunnel fire tests, the reliability of the numerical database will be significantly improved. As the database should be large enough to cover various fire scenarios, the key parameters should be varied in the experiments and numerical simulations. Table 18.2 lists a group of parameters that can be varied to form a tunnel fire database. For example, if 10 tunnel lengths, 10 fire HRRs, 10 fire locations, and 10 ventilation velocities are considered, the database will include 10^4 fire scenarios. If the database is too large, it will be time-consuming to form and train. Thus, it is important to choose key parameters and a suitable varying range to form a reasonable database.

To enforce a convenient search in the big database, all raw data should be well named, organized, and indexed with a detailed description. Since the fire process is generally complex and influenced by multiple factors, it is essential to provide

Table 18.2: Major parameters for the tunnel fire.

Type	Parameters	Remarks
Tunnel	Length	0.1–100 km
	Cross-section geometry and size	
	inclination	< 3°
Fire	Fire size or heat release rate	2–100 MW
	Initial fire location	
	Fire growth and decay rate	
	Fire spread speed	0–2 m/min
Fire protection system or Environment	Ventilation methods, design, and capacity	Longitudinal, transverse, etc.
	Distribution of Evacuation exit and emergency routes	
	Wind and altitude	

adequate input information and boundary conditions before the data process. Once the data are collected, the establishment of the database is a multi-step task: (1) data collection by searching all available literature and extracting all useful data from these documents, (2) data pre-processing by data quality check, outliers and noises removal, and filtering, and (3) data mining by extracting valuable information.

3. Tunnel Fire Prediction by IoT and AI

3.1 Internet of Things (IoT) Sensor Network and Multiple-Physical Digital Twin

The concept of the *Digital Twin* was initiated in the early 2000s and arguably first adopted by NASA in the attempt to improve the physical model simulation of spacecraft in 2010. Today, the digital twin has been applied to aircraft engines, manufacturing processes (Tao et al., 2019), and large infrastructures like the smart building and wind turbine, but its overall technology is still far from mature (Tao et al., 2019).

Some existing buildings have the building management system (BMS) or the building automation system (BAS). The BMS is a computer-based control system installed in buildings that controls and monitors the building HVAC (Heating, ventilation, and air conditioning), lighting, power systems, fire systems, and security systems (Wang, 2009). Most of these systems are only installed in the new and landmark buildings, because of the high cost and the lack of compulsory requirements. Most of the current BMS systems are not connected with each other due to the use of different Internet protocols by different manufacturers. For example, the HVAC system is not connected with the fire system, and the data are not shared between each other, even if they are made by the same company.

More recently, the building *Digital Twin* is proposed to be established upon the Building Information Model (BIM) in the design stage of a new building and include all sensors and camera information from the IoT system. Nevertheless, the current BIM software is essentially the design tools that lack the capacity for real-time data transfer and management. New approaches and standards are needed to integrate the IoT sensor network into BMS or BIM software. Optimistically, the existing fire service system has one big advantage over other building systems. That is, most of the sensors (smoke, heat, and flame detectors) are wired to the central control panel. Thus, the fire information data about ceiling temperature and smoke density could be directly recorded and processed by upgrading the current addressable system. The current addressable fire system only gives a fire alarm but does not record and provide additional information about the fire scene. Ultimately, the temperature and flow-rate data measured by HVAC sensors, as well as the CCTV camera footage, can also be shared with the fire system, although many years of research, development, and implications are needed.

The proposed fire *Digital Twin* can be regularly updated by the fire risk information, such as the fuel loads in each fire zone and the transport of high fire hazardous materials inside the building. Once a fire occurs, although the heavy smoke from the fire also quickly diminishes the visual information, the temperature and smoke IoT sensor network can continuously record and analyze the fire scene data before the arrival of firefighters. The key information of fire, such as the locations and heat release rate (or power) of the fire sources and the evacuation progress of residence, can be identified by the AI engine to support the emergency response and firefighting activities. This requires the fire *Digital Twin* to be coupled with a pre-established fire database that is used to train the AI engine. The fire database includes thousands of CFD-based fire simulations, experimental data, and empirical correlations of the specific building. With real-time information, such as data from the temperature sensors and cameras, a pre-trained AI engine can quickly recognize, visualize, and render the fire scene on the *Digital Twin*.

Finally, to improve the overall building fire safety, it is necessary to install all fire-safety related sensors and connect with other building sensors and systems to form a smart building digital twin. Figure 18.5 shows a diagram of the proposed Smart Fire Digital Twin (SFDT) for the tunnel fire (Wu et al., 2022). It has four key components:

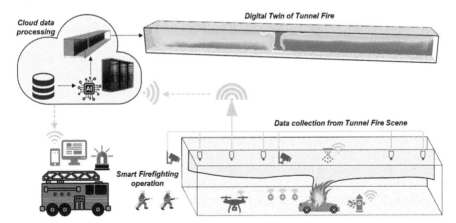

Fig. 18.5: The smart fire digital twin for a tunnel that can monitor the real-time fire behaviors, forecast the fire evolution by AI, and support smart firefighting operations.

(1) A sensor network, which is installed in the tunnel prior to fire incidents to collect the on-site temperature data.

(2) A cloud data server component, which reads the data remotely transferred from the sensor network into a server and store the data in a standard format.

(3) An AI engine component, which makes use of the available measured data to give identification, forecasting and early warning on the potential fire scenarios to the firefighters in real-time.

(4) A user interface, which could fetch the output of the AI engine for displaying the fire scene to operation center the and alarming firemen and commanders in a friendly mode.

Among these components, the AI engine enabling the digital twin to be automatic and smart in tackling the tunnel fire incident plays a central role. The arrows in the diagram indicate the direction of data flow and interaction between each component.

3.2 AI Methods for Identifying and Forecasting Tunnel Fire

The concept of AI was initially proposed on a workshop held in Dartmouth College in 1956 for dealing with computational problems related to language understanding, storage of data, and pattern matching (Russell and Norvig, 2016). Since then, AI approaches as well as other cutting-edge technologies such as remote monitoring, high-resolution sensor, high-speed computation, data-driven methods have been increasingly applied in fire safety engineering (Huang et al., 2021; Grant et al., 2015). So far, several studies have been conducted to explore the utilization of AI methods in compartment fire and tunnel fire. Most of these works applied artificial neural networks (ANNs) to train the database of fire tests (Dexters et al., 2020; Zhang et al., 2021) and numerical fire simulations (temperature data (Buffington et al., 2020; Wang et al., 2021) and fire smoke images (Lattimer et al. 2020; Wang et al., 2022; Su et al., 2021)) to identify the fire HRR, fire engineering design criterion, and the probability of flashover. Different machine-learning algorithms (Lee et al., 2004; Cabrera et al., 2020; Kou et al., 2021; Huang et al., 2021) have been proposed to train databases and pre-establish the complex relationship between sensor data and fire, so the real-time fire forecast can be achieved.

More recently, we have conducted a series of studies (Zhang et al., 2021; Wu et al., 2021; Wu et al., 2021; Wu et al., 2022; Naser, 2021) on applying AI methods to identify the real-time fire scenario and forecast critical fire events and monitor and visualize the real-time 2D fire scene in support of the smart firefighting in the tunnel. This section summarizes the methodology of AI application in identifying fire scenario information and rendering future fire scene images.

Recurrent neural networks (RNNs) are specially designed for the prediction or classification of temporal data, and its loop structure can be unfolded into a chain of cells (Fig. 18.6a). The order-dependent data are put into the structure and reserved, and then the information hidden inside the sequence of data can be interpreted as the output of the structure. Long short-term memory (LSTM) is a new type of RNN by introducing a specific internal structure that can effectively solve complex prediction problems, such as speech recognition (Hochreiter, 1997; Greff et al., 2017). LSTM cell is comprised of three gates. When data are imported into the cell, the forget

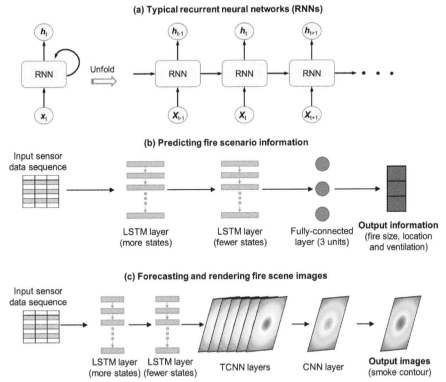

Fig. 18.6: Typical deep-learning structure diagrams for (a) Recurrent neural networks (RNNs), (b) predicting fire scenario information, and (c) forecasting and rendering fire scene images.

gate, input gate, and output gate together decide whether to block or pass it on, how to treat it, and how to output it, respectively. Then, after importing a piece of data, only useful information would be remembered, translated, and accumulated for the designed output.

One group of AI applications in fire is to identify the fire scenario information and forecast future critical fire events, and LSTM units are most effective to handle this problem. With a numerical database generated in Fig. 18.6(b), the LSTM unit in the input layer receives a vector of temperatures measured by sensors sequentially. The received data is then treated and passed on to the next state. Then, a multi-element vector (e.g., fire location, HRR, and ventilation speed) can be accurately predicted. To verify the accuracy of the AI-based fire detection, a sensitivity analysis should be carried out to optimize the database configuration and spatial-temporal arrangement of sensors (Zhang et al., 2021; Wu et al., 2021).

The other group of AI applications in smart firefighting is to forecast and render future fire scenes/images in the user interface to support the decision-making by fire commanders and firefighting operations in the field. To train massive image data, the convonutional neural network (CNN) algorithms should be used, which have been widely applied in many fields related to computer vision, such as image recognition and classification (Hayou et al., 2019). Technically, a series of convolutional and pooling layers are defined in a CNN model to pass through the input images having a larger volume to output the desired data having a smaller size (He et al., 2016). As the decoder of CNN, TCNN (temporal CNN) behaves contrarily to transform simple input data into higher-dimensional space (Dumoulin and Visin, 2016). Therefore, we proposed a deep learning model combing an LSTM model and a TCNN model, and the model is composed of LSTM, dense, and TCNN layers, as shown in Fig. 18.6(c) (Wu et al., 2021). This model is expected to generate images showing the distribution of temperature, smoke visibility, and gas concentrations in tunnels. Moreover, based on the forecasted fire scene images, critical tunnel fire information, such as the thickness and back-layering length of the fire smoke, could be estimated through post-processing.

In the future, these AI codes are expected to be further customized and optimized to handle specific fire engineering problems with a user-friendly interface. Eventually, every fire engineer and researcher will apply these AI codes to simulate complex fire scenarios fast and reveal in-depth information from the enormous database, like the CFD fire modelling tool. Today, the major users of CFD tools in the fire research and engineering application are from the fire community, rather than the mathematics and computer scientists who developed the PDE solvers of the governing equations or the CFD software developers. Therefore, it is expected that AI will also become a powerful tool for future fire engineering applications and research. A mature AI-driven fire forecast engine should be implemented into every building to identify and forecast fire scenes and support smart firefighting.

Moreover, fire engineers and researchers will be the primary user of new AI tools, rather than the AI and computer scientists. Nevertheless, developing these AI tools for the fire community needs the multidisciplinary collaboration of computing and IT communities, frontline firefighters, regulators, and policymakers. For fire engineers and researchers, it is also encouraged to have good knowledge about the AI algorithm, which helps users understand the reliability and limitations of AI's output, rather than treating AI entirely as a "black box". In other words, to ensure the physics-informed AI prediction, not only the physics-based high-fidelity CFD fire simulations are needed, but also the users should know the physics and fire dynamics to judge the AI's outputs.

4. Smart Tunnel Fire Safety Management: System Demonstration

To demonstrate the proposed smart firefighting system for tunnel, a scaled tunnel model is tested with the IoT sensor networks and AI engine trained by numerical database, as shown in Fig. 18.7. The tunnel model is scaled by 1/50, which is 1.7 m in length, 0.14 m in height, and 0.17 m in width. The back wall, ceiling, and ground of the tunnel are made of steel plates having a thickness of 3 mm. The front wall is made of 3 mm-thick fire-retardant glass to facilitate the observation of the fire scenarios (see more details in (Wu et al., 2022)).

For the numerical model, the boundary conditions and computational domain are set to be the same as the reduced-scale tunnel fire tests. Three parameters, fire location, fire size and ventilation condition in the tunnel, are considered. Based on the Froude-number scaling analysis (Li et al., 2013), the HRR and flow velocity can

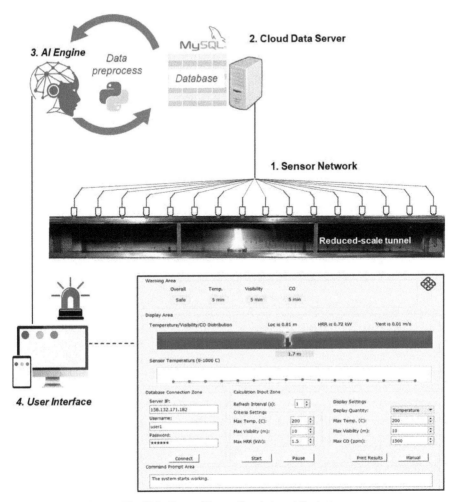

Fig. 18.7: Setup of the small-scale tunnel fire test.

Table 18.3: Major parameters for the tunnel fire.

Parameters	No. of Values	Value Range
Fire size or heat release rate (kW)	5	0.67, 1.23, 1.55, 2.45, 4.12
Fire location (m)	16	0.1–1.6 with interval of 0.1
Ventilation velocity (m/s)	5	−0.4, −0.2, 0, 0.2, 0.4

be scaled. Based on engineering practices and scaling analysis, the values of each parameter are listed in Table 18.3.

Overall, 16 fire locations, 5 HRRs, and 5 ventilation conditions produce 400 (= 16 × 5 × 5) fire scenarios. The fire scenarios are modeled with FDS version 6.7. The 3D model of small-scale tunnels is constructed, and the geometries and the surface materials of the models are set as the same as those of the scaled tunnel. The

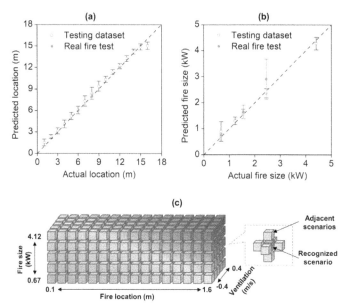

Fig. 18.8: Predictions of (a) fire location and (b) HRR on testing dataset vs. real fire test data, and (c) the evolvement of the ventilation condition predicted by the AI model (Wu et al., 2022).

fuel representing a burning vehicle was modeled with a squared burner. For doing massive numerical fire simulation, the proposed CFD models are calibrated by the laboratory experiments with burning liquid fuel flame to simulate the tunnel fire.

To demonstrate the proposed smart fire digital twin for the tunnel model, three liquid-fuel fires, producing an HRR of 0.67 kW, 1.55 kW, and 2.45 kW are tested at different locations inside the tunnel. The temperature distribution of the tunnel ceiling is measured by a datalogger GL820 produced by Graphtec Corporation. The measured temperature data are organized to form an experimental database using a similar procedure for the numerical database. Then, the experimental database is utilized to check the feasibility of predicting the information of real fire scenarios.

Figure 18.8(a,b) shows the prediction of the fire scenarios on small-scale tunnel fire tests. For comparison, the predictions on the testing dataset of the numerical data are also shown. Compared with its performance on numerical data, the trained AI model has a comparable capacity in predicting fire location and ventilation, and the deviations of the predictions on real test data are even lower. The prediction of the ventilation condition is not compared due to the missing experimental data.

To display the temporospatial distribution of the temperature on the user interface, the fire scenario should be recognized. Recognition is done by calculating the similarity of mean squared error (MSE) between the predicted fire scenario and the available ones. All the available scenarios stored in the database constituted a scenario matrix (Fig. 18.8c) having three dimensions: fire location from 0.1 m to 1.6 m along the tunnel, fire size from 0.67 kW to 4.12 kW and ventilation from –0.4 m/s (direction from right to left) to 0.4 m/s (direction from left to right). To further analyze the predicting mechanism, the input of the AI model, i.e., the measured temperature at various locations and the simulated temperature distributions at the

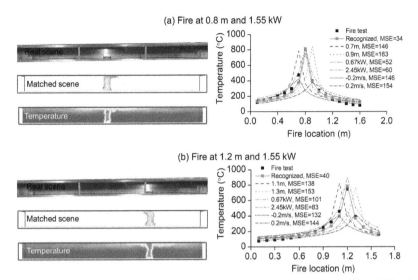

Fig. 18.9: Comparison of the test and matched fire scenes (left) and temperature profiles (right), where the fire location and the fire size in terms of HRR are: (a) 0.8 m and 1.55 kW and (b) 1.2 m and 1.55 kW.

end of the simulation at a quasi-steady state, are compared. Then, the temperature of the recognized scenario and the temperature produced by its most similar scenarios, the adjacent scenarios in all the dimensions of the scenario matrix, are selected for comparison.

Figure 18.9 illustrates the comparisons for cases where the fire is located at different positions and is burned at different HRRs. However, it is not easy to distinguish the fire size from their corresponding temperature distributions. The maximum temperature for cases having an HRR ranging from 1.23 kW to 2.45 kW is almost the same as 800°C, rather than increasing with the increase of HRR. This is because the fire sizes are so large that the fire flame impinges the ceiling of the tunnel, and the maximum temperature is essentially the flame temperature. Thus, their main difference lies in their respective distribution instead of the maximum temperature. The higher the HRR of fire is, the 'fatter' the distribution shape will be. Finally, the tested scenario is correctly identified to have an HRR of 1.55 kW (Fig. 18.9a). This does reveal that the recognition of the AI model on fire size is not simply dependent on the maximum temperature or the temperatures near the fire source. Instead, a more complicated mechanism considering the temperature distribution is developed by the AI model. A similar phenomenon is observed in the fire scenarios where the fire is located at 1.2 m (Fig. 18.9b).

After the fire scenario is recognized, the simulated fire scene, represented by the value of the HRR per unit volume (HRRPUV) and the temperature distribution, can be fetched from the corresponding fire scenario simulations stored in the numerical database. As shown on the left part of Fig. 18.9, the recognized fire scenes are much like those of the fire experiments. The fetched image of the temperature distribution is regarded as a Smart Firefighting Digital Twin of the real small-scale tunnel. Then, the digital twin is displayed on the user interface in real-time for reference.

5. Conclusions and Perspectives

This chapter reviews the state-of-the-art progress of applying AI technologies in tunnel fire safety engineering and smart firefighting. The key concepts of tunnel fire dynamics and the associated critical fire events are briefly introduced. The challenges in collecting fire data from sensors and numerical simulations are discussed. Currently, the entire engineering and scientific community is also lack of habit and standard of recording, organizing, and sharing the data, because the amount of data grows so fast that researchers lack the analytical skills to handle big data, especially the transient fire data and experimental/numerical videos. New data analytical approaches are needed with the assistance of AI methods.

A smart fire digital twin (SFDT) is proposed for the tunnel based on the deep learning algorithm pre-trained by thousands of tunnel fire simulations and real-time fire scene data from the sensor network. The BMS is still not widely used, and most of them do not share data with other systems. More recent building *Digital Twin* is proposed to be established upon the BIM, but they do not have the capacity for real-time data transfer and management. As the existing IoT sensors (smoke, heat, and flame detectors) are wired to the central control panel, there is a great advantage to form fire digital twin over other building systems.

Finally, we propose a small-scale demonstration of a smart firefighting system that can monitor and render the tunnel fire scene and forecast the fire development. Future research should include more realistic fire scenarios into the database and the training of the AI model, such as the tunnel size, shape, inclination angle, and blockage, as well as the growing/decay and moving fire sources, and more ventilation systems. The increased database will significantly increase the AI's predicting capacity. The AI model can be further integrated with the wireless sensor network, cloud server, and edge computing to form a smart firefighting system. The proposed intelligent tunnel fire safety management system is an important part of future smart buildings' multiple-physical digital twin, and it can be further upgraded for safety management in other types of buildings and hazards.

Acknowledgements

This work is funded by the Hong Kong Research Grants Council Theme-based Research Scheme (T22-505/19-N) and the National Natural Science Foundation of China (NSFC grant no. 52108480).

References

Alpert, R.L. and Science, C. 1975. Turbulent ceiling-jet induced by large-scale fires. *Combustion Science and Technology*, 11: 197–213.

Anderson, D. 1936. The construction of the Mersey tunnel (includes photographs). *Journal of the Institution of Civil Engineers*, 2: 473–516.

Baek, D., Sung, K.H. and Ryou, H.S. 2017. Experimental study on the effect of heat release rate and aspect ratio of tunnel on the plug-holing phenomena in shallow underground tunnels. *International Journal of Heat and Mass Transfer*, 113: 1135–1141.

Barbato, L., Cascetta, F., Musto, M. and Rotondo, G. 2014. Fire safety investigation for road tunnel ventilation systems: An overview. *Tunnelling and Underground Space Technology Incorporating Trenchless Technology Research*, 43: 253–265.

Beard, A. and Carvel, R. 2012. *Handbook of Tunnel Fire Safety*. ICE Publishing.

Beard, A.N. 2009. Fire safety in tunnels. *Fire Safety Journal*, 44: 276–278.

Blennemann, F. and Girnau, G. 2005. Fire protection in vehicles and tunnels for public transport. *STUVA, Köln, Mai*, 2005.

Blinov, V.I. and Khudiakov, G.N. 1961. *Diffusion Burning of Liquids*. U.S. Army Engineer Research and Development Laboratories.

Boonmee, N. and Quintiere, J.G. 2005. Glowing ignition of wood: The onset of surface combustion. *Proceedings of the Combustion Institute*, 30II: 2303–2310.

Buffington, T., Cabrera, J.M., Kurzawski, A. and Ezekoye, O.A. 2020. Deep-learning emulators of transient compartment fire simulations for inverse problems and room-scale calorimetry. *Fire Technology*, 57: 2859–2885.

Cabrera, J.M., Ezekoye, O.A. and Moser, R.D. 2020. Bayesian inference of fire evolution within a compartment using heat flux measurements. *Fire Technology*, 57: 2887–2903.

Caliendo, C., Ciambelli, P., Guglielmo, M.L. De Meo, M.G. and Russo, P. 2012. Simulation of people evacuation in the event of a road tunnel fire. *Procedia: Social and Behavioral Sciences*, 53: 178–188.

Carvel, R. 2019. A review of tunnel fire research from Edinburgh. *Fire Safety Journal*, 105: 300–306.

Casey, N. 2020. Fire incident data for Australian road tunnels. *Fire Safety Journal*, 111: 102909.

Cong, H.Y., Wang, X.S., Zhu, P., Jiang, T.H. and Shi, X.J. 2017. Improvement in smoke extraction efficiency by natural ventilation through a board-coupled shaft during tunnel fires. *Applied Thermal Engineering*, 118: 127–137.

Cote, A.E. 2008. *Fire Protection Handbook* (Vol. 2). National Fire Protection Association.

Dexters, A., Leisted, R.R., Van Coile, R., Welch, S. and Jomaas, G. 2020. Testing for knowledge: Application of machine learning techniques for prediction of flashover in a 1/5 scale ISO 13784-1 enclosure. *Fire and Materials*, 45: 708–719.

Dindarloo, S.R. and Siami-Irdemoosa, E. 2015. Maximum surface settlement based classification of shallow tunnels in soft ground. *Tunnelling and Underground Space Technology*, 49: 320–327.

Drysdale, D. 2011. *An Introduction to Fire Dynamics* (3rd Edn.) Chichester, UK: John Wiley & Sons, Ltd.

Dumoulin, V. and Visin, F. 2016. A guide to convolution arithmetic for deep learning. *ArXiv Preprint ArXiv:160307285*.

Giblin, K.A. 1997. Memorial tunnel fire ventilation test program. *ASHRAE Journal*, 39: 26.

Grant, C., Hamins, A., Bryner, N., Jones, A. and Koepke, G. 2015. Research roadmap for smart fire ighting. *NIST Special Publication 1191*.

Greff, K., Srivastava, R.K., Koutnik, J., Steunebrink, B.R. and Schmidhuber, J. 2017. LSTM: A search space odyssey. *IEEE Transactions on Neural Networks and Learning Systems*, 28: 2222–2232.

Guo, F., Gao, Z., Wan, H., Ji, J., Yu, L. and Ding, L. 2019. Influence of ambient pressure on critical ventilation velocity and backlayering distance of thermal driven smoke in tunnels with longitudinal ventilation. *International Journal of Thermal Sciences*, 145: 105989.

Gupta, V., Osorio, A.F., Torero, J.L. and Hidalgo, J.P. 2021. Mechanisms of flame spread and burnout in large enclosure fires. *Proceedings of the Combustion Institute*, 38: 4525–4533.

Haack, A. 1998. Fire protection in traffic tunnels: General aspects and results of the EUREKA project. *Tunnelling and Underground Space Technology*, 13: 377–381.

Haack, A. 2002. Current safety issues in traffic tunnels. *Tunnelling and Underground Space Technology*, 17: 117–127.

Hayou, S., Doucet, A. and Rousseau, J. 2019. On the impact of the activation function on deep neural networks training. *36th International Conference on Machine Learning, ICML 2019*. June 2019: 4746–4754.

He, K., Zhang, X., Ren, S. and Sun, J. 2016. Deep residual learning for image recognition. *Proc. IEEE Conference Computer Vision Pattern Recognition*, pp. 770–778.

Hochreiter, S. 1997. Long short-term memory. *Neural Computation*, 9(8): 1735–80. MIT Press.

Huang, X., Wu, X. and Usmani, A. 2021. Perspectives of using artificial intelligence in building fire safety. *In*: Naser, M.Z. (ed.). *Handbook Cogn. Auton. System Fire Resilient Infrastructures*, Springer.

Huang, Y., Chen, X., Xu, L. and Li, K. 2021. Single image desmoking via attentive generative adversarial network for smoke detection process. *Fire Technology*, 57: 3021–3040.

Ingason, H. and Zhen, Y. 2010. Model scale tunnel fire tests with longitudinal ventilation. *Fire Safety Journal*, 45: 371–384.

Ingason, H., Kumm, M., Nilsson, D., Lonnermark, A., Claesson, A., Li, Y.Z. et al. 2012. *The METRO Project: Final Report.* Mälardalen University Press.

Ingason, H., Li, Y.Z., Lönnermark, A. and Lonnermakr, A. 2015. *Tunnel Fire Dynamics.* (Vol. 53). London: Springer.

Jevtić, R.B. and Blagojević, M.D.J. 2014. On a linear fire detection using coaxial cables. *Thermal Science*, 18: 603–614.

Ji, J., Gao, Z.H., Fan, C.G., Zhong, W. and Sun, J.H. 2012. A study of the effect of plug-holing and boundary layer separation on natural ventilation with vertical shaft in urban road tunnel fires. *International Journal of Heat and Mass Transfer*, 55: 6032–6041.

Kou, L., Wang, X., Guo, X., Zhu, J. and Zhang, H. 2021. Deep learning based inverse model for building fire source location and intensity estimation. *Fire Safety Journal*, 121: 103310.

Kuesel, T.R., King, E.H. and Bickel, J.O. 2012. *Tunnel Engineering Handbook.* Springer Science & Business Media.

Kumar, N.V.R., Bhuvana, C. and Anushya, S. 2017. Comparison of ZigBee and Bluetooth wireless technologies-survey. *2017 International Conference Information Communication Embedded System, IEEE.* pp. 1–4.

Kurioka, H., Oka, Y., Satoh, H. and Sugawa, O. 2003. Fire properties in near field of square fire source with longitudinal ventilation in tunnels. *Fire Safety Journal*, 38: 319–340.

Lattimer, B.Y., Hodges, J.L. and Lattimer, A.M. 2020. Using machine learning in physics-based simulation of fire. *Fire Safety Journal*, 114: 102991.

Lee, E.W.M., Yuen, R.K.K., Lo, S.M., Lam, K.C. and Yeoh, G.H. 2004. A novel artificial neural network fire model for prediction of thermal interface location in single compartment fire. *Fire Safety Journal*, 39: 67–87.

Lee, S.R. and Ryou, H.S. 2005. An experimental study of the effect of the aspect ratio on the critical velocity in longitudinal ventilation tunnel fires. *Journal of Fire Sciences*, 23: 119–138.

Li, J. and Liu, J. 2020. Science mapping of tunnel fires: A scientometric analysis-based study. *Fire Technology*, 56: 2111–2135.

Li, Y.Z. and Ingason, H. 2018. Overview of research on fire safety in underground road and railway tunnels. *Tunnelling and Underground Space Technology*, 81: 568–589.

Li, Y.Z., Ingason, H. and Lönnermark, A. 2012. Numerical simulation of Runehamar tunnel fire tests. *6th International Conference Tunnel Safety and Ventilation*, pp. 203–210.

Li, Y.Z., Lei, B. and Ingason, H. 2013. Theoretical and experimental study of critical velocity for smoke control in a tunnel cross-passage. *Fire Technology*, 49: 435–449.

Liu, H., Li, J., Xie, Z., Lin, S., Whitehouse, K., Stankovic, J.A. et al. 2010. Automatic and robust breadcrumb system deployment for indoor firefighter applications, *In: Proceedings of the 8th International Conference on Mobile Systems, Applications, and Services*, pp. 21–34.

Liu, H., Xie, Z., Li, J., Lin, S., Siu, D.J., Hui, P. et al. 2013. An automatic, robust, and efficient multi-user breadcrumb system for emergency response applications. *IEEE Transactions on Mobile Computing*, 13: 723–736.

McGrattan, K., Hostikka, S., McDermott, R., Floyd, J. and Vanella, M. 2019. Fire dynamics simulator user's guide. *NIST Special Publication 1019* (6th Edn.).

Meacham, B., Bowen, R., Traw, J. and Moore, A. 2005. Performance-based building regulation: Current situation and future needs. *Building Research and Information*, 33: 91–106.

Naser, M.Z. 2021. Mechanistically informed machine learning and artificial intelligence in fire engineering and sciences. *Fire Technology*, 57(6): 2741–2784.

Ole, S. and Cruthers, H.G. 1947. Traffic tunnel and method of tunnel ventilation. http:/patentimages. storage.googleapis.com.

Omar, H.A., Abboud, K., Cheng, N., Malekshan, K.R., Gamage, A.T. and Zhuang, W. 2016. A survey on high efficiency wireless local area networks: Next generation WiFi. *IEEE Communications Surveys & Tutorials*, 18: 2315–2344.

Quintiere, J.G. 2006. *Fundamentals of Fire Phenomena.* John Wiley.

Rackauskaite, E., Bonner, M., Restuccia, F., Fernandez Anez, N., Christensen, E.G., Roenner, N. et al. 2021. Fire experiment inside a very large and open-plan compartment: x-ONE. *Fire Technology*, 58: 905–939.

Rappaport, T.S. 2002. Wireless communications—Principles and practice (The Book End). *Microwave Journal*, 45: 128–129.

Ren, R., Zhou, H., Hu, Z., He, S. and Wang, X. 2019. Statistical analysis of fire accidents in Chinese highway tunnels 2000–2016. *Tunnelling and Underground Space Technology*, 83: 452–460.

Russell, S.J. and Norvig, P. 2016. *Artificial Intelligence: A Modern Approach*. Malaysia: Pearson Education Limited.

Santos-Reyes, J. and Beard, A.N. 2017. An analysis of the emergency response system of the 1996 Channel tunnel fire. *Tunnelling and Underground Space Technology*, 65: 121–139.

Seike, M., Kawabata, N. and Hasegawa, M. 2016. Experiments of evacuation speed in smoke-filled tunnel. *Tunnelling and Underground Space Technology*, 53: 61–67.

Sevcik, E.M. 1928. *The Holland Vehicular Tunnel: A Great Engineering Achievement*. Ohio State University.

Souryal, M.R., Geissbuehler, J., Miller, L.E. and Moayeri, N. 2007. Real-time deployment of multihop relays for range extension. *Proceedings 5th Int. Conference Mobile Systems Applications and Services, ACM*, pp. 85–98.

Su, L.C., Wu, X., Zhang, X. and Huang, X. 2021. Smart performance-based design for building fire safety: Prediction of smoke motion via AI. *Journal of Building Engineering*, 43: 102529.

Takeuchi, S., Aoki, T., Tanaka, F. and Moinuddin, K.A.M. 2017. Modeling for predicting the temperature distribution of smoke during a fire in an underground road tunnel with vertical shafts. *Fire Safety Journal*, 91: 312–319.

Tao, F., Zhang, H., Liu, A. and Nee, A.Y.C. 2019. Digital twin in industry: State-of-the-Art. *IEEE Transactions on Industrial Informatics*, 15: 2405–2415.

Thomas, P.H. 1958. The movement of buoyant fluid against a stream and the venting of underground fires. *Fire Research Notes*. (Note No. 351.)

Vianello, C., Fabiano, B., Palazzi, E. and Maschio, G. 2012. Experimental study on thermal and toxic hazards connected to fire scenarios in road tunnels. *Journal of Loss Prevention in the Process Industries*, 25: 718–729.

Wang, F. and Wang, M. 2016. A computational study on effects of fire location on smoke movement in a road tunnel. *Tunnelling and Underground Space Technology*, 51: 405–413.

Wang, J., Tam, C.W., Jia, Y., Peacock, R., Reneke, P., Yujun, E. et al. 2021. P-Flash: A machine learning-based model for flashover prediction using recovered temperature data. *Fire Safety Journal*, 122: 103341.

Wang, S. 2009. *Intelligent Buildings and Building Automation*. Routledge.

Wang, Z., Zhang, T., Wu, X. and Huang, X. 2022. Predicting transient building fire based on external smoke images and deep learning. *Journal of Building Engineering*, 47: 103823.

Weng, M.C., Lu, X.L., Liu, F., Shi, X.P. and Yu, L.X. 2015. Prediction of backlayering length and critical velocity in metro tunnel fires. *Tunnelling and Underground Space Technology*, 47: 64–72.

Wu, X., Park, Y., Li, A., Huang, X., Xiao, F. and Usmani, A. 2021. Smart detection of fire source in tunnel based on the numerical database and artificial intelligence. *Fire Technology*, 57: 657–682.

Wu, X., Zhang, X., Huang, X., Xiao, F. and Usmani, A. 2021. A real-time forecast of tunnel fire based on numerical database and artificial intelligence. *Building Simulation*, 15(4): 511–524.

Wu, X., Zhang, X., Jiang, Y., Huang, X., Huang, G.G.Q. and Usmani, A. 2022. An intelligent tunnel firefighting system and small-scale demonstration. *Tunnelling and Underground Space Technology*, 120: 104301.

Yamamoto, K., Sawaguchi, Y. and Nishiki, S. 2018. Simulation of tunnel fire for evacuation safety assessment. *Safety*, 4: 12.

Yao, Y., He, K., Peng, M., Shi, L., Cheng, X. and Zhang, H. 2018. Maximum gas temperature rise beneath the ceiling in a portals-sealed tunnel fire. *Tunnelling and Underground Space Technology*, 80: 10–15.

Zhang, T., Wang, Z., Wong, H.Y., Tam, W.C., Huang, X. and Xiao, F. 2021. Real-time forecast of compartment fire and flashover based on deep learning. *Fire Safety Journal (under Review)*.

Zhang, X., Wu, X., Park, Y., Zhang, T., Huang, X., Xiao, F. et al. 2021. Perspectives of big experimental database and artificial intelligence in tunnel fire research. *Tunnelling and Underground Space Technology*, 108: 103691.

Index